BROADBAND COMMUNICATIONS

Robert C. Newman

DeVry Institute of Technology

Prentice
Hall

Upper Saddle River, New Jersey
Columbus, Ohio

Library of Congress Cataloging-in-Publication Data

Newman, Robert C.
  Broadband communications/Robert Newman.
    p. cm.
  ISBN 0-13-089321-8
   1. Broadband communication systems. I. Title.

  TK5103.4 .N49 2002
  621.382—dc21

                  2001036662

Editor in Chief: Stephen Helba
Assistant Vice President and Publisher: Charles E. Stewart, Jr.
Production Editor: Alexandrina Benedicto Wolf
Production Coordination: Carlisle Publishers Services
Design Coordinator: Diane Ernsberger
Cover Designer: Ali Mohrman
Cover photo: Corbis Stock Market
Production Manager: Matthew Ottenweller

This book was set in Helvetica, Times, Palatino, and ITC Century Book by Carlisle Communications Ltd. It was printed and bound by R.R Donnelley & Sons Company. The cover was printed by Phoenix Color Corp.

Pearson Education LTD., *London*
Pearson Education Australia PTY. Limited, *Sydney*
Pearson Education Singapore, Pte. Ltd.
Pearson Education North Asia Ltd., *Hong Kong*
Pearson Education Canada, Ltd., *Toronto*
Pearson Educación de Mexico, S.A. de C.V.
Pearson Education—Japan, *Tokyo*
Pearson Education Malaysia, Pte. Ltd.
Pearson Education, *Upper Saddle River, New Jersey*

10 9 8 7 6 5 4 3 2 1
ISBN   0-13-089321-8

This text is dedicated to all of the students who have endured my lectures, labs, projects, pop quizzes, and very lengthy exams at the DeVry Institute of Technology in Decatur, Georgia.

WHO SHOULD READ THIS BOOK?

The current telecommunications market has been growing at an astronomical pace, due largely to the expansion of Internet access to almost all sectors of society. Everyone with a computer can now have access to the Web: children are introduced to the Web at an early age. The opportunities for expansion of services seem limitless. This book is designed to provide a broad working knowledge of the broadband technologies that permeate today's society and to provide the serious student with the tools to address the numerous opportunities that abound in this field.

This text is oriented toward both the business and education community, especially toward the undergraduate student who aspires to succeed in the telecommunications industry, and the marketing and sales professional who has some basic understanding of telecommunications.

It gives the student and business professional the tools to identify and apply techniques to development and management of enterprise networking for the various applications of broadband services.

This book is not highly technical; it provides just enough detail for the reader to apply the information provided to be successful in the broadband environment. This may very well lead the reader to look at the more technical aspects of this field of study.

ORGANIZATION OF THIS BOOK

Part I includes a short introduction of the telecommunications environment and industry including the wide area network (WAN) infrastructure. Some prior knowledge of basic data communications processes is assumed, and is also a plus, as these two topics are closely interrelated. A short introduction sets the course of study. Chapter 1 provides an ideal beginning by giving significant background information in the Technology Awareness section.

Part II includes Frame Relay, Asynchronous Transfer Mode (ATM), the Synchronous Optical Network (SONET), Virtual Private Networks (VPN), and Fiber Distributed Data Interface (FDDI). These technologies are used for internetworking and provide the basis for many enterprise networks.

Part III includes Digital Subscriber Line (DSL), Integrated Services Digital Network (ISDN), and Switched Multimegabit Data Service (SMDS). ISDN and SMDS are transitional technologies and DSL is an emerging technology. These chapters are followed by the wireless/PCS environment, Fibre Channel, and considerable information on the Internet and the associated intranet and extranet environments. A considerable amount of interest is being generated in the wireless/PCS and Internet environments, particularly how they can be integrated into a seamless environment. Information is presented on the "hot" topics of WAP, GSM, CDMA, TDMA, EDGE, GPRS, and 3G.

Part IV completes the book with chapters on network management, problem solving, and troubleshooting. There is sufficient detail for the reader to be well versed in these two topic areas.

The appendices include information on the OSI model and a brief presentation on Voice over IP. Appendix E is a comprehensive broadband technologies case study/project.

Key terms are presented at the end of each chapter. Comprehensive sources were consulted to develop these definitions (Held, 1996; Newton, 1999). Review questions at the end of each chapter provide coverage of the material presented.

HOW TO UTILIZE THIS TEXT

A CD-ROM is provided with the free Instructor's Manual available to instructors using the text and includes answers (600+) to review questions found in each chapter. Also presented are suggestions in completing the activities at the end of each chapter. This CD-ROM includes all of the Visio 2000© graphics (150+) that appear in the text. These are in MicroSoft Powerpoint© format.

Many of the exercises in the book require the reader to use ancillary resources to complete the activity. Access to the Web pages of the various standards agencies and technology forums is a must. It is also beneficial that access to products such as DataPro© be available to the student. Another invaluable student resource is catalogs, such as Black Box©, which provide technical information, configurations, and prices on a number of enterprise network components.

The exercises, case studies, and projects are oriented toward practical situations and issues that relate to the current enterprise networking environment. The case studies and projects ask the reader to develop questions that must be answered in order to develop specifications for the various solutions. Brainstorming and boarding are useful techniques when conducting group exercises. These allow students to participate and provide input into the learning experience. These techniques can also be used when developing the comprehensive enterprise network design that is presented in Appendix E.

Many of the photos depicting network devices have been provided by the Black Box Corporation and can be viewed on its Website at *www.blackbox.com*.

ACKNOWLEDGMENT

I would like to thank Lawrence Bernstein (Stevens Institute of Technology, AJ) and Robert E. Morris (DeVry Institute of Technology, GA) for their invaluable review and feedback.

Comments and suggestions on this text can be sent to me at newmanrc@atl.bellsouth.net.

Robert Newman

CONTENTS ▉

PART ONE

CHAPTER 1
The Enterprise Network Basics

Chapter 1 begins with information concerning the telecommunications industry and the implications of divestiture and deregulation. It presents an overview on the network infrastructure, the functions of the network, and the various components that go to make up the overall network. This is followed by a general discussion of LANs, WANs, and MANs and how they are internetworked to form the enterprise network. This chapter includes information on several standards that apply to the industry, and concludes with a section on data communication technology awareness, which provides a general overview of many components and elements of the enterprise network and the network infrastructure.

CHAPTER 2
The WAN Infrastructure

Chapter 2 sets the stage for the rest of the text by looking at the interrelationships between the different network infrastructures. The chapter reviews leased and switched circuits and circuit switched and packet switched networks. It includes an overview of the analog and digital issues that are part of the network, and a section on multiplexing and the various interfaces prevalent in the network. Discussion also includes facility types and transmission methods across these facilities; the various routing protocols that may be encountered during transmissions; and broadband applications, issues, and concerns.

For the technically oriented reader, several other texts available provide a more in-depth look at the technical aspects of data communications. The reference section of this text provides a partial list of resources that may help in further understanding the data communications environment.

The Enterprise Network Basics

■ INTRODUCTION

The merger of the computer industry with the telecommunications industry has produced significant growth in the network technology arena. Deregulation and the divestiture process has created numerous opportunities for a significant number of new market entries. The original telephone infrastructure has migrated to a state-of-the-art broadband transport vehicle. Many industries are taking advantage of these new technologies to increase their market presence and market share.

The network communications industry is moving toward powerful gigabit communications systems that support a variety of applications that can be deployed across a wide range of disciplines. For the first time in the history of the computer and communications industry, network administrations throughout the world are embracing a set of standards for use on high capacity multimedia communications networks. This trend will lead to enhanced services to the end user and will result in significant gains in productivity with a decrease in the cost of doing business.

It is essential that the serious student and all product-oriented professionals be aware of all phases and components of the industry so that they can maximize the enormous opportunities [Cole, 2000]. This chapter attempts to provide sufficient ammunition to succeed in these endeavors. Covered in this chapter are historical legal perspectives, semitechnical presentations on the communications network infrastructure, and LAN, WAN and MAN fundamentals. The chapter concludes with a section on technology awareness, which can be used to refresh the knowledge base of the experienced networking student or provide a basis of understanding for the networking novice.

OBJECTIVES
Material included in this chapter should enable you to:

- understand the telecommunications industry history, beginning with divestiture and deregulation to today's broadband environment.

- become familiar with the terms and definitions that are commonly used in the Enterprise Networking environment.

- become familiar with the components and functions of the telephone infrastructure. Identify the various elements that are part of this infrastructure.

- become familiar with the devices that go to make up the data communications environment. Relate these components to the broadband enterprise networking environment.

- become familiar with the various standards and standards organizations that relate to the telecommunications industry.

- understand the differences between LANs, MANs, and WANs and their different applications.
- identify components that make up LAN, MAN and WAN networking environments.

divestiture
deregulation
modified final judgment (MFJ)

equal access

1.1   THE INDUSTRY

Divestiture and Deregulation History

Divestiture was basically the requirement of AT&T to divest itself of the twenty-two operating telephone companies. **Deregulation** was the process of removing regulatory authority from the telephone companies, therefore allowing for open competition. These activities were supposed to help the consumer; however, this has not always been the case. Figure 1–1 depicts the timeline from 1934 to 2001. The significant events started with the Communications Act of 1934 and became quite serious with the **Modified Final Judgment (MFJ)** of 1982 and the 1984 divestiture of AT&T. The latest activities in the industry consist of mergers of major and minor networking organizations.

The divestiture and deregulation activities of the 1980s allowed competing long-distance carriers such as MCI and Sprint to sell long-distance services on a level playing field with AT&T. This was to be known as **equal access** and was part of the MFJ decree. Equal access means that all carriers, including AT&T, could be reached by dialing a five digit code (10xxx), thus providing equal access for all carriers. City by city, subscribers were asked to choose their primary carrier. This requirement of equal access has allowed many new entries into the communications market and has provided for a wide range of new and diverse services.

Another major event was the Telecommunications Act that was passed by Congress on February 1, 1996. This document was primarily a rewrite of the Communications Act of 1934. The new act changed the rules for competition and regulation in virtually all sectors of the communications industry. This act provides for major changes in laws affecting cable television, telecommunications, and the Internet. The law's

Figure 1–1
Telecommunications
Industry Timeline

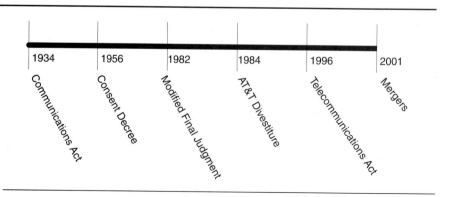

main purpose is to stimulate competition in telecommunications services. The law specifies the following:

- How local telephone carriers can compete—Local service can be offered by a variety of companies, including CATV companies, utilities such as power companies, and interexchange carriers (IECs/IXCs).
- How and under what circumstances Local Exchange Carriers (LECs) can provide long-distance services—ExBell LECs are permitted to enter long-distance markets if the move is in the public interest.
- The deregulation of cable TV rates—Incumbent Local Exchange Carriers (ILECs) and others are permitted to deliver video to homes and businesses. Rates of CATV systems with fewer than 50,000 subscribers are deregulated.
- A limit on broadcast services—A single company would be permitted to own stations that reach up to 35 percent of the viewers in the United States.

1.2 TELEPHONE NETWORK STRUCTURE

To understand the changing regulatory relationship between different phone companies and their associated regulatory agencies, it is important to understand the physical layout of the basic telecommunications infrastructure. Several major classes of telecommunications equipment must fit together to form the communications network. The network is created by the various systems, sometimes autonomously, exchanging signals across the numerous interfaces. The major components that comprise the telecommunications infrastructure are as follows:

- Customer Premises Equipment (CPE)
- Subscriber loop plant
- Local switching systems
- Interoffice trunks
- Tandem switching offices
- Interexchange trunks
- Transmission equipment

Figure 1–2 illustrates the major components of the **Public Switched Telephone Network (PSTN)**, which are necessary to support long-distance dial-up service for data communications. The PSTN consists of **Central Offices (COs)**, tandem offices, end offices, and intermediate offices. Each of these offices perform specific functions in the network hierarchy. These offices are organized into 196 **Local Access Transport Areas (LATAs)** allocated during divestiture. A LATA is usually, but not always, equivalent to the area covered by a given area code.

public switched telephone network (PSTN)
central offices (COs)

local access transport areas (LATAs)

Figure 1–2
Central Office Hierarchy

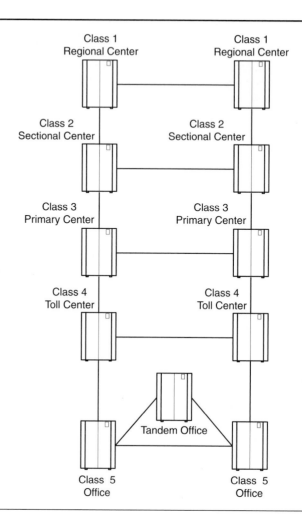

Central Offices are categorized as Class 1 through Class 5. A residential or business call is first processed in the local central office, also known as an end office, wire center, or local office. Today's network, however, is becoming flatter with fewer central office classes. For our purposes, however, we will consider the five-level structure.

Central Office Classifications

Central offices are designated as Class 1 through Class 5 based on their functionality.

- Class 1 regional center is the highest capacity switching office in the PSTN hierarchy. This is the highest level toll office in AT&T's long-distance switching hierarchy.

Figure 1–3
Local Loop Connecting a
Subscriber to a Central
Office

- Class 2 sectional center is the second highest capacity switching office in the PSTN hierarchy.
- Class 3 primary center is the third highest switching office in the PSTN hierarchy.
- Class 4 toll center is the fourth highest switching capacity office in the PSTN hierarchy.
- Class 5 office is the local switching office. It is sometimes called a wire center and is the office where your local access terminates.

Additionally, there is a Tandem Office that the telephone company utilizes to connect central offices when interoffice trunks are not available.

Note that local access is at the Class 5 office and long-distance calls go through a Class 4 office. Operator services such as information are processed in Class 4 offices. There are over 19,000 Class 5 offices and only 12 Class 1 offices.

The circuits between a residence or a business and the local CO are known as **access lines,** or **local loops**. **Trunks** connect switches and are utilized to carry multiple voice-frequency circuits using a multiplexing arrangement between local telephone exchanges. The local exchange is a facility that belongs to the local phone company where calls are switched to their proper destination and local physical connections are terminated. Figure 1–3 depicts a simple local access configuration that consists of a telephone instrument connected to a public telephone switch via an access circuit.

access lines
local loops
trunks

1.3 TELECOMMUNICATIONS NETWORK FUNCTIONS

Numerous tasks and operations must occur in the telecommunications network for it to perform its required functions. Some of these activities are simple; most are complex and require interactions with many other elements to successfully transport communications traffic. The following list includes the major categories that must be addressed.

- Network Management
- Security
- Recovery
- Addressing

- Synchronization
- Utilization
- Flow control
- Error detection/correction
- Interfacing
- Signal generation
- Handshaking
- Routing
- Formatting

Even though many of the activities and functions listed are automatic, managing the system is often time-intensive and requires a considerable amount of hands-on management from systems administrators and network technicians. Chapter 14 addresses the network management aspect of broadband communications.

1.4   TELECOMMUNICATIONS NETWORK COMPONENTS

Switched Facilities

circuit switching

Telephone calls connected on the PSTN pass over switched or shared facilities. This type of network technology is referred to as **circuit switching.** Virtually all voice telephone calls are circuit switched. Dial-up modem access is also circuit switched. The facilities in the switched environment are only utilized for the duration of the call. A switched call has three phases, namely *call setup, call transmission,* and *call termination.* A three-phase call results in a lower cost because economies of scale are achieved in sharing facilities.

Dedicated Facilities

Dedicated lines, however, like private networks, are not shared. They are essentially available 24 hours a day for the customer agreeing to pay for them. Other names for private lines are point-to-point, direct, fixed, leased, and ringdown lines.

Figure 1–4 depicts both circuit switched and dedicated circuit examples. The normal data communications devices such as modems are required to access the public network switches. The connectivity for dial-up access is temporary, whereas the dedicated access is permanent.

The terms for switched or dedicated networks, like many telecom terms, have taken a life of their own and can be used in many different ways. Private lines, for example, may refer to tie lines, long-distance lines between cities, or even dedicated local telephone lines for one user behind a Private Branch Exchange (PBX) or key system.

Many companies create or build their own networks. These private networks offer the advantages of guaranteed availability, greater reliabil-

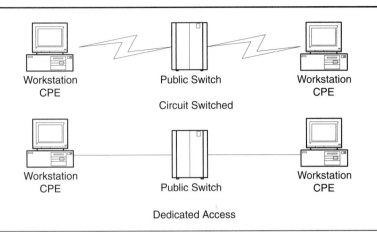

Figure 1–4
Circuit Switched and
Dedicated Access

ity, and lower cost, if properly utilized. Private networks can be created through a variety of services from a number of different vendors or actual network construction. They can be voice, data, or voice/data networks managed completely or partially by the customer. Often customers will migrate to a private network for the purpose of gaining increased control of their telecommunications services and realizing cost savings when they own right-of-ways and network infrastructures. Railroads and electric power companies often use their own right-of-ways for their communications facilities.

1.5 ELEMENTS OF A DATA COMMUNICATIONS SYSTEM

The four basic elements of a data communications system are a transmitter, a receiver, a transmission medium, and communications equipment [Stallings, 2000]. Standards relating to these elements are set forth in the **Open Systems Interconnection (OSI)** model developed by the **International Standards Organization (ISO)**. This seven-layer model is included in Appendix A.

open systems interconnection (OSI)
international standards organization (ISO)

Transmitters and receivers in this model are devices such as personal computers (PCs), workstations, printers, servers, mainframes, and minicomputers that can use and generate information. Such devices are categorized as **Data Terminal Equipment (DTE)** or **Customer Premise Equipment (CPE)**.

Data Communications Equipment (DCE) connects DTE to the network. A Data Service Unit (DSU) and modulator/demodulator (**Modem**) are examples of DCE devices. Figure 1–5 depicts a common DTE/DCE interface arrangement using modems to communicate across a PSTN. The access line is synonymous with the telephone company local loop.

data terminal equipment (DTE)
customer premise equipment (CPE).
data communications equipment (DCE)
modem

Figure 1–5
DTE/DCE Interface
Arrangement

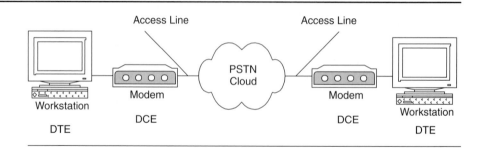

Figure 1–6
Frame and Cell Formats

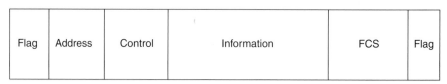

(a) HDLC frame format

Header	Information

(b) ATM cell format

frame
packet

    A **frame** or **packet** is the information unit most commonly used with these data networks. Both units contain end-user data or payload, which in turn is surrounded by additional control characters. Control characters are added by the communications logic in the transmitter and used by the communications logic in the receiver to implement the protocols and error-checking systems. Packetizing is the process of adding overhead or management data to raw user data to ensure proper delivery. Figure 1–6

protocol data units (PDU)

shows the **Protocol Data Units** (**PDU**s) for the *High-level Data Link Control (HDLC)* and *Asynchronous Transfer Mode (ATM)* layouts. The ATM PDU is

cell

called a **cell** and consists of fifty-three octets. The HDLC PDU is called a variable-length frame, standard for a number of technologies, including Frame Relay. These frames and cells are transmitted over some medium between the various communications devices.

protocol

    Before CPE devices can communicate with one another over a network, there must be a common communication "language," called a **protocol.** The type of protocol used on the network depends on the network's infrastructure and classification, that being routable or non-routable. A nonroutable protocol is local and can only be used in a single-network environment. A routable protocol functions in local environments and can communicate with remote networks. The three main protocols used in today's networks are as follows:

- NetBEUI—NetBIOS Enhanced User Interface (nonroutable)
- IPX/SPX—Internetwork Packet Exchange/Sequenced Packet Exchange (routable)
- TCP/IP—Transmission Control Protocol/Internet Protocol (routable)

Transmission **media** may be classified as guided or unguided. With guided media, the electromagnetic waves are guided along a physical path on media such as twisted pair, coaxial cable, or optical fiber. The other option, unguided media, allows for the propagation of electromagnetic signals through air and a vacuum.

media

A requirements study must be conducted to determine the appropriate type of transmission medium. If there is noise in the environment, coaxial cable or fiber might be required. There are also distance and line-of-site requirements for the various transmission media, and certain standards associated with the transmission media that may be relevant for the system designs. These standards are presented later in the chapter.

1.6   LOCAL, METROPOLITAN, AND WIDE AREA NETWORKS

Sometimes the boundaries between **Local Area Networks (LAN), Metropolitan Area Networks (MAN),** and **Wide Area Networks (WAN)** are indistinct, which makes it difficult to determine where one network ends and another begins. A useful method is to determine the type of network by examining the four prominent network properties: medium, protocol, topology, and private versus public demarcation points.

local area networks (LAN)
metropolitan area networks (MAN)
wide area networks (WAN)

- A LAN often ends where the medium changes at the demarcation point, such as wire-based networks to fiber-based networks.
- The designation can also change when there is a change in the layers that are impacted by the protocol being utilized. See Appendix A for the OSI model layers.
- A change in topology from a ring to a star often indicates a change in technologies. These are discussed later in this chapter.
- The boundary between private LANs and public WANs is the point at which LANs connect to the regional telephone network.

The communications medium could include coaxial cable, fiber-optic cable, twisted-pair wire, radio waves, or microwaves. Table 1–1 summarizes the relationship between LANs, MANs, and WANs.

Table 1–1
LAN, MAN and WAN
Comparisons

Attribute	LAN	MAN	WAN
Network coverage	Local—building or campus	Greater metro area	Large geographic area
Complexity	Minimum	Moderate	Complex
Transport speeds	Relatively fast	100 Mbps	Slow
Cost	Moderate	Low	High
Ease of service	Easy	Easy	Difficult
Error quality	Good	Good	Poor (with copper)

A primary difference between LANs, MANs, and WANs is the distance over which devices communicate with others. A LAN is local in nature: It is owned by one entity and is located in a limited geographic area and most often in a single building. Local area networks are usually faster than WANs and have a lower error rate. In larger organizations, LANs can be linked between a complex of buildings in close proximity, referred to as the campus arrangement.

Computers linked within a metropolitan area or city are part of a MAN. Metropolitan area networks have attributes of both LANs and WANs. The telecommunications facility for a MAN is usually transparent to the user.

Devices that are connected between cities or different LATAs are part of a WAN. Wide area networks usually are slower than LANs and have a higher error rate. Capital investments in WANs are high, because the outside plant infrastructure is immense.

Note that the complexion of the network changes when fiber optics are introduced into the environment. Clashes tend to become congestion dependent in low-error fiber networks.

Local Area Networks

As WANs have expanded in scope, organizations have also expanded their use of personal computers, laptop computers, personal digital assistants (PDAs), and individual workstations to support the information needs of users throughout the enterprise network. Today's organizations use small computers for word processing, order processing, messaging, sales reporting, financial analysis, engineering, and many other applications in support of the enterprise's mission. As the use of small computers has grown, the need has also grown for these computing devices to communicate, both with each other and with a centralized enterprise mainframe computer or client-server system.

Small computing devices were often used in a stand-alone manner to support applications that were local in nature. The data were often stored

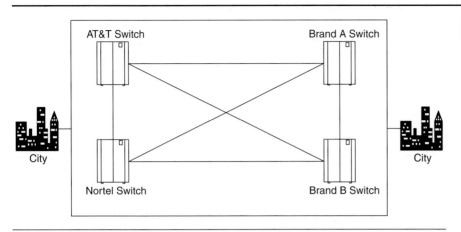

Figure 1–7
Local Area Networks

on the device's hard drive or available on a removable storage medium. Typically, additional requirements arose for various levels of access that include the following:

- Provide access to existing data that might be stored on a computer in some other division or department
- Provide access to computer peripheral devices, such as color printers, that are too expensive to be used by a single user
- Give the user the ability to exchange electronic messages with each other
- Provide access to the Internet, a corporate intranet, or extranet

During the 1980s the proliferation of Personal Computers (PCs) in the office led to the introduction of Local Area Networks (LANs). Once offices were computerized, companies began looking for effective ways to link the individual PCs, to allow for the sharing of documents and expensive peripherals. Today, LANs enable multiple-site connectivity and even telecommuting. Access to vital corporate information and e-mail is provided by file servers—computers connected to the LAN. Password features and firewalls restrict access to this information. LANs come primarily in two topologies, Ethernet/IEEE 802.3 or Token Ring. Several transmission speeds are available. Ethernet/802.3 operates at 10, 100, and 1000 megabits per second and Token Ring operates at either 4, 16, or 100 megabits per second (Mbps). Figure 1–7 depicts a logical representation of two different LANs connected to a common computer system. The computer system must provide support for the different Ethernet and Token Ring protocols.

Many of today's LANs are used to interconnect PCs and workstations with centralized computers called servers. These topologies are called client-server systems. Personal computers, shared printers, shared

Figure 1–8
Metropolitan Area Network

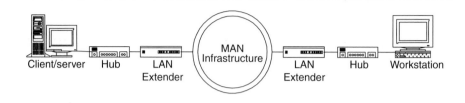

databases, factory automation, and quality control systems connected in a local area data communications system are examples of LANs.

Metropolitan Area Networks

Metropolitan area networks are connections between LANs within a city. MANs may encompass several blocks or a 50-mile radius. For example, a bank in Charlotte keeps its account records in a nearby storage facility in another part of the city. Instead of transporting the records and films between the two sites, the bank leases high-capacity phone lines to transmit images and also records data. The connections between these two sites are metropolitan area connections. MANs can be leased from a telephone company or constructed by the organization utilizing copper, fiber-optic, or microwave-based services. They may also utilize the same services as WANs such as ISDN, T-1, and T-3. These local services consist of fiber multiplexers and fiber bridges. An example of such a service is BellSouth's Native Mode LAN Interconnection (NMLI). This service uses the embedded local FDDI architecture for its transport medium. Figure 1–8 depicts a simple MAN that utilizes LAN extenders as fiber bridges. Multimode fiber connects the hub to the LAN extender and single-mode fiber is utilized to connect the LAN extender to the MAN cloud. The MAN is transparent to the user and the users may all be on the same LAN segment.

Metropolitan area networks are used to bridge the gap between WANs and LANs. Where MANs are deployed, they are usually used as a lower-cost alternative to WANs or to extend the range of an existing LAN.

Wide Area Networks (WANs)

Many networks installed by organizations use a number of public telecommunications facilities which allow enterprise computer resources to communicate over long distances. These networks might be used to provide all users at remote locations access to the resources maintained in a centrally located computer complex. Networks that link users who are widely separated geographically are called WANs.

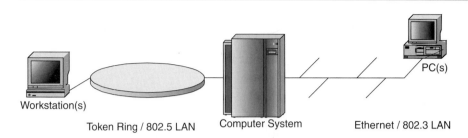

Figure 1–9
Wide Area Network
Infrastructure

Workstation(s)

Token Ring / 802.5 LAN Computer System Ethernet / 802.3 LAN

PC(s)

Wide area networks use telephone lines to connect businesses that are separated by long distances. For example, a warehouse in Denver connected to its regional sales office in Atlanta by a telephone is a WAN connection. A WAN is not confined to a limited geographical area as is a LAN. Various WAN connections are available. Selection of an appropriate WAN connection depends on factors such as quality of service needed, speed, price, compatibility with the current computer systems at any given location, and the amount of traffic between locations. The types of WAN services and technologies available include:

- Integrated Services Digital Network (ISDN)
- T-1, T-3
- Asynchronous Transfer Mode (ATM)
- Frame Relay (FR)
- Switched Multimegabit Data Service (SMDS)
- Digital Subscriber Line (DSL) services
- Wireless and Personal Communication Services (PCS)
- Fibre Channel (FC)
- Fiber Distributed Data Interface (FDDI)

Players in this environment also include the infrastructure services of the Synchronous Optical Network (SONET) and the Virtual Private Network (VPN). These backbone networks consist of many multiplexers and switching systems. Switching systems include such configurations as AT&T 5ESS and Nortel DSM100 systems. Figure 1–9 depicts a simple WAN in which numerous carriers provide seamless service across the entire network. The WAN is often used to connect similar and dissimilar LANs across multiple LATAs and states.

1.7 INTERNETWORKING

Enterprise internetworks are typically constructed using three different elements: Local Area Network data (LAN) links, Wide Area Network (WAN) data links, and network interconnection devices.

LAN data links are normally confined to a single building or within a campus of buildings. They do not cross public thoroughfares and normally operate over private cabling. Local area network facilities are generally used to create many networks that allow any device on the data link to physically communicate with any other device on the data link.

When LANs are located in widely separated locations, WAN data links are often used in conjunction with network interconnection devices to interconnect the LANs. Wide area network data links are most often used to implement point-to-point connections between pairs of network interconnection devices.

Commonly used data transfer speeds over common carrier telecommunications facilities include 19.2 kbps and 56 kbps. Digital T-1 (1.544 Mbps) and T-3 (44.736 Mbps) facilities, also available from common carriers, are widely used in enterprise internetworks. Wide area networking facilities commonly used in computer networks include:

- Common carrier telecommunications data circuits
- X.25 Packet-Switched Public Data Network circuits
- Frame Relay access
- Narrowband ISDN (N-ISDN) access
- Broadband ISDN (B-ISDN) access
- Distributed Queue Dual Bus (DQDB)/MAN access
- Switched Multimegabit Data Service (SMDS) access
- Synchronous Optical Network (SONET) facilities
- Fibre Channel connectivity

To create flexible enterprise internetworks, it is necessary to interconnect individual LAN data access links and WAN facilities. Several different types of devices can be used to accomplish this. The types of devices available for network interconnection can be divided into the following general categories:

- Bridges
- Routers
- WAN Switches
- Gateways

These devices contain firmware and software that allows them to make routing and transport decisions concerning communications traffic. Details concerning these devices are presented in Section 1.10.

1.8   BASIC DATA COMMUNICATION CONFIGURATIONS

Common equipment configurations that might be found at a central site or a remote location include host computers and client-server devices. A

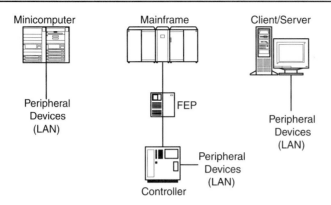

Figure 1–10
Computer Systems—Host
and Client-Server Sites

common host configuration includes a host computer, a front-end processor (FEP), and a communications controller. A smaller computer system might include a processor with a built-in Input/Output (I/O) processor. Figure 1–10 depicts examples of host and client-server sites. Both the communications controller and the I/O processor are utilized for communication interface connections. The type and number of interfaces available are dependent upon the configurations of IO slots that the processor can accommodate.

Interfaces

Most of these interfaces provide for serial communications [Elahi, 2001]. Two common interfaces are V.35 and RS232. The **RS232** interface consists of a 25-pin (DB25) standard configuration and **V.35** consists of a 34-pin standard configuration. The V.35 interface is primarily utilized for high-speed communication. These two standards define the characteristics of the DCE/DTE interface. Table 1–2 provides an overview of the various

RS232
V.35

Specification	Pinouts	Utilization
RS232	25 pins	Low-speed serial communications—connects terminals and modems
RS232	9 pins	Low-speed serial communications—often used to connect modems to laptops : IBM AT style adapter
RS449	37 pins	High-speed serial communications—connects terminals and modems
RS449	9 pins	High-speed serial communications—often used to connect modems to laptops : IBM AT style adapter
V.35	34 pins	High-speed serial communications between routers and DSUs
RS422	9/25/37 pins	Balanced interface with no physical connector—screw terminals
RS423	9/25 pins	Unbalanced interface with no physical connector—screw terminals

Table 1–2
Interface and Connector
Specifications

Figure 1–11
Ethernet LAN Example

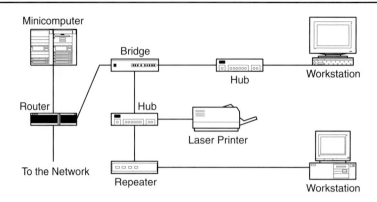

interfaces and connectors that are utilized for device connectivity in the enterprise network.

To communicate with one another, each device must be connected to a LAN. Connections between devices may be any combination of twisted-pair, coaxial cable, fiber optic, or wireless media. Twisted-pair Category 4/5 cabling and coaxial cable most commonly connect devices to a LAN. Figure 1–11 depicts a simple physical Ethernet LAN. The components are described in the LAN and MAN equipment section of this chapter.

The popularity of LANs in the 1980s led to many departments purchasing their own LANs—independent of the central computer operations staff. The compatibility of LANs from different manufacturers within a company became difficult as e-mail and file sharing emerged. This trend, referred to as decentralization, was reversed in the 1990s with the advent of the Client-Server environment. The client-server system allows the organization to provide services in both a centralized and a decentralized manner. This environment, however, has introduced system and software compatibility issues, as different protocols and topologies may well exist in the same enterprise. Thus, clients and servers must cooperatively share processing load without regard for operating system or protocol differences. Figure 1–12 illustrates how various server types communicate with different clients across the enterprise network. The clients include workstations, PCs, and terminals that utilize various operating systems.

transmission control protocol/internet protocol (TCP/IP)

The **Transmission Control Protocol/Internet Protocol (TCP/IP)** suite of protocols became a reliable choice for overcoming the operating system differences. TCP/IP is basically a networking protocol that provides communication across interconnected networks, between different computer systems and various operating systems. The TCP/IP protocol suite recognizes that the task of communications is too complex and diverse to be accomplished by a single component. The various tasks are therefore di-

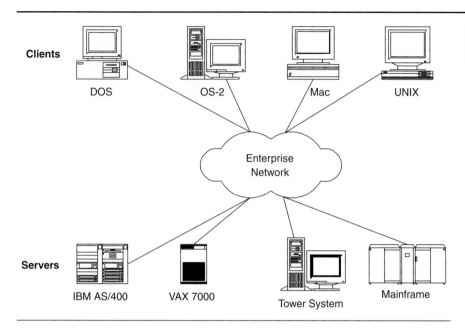

Figure 1–12
Client-Server Distributed Processing

vided into modules that may communicate with peer entities in another system. One entity within a system provides services to other entities, and in turn, uses the services of other entities. This layered protocol structure provides for an effective and efficient method for conducting communication transmissions.

1.9 STANDARDS

The standards process is important to all participants in the broadband communications industry. Without standards, it would be impossible for the multivendor, internetworking environments to exist today. Standards allow multiple vendors to manufacture competing products that will operate with a variety of networking protocols, media, and infrastructures. The end users, therefore, can be assured that hardware and software will operate as specified and will also interoperate successfully. Some of the significant standards organizations are listed in Table 1–3 along with their missions and contributions.

Of particular importance are the IEEE specifications. Table 1–4 provides common cable specifications that can be found in an enterprise network. **Baseband** is a direct transmission method that uses a bandwidth whose lowest frequency is zero.

baseband

Table 1–3
Standards Organizations

Organization	Abbreviation	Mission/Contribution
International Standards Organization	ISO	OSI seven-layer model
Comite Consultif International Telegraphique et Telephonique	CCITT	Telecommunications standards
International Telecommunications Union	ITU-T	CCITT successor
American National Standards Institute	ANSI	Information systems standards
Institute of Electrical and Electronics Engineers	IEEE	Local area network standards
Electronics Industries Association	EIA	Electrical signaling and wiring standards

Table 1–4
Media Comparisons

Specification	Medium	Speed	Segment Length	Transmission Technique
10Base5	50-ohm coax (10 mm)	10 Mbps	500 meters	Baseband
10Base2	50-ohm coax (5 mm)	10 Mbps	185 meters	Baseband
10BaseT	STP / UTP	10 Mbps	100 meters	Baseband
FOIRL	Fiber optic	10 Mbps	1000 meters	Baseband
10BROAD36	75-ohm coax	10 Mbps	1800 meters	Baseband

The enterprise data communications system utilizes the various transmission media to provide connectivity to a LAN or remote data communication devices by passing information over a WAN or MAN. It will become evident that the type of medium is an important issue in the various broadband technologies and topologies.

1.10   DATA COMMUNICATIONS TECHNOLOGY AWARENESS

Telecommunications technology is similar to Data Processing technology as it has its own set of acronyms and terms. The terms presented in this section should provide the reader with sufficient knowledge to communicate in the communications environment. A common base of technology understanding is required so that the telecommunications professional can communicate effectively with the user community. Each term identifying a device carries a specification of LAN, WAN, MAN, or multiple uses. The category of terms is divided into general technology terms, topologies, hardware devices/media, and voice systems. Sample configurations are presented where appropriate.

Technology Terms

General data processing, data communications, and network terms that students may encounter are included in this section, as is a general description of how these elements work in the broadband enterprise environment. Appendix F also provides an extensive set of acronyms.

Client Server   A **Client Server** is a computer that sits on the customer site   client server
and splits computing operations and applications between the desktop and one or more networked PCs or mainframe computer. This allows for distributed applications support and access to a central computer. In this environment, an application component called a client issues a request for services from another application called a server. There may be multiple clients sharing the services of a single server, and the client applications do not need to be aware that processing is occurring remotely. This is possible because a communication network, such as a LAN, provides the means of transporting information back and forth between the client and server component.

The client-server computing environment allows many different types of server systems to be built. The following server systems are possible in the client-server environment:

- File server
- Print server
- Database server
- Communications server
- Applications server

CPE   Customer Premise Equipment (CPE) or Customer Provided Equipment refers to the telecommunications equipment—such as key systems, answer machines, PBXs, routers, hubs, and switches—that resides on the customer's premises. These devices are on the Data Terminal Equipment (DTE) side of the DTE/DCE interface.

DCE   Data Communications Equipment (DCE) refers to devices that sit between the DTE and communications line. Modems, network interface cards, DSUs, and the interfaces on remote bridges, routers, and gateways are examples of DCE. In many cases, the long-distance carriers and Local Exchange Carriers (LECs) will only work network problems up to the DCE interface. Usually the test to be conducted by the carrier is a remote loop-back, which is used to show that the network has connectivity to the DCE device.

DTE   Data Terminal Equipment (DTE) is the source or destination device that originates or receives data over communications networks. Dumb terminals, PCs, and fax machines are examples of DTE, as are devices that have interfaces to these devices such as routers, gateways, and remote bridges. The DTE/DCE interface is important because of the physical connectivity requirements of cabling. The customer is usually responsible for network troubleshooting on the DTE side of the network interface.

operating systems

Operating Systems   Operating systems are software programs that manage the basic operations of a computer system. Examples are MS-DOS and UNIX. An operating system can be thought of as a resource manager or resource allocator. The following resources are usually managed and allocated among competing applications:

- CPU processing time
- Memory access
- Disk storage
- Input/output devices
- Retrieval and file system
- Security

OSI   Open Systems Interconnection (OSI) is the internationally accepted framework of standards for communications between different systems made by different vendors. This seven-layer standard allows for an open systems environment so that any vendor's device can communicate with any other vendor's equipment. Many vendors manufacture their products to conform to layer one and layer two of the OSI model. Appendix A provides an explanation of the standard.

Protocols   Protocols are sets of rules governing the format of message exchange. They are prevalent in the broadband environment. It is essential that one be aware of the protocol issues when designing a system and making a proposal in this environment. There are numerous protocols, including those for file transfers (Kermit), data compression (V.42bis), error control (V.42), and modulation techniques (V.34).

The reason why different networks are able to communicate over different interfaces, using different software, is the establishment of protocols. Protocols provide the rules for how communicating hardware and software components negotiate interfaces or talk to one another. Protocols may be proprietary or open and may be officially sanctioned by standards organizations or market driven. For every potential hardware-to-hardware and hardware-to-software interface, there is likely to be one or more possible protocols supported. The sum of all protocols used in a particular computer is sometimes referred to as that computer's protocol stack.

signaling system 7 (SS7)

SS7   Signaling System 7 (SS7) is an out-of-band signaling process used for supervision, alerting and addressing within the telecommunications network. It separates call setup between switches from the actual voice paths. The three basic functions of SS7 are supervising, alerting, and addressing. SS7 is an integral part of ISDN and provides for additional functions such as call forwarding, call screening, and call waiting.

SS7 is designed to be an open-ended common channel signaling (CCS) standard that can be used over a variety of digital circuit-switched networks. The overall purpose of SS7 is to provide an internationally standardized, general-purpose CCS system with the following primary characteristics:

- Suitable for operation over analog channels and at speeds below 64 kbps
- Suitable for use on point-to-point terrestrial and satellite links
- Designed to be a reliable means of transfer of information in the correct sequence without loss or duplication
- Optimized for use in digital telecommunications networks, utilizing 64 kbps digital channels
- Designed to meet present and future information transfer requirements for management, maintenance, call control, and remote control.

TCP/IP   Transmission Control Protocol/Internet Protocol (TCP/IP) is a networking protocol that provides communication across interconnected networks and between computers with diverse hardware architecture with various operating systems. TCP and IP are two of many protocols in the internet family of protocols. Virtually all computer operating systems offer TCP/IP support and most large networks rely on TCP/IP for their network traffic.

TCP/IP is a result of protocol research and development conducted on the experimental packet switched network, ARPANET, funded by the Defense Advanced Research Projects Agency (DARPA). TCP/IP is organized into five relatively independent layers:

- Application layer
- Transport layer
- Internet layer
- Network access layer
- Physical layer

Topologies
A **topology** is basically the configuration of a communication network. The physical topology describes how the network *looks*. The logical topology describes how the network *works* [Martin, 1996; Stallings, 1997].

topology

LAN Topology   A LAN is a group of data devices, such as computers, printers, and scanners, which are linked with each other within a limited area such as on the same floor or building. There are two ways of looking at the LAN technology: (1) the LAN data link technology that is used to implement a computer network, and (2) the LAN networking software that is used to provide users with local area networking facilities. The LAN data link technology is used to implement a flexible, high-speed form of networking data link. The source system generates a message and uses the facilities of a LAN data link to deliver that message to the destination system. The most commonly used LAN data link technologies are as follows:

- Ethernet
- Token Bus
- FDDI
- Token Ring
- ARCnet
- LocalTalk

A typical network user views a LAN as a collection of computing systems that are capable of communicating with one another. A user primarily interacts with high-level networking software that allows the use of the networked computers. The following are examples of the most commonly used networking software systems:

- NetWare
- TCP/IP
- AppleTalk
- DECnet
- VINES
- LANtastic
- System Network Architecture (SNA)
- LAN Manager

bus
network interface cards
(NIC)

Bus Topology   A **bus** is an electrical connection that allows two or more wires or lines to be connected together. All **Network Interface Cards** (**NIC**s) receive all the same information put on the bus, but only the card that is "addressed" will accept the information. With the bus topology, each system is directly attached to a common communication channel, where signals that are transmitted over the channel comprise the messages. As each message passes along the channel, each system receives it and examines the destination address that is contained in the message. If the destination address is for that system, the message is accepted and processed; otherwise, it is ignored. Figure 1–13 illustrates a logical bus topology.

ring

Ring Topology   A **ring** is a LAN in which all the PCs are connected through a wiring loop from workstation to workstation, forming a circle or ring. Data are sent around the ring to each workstation in the same di-

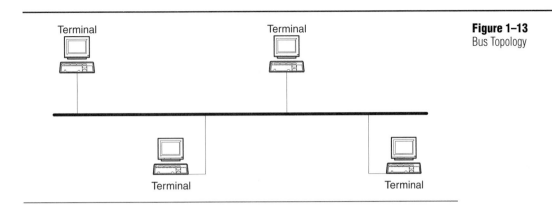

Figure 1–13
Bus Topology

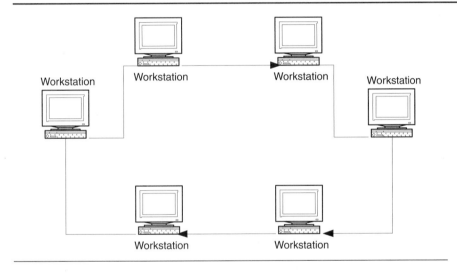

Figure 1–14
Ring Topology

rection. Each system acts as a repeater for all signals it receives and then retransmits them to the next system in the ring at their original signal strength. All messages transmitted by any system are received by all other systems, but not simultaneously. The system that originates a message is usually responsible for determining that a message has made its way all the way around the ring and then removing it from the ring. Figure 1–14 shows a logical ring topology.

Star Topology   A **Star** is a topology in which all phones or workstations are wired directly to a central service unit. The central service unit establishes, maintains, and breaks connections between workstations. With the star topology, all transmissions from one system to another pass through the central point, which may consist of a device that plays a role in managing

star

Figure 1–15
Star Topology

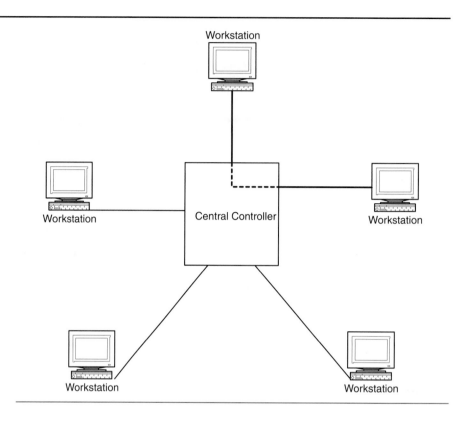

and controlling communication. The device at the center of a star may act as a switching device. When one system wishes to communicate with another system, the central switch may establish a path between the two systems that wish to communicate.

The star topology has been utilized for many years in dial-up telephone systems, in which individual telephone sets are the communicating systems and a Private Branch Exchange (PBX) acts as the central controller. Figure 1–15 shows a logical star topology.

MAN

MAN Topology A **MAN** is a group of data devices, or LANs, linked with each other within a city or large campus area. MANs operate in a manner similar to LANS, but over longer distances. MANs can be used to bridge the gap between WANs and LANs. A MAN could be implemented over a Fiber Distributed Data Interface (FDDI) ring or LEC facilities with LAN extenders and fiber bridges. Figure 1–16 shows two client-server systems that are using the FDDI technology to extend their access across a metropolitan area. The FDDI network is transparent to the users. Other devices can be utilized in place of the gateways.

Figure 1–16
Metropolitan Area Network
Topology

WAN Topology   A **WAN** is a group of data devices, or LANs, that can   WAN
communicate with each other from different cities, different LATAs,
different states, and different countries. Its primary function is to tie
together users who are widely separated geographically. Many of today's networks employ public telecommunications facilities to provide users with access to the resources of centrally located computer
complexes and to permit fast interchange of information among users.
Intelligent terminals, personal computers, workstations, minicomputers, and other forms of programmable devices are all part of these
large networks.

Portal   An encapsulation facility allows two systems that conform to a   portal
given network architecture to communicate using a network that conforms to some other network architecture. The encapsulation facility at
each end is referred to as a **portal.** A pair of portals, one on each end of the
network, can be viewed as implementing what is sometimes called a tunnel. The tunnel transports messages through a network conforming to a
foreign network architecture. Unlike a gateway, an encapsulation facility
does not perform actual protocol conversions. Tunnels will be discussed
in Chapter 6.

Hardware Devices/Media

The devices and hardware that comprise the physical design of a network are considerable and varied in functionality and cost. The systems
designer has a wealth of choices when designing a communications
network, and thus numerous pitfalls to avoid. This section comments
on the general categories of network components, including transport
media, cabling, central office components, and general LAN/MAN/WAN
devices [Institute of Electrical and Electronics Engineers (IEEE),
www.ieee.org, 2000; Palmer and Sinclair, 1999].

Transport Media   Components of transport media include coaxial cable, copper wire, fiber, radio waves, and wave guide. The wireless topics will be discussed in Chapter 11. Specifications for copper wire include
10Base2, 10BaseT, and 10Base5. Figure 1–17 shows two categories of
twisted pair cable.

Figure 1–17
UTP Cable (Reproduced
with permission from Black
Box Corp.)

Category 5 cable   Category 5E cable

Figure 1–18
Coaxial Cable (Reproduced
with permission from Black
Box Corp.)

Copper Wire

10Base2

10Base2 is the IEEE 802.3 Ethernet standard for data transmission over thin coaxial. It is commonly called thin Ethernet, or Thinnet, because the cable diameter is half the size of 10base5. 10base2 LANs run at 10 Mbps.

10BaseT

10BaseT is known as twisted pair Ethernet because it uses the same type of wire as used for connecting telephone systems in an office building. It operates at 10 Mbps and 100 Mbps. It is available in both Unshielded Twisted Pair (UTP) and Shielded Twisted Pair (STP).

10Base5

10Base5 is a transmission specified by IEEE 802.3 that carries information at 10 Mbps using 50-ohm coaxial cable. It is sometimes called Thicknet Ethernet.

coax

Coaxial Cable   Coaxial **(Coax)** cable is composed of an insulated central conducting wire wrapped in another cylindrical conducting wire. This package is usually wrapped in another insulating layer and an outer protective layer (See Figure 1–18). It is typically used on 10Base2 Ethernet implementations and for connections to IBM 327x type terminals.

Multi-mode fiber ST connectors   Single-mode fiber SC connectors

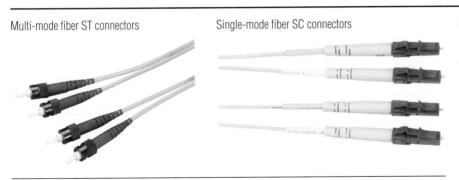

Figure 1–19
Multi-Mode and Single-Mode Fibers (Reproduced with permission from Black Box Corp.)

Spread spectrum   Infrared system

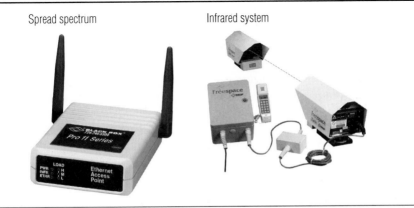

Figure 1–20
Spread Spectrum and Infrared System Components (Reproduced with permission from Black Box Corp.)

Fiber   **Fiber** is a shortened way of saying fiber optic. An optical fiber can be used to carry data signals in the form of modulated light beams. An optical fiber consists of a thin cylinder of glass, called the core, surrounded by a concentric layer of glass, called the cladding. This package is all wrapped in a protective sheath. Fiber is manufactured in single-mode and multi-mode versions. Fiber-optic cables have the potential for supporting high transmission rates. The emitter for single-mode fiber is a laser and the emitter for multi-mode is a Light-Emitting Diode (LED). Figure 1–19 shows both multi-mode and single-mode fiber with ST (round) and SC (square) connectors.

fiber

Radio Waves   Microwave is deployed in two frequency ranges: satellite and terrestrial. Wireless LAN technologies are implemented using infrared, spread spectrum, and narrowband microwave (radio). Figure 1–20 shows devices that are part of an infrared and spread spectrum configuration. Additional details are presented in Chapter 11.

In an infrared LAN an individual cell is limited to a single room, because infrared light does not penetrate walls. Infrared can consist of either diffused or directed beam infrared. Its modulation technique is Amplitude

Shift Keying (ASK) and the access method is Carrier Sense Multiple Access (CSMA) or token passing. The data rate is from 1 Mbps to 10 Mbps and its range is from 50 feet to 200 feet, which is based on the physical layout.

A spread spectrum LAN makes use of the spread spectrum transmission technology. No licensing is required by the Federal Communications Commission (FCC) for use in the United States. Spread spectrum is available in either frequency hopping or direct sequence. The modulation technique is Frequency Shift Keying (FSK) and Quadrature Phase Shift Keying (QPSK), respectively, and the access method is Carrier Sense Multiple Access (CSMA). The data rate is 1 Mbps to 20 Mbps and its range is from 100 feet to 800 feet.

Narrowband microwave (radio) LANs do not use spread spectrum. Some of these products require FCC licensing. The modulation technique is FS/QPSK and the access method is ALOHA or CSMA. The data rate is 10 Mbps to 20 Mbps and the range is 40 feet to 130 feet.

Central Office Equipment   A **Central Office (CO)** is utilized as a centralized location for connecting subscriber's telephone lines. Several different hardware devices are utilized in the COs. These include the WAN switches, channel banks, digital access cross-connect switches (DACS), and multiplexers.

Central Office Switches   Central office switches are designated as Class 1 through Class 5, depending upon their capabilities.

Class 1 regional center is the highest capacity switching office in the PSTN network hierarchy.

Class 2 sectional center is the second highest capacity switching office in the PSTN network hierarchy.

Class 3 primary center is the third highest switching office in the PSTN network hierarchy.

Class 4 toll center is the fourth highest switching capacity office in the PSTN network hierarchy.

Class 5 office is the local switching office

A WAN Switch is located in a CO and is utilized to serve a large number of communication facilities. These switches take the form of equipment such as an AT&T 5ESS or a Nortel DSM10. Frame Relay and SMDS Statistical Time-Division Multiplexed (STDM) switches also fit in this category.

Channel Banks   The original channel bank combined twenty-four analog signal sources into a single DS-1 (1.544 Mbps) bit stream to transmit across either a public or private network. The newer D-4 (superframe) channel bank provides for both voice and digital data traffic. A Data Service Unit (DSU) can be plugged into the D-4 channel bank to carry data at up to 56 kbps digitally.

Subscriber Loop Carrier (SLC96) is basically a channel bank that converts and multiplexes ninety-six voice signals into digital bit streams. It consists of a Central Office Terminal (COT) and the Remote Terminal (RT). Five T-1 lines connect the COT and RT, with four primary T-1 lines carrying the bits for the ninety-six channels (4 * 24), and a spare T-1 line that is used for backup in the event of a failure. This method of redundant services is called protection switching. The SLC96 COT is colocated with the central office equipment and the RT is located near the customer's premises.

Digital Access Cross-connect Switch The **Digital Access Cross-Connect Switch (DACS)** is a system that uses a time division multiplexing scheme to switch and cross-connect digital bit streams. The number of input ports equals the number of output ports, which allows for a non-blocking connectivity. This means that all circuits connected to the system can operate simultaneously with any overload. These devices are used by the telephone companies (telcos) for the following functions:

digital access cross-connect switch (DACS)

- Administration and testing
- Reconfiguration of ports on the system
- Rearranging of channels within a digital span
- Load balancing
- Network recovery and switching

The DACS also allows for linking traffic across multiple digital links and the insertion and deletion of bit streams into a channel (drop and insert). These devices were developed for use in a central office or Point Of Presence (POP) to gain access to every digital bit stream at the DS-0 and DS-3 levels. A channel on a T-1, called a digroup, can be multiplexed onto a channel from a different T-1, or from one T-3 to another T-3. The DACS allows a considerable amount of flexibility in the manipulation of the central office facilities in the event of a major network failure. Figure 1–21 shows how multiple, partially filled T-1s can be multiplexed into full (24 channel) T-1s. Four partially filled T-1s can be concentrated into two T-1s.

Multiplexers The next step above the channel bank is the **Multiplexer (MUX)**. The MUX provides flexibility, in that services can be channelized into twenty-four fixed time slots or non-channelized. The MUX supports multiple interfaces via the use of a wide range of plug-in cards which provide subrate speeds of less than the basic 64 kbps or super-rate channel capabilities above 64 kbps.

multiplexer (MUX)

A T-1 MUX handles signals from different types of sources, including voice, data, video, and fax. A T-1 MUX can also support several modulation standards such as Pulse Code Modulation (PCM) and Adaptive Differential PCM (ADPCM). The MUX may also allow for high-compression voice multiplexing at 8 kbps and 16 kbps, allowing a single 64 kbps channel to be used by up to eight voice devices.

Figure 1–21
Concentration with a DACS

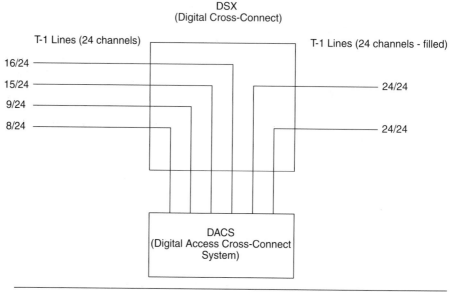

Data Processing Equipment Data processing equipment includes communication controllers, disk drives, and front-end processors. These devices are normally part of a mainframe computer system.

Controllers A controller is a device that controls the operation of another piece of equipment. Specifically, in data communications, a controller resides between a host and terminals and relays information between them. Controllers can be housed in the host, be stand-alone, or reside on a file server. Interfaces and interface cards that provide OSI layer 1 physical connectivity reside in these devices.

Disk Technology Servers require enormous amounts of disk storage. Web Servers store thousands of web pages and associated image, sound, and video files. Storage devices can include Small Computer System Interface (SCSI) disk drives, Fibre Channel, and Redundant Array of Inexpensive Disk (RAID) drives. High-end PC servers use SCSI drives, which are somewhat faster, but more expensive than the standard Enhanced Integrated Drive Electronic (EIDE) disk drive (see Figure 1–22). For extremely high-speed disk access, high-end servers are moving toward Fibre Channel disk access. Whereas SCSI hosts must be within about 12 meters of their disk drives, Fibre Channel allows disk drives to be hundreds of meters and even kilometers away from their hosts. RAID uses multiple disk drives controlled by a single controller board.

front-end processor (FEP)

Front End Processor The **Front-End Processor** (**FEP**) acts as a traffic cop of the mainframe data communications world. It typically resides in

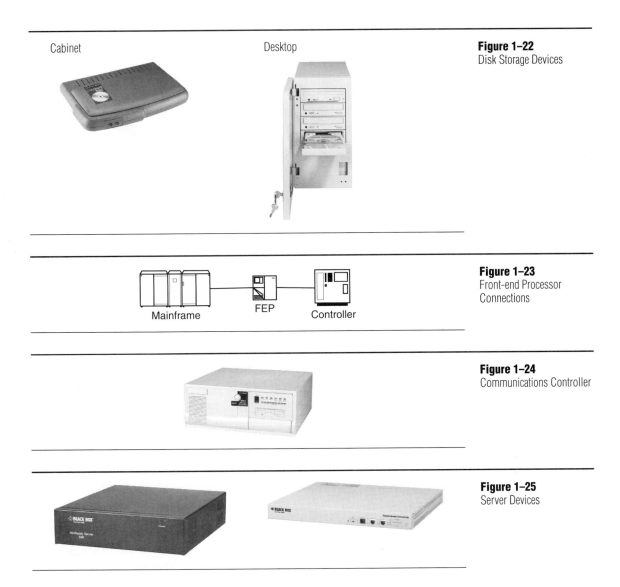

Cabinet   Desktop

Figure 1–22
Disk Storage Devices

Figure 1–23
Front-end Processor
Connections

Mainframe   FEP   Controller

Figure 1–24
Communications Controller

Figure 1–25
Server Devices

front of the mainframe computer and is designed to handle the telecommunications burden so the mainframe computer can efficiently process its programmed functions. Figure 1–23 shows the connectivity of a FEP between a mainframe processor and a communications controller (see Figure 1–24). Cable distance is a major concern in this connectivity. Smaller systems use Input/Output Processor (IOP) modules to accomplish this same task.

Servers   A server has a hardware and software component. It is a shared computer on the LAN and a program that provides some service to the other client programs. The server (Figure 1–25) is a component of the

Figure 1–26
Local and Remote Bridges
(Reproduced with
permission from Black Box
Corp.)

Local Bridge   Remote Bridge

client-server environment and connects to a hubbing device. Servers are gatekeepers to systems and information files. The following server systems make the client-server environment possible:

- File server
- Print server
- Database server
- Communications server
- Application server

LAN and MAN Equipment   Equipment described in this section is primarily utilized in a LAN, but some of these devices can be used interchangeably in either topology. Devices include bridges, firewalls, gateways, hubs, LAN switches, Multistation Access Units (MAUs), Network Interface Cards (NICs), repeaters, routers, and transceivers [Gelber, 1997].

bridges

Bridges   **Bridges** connect multiple LANs. They have less intelligence than routers, because they use simple packet filtering to determine whether traffic stays on the LAN or is passed off to an adjacent LAN. Segmenting is a common function for a bridge. There are both local and remote bridges (See Figure 1–26). See the following discussion on LAN switches for increased functionality.

A bridge forwards frames from one LAN segment, but is more flexible and intelligent than a repeater. A bridge interconnects separate LAN or WAN data links rather than only cable segments. Some bridges learn the addresses of the stations that can be reached over each data link they bridge, so they can selectively relay only traffic that needs to flow across each bridge. The bridge operates in the OSI Layer 2 Medium Access Control (MAC) sublayer, and is transparent to software operating in the layers above the MAC sublayer.

A bridge can interconnect networks that use different transmission techniques and/or different MAC methods. A bridge might be used to interconnect an Ethernet LAN with a Token Ring or a FDDI LAN. Examples of these multiple-protocol bridges are translational bridges and source-route transparent bridges. A pair of remote bridges with a telecommunications facility between them can be used to interconnect two LANs that are situated in different geographical locations. Figure 1–27 depicts two Ether-

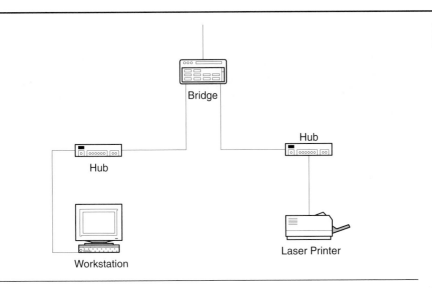

Figure 1–27
Local Bridge Configuration

Bridge

Hub

Hub

Workstation

Laser Printer

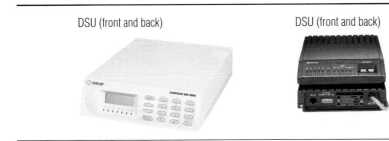

DSU (front and back)

DSU (front and back)

Figure 1–28
Data Service Units
(Reproduced with
permission from Black Box
Corp.)

net LAN segments—one segment has a workstation and the other segment has a printer. Access to the hubbing devices is through the local bridge, whose access is based on the proper addressing scheme.

Data Service Units   The primary function of a **Data** (digital) **Service Unit (DSU)** is to convert a unipolar signal from the customer's DTE into a bipolar signal for the network (DCE) and vice versa. The Channel and Digital Service Units (CSU/DSUs) have evolved to the point of being combined, so they are not generally available separately. This combination allows for the transmission of high-speed data communications rates from 56 kbps to the T-1 rate of 1.544 Mbps, with additional capabilities for fractional T-1 (24 channel) connectivity. Some of the latest technology equipment, such as multiplexers and channels, integrates the functions of the CSU/DSU into the chassis. Figure 1–28 depicts two models of DSUs.

data service unit (DSU)

Figure 1–29
Firewall Devices
(Reproduced with
permission from Black Box
Corp.)

Firewall   Firewall

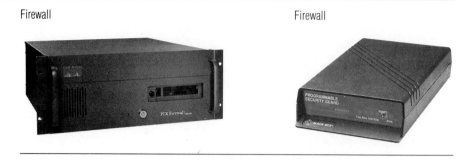

firewall

Firewalls A **firewall** is a computer with special software installed between the Internet and a private network for the purpose of preventing unauthorized access to the private network. All network packets entering the firewall are filtered, or examined, to determine whether those users have authority to access requested files or services and whether the information contained within the message meets the criteria for forwarding over the internal network.

A filter is a program that examines the source address and destination address of every incoming packet to the firewall server. Routers are other network access devices that are capable of filtering data packets. Filter tables are lists of addresses whose data packets and embedded messages are either allowed or prohibited from proceeding through the firewall server into the corporate network. Filter tables can also limit the access of certain IP addresses to certain directories. Figure 1–29 shows rack-mount and table-mount versions of firewall devices.

gateway

Gateways A **gateway** is an entrance and exit into a communications network. Technically, a gateway is a translation device that converts electrical signals from one type of network to another, or converts code from one type of application to another. Gateways operate at all layers of the OSI model.

A gateway is a fundamentally different type of device than a bridge, repeater, router, or switch and can be used in conjunction with them. A gateway makes it possible to communicate with an application program that is running in a system conforming to some other network architecture.

A gateway performs its function in Layer 7 (Application) of the OSI model. The function of a gateway is to convert from one set of communication protocols to some other set of communication protocols. Protocol conversion includes message format conversion, address translation, and protocol conversion.

Internet gateways offer a LAN-attached link for client PCs to access a multitude of Internet-attached resources including e-mail, FTP/Telnet, news groups, Gopher, and the World Wide Web. The Internet gateway translates between the LAN's transport protocol and TCP/IP. Figure 1–30 shows a sophisticated gateway device.

Gateway

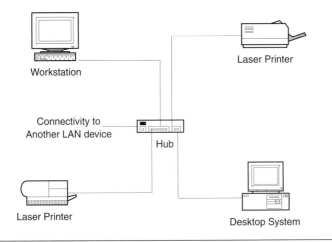

Figure 1–30
Gateway Device

12 port hub (front and back views)   Stacked hubs (12 and 24 ports)

Figure 1–31
Standalone and Stacked
Ethernet Hubs

Figure 1–32
Ethernet Hub

Laser Printer

Workstation

Connectivity to
Another LAN device

Hub

Laser Printer

Desktop System

Hubs   A **hub** is the intelligent wiring delivery point to which print-   hub
ers, scanners, PCs, and so forth are connected within a segment of an
Ethernet/802.3 LAN. Hubs enable LANs to be connected to twisted-pair
cabling instead of coaxial cable in the 10Base2 topology. Since it operates in
a half duplex mode, only one device at a time can transmit through a hub.
Speed is usually 10/100 Mbps. A simple hub (Figure 1–31) is an OSI Layer
1 device. Wiring hubs are useful for their centralized management capabilities and for their ability to isolate nodes from disruption. Figure 1–32

Figure 1–33
Token Ring MAU

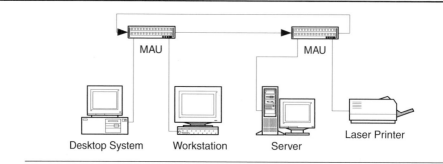

Figure 1–34
Multistation Access Units
(MAU Reproduced with
permission from Black Box
Corp.)

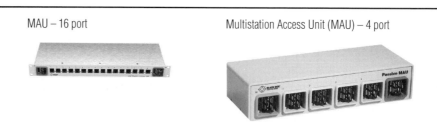

MAU – 16 port   Multistation Access Unit (MAU) – 4 port

depicts a hub that is used to connect two printers, a desktop system, and a workstation with another LAN device. The desktop system and the workstation can access either printer.

Modem   A modulator/demodulator (**modem**) is a DCE device that modulates and demodulates signals. Modulation techniques include amplitude modulation, frequency modulation, and phase modulation. The modem provides an interface between digital devices and analog circuits and equipment. Modems are designed for different applications and are available in a number of configurations and speeds. Many PCs are currently being shipped with a 56 kbps modem card for Internet access.

multistation access unit
(MAU)

Multistation Access Unit   A **Multistation Access Unit** (**MAU**) is the intelligent wiring delivery point to which printers, scanners, PCs, and so forth, are connected within a segment of a Token Ring/802.5 LAN. Connectivity is usually via coaxial cable; however, shielded twisted-pair cable is currently being utilized. The transmission speed is usually 4/16 Mbps.

A MAU (Figure 1–34) is a small electronics device that usually contains eight ports for device connectivity plus a ring-in and a ring-out port for additional MAU devices which allow cascades to form larger rings. Figure 1–33 shows two MAUs cascaded into one ring topology. The server and the two computers can access the printer.

network interface card
(NIC)

Network Interface Card   A **Network Interface Card** (**NIC**) is an adapter that connects a device to the network. The NIC, usually a PC extension board, provides the functionality needed by the connected device to share

Network Interface Card (NIC)

Network Interface Card (NIC)

Figure 1–35
PC Network Interface Cards
(Reproduced with
permission from Black Box
Corp.)

Repeater for coaxial cable

Fiber Extender

Figure 1–36
Coaxial Cable and Fiber
Repeater Devices
(Reproduced with
permission from Black Box
Corp.)

a cable or medium with other stations. Some types of computing devices
that are designed for use on specific types of networks, such as a network
printer, have the functions of a NIC integrated directly into the device.
Figure 1–35 shows two different types of network interface cards.

Network Termination 1 **Network Termination 1 (NT-1)** provides
functions relating to the physical and electrical termination of the local
loop between the carrier and the user ISDN CPE. NT-1 functions are required for both BRI and PRI ISDN services. Additional details will be provided in Chapter 9.

network termination 1
(NT-1)

Repeater The simplest facility used for network interconnection is
the **repeater** (Figure 1–36). The major function of a repeater is to receive a
signal from one LAN cable segment and to retransmit it, regenerating the
signal at its original strength over one or more cable segments. A repeater
operates in the OSI model Layer 1 (Physical) and is transparent to all the
protocols operating in the layers above that layer.

repeater

Using a repeater between two or more LAN cable segments requires
that the same physical layer protocols be used to send signals over all the
cable segments. Repeaters are used in bus-structured LANs, such as Ethernet, to connect individual cable segments to form a larger LAN.

Figure 1–37
Ethernet LAN Repeater

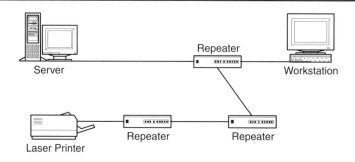

Figure 1–38
Router Devices

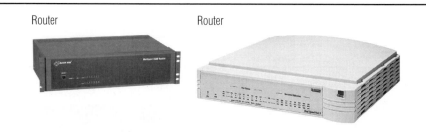

Repeaters are generally transparent to the networking software operating in the individual computers attached to a LAN. The main characteristic of a repeater is that all the signals that are generated on the cable segment on one side of the repeater are propagated to the cable segment or segments on the other side of the repeater. A repeater implements no form of filtering capability.

The 5-4-3 rule must be observed when deploying repeaters. This means that between two end devices on a five-segment LAN, four repeaters can be utilized, with three segments that have devices connected. Figure 1–37 shows three repeaters with devices attached in two Ethernet segments.

routers

Routers **Routers** (Figure 1–38) connect multiple LANs. They are more functional than bridges in that they can handle more protocols, offer traffic control capabilities, and perform better in highly meshed networks. Routers are the "central offices" for the Internet. Filtering and firewall functions are available in a router.

Routers provide the ability to route messages from one system to another where there may be multiple paths between them. A router performs its function in the OSI model Layer 3 (Network). Routers typically have more intelligence than bridges and can be used to construct enterprise internetworks of almost arbitrary complexity. Interconnected routers in an internet all participate in a distributed algorithm to decide on the optimal path over which each message should travel from a source system to a destination system.

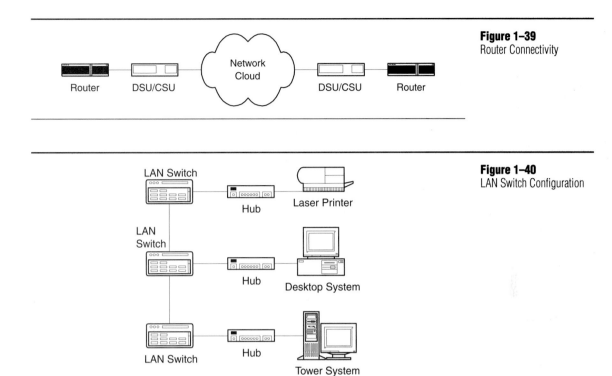

Figure 1–39
Router Connectivity

Figure 1–40
LAN Switch Configuration

In the internetworking environment, a router is often called an inter-mediate system. By contrast, systems that originate data traffic and serve as the final destination for that traffic are called end systems. In general, a router performs the routing function by determining the next system to which a message should be sent. It then transmits the message to the next system over the appropriate link to bring the message closer to its final destination. Figure 1–39 illustrates two user networks that are connected through a network cloud utilizing two DSUs.

Switches   In addition to repeaters, bridges, and routers, a variety of different types of switching facilities can be used in constructing inter-nets. The main purpose of interconnecting LAN data link segments using interconnection devices is to allow a greater degree of sharing a commu-nication medium. Network interconnection devices allow a larger num-ber of different network devices to be attached to the same LAN.

A LAN switch functions like a bridge and acts as a backbone inter-face device. Like a bridge, a LAN switch can be used to segment LANs to improve performance and enhance security. A LAN switch provides switching through hardware, which provides improved performance over a bridge that switches through software. Speeds available are 10 Mbps and 100 Mbps. A LAN switch operates at Layer 2 of the OSI model. Figure 1–40 depicts three LAN switches that are utilized as backbone

Figure 1–41
LAN Switches

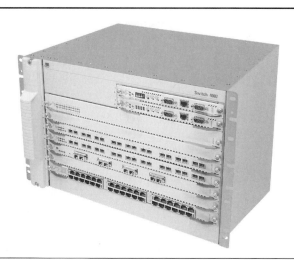

devices on three different locations in a building. These LAN switches could be connected with multimode fiber, coaxial cable, UTP, or STP. LAN switches (Figure 1–41) are being substituted for bridges.

transceiver

Transceivers   A **transceiver** is the attachment hardware in 802.3 networks that connects the controller interface to the transmission cable. The transceiver contains the carrier-sense logic, the transmit-receive logic, and the collision-detect logic. It is called the Attachment Unit Interface (AUI), however it is also called the medium attachment unit by some users. It provides MAC services such as jabber inhibit, and heartbeat for a LAN station. It provides connectivity for a number of different physical media, such as coax, twisted pair, and fiber (Figure 1–42).

Voice Systems

private branch exchange (PBXs)

Voice systems include **Private Branch Exchanges** (**PBXs**), key systems, and hybrid systems. Table 1–5 provides a comparison of these systems' features [Green, 1996].

Private Branch Exchange   In the typical PBX environment, a T-1 digital trunk interface, which acts like a channel bank, resides in a shelf in the PBX system. As an integral part of the PBX, this T-1 interface is designed primarily around voice circuits; however, newer PBX generations use this interface to provide video and data services. The PBX offers some features that are not accessible to the multiplexer or network, including the following [FitzGerald and Dennis, 1999]:

- Queuing a channel on a T-1
- Station Message Detail Recording (SMDR)

Fiber Transceiver

Token Ring Media Converter

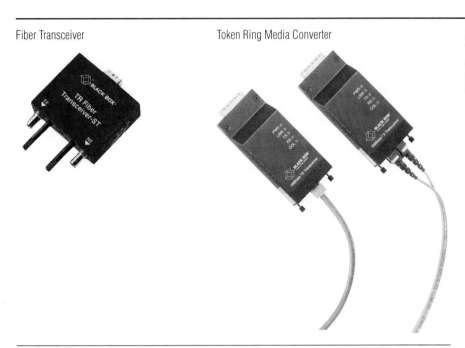

Figure 1–42
Transceiver Device
(Reproduced with
permission from Black Box
Corp.)

Table 1–5
PBX, Key System, and
Hybrid System
Comparisons

Feature	PBX	Key System	Hybrid System
Generic program	Stored in RAM	Stored in ROM	Stored in RAM or ROM
Networking	Full	Tie line	Tie line or not available
Trunk access	Pooled	Button selection	Pooled or button selection
Least-cost routing	Programmable	Programmed in ROM	Programmable or programmable in ROM
Automatic Call Distribution (ACD)	Available	Not available	Available with reduced features
Attendant call transfer	Transfer to station	Announce over intercom	Transfer or announce
Dial code restrictions	Full	Limited or none	Limited to full
Traffic usage measurement	Available	None	Limited or none

- Call forwarding
- Call transfer
- Conferencing
- Hunting
- Least-cost routing
- Redialing

Digital PBXs can switch data as easily as voice, but office automation does not function well in a circuit switched environment, and the bandwidth required is often greater than a PBX supports.

Key Systems   Many key systems are designated by the capacity in central office lines and stations. For example, an 824 system could accommodate a maximum of eight lines and twenty-four stations. In a key system each line is terminated via a button on a telephone instrument. The number of buttons that can be physically terminated on the telephone tends to put an upper limit on the feasible size of a key system. At some point it is awkward to select lines manually for outgoing calls and to use the intercom for announcing incoming calls. A hybrid, in which the attendant transfers calls and the user dials an outgoing call access code, is usually required. Many key systems require proprietary telephone sets, and may not use the standard 2500 telephone set. Because key system attendants must announce each call to the called party, most modern systems have an internal paging feature that allows the attendant to hold a two-way conversation over a speaker/microphone that is built into the telephone set.

Hybrid Systems   Private branch exchanges are rarely economical in small line sizes, and large organizations require features and capacity that key systems lack. For companies in between, the market offers a combination PBX and key system that is called a hybrid. A hybrid system has characteristics of both the PBX and key system and can usually be configured as either.

■ SUMMARY

Divestiture and deregulation specifies the rules that govern the relationships between the various communication vendors and the customer base. The communications professional must be aware of the current ever-changing regulatory environment.

It is essential that a system designer, marketer, or system administrator possess a general knowledge and awareness of basic data communication architecture and components so that a high level of communication can be established with the user.

Merging of different enterprise networks can result in a number of different protocols and different topologies that must interact successfully. The broadband professional must possess or develop the skills to make this merger happen successfully.

Key to the development of successful LAN programs was the development of the devices called routers, bridges, switches, and gateways for sending data between LANs. This allowed for transmission and access to LANs that had both the same and different protocol suites.

Several hardware devices and software elements that are components of the LAN, MAN, and WAN architectures can impact the performance and integrity of the enterprise network. It is essential that anyone working in this environment be cognizant of this impact and understand how to effectively utilize these network elements.

Key Terms

10Base2	International Standards Organization (ISO)
10Base5	Local Access Transport Areas (LATAs)
10BaseT	Local Area Network (LAN)
Access Lines	Local Loops
Baseband	MAN
Bridges	Multistation Access Unit (MAU)
Bus	Media
Cell	Metropolitan Area Network (MAN)
Central Office (CO)	Modem
Circuit Switching	Modified Final Judgment
Client-Server	Multiplexer (MUX)
COAX	Network Interface Cards (NIC)
Customer Premises Equipment (CPE)	Network Termination 1 (NT-1)
Digital Access Cross-Connect Switch (DACS)	Open Systems Interconnection (OSI)
Data Communications Equipment (DCE)	Operating Systems
Data Service Unit (DSU)	Packet
Deregulation	Private Branch Exchange (PBXs)
Divestiture	Portal
Equal Access	Protocol
Fiber	Protocol Data Unit (PDU)
Firewall	Public Switched Telephone Network (PSTN)
Frame	Repeater
Front-end Processor (FEP)	Ring
Gateway	Router
Hub	RS232

Signaling System 7 (SS7)

Star

Topology

Transceiver

Transmission Control Protocol/Internet Protocol (TCP/IP)

Trunks

V.35

WAN

Wide Area Network (WAN)

REVIEW QUESTIONS

1. Explain divestiture and equal access.

2. What are the major stipulations of the Telecommunications Act of 1996?

3. What major components comprise the telecommunications infrastructure?

4. What is a LATA and how many are there? What is the significance of a LATA?

5. What are the central office categories? What are their different functions?

6. What is the difference between an access line and a trunk?

7. What tasks are performed in the telecommunications network? Is there a significant amount of energies required for these functions?

8. What are the three phases of a switched call?

9. What is the difference between a dedicated call and a switched call?

10. What are the four basic elements of a data communications system? What are their functions?

11. Explain the DTE/DCE concept. Why is it important?

12. Explain the differences between the various protocol data units. What is the difference between a frame and a packet?

13. What is the difference between guided and unguided media? Give examples of each.

14. Four network properties can be utilized to determine the demarcation point between a LAN and a WAN. Describe each.

15. Compare and contrast LANs, MANs, and WANs.

16. Discuss the WAN services and technologies that can be utilized over a WAN.

17. Describe the structure of a MAN. What is the major function and usage of a MAN?

18. What led to the development of local area networks? What are the technologies that are utilized in the LAN environment?

19. What are the most commonly used networking facilities utilized in internetworking?

20. Describe the configurations of the basic data processing systems.

21. Discuss the RS232 and V.35 interfaces. How are they used? How many pins are utilized in each of these interfaces?

22. Why does the network need to utilize TCP/IP?

23. What is the role of agency standards in networking?

24. What are the different IEEE media specifications? How are they utilized?

25. What is a client-server system? Give examples.

26. What is the relationship between CPE and DTE?

27. What is the relationship between DTE and DCE?

28. What is a protocol? Give examples. What is a protocol stack?

29. How is SS7 utilized in the network?

30. What are the most common LAN data link technologies?

31. Give examples of the most commonly used networking software systems.

32. What is the difference between a bus, a ring, and a star topology?

33. What is a portal? Where would a portal be found?

34. What are the different varieties of transport media? Describe the wire-based media.

35. What is a channel bank? How is it different from a SLC96?

36. What is the function of a DACS? What is a POP? How does it relate to a DACS?

37. What is the function of a central office multiplexer?

38. Describe the configuration and connectivity of a mainframe, FEP, and communications controller.

39. What is the function of a bridge? How does this differ from a LAN switch?

40. Describe the use of a DSU and modem.

41. What is the difference between a firewall, gateway, and router? What is the primary function of a router?

42. Describe the utilization of a MAU and hub. How are they similar and different?

43. What is a network interface card? How is it used?

44. What is the function of a repeater? Is there a limitation on their usage?

45. What are the three major categories of voice systems? Describe each.

ACTIVITIES

1. Use the Internet and other online sources, trade journals, and texts to research any component that has been presented in this chapter. If the subject is hardware or software, show how it interacts with the network. If not, describe how it is relevant to the enterprise network.

2. Utilize the sources developed in activity 1 to price out the components that are necessary to procure the network devices.

3. Based on the student's expertise level, utilize either the logical or the physical representation technique to illustrate a WAN, MAN, or LAN network. The components of this activity would include the following steps:

- Identify a customer or potential customer and develop a design that depicts a simple LAN, MAN, or WAN network.

- Identify the various components including routers, hubs, switches, repeaters, MAUs, and workstations.

- Show how wiring and physical connectivity is important to the network design.

- Identify DCE devices and network components such as DSUs, modems, central offices, carriers, and facilities. They should include the appropriate clouds such as IXCs and LECs.

This activity can be quite simple in nature or can be complex depending on the amount of time available for the task. Note that a logical design might be sufficient if specific details such as device interface connectivity are not required. A physical design would be required, however, if it were to be an attachment to an order or a Request For Proposal (RFP). These activities can also be the source for both individual and group projects.

A number of Websites are useful in developing these network designs. These sites include sample applications and the physical network design including the components necessary to effect a solution. The following sites are available:

- www.3com.com/technology/index.html
- www.cisco.com/
- www.nortelnetworks.com/products/index2.html

4. Arrange for an IXC or LEC to conduct a tour of a local networking facility. Sketch the layout of this facility.

2

The WAN Infrastructure

■ INTRODUCTION

Wide area data links are typically used to provide point-to-point connections between pairs of systems that are typically located some distance from one another. Historically, most telecommunications facilities offered by common carriers provide an analog communications channel designed for the purpose of carrying telephone voice traffic. This is commonly referred to as Plain Old Telephone Service (POTS).

Data communications usually involve a computer with one or more terminals connected by communications lines or a number of computers interconnected with the Internet. The communications lines might be standard telephone lines, dedicated high-speed data communications lines, or special arrangements that use telephone lines for services such as Integrated Services Digital Network (ISDN) or Asymmetrical Digital Subscriber Line (ADSL).

Telecommunications carriers now also provide specialized digital telecommunications circuits optimized for data transmission rather than voice transmission. There are various levels of digital service available, corresponding to different data rates. Common digital signal levels include *DS0* (64 kbps), *DS1* (1.544 Mbps), and *DS3* (44.736 Mbps).

A new trend is emerging during which these private line and dedicated facilities are being replaced by the new broadband technologies. It is therefore essential that the network professional know the components and operation of the private line environment and how to migrate it to the new broadband technologies.

This chapter will give the reader the tools to identify new technologies that can be utilized to replace the older private line and dial-up technologies. By understanding how the current environment works, the reader can use this information to develop a migration strategy to the newer broadband technologies.

OBJECTIVES
Material included in this chapter should enable you to:

- understand the basic technology of dial-up and dedicated services, learn how to migrate them to broadband services, and identify broadband applications.

- become familiar with terms, definitions, and standards of the broadband network infrastructure.

- understand the various transmission methods and how they operate within the broadband environment.

- identify and utilize criteria to specify configurations and applications in the broadband arena.

- understand how the various routing protocols play a major role in broadband networks. Identify the pros and cons for each routing protocol.

- become familiar with circuit switching and packet switching technologies. Identify the pros and cons for circuit switching and packet switching networks.

2.1   MIGRATION TO BROADBAND SYSTEMS

Overlaying the voice network of switches, transmission equipment, and Customer Premise Equipment (CPE) is a vast array of communications services. The Public Switched Telephone Network (PSTN) is only one of the many services that ride these facilities, which is not a single network, but many separate networks under multiple ownership that come together in the local exchange network. The power of the PSTN lies in the fact that the network's structure is transparent to the user.

Although the PSTN is orderly and easy to use, the same cannot be said for data networks. Data networks use the same backbone facilities and infrastructure as voice, including the copper cable local loop and the fiber-optic transmission facilities. Communication between different networks on these facilities, however, is not assured. The data equivalents of the PSTN are value-added networks that use packet switching on Switched Multimegabit Data Service (SMDS) or Frame Relay technologies.

These technologies, as discussed in Chapters 3 and 10, enable users to send traffic across a public network over Permanent Virtual Circuits (PVCs). A **virtual circuit** is defined in the network software; the path does not occupy a fixed hardware circuit as does a private line dedicated circuit. Another type of circuit, called a Switched Virtual Circuit (SVC), which would allow a user to dial an end-to-end data connection over a public data network, is not available at this time. The primary reasons why SVC services have not been implemented are the lack of demand for the Local Exchange Carriers (LECs) to implement this type of service and the exorbitant infrastructure costs for the carriers.

A new form of communications network is replacing the current telecommunications environment. With a few exceptions such as Frame Relay, today's network services are composed of fixed-bandwidth circuits. These fixed bandwidth circuits are not efficient because they are underutilized when there is no traffic and blocked when the traffic exceeds the fixed bandwidth. Some services, such as voice, work effectively in a fixed bandwidth environment; however, many of today's applications require a lot of bursty-oriented bandwidth and then no bandwidth at all.

Switching is at the heart of many applications. It is possible to connect to anyone anywhere with a PSTN connection but this is insufficient with these new applications. A new switching and multiplexing system is required to enable users to connect to anyone they choose, get the bandwidth they need when they require it, and only pay for the time that the connection is utilized. A technology that can fulfill these broadband switching requirements is Asynchronous Transfer Mode (ATM), which will be discussed in Chapter 4.

For the reader to understand how this transition will take place between the current private line dedicated circuits, dial-up circuit switch-

virtual circuit

ing, and the broadband switching technologies, information on the various elements and components of this networking environment will be presented. This discussion begins with details concerning network configurations and network design.

2.2 NETWORK CONFIGURATIONS AND DESIGN

The design of most networks, including wide area networks, incorporates a variety of different topologies. **Topology** refers to the physical design and connectivity of a communications network. A topology can be defined as the physical arrangement of nodes and links to form a network, including the connectivity pattern of the network elements.

topology

The basic function of a communications network is to provide access paths by which an end user at one location can assess end users at other locations. The designer must design the network to meet the needs of the users rather than base it on some particular topology or technology. *Top-down Network Design* [Oppenheimer, 1999] by CISCO Press is an excellent book that takes a logical network design approach and transforms it into a physical network design.

The various technologies presented in this book include a majority of the current broadband offerings, including Frame Relay, Asynchronous Transfer Mode (ATM), Digital Subscriber Line (DSL), Fiber Distributed Data Interface (FDDI), and wireless technology.

There are three basic types of communications networks: *centralized, decentralized,* and *distributed.*

- A centralized network is a computer network with a central processing node through which all data and communications flow. This type of network requires dedicated hard-wired devices that operate using high-speed parallel or serial transmission techniques.
- Decentralized networking implies some processing distribution function using a host processor to control several remotely located processors. The host processor can off-load activities to other processors, but it still maintains control. Data are often transported over serial links in a synchronous or asynchronous transmission mode.
- In distributed processing, a computer or node in the network performs its own processing and stores some of its data while the network manages communications between the nodes. This type of network could be considered a peer-to-peer network spread over a large geographical area. It usually consists of many different types of hardware and software components. In distributed networks, all nodes in the network share responsibility for application processing, which requires that data format conversion and communications protocol-handshaking capabilities reside in each node.

Figure 2–1
Access Line and POP
Connectivity

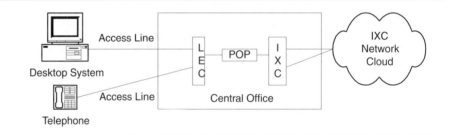

2.3   ACCESS CIRCUIT TYPES

Whether the communications network topology is decentralized or distributed, there must be a method for providing connectivity between the processors or nodes. In today's environment, this connectivity takes the form of an access circuit or line. These access facilities are available from either Incumbent Local Exchange Carriers (ILECs), which are the Telcos, or Competitive Local Exchange Carriers (CLECs).

circuit

In telecommunications, a **circuit** is a discrete (specific) path between two or more points along which signals can be carried. Unless otherwise qualified, a circuit is a physical path consisting of one or more wire pairs (two wire/four wire) and possibly intermediate switching points. An

access line

access line or local loop is the local access connection or telephone line between a customer's premise and a carrier's Point Of Presence (POP),

central office (CO)

which is in the carrier's **Central Office (CO)**. A POP generally takes the form of a switch or router. It can be a meet point for Internet service Providers (ISPs), where they exchange traffic and routes. Figure 2–1 de-

interexchange carrier
(IXC)

scribes the access line (local loop) and POP environment. The **Interexchange Carrier (IXC)** communicates with the LEC through the POP, and this POP device is often collocated with the carrier's equipment.

Dedicated (Leased Line) Circuit
Dedicated lines often have the following characteristics:

- Fixed pricing
- Available 24 hours/day, 7 days/week
- Mileage sensitivity
- Managed by the user
- Secure service
- Good quality

Leased line service is available in numerous speeds. It is essential that the user be aware of these speeds, because a migration to a broadband product may be speed sensitive. Service is usually available in two categories,

Digital Data Service (DDS)	2.4 kbps, 4.8 kbps, 9.6 kbps, 19.2 kbps, 28.8 kbps, 56 kbps	**Table 2–1** Circuit Speeds
Digital Signal Hierarchy (DSx)	DS0 (64 kbps) DS1 (1.544 Mbps) DS3 (44.736 Mbps)	

Digital Data Service (DDS) and high-speed services. Table 2–1 provides a list of generally available speeds.

A network is an arrangement of circuits. In a dial-up or *switched* connection, a circuit is reserved for use by one user for the duration of the calling session. In a *dedicated* or leased line arrangement, a circuit is reserved in advance and can only be used by the owner or renter of the circuit. It is advantageous for a user to lease a dedicated line if security is an issue and if the circuit will be in use for most of the time. Occasional use would dictate a dial-up line; however, dial-up quality is usually less than dedicated quality.

Switched or dial-up circuits are provided by the standard public telephone network, and using them for data is similar to making a telephone call. A temporary connection is built between Data Terminal Equipment (DTE) as if there was a direct connection between the two devices. The circuit is set up on demand and discontinued when the transmission is complete. This technique is called circuit switching. An obvious advantage of circuit switching is flexibility. Usage charges are based on the duration of the call and distance, just like in a standard telephone call.

One factor that may impact the quality of a dial-up circuit is the availability of the actual network facilities used in routing the call when the connection occurs. Circuits may be variable in quality, and therefore can be good in one situation and marginal in another. Because of this variability, the data transmission speed that can be achieved dependably is less than with dedicated facilities. When a subscriber uses switched circuits, it exposes itself to security risks. It is possible for unauthorized persons to dial in and access the computer facilities if security procedures are not adequate.

Figure 2–2 provides a general model of the difference between a private line, which is dedicated to the user, and switching, which provides shared facilities for the numerous users.

WAN switches are essentially components of public or private communications networks. A switching system essentially contains a specialized computer that performs routing of signals over transmission routes. Examples of these WAN switches include AT&T #5ESS and Nortel DMS10. A communications network consists of numerous similar and dissimilar types of switches with various types of transmission links to connect them for seamless communication. Transport of signals through

Figure 2–2
Dedicated/Switched Lines

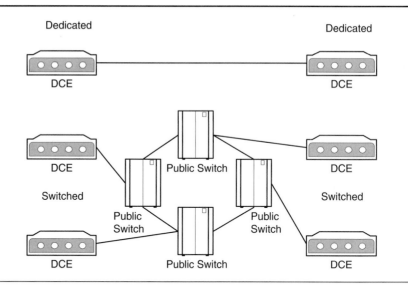

circuit switching
packet switching

the switched telephone network is accomplished via **circuit switching** or **packet switching.**

2.4   CIRCUIT SWITCHING

Circuit switching is like having your own highway (facility) on which your conversation can travel. As long as the connection stays open, no one else can use it. Most voice telephone calls are circuit switched. The process is as follows:

1. A request is made to the network from a user when a number is dialed.
2. A circuit path is established end-to-end through the telephone facility infrastructure.
3. The conversation or data transmission is completed.
4. The circuit path is released back to the infrastructure.

Note that the user has complete use of the facility as long as the connection is established. If the user hangs up and redials, the path will probably be a different one. This is what happens when you get a facility that is of poor quality and redial and then get a good-quality facility.

Figure 2–3 provides an overview of the circuit switched environment. Each of the end devices connects to the network via some DCE device such as a modem. This connectivity is through a local CO, or end of-

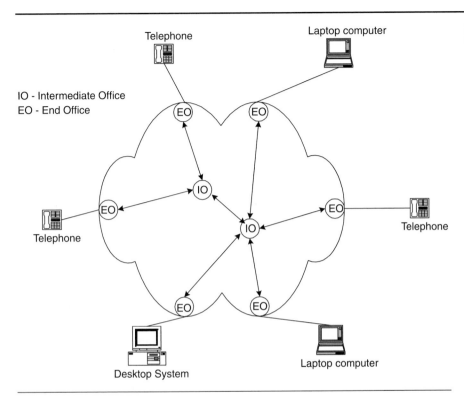

Figure 2–3
Circuit Switched Network

IO - Intermediate Office
EO - End Office

fice, wire center, or Class 5 office. The circuit switched network establishes a path through adjacent COs until connectivity is established to the other user. These COs, called nodes, can be classified as end nodes or intermediate nodes. The intermediate nodes connect other intermediate nodes and end nodes.

The connection path is established before transmission begins, which means that the channel capacity must be reserved in advance between the two end locations before any data can flow. All switches or nodes in the path must have available internal switching capacity to handle the requested connection. The switches have the intelligence to make these allocations and devise a route to provide this connectivity through the network. This intelligence is provided by the Signaling System 7 (SS7) network.

Figure 2–4 illustrates how a call is routed over a circuit switched network. Caller 1 dials a telephone number and accesses Switch A, which causes a path to be established through the circuit switched network (Switches A, C, D, F) to the device that is called behind Switch F. This path is utilized by callers 1 and 2 until one of the users disconnects (hangs up).

Figure 2–4
Circuit Switched Call
Routing and Setup

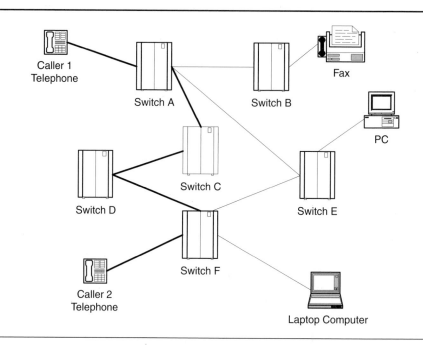

Circuit switching can be inefficient, because channel capacity is dedicated for the duration of the connection, whether there is a presence of data traffic. The channel may be idle during most of the time when there is computer-to-computer communication. This efficiency improves for voice traffic, but it still does not approach 100 percent. Circuit switching was designed for voice traffic, but is now being used more for data transmissions.

Circuit-switching technology has been driven by voice traffic applications. Two primary requirements for voice traffic are no transmission delay and no variation in delay. A constant transmission rate must be maintained, because transmission and reception occur at the same signal rate for human conversations. The quality of the signal must also be sufficiently high to provide intelligibility.

Circuit switching, with its widespread implementations, is well suited for the transmission of analog voice signals. Even though it is inefficient, circuit switching will remain an attractive choice for both wide and local area networking for some time. The main feature of circuit switching is transparency, which means that it appears as if there is a direct end-to-end connection with no special requirements of the end stations.

Figure 2–5
Packet-Switching Network

2.5   PACKET SWITCHING

Packet switching is like everyone having their own railway cars, but sharing the same track (facility). The track (facility) takes the cars to their proper destination. Packet switching is popular because most data communications consist of short bursts of data with intervening spaces that usually last longer than the actual burst of data. Packet switching takes advantage of this characteristic by interweaving bursts of data from many users to maximize use of the shared communication network.

Packet Data Networks (PDNs) are based on packet switching technology in which messages are broken down into fixed-length components called **packets** and sent through a network individually. A packet consists of the data to be transmitted and certain control information. Packet switching networks are designed to provide several alternative high-speed paths from one node to another. Figure 2–5 shows the different paths or routes available for a packet to traverse the packet switching network.

packet

As an example, messages from the Orlando node to the New Orleans node would normally pass through the Jackson node; however, if the facility between these nodes had a failure, then the message could be routed through the Atlanta or Columbia node. This configuration provides a redundant, fail-safe capability and implies that the route that a specific message takes is in part a function of the condition and traffic capacity of the various network links on the network when the message is sent. The situation and the route taken by the message are dynamic.

Numerous packet switches are interconnected to form a WAN. A switch usually provides for multiple input/output (I/O) connections, making it possible to form many different topologies and to connect to several switches and computers. A WAN is usually not symmetric in design, because the interconnections among packet switches, and the capacity of each connection, are based on the expected traffic and the ability to provide redundancy in the event of failure.

A WAN allows many computers to send packets simultaneously, which is accomplished through a fundamental paradigm used with wide area packet switching called **store and forward.** This store and forward operation requires the switches to buffer packets in memory. The store operation occurs when a packet arrives, and is accomplished by placing a copy of the packet in the switch's memory and notifying the processor of the activity. The forward activity requires the processor to examine the packet and determine over which interface it should be sent on the path to its final destination.

store and forward

A system that uses the store and forward paradigm can transport packets through the network as fast as the packet switch hardware and software allows. If multiple packets are sent to the same output device, the packet switch can hold the packets in memory until the output device is ready. Also, if the output device is busy, the processor places the outgoing packet in a queue that is associated with the device. To summarize, the store and forward technique allows a packet switch to buffer a short burst of packets that arrive simultaneously and transport them successfully to their destination.

Packets and Packetizing

Network layer protocols are responsible for providing end-to-end addressing schemes and for enabling internetwork routing of network layer data packets. The term *packets* is usually associated with network layer protocols, whereas the term *frames* is usually associated with data link layer protocols. In addition to the original data, each packet contains an address field, which is added by the packet assembler/dissassembler to give the destination's identification. Other checking or control fields may also be added to the packet to ensure data integrity. A generic form of a

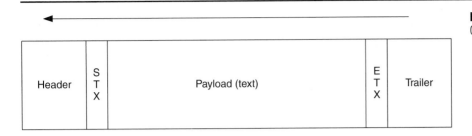

Figure 2–6
Generic Format of a Packet

data packet is shown in Figure 2–6. The three principal elements of the packet include:

- Header—control information such as synchronizing bits, address of the destination device, address of the transmitting device, and length of the packet
- Payload—the data to be transmitted
- Trailer—end of packet, error detection, and correction bits

When packet switching is used, messages are segmented into packets of a predetermined size before they are transmitted, a process called packetizing, and the packets are commonly less than 1,000 bytes long. This packetizing function is performed by hardware or software called the **Packet Assembler/Disassembler (PAD)**. The PAD may exist in the user's DTE or in the PDN's node. At the receiving end, another PAD assembles the message from the packets. Since the packets do not follow each other in sequence, they may arrive at the destination out of order and must be reassembled correctly. Throughput can be impacted by out-of order, flawed, or missing packets.

packet assembler/ disassembler (PAD)

Switching Techniques

The two approaches used to transport the packets are datagram and virtual circuit. Datagram is a connectionless service, whereas virtual circuit is a connection-oriented approach. In the datagram approach, each packet is treated independently, with no reference to packets that have already gone. Since each packet contains the destination address, each node looks at this address and makes a decision for the route to the next node. Thus, all packets of a particular message may not follow the same route and could arrive at the destination out of order, or not at all, due to a number of circumstances. The destination node must determine if all packets have been received and if they are in the proper order. In this technique, each globally addressed message packet, which is treated independently, is referred to as a **datagram.**

datagram

In the virtual circuit approach, the packet switched network establishes what appears to be one end-to-end circuit between the sender and receiver. This appears very similar to the circuit switching approach where a path is chosen before any transmission takes place. Each packet contains a virtual circuit identification, instead of a destination address, and data. Each node of the preestablished route knows where to direct such packets; therefore, no routing decisions are required. As in circuit switching, all messages for that transmission take the same route over the virtual circuit which has been set up for that particular transmission. The two DTE devices believe that they have a dedicated point-to-point circuit. At any time, each DTE can have more than one virtual circuit to any other DTE and can have virtual circuits to more than one DTE.

The main characteristic of the virtual circuit technique is that the route between stations is set up prior to data transfer. This, however, does not mean that there is a dedicated path, as in circuit switching. A packet is still buffered at each node and queued for output over a line, while other packets on other virtual circuits may share the use of the line. The difference from the datagram approach is that the node does not need to make a routing decision for each packet. It is made only once for all packets using that virtual circuit.

The following process is depicted in Figure 2–7.

1. Information is packetized at the sending desktop system location by way of a PAD function.

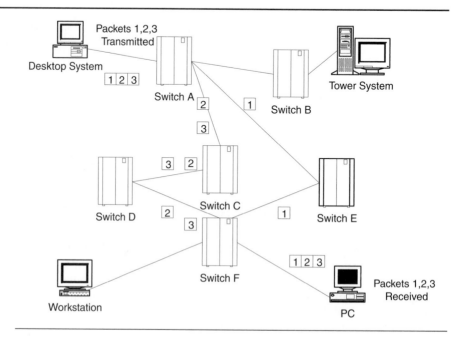

Figure 2–7
Packet Switched Network

2. Three packets of information (1,2,3) are provided to the packet switched network at switch A.
3. The packets are routed through the network (Switches A, C, D, F) to the destination based on certain conditions and attributes. They may take different paths to their final destination.
4. The packets (1,2,3) arrive at the receiving location where they are reassembled into usable information via another PAD function. (The process of datagram or connectionless switching).

Note that the user has no input into the path that the packets take. The packet switching network makes all decisions based on several conditions. Possibly the packets will arrive at the destination out of order and be reassembled into the correct order.

Circuit Switching Versus Packet Switching

Packet switching has several advantages over circuit switching:

- A packet switching network can perform data-rate conversion.
- Line efficiency is greater because a single node-to-node link can be dynamically shared by many packets over time.
- A heavy load on a circuit switch may cause blocking; however, the packet switched network will still accept the packets at a reduced rate.
- Priorities can be set for packets that have been queued for transmission.

For the technically inclined, an in-depth technical discourse on circuit switching and packet switching can be found in *Data & Computer Communications* [William Stallings, 1997].

2.6　ANALOG VERSUS DIGITAL OVERVIEW

The terms **analog** and **digital** are used to convey (1) information or data, (2) signals and signaling, and (3) transmission [Miller, 2000]. Analog refers to a continuously variable waveform, whereas digital is discretely variable. analog
digital

Analog and Digital Comparisons

Information that is being transported across the network is represented as either analog or digital traffic. An **analog signal** is an electrical signal with continuously varying waveform. The term **analog data** refers to data—such as voice, video, or pictures—in which the energy levels can vary continuously over time, taking on all possible values over an interval range. analog signal
analog data

Analog signals have three basic characteristics that can be used to convey information—amplitude, frequency, and phase. **Amplitude** is the amplitude

Figure 2–8
Continuous and Discrete
Signals

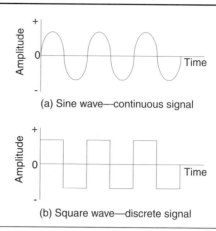

(a) Sine wave—continuous signal

(b) Square wave—discrete signal

maximum value or strength of the signal over time, typically measured in volts. **Frequency** is the rate in cycles per second (Hertz, hz) at which the signal repeats. **Phase** is a measure of the relative position in time within a single period of a signal, which is a shift in the signal by a certain number of degrees (e.g., 180 degrees).

    A **digital signal** is one that can assume one of several discrete states. Two different states can be used to represent the two binary digits (1 and 0). Information to be transmitted is converted into a digital code consisting of a specific 1s and 0s pattern. **Digital data** refers to data, such as alphanumeric characters, that can only take on a specific, finite set of values.

    Figure 2–8 shows an example of a continuous (sine wave) signal and a discrete (square wave) signal. Note that both signals oscillate about the zero-voltage reference line.

frequency
phase

digital signal

digital data

Analog/Digital Coding and Decoding

coder/decoder (CODEC)

A **Coder/Decoder** (**CODEC**) converts voice signals from their analog form to digital acceptable, modern digital PBXs, and digital transmission systems. It then converts those digital signals back to their original analog state at the receiving end. The CODEC is actually in the telephone set for some products, which means that the telephone converts the analog voice signal and sends out a digital signal. The normal Analog-to-Digital (A/D) and Digital to Analog (D/A) processes are as follows:

- Analog signals from the telephone at the transmitting subscriber site are sent over twisted-pair copper wires to the Central Office (CO), where the A/D conversions are made.
- The digitized signal is sent over the PSTN to the CO at the destination and converted back to the analog signal by the D/A CODEC.
- The D/A CODEC output analog signal is sent via twisted pair wire to the receiving subscriber's telephone.

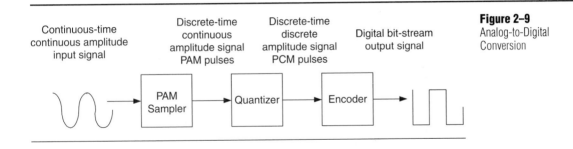

Figure 2–9
Analog-to-Digital
Conversion

Pulse Code Modulation

The system for generating and transmitting telephone signals digitally is called **Pulse Code Modulation (PCM)**. PCM is used to create a digital signal from an analog signal. A PCM channel band samples the voice signal 8,000 times per second (125 milliseconds, ms), converts each sample to an 8-bit word, and transmits it over a T-1 circuit interspersed with digital signals from the other twenty-three channels. This T-1 (1.544 Mbps) is known as the basic building block of the North American digital hierarchy (DS-1).

pulse code modulation (PCM)

Speech signals originate at the transmitter button of a telephone set as an analog signal. To carry the information in the analog signal over a digital transmission facility, it is necessary to convert the analog signal to a digital signal. The steps in this process include sampling, quantizing, and encoding.

If voice data are limited to frequencies below 4,000 Hz, a conservative procedure for intelligibility, then 8,000 samples per second are sufficient to completely characterize the voice signal. The first step in this process is called Pulse Amplitude Modulation (PAM), which is a quantizing process of the analog signal. Figure 2–9 shows the steps necessary to take an analog signal and develop a digital signal using PAM and PCM. On reception, the process is reversed to reproduce the original analog signal.

To summarize, the following three steps are necessary for the A/D conversion process:

1. Sampling—The analog signal is sampled periodically. Each sample generates a PAM signal, with an amplitude equal to that of the analog signal.
2. Quantizing—The amplitude (height) of the PAM signal is measured to derive a number that represents its amplitude level.
3. Encoding—The decimal (base 10) number derived in the quantizing step is then converted to its equivalent 8-bit binary number.

Repeaters and Amplifiers

Using digital signals to represent voice or data is more efficient than using analog signals. Digital signals are easier to interface with each other, the electronics are cheaper, and the digital signals are less susceptible to noise. *Noise* is any unwanted signal of sufficient amplitude that it interferes with the communications process. If an unwanted signal is induced into the medium carrying the desired signal, the unwanted signal combines with the desired signal. If this combined signal passes through an amplifier, the resulting combined signal is magnified, producing an unintelligible transmission and thus requiring a retransmission. Analog signals can only be carried a certain distance by a transmission medium before the signal gets so attenuated that it must be amplified. **Attenuation** is the loss of signal strength during transmission due to resistance of the media.

attenuation

When digital signals are used to convey voice, data, or video, the signal is continuously varying from one discrete voltage level to another. The two states of voltage levels are used to represent a digit 1 or digit 0. The digital signal is sent down the transmission medium by the appropriate voltage level. Instead of amplifiers that are used in analog transmission, **repeaters,** or regenerators, are used to re-create the digital signal and pass it along the medium. The distance between repeaters depends on the signal carried and the medium utilized.

repeater

Digital signal regenerators strip impairments out of a signal by regenerating clean 1s and 0s. The decoder at the receiving end of a circuit uses an 8-bit pattern of 1s and 0s to reconstitute the original analog signal that had been coded by a CODEC at the transmitting end of a circuit. The PSTN has evolved into an all-digital network; however, there are still analog components remaining in the enterprise network environment. The circuitry that connects the ubiquitous telephone to the local central offices is mostly analog, which means that the inputs and outputs from the central offices are often analog.

Analog Versus Digital Transmission

Remember that analog information has a continuous waveform representation and that digital information is represented by discrete pulses. The continuous signal may represent speech and the discrete signal may represent binary 1s and 0s. Originally, all transmission was analog, but most new network facilities are now digital. Analog data can be transmitted in the form of analog or digital signals. Speech data are normally carried in the form of electrical analog signals in most local telephone loops; however, speech data can be digitized and carried in the form of digital signals in a digital network.

The primary reasons why digital transmission is superior to analog transmission are as follows:

- Better data integrity
- Higher capacity, so easier integration
- Better security and privacy
- Lower cost

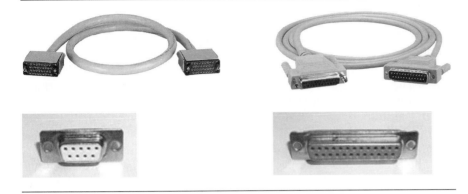

Figure 2–10
Interfaces and Cable
Assemblies

2.7   INTERFACES

Point-to-Point networks involving the simple interconnection of two pieces of equipment are relatively simple to establish. They may use either a digital line with a **Data Service Unit (DSU)** or an analog line with a **modem.** Common interface standards for these devices include the RS232, RS449, X.21, and V.35 specifications. Figure 2–10 shows examples of the RS232 and V.35 interface connectors and cable assemblies.

data service unit (DSU)
modem

Physical interface standards typically specify certain characteristics about the connection between **Data Terminal Equipment (DTE)** and **Data Communications Equipment (DCE)**. The most commonly used serial DTE/DCE interface standard in the United States is the RS232-D Electronic Industries Association (EIA) standard.

data terminal equipment (DTE)
data communications equipment (DCE)

If the RS232-D standard is fully implemented, the DTE and DCE are connected by a DB25 cable to each device using a 25-pin connector. Each pin has a specific function in establishing communication between the devices. One drawback of the RS232 standard is its limited bandwidth and distance.

Another standard, RS449, was designed to replace the RS232, and increase both the bandwidth and distance specifications. The RS449 standard, unlike RS232, differentiates between operational and electrical specifications. RS449 defines pin functions, but relies on the electrical standards, RS422 and RS423, for the electrical specifications. RS449 defines a 37-pin connector, which includes those of the RS232, and additional pins for testing. However, sheer market power has retained the RS232 as the preferred interface.

The two electrical standards, RS422 and RS423, correspond to balanced circuits and unbalanced circuits, respectively. An unbalanced circuit uses one line for signal transmission and a common ground. A balanced circuit uses two lines for signal transmission. Balanced signals are less susceptible to noise and allow higher transmission rates over longer distances.

The X.21 interfaced standard is defined by ITU-T and uses a 15-pin connector. Like RS449, it allows for both balanced and unbalanced circuits. There are several significant differences between X.21 and the RS standards. The first is that X.21 was designed as a digital signaling interface. The second involves control circuits in the interface. RS standards define specific circuits for control functions. However, X.21 puts more logic circuits (intelligence) in the DTE and DCE devices that can interpret control sequences and reduce the number of connecting pins in the interface.

V.35

The ITU-T **V.35** is a high-speed digital interface standard, which has been superceded by ITU-T standards V.36 and V.37. The standard describes synchronous DCE devices that operate at data signaling rates of 56 kbps, 64 kbps, and higher. This interface consists of a rectangular connector with thirty-four circuit pins. It is frequently utilized in Frame Relay and ATM installations.

user-to-network interface (UNI)

An important interface concept is the **User-to-Network Interface (UNI)**. This interface provides a physical point of separation between the responsibilities of the carrier and the customer. Any equipment and wiring on the user side of the UNI is the responsibility of the customer (unless a maintenance agreement makes equipment and wiring the responsibility of the carrier). Any equipment or lines on the network side of the UNI are the responsibility of the carrier. This is called the **Network-to-Network Interface (NNI)**.

network-to-network interface (NNI)

Another important interface concept is DTE/DCE. It provides a functional separation between different types of equipment, regardless of what side of the UNI they happen to be located. A DSU is an example of DCE, which may be owned by the customer and sit on the user side of the UNI. A computer is an example of DTE, but its modem is DCE. Both are located on the user side of the UNI. At the other end of a dedicated or dial-up line, or on the network side, there are DCE devices owned by the end users in the form of modems and DSUs. Figure 2–11 shows both the UNI/NNI relationship and examples of the DTE/DCE interface.

Figure 2–11
NNI/UNI

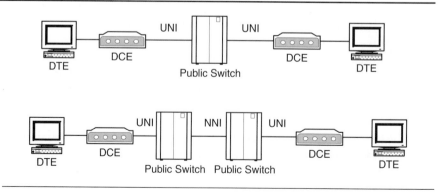

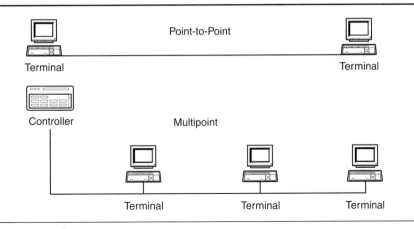

Figure 2–12
Point-to-Point and
Multipoint Configurations

Physical network wiring configurations are point to point and multipoint. *Point to point* configurations have only one connection between the two terminals. *Multipoint* terminals share the connection with the other terminals and require a controller to handle polling and prioritization. IBM synchronous/SDLC configurations can operate in a multipoint configuration. Figure 2–12 depicts both configurations.

2.8 MULTIPLEXING

Circuits must be combined or aggregated for transmission efficiency. This process, called multiplexing, utilizes several different techniques to accomplish this task. **Frequency Division Multiplexing (FDM)**, **Time Division Multiplexing (TDM)**, and **Statistical Time Division Multiplexing (STDM)** are examples of such techniques.

Multiplexing circuits saves facilities and resources—a T-1 MUX offers twenty-four DS-0 (64 kbps) channels, whereas a T-3 MUX offers twenty-eight DS-1 channels or 672 DS-0 channels. Multiplexers can feed other multiplexers to aggregate circuits for long-haul transmission. Figure 2–13 depicts three levels of multiplexing. Low-speed circuits are multiplexed with DS-0s, DS-0s are multiplexed with DS-1s, and DS-1s are multiplexed with DS-3s, which produces a facility containing 672 DS-0 circuits.

frequency division multiplexing (FDM)
time division multiplexing (TDM)
statistical time division multiplexing (STDM)

Frequency Division Multiplexing
Frequency Division Multiplexing (FDM) is an analog technology that achieves the combining of several digital signals onto one medium by sending signals in a number of distinct frequency ranges over that medium. FDM breaks the available bandwidth into separate full-time

Figure 2–13
Multiplexing/Aggregating
Signals

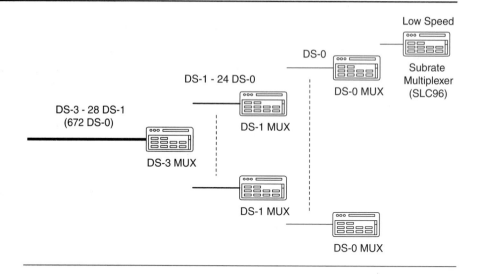

Figure 2–14
Frequency Division
Multiplexing

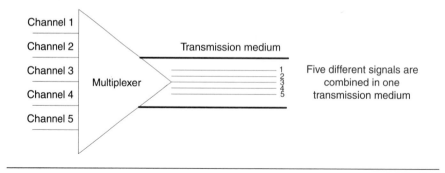

channels. Figure 2–14 depicts a multiplexer using FDM to aggregate five channels over one physical medium. An example of FDM is broadcast and cable television.

Time Division Multiplexing

Time Division Multiplexing (TDM) is a digital technology that involves sequencing groups of bits from a number of individual input streams, one after another, in such a manner as to associate them with the appropriate destination receiver. In TDM, the entire channel is allocated for short periods of time, using time slots instead of separate full-time channels, as is the case with FDM. Each time slot is assigned to a specific input channel. Figure 2–15 provides an example of five input channels that will be providing inputs to the multiplexer.

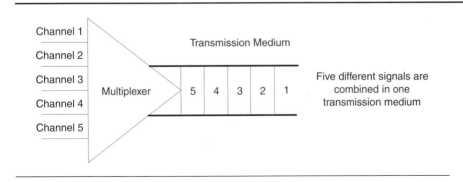

Figure 2–15
Time Division Multiplexing

Statistical Time Division Multiplexing

In a synchronous time division multiplexer, many of the time slots in a frame generally contain no bits. In a typical application where a number of terminals share a computer port, usually some terminals are not transferring any data. An alternative to TDM is Statistical TDM (STDM). As with TDM, the statistical multiplexer has several I/O lines on one side, each with an associated buffer, and a higher-speed multiplexed line on the other.

STDM is similar to TDM except more efficient, because all of the time slots contain data. STDM takes advantage of the sporadic nature of terminal users and allocates bandwidth to each terminal on the basis of demands and needs. A STDM does not assign specific time slots to input channels. Instead, it transmits the input's address along with each bit stream of data. Additional overhead, however, is a factor because of the additional bits for the addresses, but the amount is negligible and incurs no deterioration of response time.

The term *statistical* refers to the method by which time slots are allocated. A statistical multiplexer decides how many time slots to allocate in the next second based on the amount of data sent by a given device in the last second. Complicated algorithms produce constant calculations, so the time-slot utilization is dynamic based on the user's most recent demands and probable future needs.

Because the STDM device utilizes a buffer to accommodate high traffic rates, delays can occur when there is considerable buffering of data. Such a situation occurs when all users try to transmit simultaneously. Some STDM models have the ability to stop terminals from transmitting when there is a shortage of time slots. Figure 2–16 provides a comparison of a TDM and STDM bit stream. Note that STDM transmits no empty slots, and has additional capacity that could be utilized by additional traffic.

Figure 2–16
TDM Comparison with
STDM

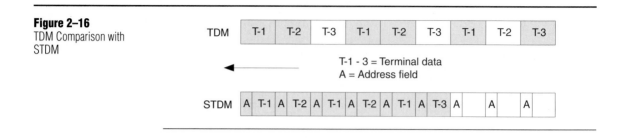

Figure 2–17
Transmission Modes

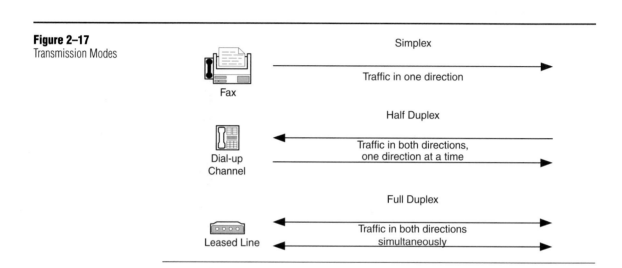

2.9 TRANSMISSION METHODS

Simplex, Half, and Full Duplex

Transmission methods are either simplex, half duplex, or full duplex. It will be necessary to know the application so that the transmission method can be determined. **Simplex** is transmission in one direction; **half duplex** is transmission in both directions, one direction at a time; and **full duplex** is transmission in both directions at the same time. Examples of these methods are as follows:

simplex
half duplex
full duplex

- Simplex—alarm systems, printers
- Half duplex—dial-up channel
- Full duplex—leased line

Figure 2–17 provides an example of each method. Fax traffic is simplex because it is normally in one direction. A half-duplex dial-up channel carries information in both directions; however, a turnaround is required to

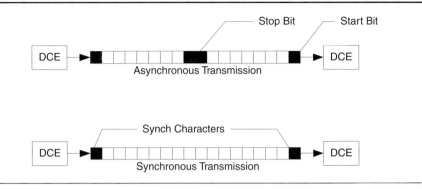

Figure 2–18
Asynchronous vs.
Synchronous Transmission

ensure that only one device is transmitting during any given time interval. A full-duplex circuit with multiple channels can transmit on one channel and simultaneously receive on the other.

Asynchronous versus Synchronous Transmission

Serial transmission can be accomplished using either a synchronous or asynchronous technique. It is necessary for the DCE devices to establish and maintain some type of timing between them so that detectable events are detected, produced, and transmitted accurately. The two alternatives for establishing and maintaining timing are asynchronous and synchronous transmission. **Asynchronous transmission,** or start-stop transmission, establishes synchronization by including start and stop bits with the data bits. Each character in a message is transmitted as an individual entity, without regard to when the previous character was transmitted.

asynchronous transmission

With **synchronous transmission,** timing is provided by a clocking signal supplied by the DCE device or the carrier. All characters in a synchronous message are sent contiguously, one after another, with framing characters at the beginning and end of the entire message block. Figure 2–18 shows the difference between asynchronous and synchronous packages. The asynchronous example depicts two groups of 7 bits, each with a start and stop bit. The synchronous example depicts a 16-bit message with a synch character at the beginning and end of the message.

synchronous transmission

Two-Wire and Four-Wire Facilities

Most local loops used for connection to the PSTN to supply switched, dial-up service are physically described as two-wire circuits. One of these wires is used as a ground wire for the circuit, and the other is utilized for data signaling. Dial-up and switched circuits generate a dial tone and are connected to a CO switch.

A four-wire circuit consists of two wires capable of simultaneously carrying a data signal, each with its own ground. Four-wire circuits are known as private lines or leased lines and bypass telephone company switching equipment. They have no dial tone and are always operational between the points designated by the subscriber.

2.10 COMMUNICATIONS ENTITIES

Access Requirements

Each vendor has a Point Of Presence (POP) where connections are made to other vendors/telcos/Interexchange Carriers (IXCs), Local Exchange Carriers (LEC), Competitive Local Exchange Carriers (CLECs), and Incumbent Local Exchange Carriers (ILECs). This access is one of the deregulation requirements discussed in Chapter 1.

point of presence (POP)

local exchange carrier (LEC)

The **Point Of Presence** (**POP**) is a physical place where a carrier has a presence for network access and generally takes the form of a switch or router. A large Interexchange Carrier (IXC) will have a number of POPs, at which they interface with the **Local Exchange Carrier** (**LEC**) networks to accept originating traffic and deliver terminating long-distance traffic. The basis on which the interface is accomplished can include switched and dedicated connections. Similarly, providers of X.25, Frame Relay, and ATM services have specialized POPs, which may be collocated with the circuit switched POP for voice traffic. A POP is also a meet point for Internet Service Providers (ISPs), where they exchange traffic and routes.

Connectivity between LECs is established with IXCs, such as AT&T, MCI, Sprint, and other carriers. Each carrier has a facility POP for interfacing with each other and the LECs. The term *point of presence* is beginning to change to *Point Of Interface* (POI). POP implies that a switching system is at that location; however, the POI can be in a closet in a basement or hotel, connected via a dedicated trunk to another part of the country. Regardless of whether the interface point is a POP or a POI, it is transparent to the subscriber.

Figure 2–19 shows a POP environment where the various IXC, LEC, CLEC, and telco entities interface with each other.

local access transport area (LATA)

Recall from Chapter 1 that LATA boundaries determine the rules for this connectivity. As part of divestiture, 165 **Local Access Transport Areas (LATAs),** were defined in the United States. The LECs basically provided telephone service within the LATAs, and interLATA telephone traffic was provided by the long-distance carriers or IXCs such as AT&T, MCI, and Sprint. The Telecommunications Act of 1996 effectively eliminated the differences in the areas and the type of traffic the LECs and IXCs could provide. The Act of 1996 defined provisions for ILECs to enter interLATA transport and for IXCs to provide intraLATA service.

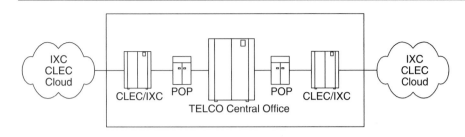

Figure 2–19
POP between IXC and Local Access

Routing and Routing Protocols

TCP/IP consists of the Transmission Control Protocol, specified in RFC793, and the Internet Protocol, specified in RFC791 [*Guide to TCP/IP*, 1998]. TCP operates at OSI Layer 4 and was designed for point-to-point communications between computers on the same network. IP operates at OSI Layer 3 and provides for communications between computers linked to different networks or to WANs. On a TCP/IP network, the packets of routed protocols contain internetwork layer addressing information, which allows user traffic to be directed from one network to another [Huitema, 1999].

TCP/IP

Address Resolution Protocol (ARP) and Reverse Address Resolution Protocol (RARP) are both Internet Protocol (IP) routed. ARP maps IP addresses to MAC addresses, which are used to send packets to their destinations. This means that ARP is used to obtain the physical address when only the logical address is known. The process includes two steps: (1) An ARP request with the IP address is broadcast onto the network; and (2) the node on which the IP address resides responds with the hardware address. RARP is similar to ARP in that it binds MAC addresses to IP addresses; however, it is primarily used for diskless workstations. The workstation uses the protocol to obtain its IP address from a server.

Routers connect two or more network segments and use the ARP in conjunction with their routing tables to transport data packets. A router requires an IP address for every network segment to which it is connected and a separate network interface for each network segment. Figure 2–20 provides an example of this router IP address and interface configuration. Note that each device on the network has both an IP and a MAC address. There are two segments in this example as indicated by the numbers 4 and 5 in the IP addresses.

When devices send packets to destinations not on their segment, the router connected to the segment on which the packet originated recognizes that the destination host is on a different subnetwork. The router's responsibility is to determine which network should receive the packet by

Figure 2–20
Router Configuration with
Two Segments

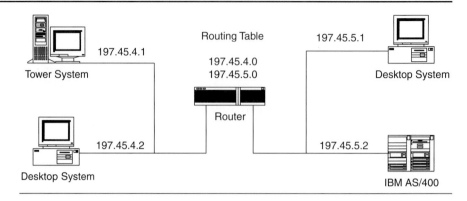

referencing its routing table to determine which interface is connected to the destination network. If the destination address is not in the routing table, an ARP request must be issued to locate the path to the destination. This ARP request is forwarded through the network to as many routers as is necessary to resolve the device address requested. This process continues until the hop count or Time To Live (TTL) reaches its maximum or the destination is located, before discarding the packets.

When multiple routers exist on a network, they must be able to share routing information in order to update each other's routing tables. Routing protocols carry routing table information and do not transmit normal network traffic. Routers learn about paths to other networks in one of two ways: static or dynamic configurations. A *static* configuration is a manual process during which a network administrator must enter the paths used to traverse one segment to another into the router's routing table memory. With *dynamic* configurations, the routers communicate with each other and create and maintain routing tables automatically. Static routing configurations are time consuming and error prone, but in certain situations are required. Following is a list of dynamic routing protocols [*Internetworking Technologies Handbook*, 1998]:

- Routing Information Protocol (RIP)
- Border Gateway Protocol (BGP)
- Exterior Gateway Protocol (EGP)
- Interior Gateway Routing Protocol (IGRP)
- Enhanced Interior Gateway Routing Protocol (EIGRP)
- Open Shortest Path First protocol (OSPF)

Routing protocols are available in two categories—*Interior Gateway Protocols (IGPs)* and *Exterior Gateway Protocols (EGPs)*. IGPs send routing tables and routing information between routers on the same internal network, whereas EGPs send routing information between networks.

This book will not explore the details for each of these routing protocols, although anyone who configures router networks will need in-depth training before undertaking such an endeavor. Certification textbooks and courses are available to help prepare the network administrator for such activities. One example textbook is the *CCNA Guide to Cisco Networking Fundamentals* [Kurt Hudson, 2000]. A general knowledge of the various routing protocols is necessary when discussing network routing. Following is a brief description of the various dynamic routing protocols.

Routing Information Protocol

Routing Information Protocol (RIP) is a distance-vector protocol and until recently was the most commonly used interior gateway protocol. A *distance-vector* protocol calculates distances between destinations by counting hops between routers. This protocol uses frequent broadcasts of the entire routing table on all interfaces to update neighboring routers. Nonadjacent routers learn of the updates secondhand. If multiple paths to the destination exist, the path with the fewest number of hops is selected for the transmission. IP RIP is formally defined in RFC1085 (RIP) and RFC1723 (RIP2).

The basic features and characteristics of RIP include routing metrics, routing stability, routing updates, and routing timers:

- RIP uses a single metric (*hop count*) to measure the distance between the source and destination network. Each hop in a path from source to destination is assigned a hop-count value, which is typically 1.
- RIP implements the split-horizon and hold-down mechanisms to prevent incorrect routing information from being propagated over the network. A hop-count limit prevents routing loops from continuing indefinitely. A *split horizon* is the view that a router has of a WAN interface in a partial-mesh environment where an incoming packet may need to be sent out on the same interface over which it was received to reach its ultimate destination.
- RIP sends routing update messages at regular intervals and when the network topology changes. When a router receives a routing update that includes changes to an entry, it updates its routing table to reflect the new route. The metric value for the path is increased by one, and the sender is indicated as the next hop. RIP routers maintain only the best route to a destination. After updating its routing table, the router immediately begins transmitting routing updates to inform other network routers of the change.
- RIP uses a routing update timer, a route timeout, and a route flush timer. The routing update timer clocks the interval between periodic routing updates, which is usually 30 seconds. When the route-timeout timer expires, the route is marked invalid, but is retained in the table until the route flush timer expires.

Two primary difficulties that can occur with RIP are the possibilities of (1) sending traffic through an inefficient path and (2) the possibility of a routing update taking a long time to reach convergence. *Convergence* is the point at which the internetworking devices share a common understanding of the routing topology.

Border GateWay Protocol

Border Gateway Protocol (BGP) provides a list of routers, reachable addresses, and cost metrics for path determination. BGP performs interdomain routing in TCP/IP networks. BGP is an exterior gateway protocol, which means that it performs routing between multiple autonomous systems (domains) and exchanges routing and reachability information with other BGP systems. Instead of transmitting the entire routing table, BGP only sends updates for the affected parts of the routing tables. BGP is described in RFC1654 and RFC1771.

BGP was developed to replace its predecessor, the now-obsolete Exterior Gateway Protocol (EGP), as the standard exterior gateway routing protocol used in the global Internet. BGP solved serious problems with scalability for Internet growth.

BGP performs three types of routing: inter-autonomous system routing, intra-autonomous system routing, and pass-through autonomous system routing.

- Inter-autonomous system routing occurs between two or more BGP routers in different autonomous systems.
- Intra-autonomous routing occurs between two or more BGP routers located within the same autonomous system.
- Pass-through autonomous routing occurs between two or more BGP routers that exchange traffic across an autonomous system that does not run BGP.

The primary function of a BGP system is to exchange network-reachability information, including that about the list of autonomous system paths with other BGP systems. Each BGP router maintains a routing table which lists all feasible paths to a particular network. The router does not refresh the routing table; rather, routing information received from peer routers is retained until receipt of an incremental update.

BGP devices exchange routing information upon initial data exchange and after incremental updates. When a router first connects to the network, BGP routers exchange their entire BGP routing tables. BGP routers do not send regularly scheduled routing updates, but only the portion of the routing table that has changed.

BGP uses a single routing metric to determine the best path to a given network. This metric consists of an arbitrary unit number that specifies the degree of preference of a particular link. The value assigned to a link can be based on any number of criteria, such as speed, delay, cost, and number of paths.

Exterior Gateway Protocol

Exterior Gateway Protocol (EGP) is a particular instance of an exterior gateway protocol and should not be confused with the EGP category. EGPs once transferred entire routing tables between networks during the transactions. The tables consisted of routers, addresses, and cost metrics associated with each router. EGP has been replaced by BGP.

Interior Gateway Routing Protocol

Interior Gateway Routing Protocol (IGRP), like EIGRP, is a proprietary protocol developed by Cisco Systems. IGRP is a distance-vector protocol like RIP, but it has a maximum hop count of 255 and uses additional factors other than hop count when determining the best path. Distance-vector routing protocols require each router to send all or a portion of its routing table in a routing update message at regular intervals to each of its neighboring routers. As routing information proliferates through the network, routers can calculate distances to all nodes within the internetwork.

IGRP uses a combination of metrics, including internetwork delay, bandwidth, reliability, and network load. Network administrators can set the weighting factors for each of these metrics. A wide range for IGRP metrics allows satisfactory metric setting in internetworks with widely varying performance characteristics. The metric components are combined in a user-definable algorithm, which allows the network administrator to influence the route selection. To provide additional flexibility, IGRP permits multipath routing.

IGRP provides features designed to enhance its stability, including split horizons, hold-downs, and poison-reverse updates.

- Split horizons derive from the premise that it is never useful to send information about a route back in the direction from which it came. The split horizon rule helps prevent routing loops that can cause a severe reduction in network performance.
- Hold-downs are used to prevent regular update messages from inappropriately reinstating a route that might have gone bad. When a router goes down, neighboring routers detect this via the lack of regularly scheduled update messages. These routers then calculate new routes and send routing update messages to inform their neighbors of the route change. Hold-downs tell routers to hold down any changes that might affect routes for some period of time.
- Split horizons should prevent routing loops between adjacent routers, but poison-reverse updates are necessary to defeat larger

routing loops. Increases in routing metrics usually indicate routing loops. Poison-reverse updates are then sent to remove the route and place it in hold-down.

IGRP maintains a number of timers and variables containing time intervals, which include an update time, an invalid timer, a hold-time period, and a flush timer.

- The update timer specifies how frequently routing update messages should be sent.
- The invalid timer specifies how long a router should wait in the absence of routing update messages about a specific route before declaring the route invalid.
- The hold-time variable specifies the hold-down period.
- The flush timer indicates how much time should pass before a route should be flushed from the routing table.

Enhanced Interior Gateway Routing Protocol

Like IGRP, Enhanced Interior Gateway Routing Protocol (EIGRP) is a proprietary protocol developed by Cisco Systems. EIGRP has a maximum hop count of 224 and provides the same features as IGRP, except EIGRP does not update routers that are not affected by topology changes. This provides for a faster convergence time, which is the time it takes to update all routers on a network. EIGRP provides compatibility and seamless interoperation with IGRP routers. Enhanced IGRP also integrates the capabilities of link-state protocols into distance-vector protocols.

EIGRP includes support for a variable-length subnet mask, support for partial updates, and support for multiple network layer protocols. Fast convergence is accomplished because EIGRP stores all of its neighbor's routing tables so that it can quickly adapt to alternate routes. If no alternate route exists, EIGRP queries its neighbors to discover an alternate route. EIGRP makes no periodic updates, but sends partial updates only when the metric for a route changes.

Enhanced IGRP employs the following four key technologies that combine to differentiate it from other routing technologies.

- Neighbor discovery/recovery is used by routers to dynamically learn about other routers on their directly attached networks.
- Reliable Transport Protocol is responsible for guaranteed, ordered delivery of EIGRP packets to all neighbors.
- Dual finite-state machine embodies the decision process for all route computations by tracking all routes advertised by all neighbors.
- Protocol-dependent modules are responsible for network-layer protocol-specific requirements.

Enhanced IGRP relies on the following fundamental concepts:

- When a router discovers a new neighbor, it records the neighbor's address and interface as an entry in the neighbor table.
- The topology table contains all destinations advertised by neighboring routers. A topology table entry for a destination can exist in either an active or passive state.
- Enhanced IGRP provides support for both internal and external routes.

Open Shortest Path First Protocol

Open Shortest Path First (OSPF) is an open protocol specified in RFC1247. It is a link-state routing protocol for TCP/IP networks that calls for the sending of *Link-State Advertisements (LSAs)* to all routers with the same hierarchical area. Information on attached interfaces, metrics used, and other variables is included in OSPF LSAs. As OSPF routers accumulate link-state information, they use the SPF (Dijkstra's) algorithm to calculate the shortest path to each node.

OSPF is an efficient IGP that attempts to determine the optimum path between two points by constructing a topology of the entire network. OSPF considers many factors such as route speed, traffic levels, link reliability, and security in this path determination. Routers using link-state protocols flood routing update information across the network to each router, which means that the routers receive the information firsthand.

Unlike RIP, OSPF can operate with a hierarchy. The largest entity within the hierarchy is the Autonomous System (AS). An autonomous system is a collection of networks under a common administration which shares a common routing strategy. Even though OSPF is an interior gateway routing protocol, it is capable of receiving routes from, and sending routes to, other autonomous areas. An AS can be divided into several areas, which are groups of contiguous networks and attached hosts. Routers with multiple interfaces can participate in multiple areas. These routers, which are called area border routers, maintain separate topological databases for each area. Each area behaves like an independent network and the router's database includes only the state of the area's links. The flooding protocol stops at the boundaries of the area, and the routers compute only the routes within that area.

A topological database is an overall picture of networks in relationship to routers. The topological database contains the collection of LSAs received from all routers in the same area. An area's topology is invisible to entities outside the area, which results in less routing traffic in the network. Autonomous System (AS) border routers running OSPF learn about exterior routes through exterior gateway protocols such as BPG and EGP.

Figure 2–21
OSPF Hierarchy with
Multiple Areas

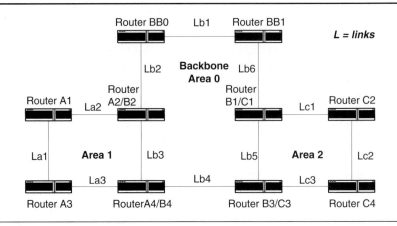

An OSPF backbone is responsible for distributing routing information between areas. It consists of all area border routers, networks not wholly contained in any area, and their attached routers. The backbone itself is also an OSPF area, so all backbone routers use the same procedures and algorithms to maintain routing information within the backbone. Figure 2–21 shows an example of an internetwork consisting of an AS with three areas. OSPF Area 1 and Area 2 are accessible to each other through the OSPF backbone Area 0. The backbone area includes the border routers BB0 and BB1 from the area border routers A2/B2, A4/B4, B1/C1, and B3/C3, and from the links Lb1, Lb2, Lb3, Lb4, Lb5, and Lb6.

2.11   BROADBAND ACCESS APPLICATIONS

An initial step for identifying situations where broadband services may be deployed is to identify traditional dedicated, leased line, and dial-up applications. These applications may involve small to midsize customers with multiple sites who might connect their sites, separated by distance, with a dedicated and nonsharing link, or users who have historically used dial-up facilities to access remote sites. In many situations, the speeds of the circuits may not exceed 19.2 kbps. This is important to know, since the lowest access line rate for Frame Relay and a number of the other broadband technologies is 56 kbps.

Application examples that are feasible might include the following:

- Home and small-business access for telecommuting employees or self-employed persons who require a guaranteed access when needed. An example is a connection to a network management system for Tier II support.

Table 2–2
Interface Standards

Standard	Description
RS232-D	A set of standards specifying electrical, functional, and mechanical interface specifications for a 25-pin connector.
RS449	Originally designed to replace RS232 to improve transmission capabilities. They are not compatible, since RS449 has a 37-connector with different functions. It uses twisted-pair cable.
X.21	The interface between DTE and DCE for synchronous operation on public switched networks.
V.35	An ITU-T standard for trunk interface between a network device and packet network that defines signaling for data rates greater than 19.2 kbps. The physical connector is a 34-pin rectangular M-block connector. V.36 is intended to replace V.35. RS422 and RS423 are the electrical specifications associated with V.35.

- Connectivity in the geographic areas not yet served by a broadband technology such as Frame Relay.
- A secure facility that is available on a 7-day x 24-hour basis for those who do not want to share facilities.
- A redundant/backup facility in the event of a primary facility failure where the customer must communicate if or when the primary facility fails. An example is a bank or store where bankcards and debit cards are utilized for payment.
- Any requirement that can be satisfied by the various access products for connectivity to a remote location.
- The network organization must show added value in the form of excellent service and rapid trouble isolation response. A 7 by 24 network management facility is a must to accomplish this feat.

2.12 STANDARDS

Several standards in the wide area network environment are associated with physical interfaces, transmission protocols, and routing protocols. A significant number of standards are also associated with the various broadband technologies presented in this book. The standards for each of the broadband technologies are discussed throughout this book and are then summarized in Appendix C. The primary device interface standards are listed in Table 2–2.

2.13 ISSUES AND CONSIDERATIONS

A primary issue, when upgrading from dial-up and leased line implementations to broadband access, is the speed of the current access circuits.

Although the upgrade has certain advantages for dedicated access, cost justification for this upgrade will probably be required. There may also be advantages for a dial-up type access such as ISDN.

Dedicated Access

The primary attributes of a dedicated private line are as follows:

- Speeds higher than dial-up
- Permanent circuit path
- Quality higher than dial-up
- Available continuously
- More expensive than dial-up
- Speeds: 56 kbps, 64 kbps, 1.544 Mbps, 44.736 Mbps

These attributes must be considered when deciding which broadband technology meets the cost-benefit considerations for each enterprise network. Upcoming chapters discuss each of these broadband technologies in detail.

■ SUMMARY

Wide area data links have historically been utilized to connect pairs of systems that were located at some distance from each other. Asynchronous transmission was widely used, particularly for connections between terminals and host computers. Transmission methods included simplex, half-duplex, and full-duplex alternatives.

The OSI network layer performs major functions such as routing, network control, congestion control, and collection of accounting data. Numerous routing algorithms and protocols exist, including BGP, IEGRP, RIP, and OSPF. Routing algorithms for both static and dynamic options are utilized.

Transport across the WAN is accomplished by circuit switching and packet switching. Circuit switching creates a connection with a dedicated path for the call duration and then releases it when the call is disconnected. All packets are forwarded, in the proper sequence, over this predetermined path. In packet switching, a message is broken into fixed-length packets and transmitted to the destination through different packet switching network nodes. The message is reassembled at the destination.

Numerous hardware devices are utilized in the WAN. These devices fit in the categories of multiplexers, amplifiers, and repeaters. Multiplexers include Frequency Division Multiplexers (FDMs), Time Division Multiplexers (TDMs), and Statistical Time Division Multiplexers (STDMs). Amplifiers are utilized for analog signals, and repeaters are used for digital signals.

Of particular importance in the WAN are the interface specifications which include the user-to-network (UNI) and the network-to-network (NNI) interface. Both the data communications equipment (DCE) and data terminal equipment (DTE) interfaces determine the type of connectors and other physical media requirements.

The WAN environment includes independent entities, such as local exchange carriers, (telephone companies), and interexchange carriers which provide services across both LATA and state boundaries. The organizations interface with each other through point-of-presence (POP) devices.

Certain current applications use dial-up and dedicated facilities to accomplish their missions. These applications can be converted from this older technology to a new broadband technology for increased bandwidth and efficiency.

Key Terms

Access Line	Local Access Transport Areas (LATA)
Amplitude	Local Exchange Carrier (LEC)
Asynchronous Transmission	Modem
Attenuation	Network-to-Network Interface (NNI)
Analog	Packet
Analog Data	Packet Assembler/Dissassembler (PAD)
Analog Signal	Packet Switching
Central Office (CO)	Point of Presence (POP)
Circuit	Pulse Code Modulation (PCM)
Circuit Switching	Repeater
Coder/Decoder (CODEC)	Simplex
Datagram	Statistical Time Division Multiplexing (STDM)
Data Communications Equipment (DCE)	Store and Forward
Data Terminal Equipment (DTE)	Synchronous Transmission
Data Service Unit (DSU)	TCP/IP
Digital	Time Division Multiplexing (TDM)
Frequency Division Multiplexing (FDM)	Topology
Full Duplex	User-to-Network Interface (UNI)
Half Duplex	V.35
Interexchange Carrier (IXC)	Virtual circuit

REVIEW QUESTIONS

1. Provide an overview of the pre-broadband network environment.

2. Describe a virtual circuit.

3. Describe the relationship between an access line, local loop, and a POP.

4. What is the difference between a topology and a technology? Give examples of each.

5. Describe the differences between distributed, decentralized, and centralized data centers.

6. What are the characteristics of a dedicated line?

7. Provide a list of speeds that are allowed on DDS and DSx circuits.

8. What is the difference between switched and dedicated communications service?

9. Describe circuit switching and the process that is required to use a facility.

10. Describe the process of routing a call in a circuit switched network.

11. Describe the packet switching network.

12. What is a packet and what is packetizing? How is a PAD utilized?

13. Define datagram. Where is it used?

14. How is circuit switching different from packet switching? What are the advantages and disadvantages?

15. Explain the store and forward operations involved in packet switching.

16. Describe the elements of a packet.

17. What is the difference between an analog signal and a digital signal?

18. Describe the three basic characteristics of the analog signal that can be used to convey information.

19. What is the difference between analog data and digital data?

20. What is the function of a CODEC? How does it work?

21. Describe PCM. How does it relate to PAM?

22. Where would one use a repeater or an amplifier? What is the difference between the two devices?

23. Describe the UNI and NNI specifications. What are the differences? Where do they fit in the broadband network?

24. What is the difference between Point-to-Point and MultiPoint?

25. What is the difference between DTE and DCE? Provide examples where this applies.

26. What is the difference between FDM, TDM, and STDM?

27. When multiplexing, how many channels can be derived from a DS-0, DS-1, or DS-3?

28. Describe the combining process of frequency division multiplexing. What are examples of FDM?

29. Describe the signal combining process of time division multiplexing.

30. What is the difference between full-duplex and half-duplex transmission?

31. What is the difference between synchronous and asynchronous transmission?

32. Describe the differences between two-wire and four-wire transport facilities.

33. What is the relationship between a LEC, IXC, and POP? How does this relate to LATAs?

34. What is the difference between an ARP and a RARP? Why are they both necessary?

35. Describe the differences between dynamic and static routes.

36. What are the primary dynamic routing protocols?

37. What is the difference between IGP and EGP?

38. Describe RIP and compare it with OSPF.

39. What are the major differences between EIGRP and BGP?

40. Give examples of potential broadband access applications.

1. Create a list of the highlights and differences of dedicated services and switched services. Identify the specific vendor opportunities to compete for existing customer business with dedicated or switched product suites. What would be the difficulties involved in replacing either network configuration with broadband services?

2. Develop an application that could utilize a packet switched topology. Show how this would compare with a circuit switched network.

3. Develop a comparison table of the dynamic routing protocols. Identify where they would be utilized and the features of each protocol.

4. Contact the local ILEC or CLEC and obtain brochures and information on services that can address the various industries that use telecommunications in their daily operations.

5. Arrange for a carrier representative to present information on the WAN infrastructure.

6. Arrange for a tour of a network control center or central office. Sketch the basic components of the center.

7. Identify the various ILECs and CLECs that provide services in your area. Determine the various products and services that are offered by these organizations.

CASE STUDY / PROJECT

This project can consume a considerable amount of time and might best be assigned as a group project. Part of it, however, can be conducted in class. Using brainstorming in class after some initial research by the students would be effective in identifying the questions that need to be answered by the students.

Situation

Several companies provide for purchasing products and services through a mail-order house. These include both major corporations and small mom-and-pop operations. A considerable number of these are advertised on television.

WearhouseRUs (WRU) is a large mail-order house that takes orders for clothing items. WRU has locations in seven major cities throughout the South. These orders are transmitted daily to the host computer that is located in Atlanta, Ga. These orders are processed at the host location and submitted to the various clothing suppliers for shipment directly to the customer. Competition in this industry has become intense with the advent of e-Commerce and the Internet. It is now possible to access the Internet and through a number of Websites, order almost any clothing item for delivery to the doorstep. In order to compete, WRU must provide a level of service comparable with the Internet clothing providers.

Figure 2–22
CaseStudy WRU Network
Environment

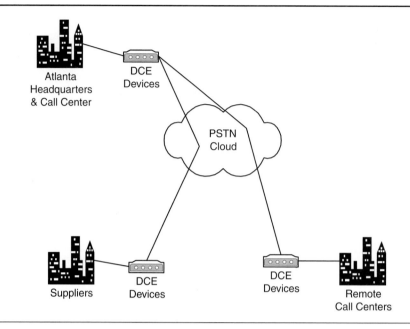

WRU has been using the PSTN for connecting to its computer center host processor to transmit the daily transaction information (Figure 2–22). The elapsed time for this transport is approximately 25 minutes for each of seven locations. The host computer site has one dial-up 19.2 kbps modem. Each store site has a 19.2 kbps modem. The order transmission to the clothing suppliers is also accomplished via the 19.2 kbps modem. Each of the clothing suppliers provides the modem for the transmission to WRU. The telecommunications manager is seeking to improve the overall order process. The local ILEC has contacted WRU and proposed a leased line solution to replace the dial-up operation.

Details

The call centers in the various WRU locations are open from 9 a.m. to 9 p.m., Monday through Friday. The centers are located in Miami, Charlotte, Atlanta, Birmingham, New Orleans, Nashville, and Louisville. The Computer center and headquarters is located in Atlanta, Ga. After closing each day, the host computer dials each site in turn and uploads the daily transactions from the servers. Approximately 25 minutes of transmission time is required for each of the seven data transmissions from the remote locations to the host computer. The order transmission to the clothing suppliers takes approximately 30 seconds per order for each of the 500 orders. Each order consists of an average of 500 characters. There are twenty clothing suppliers that are located throughout the United States. The suppliers have several different operating schedules. Numerous retransmissions are required due to the difficulties in the network.

Requirements

It is your responsibility to evaluate the current situation at WRU and provide a recommendation to higher management. The first step is to make a list of questions that will help clarify the current situation at WRU. This may require researching the clothing industry. Look at the state-of-the-art telecommunications and make recommendations that might improve the situation. Look at additional applications that might be utilized on the upgraded facilities. Remember that the ILEC has proposed a leased line solution.

PART TWO

CHAPTER 3
Frame Relay/X.25

Chapter 3 begins with a discussion of the Frame Relay technology and a comparison of X.25 with Frame Relay. Discussion continues with the architecture of switched and permanent virtual circuits. Throughput and capacity, which includes congestion control, is detailed, including information about Forward Explicit Congestion (FECN), Backward Explicit Congestion Notification (BECN), Committed Information Rate (CIR), and Discard Eligibility (DE). Topics also include various components and attributes, and the various interfaces such as NNI and UNI. The chapter concludes with standards issues and the advantages and disadvantages of the Frame Relay technology.

CHAPTER 4
Asynchronous Transfer Mode

Chapter 4 introduces the Asynchronous Transfer Mode (ATM) with an overview of the technology, its characteristics, and its architecture. The various ATM cell functions are described and include discussions on the virtual channel identifier and virtual path identifier. The NNI/UNI/PNNI interfaces are discussed, as is access connectivity. The chapter also presents considerable information on the components of ATM, which include the ATM Layer and the ATM Adaptation Layers (AALs), and the various classes of service provided by ATM. Another section covers capacity and throughput and how it applies to both public and private ATM backbone networks. The chapter concludes with a discussion on Quality of Service (QoS), ATM applications, and advantages and disadvantages of the ATM technology.

CHAPTER 5
Synchronous Optical Network

Chapter 5 begins with an overview of the technology and the infrastructure and speeds offered by the Synchronous Optical Network (SONET). Discussion continues with the SONET frame, the multiplexing process, and the SONET components of path overhead, line overhead, and section overhead. This discussion also concerns the Synchronous Payload Envelope (SPE) and the various SONET pointers. Other topics are end-user topology, and the various equipment elements for SONET, including T-1 multiplexers, digital cross-connect systems, digital switches, regenerators, and optical media. The chapter concludes with information on the various SONET standards such as quality of service (QoS), SONET applications, and the advantages and disadvantages of the SONET technology.

CHAPTER 6
Virtual Private Network

Chapter 6 presents the concept of a Virtual Private Network (VPN) and the VPN environment.

The chapter looks at the various VPN types, including details about access VPNs, intranet VPNs, and extranet VPNs. A considerable amount of text is oriented toward tunneling, security, and connectivity issues. A discussion presents details concerning VPN hardware and software that includes firewalls, gateways, and routers. The software section includes information about Point-to-Point Tunneling Protocol (PPTP), Layer 2 Forwarding Protocol (L2F), Layer 2 Tunneling Protocol (L2TP), and Internet Security (IPSec). VPN implementation scenarios are also presented, as are VPN standards and applications. The chapter concludes with a discussion of issues and considerations, and advantages and disadvantages, of a VPN installation.

CHAPTER 7
Fiber Distributed Data Interface

Chapter 7 begins with an overview of the Fiber Distributed Data Interface (FDDI) technology and topology which includes a discussion of the various device types utilized in the network. The chapter also compares FDDI and Token Ring, and the FDDI physical specifications, which also includes a presentation on the physical media utilized by FDDI. Other topics include FDDI hardware port types and ring connectivity, FDDI architecture, protocol architecture, ring scheduling, and ring operation. The FDDI frame and token format and the other FDDI standards are described. The chapter concludes with a discussion of FDDI-I and FDDI-II, FDDI applications, and the advantages and disadvantages of FDDI.

Frame Relay/X.25

■ INTRODUCTION

Frame Relay is a point-to-point Permanent Virtual Circuit (PVC) technology that offers WAN communications over a fast, reliable, digital, packet switching network. It was developed out of X.25 and ISDN technology. The premise of Frame Relay is that modern communication systems, which are relatively error free, do not require the extensive and resource-consuming operations that are required of older networks for error correction.

There is an entirely different set of acronyms to learn in the Frame Relay environment. Terms such as FRAD, PAD, CIR, FECN, BECN, LMI, PVC, SVC, and DLCI are all commonplace in discussions. Each of these acronyms will be explained in detail.

Frame Relay is a telecommunications service designed for cost-efficient data transmission for intermittent traffic between LANs and between end points in a WAN. Frame Relay puts data in a variable-size unit called a frame, and leaves any necessary error correction (retransmission of data) up to the end points, which results in increased throughput of data transmission.

For most services, the network provides a PVC, which means that the customer sees a continuous, dedicated connection without having to pay for a full-time leased line, while the service provider figures out the route each frame travels to its destination and can charge based on usage. An enterprise can select a level of service quality through a Committed Information Rate (CIR)—prioritizing some frames and making others less important.

Frame Relay is offered by a number of service providers, including AT&T, Sprint, MCI, and various Telcos. Frame Relay is provided on fractional or full T-1 carriers. Frame Relay complements and provides a midrange service between ISDN, which offers bandwidth at 128 kbps, and Asynchronous Transfer Mode (ATM), which operates in a somewhat similar fashion to Frame Relay, but at speeds from 155 Mbps (OC-3) or 622 Mbps (OC-12).

Frame Relay services are quickly growing in popularity. They are relatively inexpensive and allow the customer to specify the bandwidth

OBJECTIVES
Material included in this chapter should enable you to:

- become familiar with the terms and definitions that are part of the Frame Relay environment.

- understand the function and interaction of the Enterprise Network components with the Frame Relay infrastructure.

- understand the differences between Frame Relay and X.25.

- identify components that comprise the Frame Relay networking environments.

- understand the significance of congestion control in the Frame Relay environment.

- identify applications that fit in the Frame Relay environment.

- determine the best fit of fast-packet services for a particular situation and environment by comparing Frame Relay with the other available technologies.

- successfully identify a Frame Relay network solution.

Figure 3–1
Switching Tree Structure

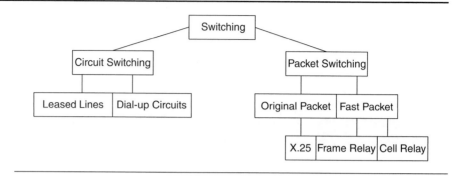

needed. Frame relay is often used to interconnect LANs over WAN facilities. It provides a dedicated connection during the transmission period. Under certain circumstances, Frame Relay can be used for voice and video transmission, and is a packet-switching technology.

Figure 3–1 depicts the switching tree structure for packet and circuit switching (see Chapter 1). The two subdivisions of packet switching are original packet switching and fast-packet switching. Fast-packet switching is further subdivided into Frame Relay and Cell Relay (ATM). Frame Relay is categorized as a fast-packet service.

3.1   FRAME RELAY TECHNOLOGY

What Is It?

virtual circuit

Frame Relay is a simple connection-oriented, **virtual circuit,** packet service. It can provide both Switched Virtual Circuits (SVCs) and Permanent Virtual Circuits (PVCs). Frame Relay is basically a multiplexed interface to a packet switched network. It provides a means for statistically multiplexing many logical data conversations over a physical transmission link.

Frame Relay is a high-performance WAN protocol that operates at the physical and data link layers of the OSI model. It was originally designed for use across ISDN interfaces. As mentioned, it is an example of a packet switching technology. Packet switched networks enable end stations to dynamically share the network medium and the available bandwidth. Variable-length packets are used for more efficient and flexible data transfers. These packets are then switched between the various network segments up to the destination.

Frame Relay services provide bandwidth on demand. A user may use 64 kbps for an application at a given point in time and may use 128 kbps for a different application at another instant. Frame Relay allows the user to obtain dynamic allocation of bandwidth capacity. The main approaches used are as follows:

- Circuit switching using Time Division Multiplexing (TDM)—In the TDM approach, time slots are reserved for each user device. This method is fast and uses no error checking.
- Unchannelized T-1—This is a modification of the TDM approach which uses proprietary schemes for using traffic in a nonslotted manner.
- Statistical Time Division Multiplexing (STDM)—Each user is identified by a logical channel number and is provided with a virtual circuit.

Statistical multiplexing techniques control network access in a packet switched network. The advantage of this technique is that it accommodates more flexibility and more efficient use of bandwidth. Frame Relay is described as a streamlined version of X.25 that offers fewer robust capabilities than X.25. The design approach of Frame Relay focuses on eliminating some of the problems of earlier protocols, such as X.25, in dealing with errors.

Historically, the network switches performed error-checking operations on all traffic. The premise of Frame Relay is that modern communications systems do not require the additional operations for error correction. In the event that an error occurs, Frame Relay assumes that end-user machines will handle the detection and resolution of errors. The idea is to notify the user about actual or potential congestion problems, and for the user to respond accordingly. Frame Relay assumes that the user machine supports end-to-end acknowledgment of traffic, which in the past was the network's responsibility.

Frame Relay was developed due to major trends in the communications industry [Frame Relay Forum, 2000]. The trends include:

- Increased need for speed across the network platforms within the end-user and the carrier networks
- Improved transmission facilities
- Increased intelligence of the devices attached to the network
- Increased need to connect LANs and WANs and the Internetworking capabilities

3.2 FRAME RELAY VERSUS X.25

Frame relay is based on the older X.25 packet switching technology, which was designed for transmitting analog data such as voice conversations. X.25 was the first universal data communications protocol. It stimulated growth in data communications traffic, largely due to its reliability and

Table 3–1
OSI Model Layers 1 to 3

OSI Model Layer	Frame Relay	X.25
3—Network	Handled at higher layers	Packet switching, orderly communications, connection reliability
2—Data Link (MAC / LLC)	LAPF; Data transfer, frame composition, frame size checking, packet switching, reliability checking	Data transfer, error checking and correction, frame sequence, flow control, frame composition
1—Physical	Physical connectivity	ITU-T X.21 physical connectivity

robustness [International Standards Organization (ISO), 2000]. X.25 was developed when the quality of telecommunications lines was much worse than today and communication errors were more common.

Unlike X.25, which was designed for analog signals, Frame Relay is a fast-packet technology, which means that the protocol does not attempt to correct errors. When an error is detected in a frame, it is simply "dropped" (thrown away). In Frame Relay, the end-points are responsible for detecting and re-transmitting dropped frames. For X.25, the network switches performed error-checking operations on all traffic, including retransmissions in the event of errors and transmissions of messages to the user in the event of failed retransmission. Today, the incidence of error in modern digital networks is extraordinarily small relative to analog networks.

X.25 and Frame Relay both perform error detection, but X.25 will request a retransmission, whereas Frame Relay will discard a bad frame. Frame Relay operates at Layers 1 and 2 of the OSI model. Since X.25 operates at OSI Layers 1 to 3, additional overhead is incurred, which increases the throughput time for X.25 and makes Frame Relay efficient and more attractive for the network user. Table 3–1 compares three layers of the OSI model for Frame Relay and X.25.

One of the greatest merits of the simple Frame Relay data transfer protocol is that it provides a high degree of transparency to the higher layer protocols that are carried. This contrasts with X.25, where the scope for destructive interference with higher layer protocols often causes problems and can seriously impair performance and throughput.

Key features of X.25 that result in considerable overhead are as follows:

- Call control packets, used for setting up and clearing virtual circuits, are carried on the same channel and same virtual circuit as data packets, and these involve in-band signaling.
- Multiplexing of virtual circuits occurs at layer 3.
- Both layers 2 and 3 provide flow and error control mechanisms.

Following are some key features of Frame Relay that improve performance over X.25.

- Call control signaling is carried on a separate logical connection from user data. Intermediate nodes need not maintain state tables or process messages relating to call control on an individual per-connection basis.
- Multiplexing and switching of logical connections takes place at layer 2 instead of layer 3, eliminating an entire layer of processing.
- There is no hop-by-hop flow or error control. Both are the responsibility of the higher layers.

Another difference between Frame Relay and X.25 is that Frame Relay uses only the physical layer and the Link Access Procedure Frame (LAPF) mode bearer services. The physical layer consists of interfaces similar to those in X.25. The Layer 2 LAPF is designed for fast communications services without the overhead of X.25, but it also includes an optional sublayer for situations requiring high reliability.

Frame Relay has several elements in common with X.25. Both use packet switching over virtual circuits. Also, as with X.25, the virtual connections can be switched or be permanent. In Frame Relay, DTE might be a router, bridge, or computer that is connected to a DCE, which is a network device that connects to a Frame Relay WAN. Instead of using a PAD to convert packets, as in X.25, Frame Relay uses a **Frame Relay Assembler/Disassembler (FRAD)**, which is often a module in a router, switch, or chassis hub.

> frame relay assembler/disassembler (FRAD)

To summarize, Frame Relay does the following:

- Utilizes less overhead, and therefore is faster than X.25
- Assumes that the network is error free
- Assumes that intelligent devices are on each end of the connection

3.3 FRAME RELAY PROTOCOL ARCHITECTURE

There are two separate planes of operation for the Frame Relay protocol: a **control plane** and a **user plane.** The control plane is involved in the establishment and termination of logical connections, and the user plane is responsible for the transfer of user data between subscribers. The control-plane protocols are between a subscriber and the network, whereas user plane protocols provide end-to-end functionality. Figure 3–2 depicts the components of both the control and user planes that are used to support the frame mode bearer service.

> control plane
> user plane

Control Plane
The control plane for frame mode bearer services is similar to that for common channel signaling for circuit switched services, in that a separate

Figure 3–2
User / Control Plane

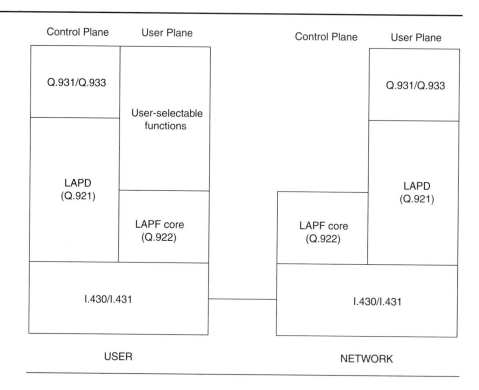

link access procedure
(LAPD)

logical channel is used for control information. At the data link layer, **Link Access Procedure (LAPD)**(Q.921) is used to provide a reliable data link control service, with error and flow control between the user and the network over the D channel. This data link service is used for the exchange of Q.933 control signaling messages.

User Plane

For the actual transfer of information between end users, the user plane protocol is LAPF (Q.922). Q.922 is an enhanced version of LAPD (Q.921). Only the core functions of LAPF are used for Frame Relay. The core functions of LAPF in the user plane constitute a sublayer of the data link layer. This provides the bearer service of transferring data link frames from one subscriber to another, with no flow or error control. Based on the core functions, a network offers frame relaying as a connection-oriented link layer service which allows preservation of the order of frame transfer from one edge of the network to another with a small probability of frame loss.

3.4   FRAME RELAY NETWORKS

One of the biggest impediments to high throughput has been the necessity to guarantee the delivery of data. Public networks were the culprits because of their noisy and unreliability due to the use of copper as the primary transmission medium. The introduction of fiber optics has eliminated a considerable number of the barriers to reliable data transmission. Frame Relay is one of the first services to take advantage of the high-quality digitized public carrier transmission lines.

Like X.25 networks, Frame Relay networks are generally operated by either a common carrier or by a private telecommunications service provider. An organization usually contracts with a Frame Relay service provider to implement point-to-point connections between pairs of DTE devices.

Local and long distance companies are combining to build Frame Relay networks. Shared by multiple users, they have the benefits of a dedicated network. Frame Relay is a virtual, value-added, private network service where end users are not required to lease dedicated lines. It is an alternative to businesses creating their own private data network. Frame Relay is mainly used for LAN-to-LAN data transmissions across wide areas and is cost effective for organizations with more than four sites.

Frame Relay uses multiple virtual connections over a single cable medium. Each virtual connection, which is logical rather than physical, provides a data path between two communicating nodes. Two types of virtual connections exist within Frame Relay: **Switched Virtual Circuit (SVC)** and **Permanent Virtual Circuit (PVC)**.

switched virtual circuit (SVC)
permanent virtual circuit (PVC)

Switched Virtual Circuits

Switched Virtual Circuits (SVCs) are temporary connections used in situations that require only sporadic data transfer between DTE devices across the Frame Relay network. A SVC is designed to allow the network or T-carrier provider to determine the data throughput rate. It can be adjusted according to the needs of the application and the current network traffic conditions. Multiple SVC connections can be supported on a single cable from point to point. Communication across a SVC consists of four operational states:

- Call Setup—The virtual circuit between two DTE devices is established.
- Data Transfer—Data are transmitted between the two DTE devices.
- Idle—The connection between the two DTE devices is active; however, no data is transferred.
- Call Termination—The virtual circuit between the two DTE devices is terminated.

After the virtual circuit is terminated, the DTE devices must establish a new SVC if additional data transmission is required. Actual deployment of SVCs is minimal in today's Frame Relay networks.

The main benefit of a SVC arrangement is that individual data links need only be established when needed and may be cleared afterwards.

This simplifies the network and its management, and in addition has the effect of automatically optimizing the routing of connections each time they are newly established.

Permanent Virtual Circuits

Permanent Virtual Circuits (PVCs) are permanently established connections that are used for frequent and consistent data transfers between DTE devices across the Frame Relay network. Communication across a PVC does not require the call setup and termination states that are used with SVCs. PVCs always operate in one of the following two operational states:

- Data Transfer—Data are transmitted between the DTE devices over the virtual circuit.
- Idle—The connection between DTE devices is active, but no data are transferred.

DTE devices can begin transferring data whenever they are ready because the circuit is permanently established. Signal transmission is handled at the physical layer, and virtual connections are part of the LAPF layer. A single cable medium can support multiple virtual connections to different network destinations.

3.5   FRAME RELAY FRAME

Frame Relay operates at the data link layer of the OSI model rather than at the network layer. A frame can incorporate packets from different protocols such as Ethernet and X.25. It is variable in size (8 k byte increments) and can be as large as 4,096 bytes or greater. Figure 3–3 depicts the general Frame Relay frame format.

frame check sequence (FCS)

The frame begins and ends with one-octet flag fields that have the binary value 01111110. The start flag signals the beginning of the frame and allows for synchronization. The **Frame Check Sequence (FCS)** field allows each switch to check for errors in the Frame Relay header. If it finds an error, the switch discards the frame.

Figure 3–3
Frame Relay Frame Format

Flag (8)	Header (16)	User Information	FCS (16)	Flag (8)

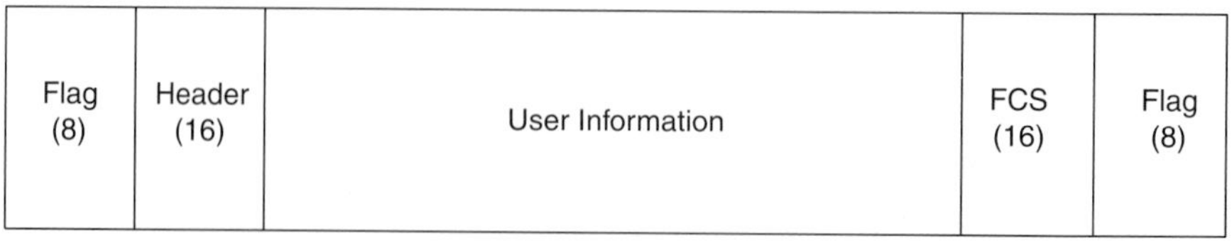

DLCI(6)	C/R(1)	EA(1)	DLCI(4)	FECN(1)	BECN(1)	DE(1)	EA(1)

Figure 3–4
Frame Relay Header Format

Header Format

The Frame Relay header (Figure 3–4) consists of two bytes (16 bits) that contain the following fields:

- Data Link Connection Identifier (DLCI)—10 bits total
- Command / Response (C/R)—1 bit
- Address Extension (AE)—2 bits total
- Discard Eligible (DE)—1 bit
- Forward Explicit Congestion Notification (FECN)—1 bit
- Backward Explicit Congestion Notification (BECN)—1 bit

The DCLI field identifies a specific virtual circuit. The AE bit is set to one if the octet it ends is the last octet in the address field; otherwise, it is set to zero. For transmissions within the CIR, the DE bit is set to zero. For frames going faster than the **Committed Information Rate (CIR)**, the DE bit is set to one. The BECN field is set to tell the station that receives the frame to slow down. The FECN tells the station at the other end to slow down. The C/R field allows the two communicating parties to indicate whether the message is a command or a response.

committed information rate (CIR)

3.6 THROUGHPUT AND CAPACITY

A potential problem associated with Frame Relay is that carriers tend to oversell capacity, calculating that the network will not be used by everyone at the same time. When the carrier's network is oversubscribed, frames can be dropped during peak periods of usage to avoid congestion. If congestion conditions exist in the network, the carrier's switches will warn customer devices, so they can step down their transmission rates. If they fail to do so, their discard eligible data will be dropped. Required for Frame Relay service, a CIR is utilized because of the bursty nature of frame relay traffic. The usual charge for CIR is allocated in 8 k increments.

CIR refers to the average maximum transmission speed of a user over a link to the Frame Relay Network. The customer is always free to burst up to the maximum circuit and port speed. This amount may be subject to an additional surcharge. Excess bursts can be marked by the carrier as discard eligible and subsequently discarded in the event of network congestion. A CIR of zero indicates that all frames are discard eligible in the event of congestion in the network.

When the sum of the data arriving over all virtual circuits exceeds the access rate, the situation is called *over subscription.* This can occur when the CIR is exceeded by burst traffic for the virtual circuits. Over-subscription results in dropped packets. In such a case, the dropped packets must be retransmitted. A reduction in the number of PVCs per virtual circuit can improve this over subscription situation. A network management system can detect this type of situation.

A user that subscribes to Frame Relay service leases a circuit of a certain port speed to the carrier's CO. The port speed is the maximum rate at which data can be transmitted. The subscriber contracts with the carrier for a certain CIR measured in bits per second. This amount of CIR can be from zero to the speed of the circuit and is usually allocated in 8 kbps increments. The basic mechanism for relieving congestion in a crowded Frame Relay network is to discard frames, which can result in slow response time and degraded service.

Committed Information Rate and Discard Eligibility

Frame Relay handles congestion by discarding traffic to avoid these problems. The network may use the DE bit to aid in determining what traffic to discard. The committed information rate can be utilized in conjunction with the DE bit to accomplish this task. The end user estimates the amount of traffic that will be sent to the network during the normal period of time. This measured average traffic is called CIR. The rate is agreed upon by both the user and the carrier and becomes part of the service contract.

The network measures the traffic during a time interval and, if it is less than the CIR value that has been set in the Frame Relay switch, the network will not alter the DE bit. However, if the rate exceeds the CIR value during the time interval, the network will allow the traffic to pass unless the network is congested; in which case, the congested network will set the DE bit to one. It is likely that this traffic will be discarded.

committed burst rate (B_c)
excess burst rate (B_e)

The factors that comprise the CIR are **Committed Burst Rate (B_c)** and **Excess Burst Rate (B_e).** Some references use size instead of rate in their documentation. These service parameters are policed at the point of entry to the network, and they can be set independently for each direction of transmission to cater efficiently for applications that send more information in one direction than the other.

Committed Burst Rate (B_c) and Excess Burst Rate (B_e)

The Committed Burst Rate (B_c) describes the maximum amount of data that a user is allowed to offer to the network during some time interval. This is established during call setup or pre-provisioned with a PVC.

The Excess Burst Rate (B_e) describes the amount of data that a user may send that exceeds the committed burst rate during the time interval. The excess burst rate also identifies the maximum number of bits that the network will attempt to deliver in excess of the committed burst rate during this time interval.

The CIR describes the information transfer rate that the network must commit to in order to support user traffic during normal operations. The full implementation of CIR, B_e, and B_c is not supported by all networks. Alternatively, the provider may simply agree to provide the user with an access rate interface of DS-0 or DS-1, with the stipulation that the user can send data at the access rate, but only for a certain period of time.

Most public network providers offer Frame Relay at two prices: either a fixed monthly fee based on an access rate in bits or usage basis measured in volume of bits.

Congestion Control

Frame Relay does not perform error detection and correction on each hop between pairs of switches as does packet switching. Nor does it have very good flow control. It places lower processing burdens on switches than X.25, which allows for much higher speeds and much lower costs and reduces delay (latency) at each switch. Frame Relay reduces network overhead by implementing two simple congestion-notification mechanisms, **Forward Explicit Congestion Notification (FECN)** and **Backward Explicit Congestion Notification (BECN)**, rather than explicit, per-virtual-circuit flow control. Frame Relay is typically implemented on reliable network media, so data integrity is not sacrificed because flow control can be left up to higher layer protocols.

forward explicit congestion notification (FECN)
backward explicit congestion notification (BECN)

FECN, BECN, and Discard Eligibility

A two-byte (16 bits) control field is used for network addressing and control. Frame relay has 3 bits in the frame definition to handle flow control. FECN and BECN are each controlled by a single bit contained in the Frame Relay frame header. The Frame Relay frame also contains a **Discard Eligibility (DE)** bit, which is used to identify less important traffic that can be dropped during periods of congestion. BECN is sent back to the original source user to throttle back its transmission to prevent network congestion, whereas FECN warns the destination recipient of impending network congestion. If the DE bit is set, permission to discard frames in excess of the CIR is given to relieve network congestion.

discard eligibility (DE)

The FECN bit is part of the address field in the Frame Relay frame header. The FECN mechanism is initiated when a DTE device sends Frame Relay frames into the network. If the network is congested, DCE devices (switches) set the value of the frames' FECN bit to one. When the frames reach the destination DTE device, the address field indicates that the frame experienced congestion in the path from the source to the

destination. The DTE device can relay this information to a higher layer protocol for processing. Depending on the implementation, flow control may be initiated, or the indication may be ignored.

The BECN bit is part of the address field in the Frame Relay header. DCE devices set the value of the BECN bit to one in frames traveling in the opposite direction of frames with their FECN bit set. This informs the receiving DTE device that a particular path through the network is congested. The DTE device can then relay this information to a higher layer protocol for processing. Depending on the implementation, flow control may be initiated, or the indication may be ignored.

The discard eligibility bit is used to indicate that a frame has lower importance than other frames. The DE bit is part of the address field in the Frame Relay frame header. DTE devices can set the value of the DE bit of a frame to one to indicate that the frame has a lower importance than other frames. When the network becomes congested, DCE devices will discard frames with the DE bit set before discarding those that do not. This reduces the likelihood of critical data being dropped by Frame Relay DCE devices during periods of congestion.

Frame Relay Error Checking

cyclic redundancy check (CRC)

Frame Relay uses a common error-checking mechanism known as **Cyclic Redundancy Check (CRC)**. The CRC compares two calculated values to determine whether errors occurred during the transmission from source to destination. The CRC is stored in the Frame Check Sequence (FCS) field of the Frame Relay frame. Frame Relay reduces network overhead by implementing error checking rather than error correction. Frame Relay is typically implemented on reliable network media, so data integrity is not sacrificed as error correction can be left up to higher layer protocols that run on top of the Frame Relay protocol stack.

3.7   FRAME RELAY COMPONENTS

To access a Frame Relay network, a DTE must either incorporate a frame relay interface card or alternatively use some other standard interface and an external conversion device. A frame relay packet assembler/disassembler (FRAD) is required at each end of the Frame Relay circuit. The FRADs provide for the conversion of continuous bit stream oriented data signals into a frame format. Each of the FRADs represents an addressable node on the

packet assembler/ dissembler (PAD)

Frame Relay network. (Note: X.25 requires a **Packet Assembler/Dissembler (PAD)** at each end.) The FRAD function can take place in several CPE devices, namely a router, switch, bridge or gateway. FRADs are also available as stand-alone units, which may be confusing because both the software and the hardware have the same name (FRAD). Figure 3–5 shows a PAD in an X.25 configuration and a FRAD in a Frame Relay configuration.

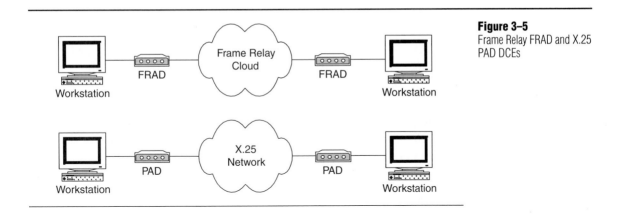

Figure 3–5
Frame Relay FRAD and X.25 PAD DCEs

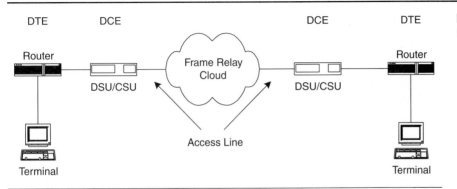

Figure 3–6
CPE / DTE Interfaces

Figure 3–6 shows the DCE/DTE interface points in the Frame Relay environment. The DSUs provide for the DCE interface specifications and the router provides for the DTE interface specifications. The access line or local loop provides connectivity from the DCE device (DSU) to the LEC or telco.

Frame Relay Switches

Frame Relay switches are available in basically two versions—Central Office switch and customer network switch. As an example, the Cascade 9000 B-STDX is utilized by a number of carriers to provide Frame Relay and SMDS traffic across the public network. Obviously, the CO-based switch is larger and more robust than the customer switch. The user-based Frame Relay switch is available in a desktop or rack-mounted version. Frame Relay switches are available from numerous vendors.

3.8   ADDRESSING

data link connection
identifier (DLCI)

link access procedure
balanced (LAPB)

The address field contains the **Data Link Connection Identifier (DLCI)**, which is used to identify the virtual connection over which the frame is transported. The DLCI has the same function in Frame Relay as the Logical Channel Identifier (LCI) field in the X.25 **Link Access Procedure Balanced (LAPB)** header. Data transfer involves the following three steps:

1. Establish a logical connection between two endpoints, and assign a unique DLCI to the connection.
2. Exchange information in data frames. Each frame includes a DLCI field to identify the connection.
3. Release the logical connection.

Data Link Connection Identifier

The DLCI provides the identification mechanism that identifies a particular virtual circuit and a particular station at the remote end of the virtual circuit. A computer connected to a Frame Relay network can communicate with any number of other computers by placing the appropriate DLCI value in each frame it transmits. Each end of a virtual circuit will have its own DLCI. The DLCI is identical to a virtual circuit number in a network layer protocol. The Frame Relay network is responsible for mapping from the source user's DLCI to the destination's DLCI.

Frame Relay DLCIs have local significance, which means that the values themselves are not unique in the Frame Relay WAN. Two DTE devices connected by a virtual circuit may use a different DLCI value to refer to the same connection.

The DLCI address field can be two, three, or four octets in length, thus allowing for DLCI value lengths of 10, 17, or 24 bits long. The DLCI values are consistent with the service access point identifier (SAPI) and Terminal Endpoint Identifier (TEI) values used in ISDN LAPD. This allows LAPD and Frame Relay frames to be multiplexed on the ISDN D channel. Table 3–2 provides a list of the permissible DLCI values.

The DLCI can also be used to provide the protocol identification function. One computer can establish any number of virtual circuits with another computer attached to the Frame Relay network. Each virtual circuit connecting the two computers is assigned a different DLCI and can be used to carry packets associated with a different Network layer protocol.

The DLCIs are premapped to a destination node. This simplifies the process at the routers, because they need only to look at their routing table, check the DLCI in the table, and then route the traffic to the proper port based on this address.

DLCI Values	Function	**Table 3–2** DLCI Values
Two-octet address format		
0	In-channel signaling	
1–15	Reserved	
16–991	Assigned using Frame Relay connection procedures	
Three-octet address format		
0	In-channel signaling	
1–1,023	Reserved	
1,024–63,487	Assigned using Frame Relay connection procedures	
Four-octet address format		
0	In-channel signaling	
1–131,017	Reserved	
131,072–8,126,463	Assigned using Frame Relay connection procedures	

Local Management Interface (LMI)

To supplement the basic frame information, several **Local Management Interface (LMI)** extensions must be included as a header in the data portion of the frame. Frame relay LMI extensions include global addressing, virtual-circuit status messages, and multicasting. Its main job is to provide status and configuration about PVCs at a User-to-Network Interface (UNI). This extension helps to synchronize communications between the DCE and the DTE, and to ensure that a complete connection exists before data are transmitted. A network can also use the LMI to inform the user CPE about the addition and deletion of PVCs, and advise if they are active or inactive. Optional LMI features are available for multimedia applications, global addressing, and traditional XON/XOFF flow control.

local management interface (LMI)

The LMI global addressing extension gives Frame Relay DLCI values a global rather than a local significance. DLCI values become DTE addresses that are unique in the Frame Relay WAN. The global addressing extension adds functionality and manageability to Frame Relay internetworks. Individual network interfaces and the end nodes attached to them can be identified by using standard address resolution and discovery techniques. In addition, the entire Frame Relay network appears to be a typical LAN to routers on its periphery.

LMI virtual circuit status messages provide communication and synchronization between Frame Relay DTE and DCE devices. These messages are used to periodically report on the status of PVCs, which prevents data from being sent into the bit bucket.

The LMI multicasting extension allows multicast groups to be assigned. Multicasting saves bandwidth by allowing routing updates and address-resolution messages to be sent only to specific groups of routers.

Figure 3–7
Frame Relay DLCI / PCVs

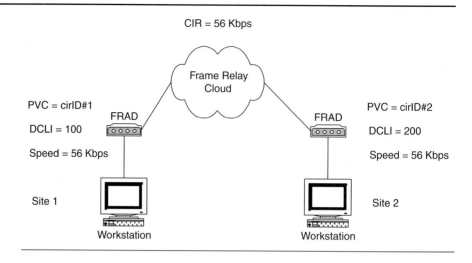

The extension also transmits reports on the status of multicast groups in update messages.

Connectivity

Connections are made from the customer locations to the Frame Relay cloud via Permanent Virtual Circuits (PVCs). A PVC is similar to a dedicated line, in that communication devices are not concerned with route management and error checking. A PVC assignment is required in the Frame Relay switch and the name usually consists of the circuit ID. Also required is a DLCI which is used in addressing the packet. The Frame Relay network depicted in Figure 3–7 shows two sites operating at 56 kbps with a CIR of 56 kbps. Site 1 has a DLCI of 100 and a PVC name of cirID#1. Site 2 has a DLCI of 200 and a PVC name of cirID#2. This information is coded into the respective FRAD devices. The CIR is coded into the CO Frame Relay switch along with the connectivity information. The DLCIs have local significance because they are both served out of the same CO-based switch.

3.9 NETWORK INTERFACES

As in packet switched data networks, the switches within a given carrier's network are commonly provided by a single manufacturer, therefore it is not necessary to use a standardized interface between the nodes within the network. Certain situations, however, require switching across two different WANs or subnetworks, supplied by different manufacturers.

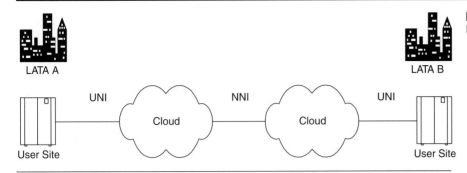

Figure 3–8
UNI to NNI Relationship

LATA A   LATA B

UNI   NNI   UNI

Cloud   Cloud

User Site   User Site

Two concepts that apply to the WAN are the **User-to-Network Interface (UNI)** and **Network-to-Network Interface (NNI)**. The UNI defines standards for interoperability between end-user equipment and the network equipment. The NNI defines standards for interoperability between the network vendor's equipment. The relationship of the NNI to the UNI is depicted in Figure 3–8.

user-to-network interface (UNI)
network-to-network interface (NNI)

NNI and UNI

A PVC operating across more than one Frame Relay network is called a multinetwork PVC. Each piece of the PVC provided by each network is a PVC segment. The multinetwork PVC is the combination of the relevant PVC segments. Full internetworking operations between Frame Relay require that the procedures set forth in ANSI T-1.617 Annex D be used at the UNI and the NNI interfaces. This means that a user sends a Status Enquiry (SE) message to the network and the network responds with a Status (S) message. The SE is used to query the receiver about the status of PVC segments. The SE then provides information about PVC segments to the user. These messages are sent across the network in HDLC Unnumbered Information (UI) frames.

NNI Operations

NNI operations take place through exchange of the S and SE messages that contain information about the status of PVCs. These operations include the following:

- Notification of the adding of a PVC
- Detection of the deletion of a PVC
- Verification of links between Frame Relay nodes
- Notification of a PVC segment availability/unavailability
- Verification of Frame Relay nodes
- Notification of UNI and NNI failures

Figure 3–9
Frame Relay Multisegment
Network NNI/UNI

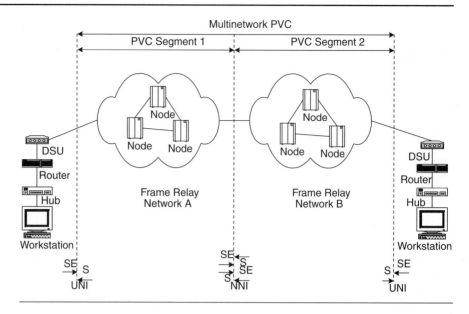

Figure 3–9 depicts a multinetwork PVC which includes two Frame Relay network segments with two PVC segments that connect two different Ethernet LANs. It also shows the direction of flow for both the Status (S) and Status Enquiry (SE) messages. The figure also shows the DCE interface at the DSU and the DTE interface at the router for each PVC segment. The requirement for the NNI could be either the need to cross LATA boundaries or access facilities between different carriers.

3.10 FRAME RELAY STANDARDS

To fully understand how Frame Relay works, one must understand the underlying standards for the technology. The following ITU-T recommendations define Frame Relay.

- Recommendation I.233 describes the Frame Relay service.
- Recommendation I.122 defines the framework of recommendations that specify Frame Relay.
- Recommendation Q.922 defines the core aspects of Frame Relay, including the data link procedures for frame format, the address field, and so forth.
- Recommendation I.370 defines the congestion management procedures.
- Recommendation Q.933 defines the signaling procedures to set up switched virtual connections. It is not utilized for PVC services.

Layers 4–7	Higher Layer Information
Network Layer 3	Q.933 Signaling
Data Link Layer 2	Q.922 Core aspects
Physical Layer 1	Connectivity

Table 3–3
Protocol Stack for Frame Relay

Table 3–3 shows the relevant OSI layered structure for Frame Relay.

In 1988, the CCITT approved Recommendation I.122, Framework for Additional Packet Mode Bearer Services, which is part of a series of ISDN-related specifications. ISDN developers had been using a protocol called LAPD to carry signaling information on the D channel of ISDN as defined by CCITT Recommendation Q.921. LAPD has characteristics that could be useful in other applications, such as the provision for multiplexing virtual circuits at Layer 2 in the frame level (as opposed to Layer 3 in the X.25 networks). Therefore, I.122 was written to provide a framework outlining how such a protocol might be used in applications other than ISDN signaling.

Frame Relay Requests for Comments and Management Information Bases

Request for Comments (RFC) 1294 Multiprotocol Interconnect over Frame Relay establishes how the protocols are encapsulated within the Frame Relay frame and transported across the network. It stipulates that the protocols running over Frame Relay must encapsulate their Protocol Data Units (PDUs) within the ITU-T **Q.922 Annex A** frame. It also stipulates fragmentation guidelines. RFC document 1315 is the Management Information Base (MIB) for Frame Relay UNI. The objects are organized into three object groups: (1) data link connection management, (2) circuit group, and (3) error group. These groups are stored in tables in the MIB and can be accessed by Simple Network Management Protocol (SNMP). Appendix C also identifies several RFCs involved in the Frame Relay technology.

Q.922 Annex A

X.25 Standards

Four protocols of particular importance in maintaining an X.25 network are as follows:

- X.3 protocol—Specifies how a PAD converts a packet to be sent in the X.25 format, and how to strip the X.25 information out when the packet reaches the destination network.
- X.20 protocol—Defines start and stop transmissions between the DTE and DCE devices.
- X.28 protocol—Specifies the interface between the DTE and the PAD.

- X.29 protocol—Specifies how control information is sent between the DTE and the PAD, the form in which the control information is sent.

Information concerning X.25 is available in the following RFC documents:

- RFC 1382 SNMP MIB Extension for the X.25 Packet Layer
- RFC 1381 SNMP MIB Extension for the X.25 LAPB

Appendix C also identifies several RFCs involved in the X.25 technology.

3.11   FRAME RELAY ADVANTAGES AND DISADVANTAGES

Frame Relay Advantages
Utilizing the Frame Relay technology involves several advantages:

- A long-distance provider, not the end-user, manages the network. This allows companies to concentrate on their main business, instead of maintaining their networks. Network Management, if performed correctly, is an expensive undertaking. As an example, the Banking industry's primary function is financial operations, and may not be equipped to support a network management function. It is more effective and efficient for an organization that is oriented toward network management to undertake this function.
- Less hardware is required at each site than that for private networks. Because Frame Relay utilizes PVCs, fewer routers and DCE devices are required at the central site. The switches are located in the carrier's central offices. The air-conditioning and power facilities that are required for the infrastructure is all provided by the carrier.
- Adding capacity is easier on Frame Relay than on private lines. Because logical circuits are multiplexed using PVCs, additional circuits can be added with simple software changes by the carrier. Private lines require a DCE device at each end of the circuit.
- Variable-size frames handle intermittent data traffic between LANs and WANs. Different frame size capacity handles the varying requirements of a versatile and dynamic network. Some of the frames may be short requests with longer frames as a response.
- Speed is fast because error correction is left up to endpoints— This increases speeds of overall transmission since Frame Relay operates at OSI layers 1 and 2, and therefore, throughput is greater.
- Frame Relay is based on older protocol X.25; however, it does not carry the overhead required by X.25, which was designed for analog signals. Frame Relay is a fast-packet technology, and so does

not attempt to correct errors. It is not necessary for this additional function due to the improved quality of the network.

- Transmission is on a dedicated path. Some applications require a dedicated transmission path due to the time-sensitive nature of the data. Although Frame Relay is not ideally suited for either streaming audio or video, which requires a steady flow of data, if provisioned properly, these applications can be handled reliably.
- Customer sees a dedicated circuit through a PVC. Effectively, the user has a private line access, which has all the security attributes associated with the private line.
- Frame Relay is similar to a private line because it is permanently connected. Faster access is the result as setup and tear down are not required.
- Numerous speed options are available. Fractional rates of 56 kbps to 44.736 Mbps are available for DS-1 and DS-3. (i.e. 64 kbps–1.544 Mbps in increments of 64 kbps)
- A CIR is established for each PVC. This allows the customer to burst up to this rate and be guaranteed a successful transmission. The customer can decide what level of risk is acceptable for the application.
- A DLCI is established for each PVC. This is basically the address that is coded in the user's DTE devices. It can be utilized to identify the circuit and the location of the devices. A database can be created for network management utilizing a structured DLCI numbering scheme.
- Frame Relay is cost effective for one-to-many customer terminations. Consider a bank with numerous branches as opposed to many-to-many, where all sites need to communicate with all other sites (SMDS topology).

It appears that these advantages would always suggest a Frame Relay solution; however, this is not the case. At times it is not the appropriate network solution.

Frame Relay Disadvantages

- Multimedia transmissions. The prime limitation to Frame Relay has been the ability to transmit time-sensitive (brittle) applications due to the burst transmission characteristics of the system. When framed voice and video do not arrive in a predictable timed fashion, problems can occur in the conversion process that translates the voice and video signals back to their original form.
- Analog to digital (A/D) conversions are a problem. Digital conversion, compression, and multiplexing technologies have not advanced sufficiently to satisfactorily handle conversion of the common 56 to 64 kbps analog telephone capabilities.

- International access may not be available. Several countries still do not offer Frame Relay access or service capabilities.
- The technology has certain limitations. It has not yet been determined if Frame Relay is a viable alternative to the circuit switched dial-up telephone network for voice communications.
- Carrier failures are a reality. In several instances major providers of the Frame Relay service have experienced failures in the backbone network. This requires that the user have a backup plan with an alternative service when, and if, this occurs.

3.12   FRAME RELAY APPLICATIONS

Numerous organizations can utilize Frame Relay to enhance the performance of their network applications. Industries and applications that are supported are as follows:

- Education—distance learning; Computer-based training from remote sites to a central host computer center.
- Government—enhanced customer service and electronic transaction processing; Electronic tax filings and other governmental filing requirements; Electronic payment processing for government programs.
- Health Care—telemedicine, electronic filing and record consolidation; Filling of prescription from remote requests from customers; Filing of insurance forms from hospitals and other health care providers; Centralization of a patient's records in a single database.
- Legal—electronic research; Centralized law library access.
- General business—LAN connectivity; Business-to-business electronic commerce, data center consolidation, network cost reduction through leased line replacement, host-to-host network connectivity. Web-based applications, groupware software, E-mail, and Web-based customer support applications.
- Manufacturing—remote operational management; Just-in-time management of remote production and distribution centers.
- All—Business e-mail with fairly large file attachments; large company backbone for LANs to ISP; ISP.
- Any private line application where there is a one-to-many network.

Figures 3–10 and 3–11 illustrate a before and after network design that converts a private line network to a Frame Relay network, the classical application for Frame Relay.

Note the marked reduction in the number of DCE devices (DSUs) at the host location with a Frame Relay implementation.

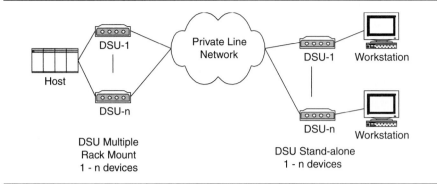

Figure 3–10
Private Line Network with
Multiple Leased Lines

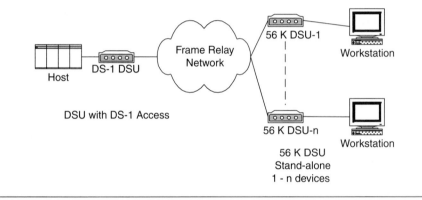

Figure 3–11
Frame Relay Network with
Multiple PVCs

3.13 TECHNOLOGY ALTERNATIVES

Competitors for Frame Relay are X.25, ISDN, leased lines, fractional T-1, ATM, and DSL. Obviously Frame Relay is superior to X.25 in that throughput is increased with Frame Relay since Layers 1 and 2 are used instead of Layers 1 to 3 for X.25. There is also a reduction in the number of DCE devices when implementing Frame Relay.

ISDN competes with Frame Relay; however, it is often used as a backup technology in the event that the Frame Relay network fails. Frame Relay has the capability of faster speeds than ISDN.

A leased line network competes with Frame Relay when it is local. Since leased lines are distance sensitive, costs are usually higher for leased lines when mileage between the endpoints is great. Cost savings are realized due to mileage cost and a reduction in the requirements of DCE devices at the host site. Fractional T-1 service in increments of 64 kbps has become an alternative in today's networks. This service requires a DSU device that can be optioned for the twenty-four DS-0 channels.

ATM can compete with Frame Relay at the lower speeds of DS-1 and DS-3. Frame Relay may not be offered at speeds greater than DS-1 in many areas. ATM, which can operate at speeds greater than Frame Relay, may not be available for public usage by all carriers in all areas.

ADSL competes with Frame Relay because it uses the embedded copper that is the customer's current access lines. DSL service, however, is not as available as Frame Relay service. Frame Relay may very well have a better grade of service and be less subject to outages than ADSL.

3.14   ISSUES AND CONSIDERATIONS

Frame Relay is a robust service. The most likely point of failure is in the access circuits. Once the traffic reaches the carrier's backbone network, disruption in unlikely. The major LECs offer Frame Relay in their serving areas. LEC Frame Relay must connect to an IEC to bridge LATA boundaries. Some LECs and IECs have worked out NNI arrangements to make the service area boundaries transparent to the customer.

A network manager should look at the alternative service features, access port and circuit speed issues, CIR and pricing elements, network management capabilities, and network connectivity when considering a service.

Service Features
Data communications features for Frame Relay and X.25 should be examined and compared with the services available from the various data communications network alternatives. Features to be considered are as follows:

- Virtual circuit or datagram services—users with very short messages may require a datagram service.
- Security—networks should offer password security and encryption.
- Protocol conversion—communication between terminals using unlike protocols may be supported by the network.
- Billing service—consider whether detailed call accounting is required or whether message charges are bulked to a user number.
- Multidestination message service—consider the importance of broadcasting messages to many stations simultaneously.
- Closed user groups—these are private networks within the network, designated for the exclusive use of users who gain access only with proper authentication.

Access Circuit and Port Speed

It is necessary to determine the required speed of the access circuit and port speed into the carrier's network. Speeds of 56/64 kbps may not offer enough bandwidth for the access circuit. It may be necessary to provide a T-1, fractional T-1, or even a DS-3 access circuit to provide sufficient throughput. It might be beneficial to determine if access can be shared with some other services that are routed to the same carrier.

Committed Information Rate

Selection of the CIR is one of the most important factors in determining the success of Frame Relay. It is necessary to review the expected maximum data throughput requirement of the network and the time of day that bursts are likly to occur. Review how the carrier handles bursts over the CIR. Determine whether the network buffers or discards packets when congestion occurs.

Pricing Elements

Frame Relay has three pricing elements that need to be researched:

- CIR—the throughput the carrier guarantees will be transported
- Port speed—the speed of the access port into the carrier's network. It is in multiples of 64 kbps
- Access circuit—the cost of the access circuit provided by the ILEC or CLEC, 56 kbps to 44.736 Mbps

Network Management Information

A major advantage of Frame Relay over fixed networks is the amount of network information that is available. Determine what reports are available, how often they are produced, and how they are obtained. This information may be available online. Most carriers have some type of network management system for monitoring the performance of the Frame Relay network.

Network Connectivity

The location of the carrier's Point Of Presence (POP) is of concern to designers. In locations without a POP in the same city, the cost of the local access circuit can be high enough that Frame Relay is not cost effective. Some of the major Interexchange Carriers (IECs) have signed NNI agreements to make local access to IEC Frame Relay networks more economical. A company with both inter- and intraLATA connections on the same network can often provide local access circuits at a lower cost.

■ SUMMARY

A technology that has gained acceptance in the WAN community is Frame Relay, a fast-packet switching protocol that supports variable-length frame structure of up to 4,096 bytes. Frame Relay began as a concept of a simple packet access to ISDN. In contrast to conventional packet switching, fast-packet protocols such as Frame Relay are used in operating environments that include digital broadband systems.

Frame Relay was developed to reduce the overhead of packet switching and provide more efficient data transmission. This Layer 2 protocol divides messages into variable-length frames for transmission through the network at multimegabit speeds. Frame Relay has approximately 25 percent of the overhead of basic packet switching.

Because the Frame Relay protocol has been streamlined to make it fast and efficient, it handles circuit congestion differently. It accomplishes this control by utilizing the Committed Information Rate (CIR) concept. The basic mechanism for relieving congestion is to discard frames.

Using Frame Relay to interconnect the LANs of a corporation can provide a one-to-many connectivity through a single access circuit, using multiple virtual circuits. Frame Relay's high throughput makes it suitable for LAN and WAN data and imaging network traffic, but less satisfactory for voice and real-time video.

The majority of Frame Relay networks deployed today are provisioned by service providers who intend to offer transmission services to customers. This is often referred to as a public Frame Relay service. Frame Relay is implemented in both public carrier provided networks and private enterprise networks.

The benefits and advantages of frame relay services include the following:
- Standards based
- Increased utilization and efficiency
- Savings through network consolidation
- Improved network up-time
- Improvements in response time
- Easily modifiable and fast growth

Key Terms

Backward Explicit Congestion Notification (BECN)   Committed Information Rate (CIR)

Committed Burst Rate (B_c)   Control Plane

Cyclic Redundancy Check (CRC)

Discard Eligibility (DE)

Data Link Connection Identifier (DLCI)

Excess Burst Rate (B_e)

Forward Explicit Congestion Notification (FECN)

Frame Check Sequence (FCS)

Frame Relay Assembler/Disassembler (FRAD)

Link Access Procedure Balanced (LAPB)

Link Access Procedure (LAPD)

Local Management Interface (LMI)

Network-to-Network Interface (NNI)

Packet Assembler/Dissassembler (PAD)

Permanent Virtual Circuit (PVC)

Q.922 Annex A

Switched Virtual Circuit (SVC)

User-to-Network Interface (UNI)

User Plane

Virtual Circuit

REVIEW QUESTIONS

1. Describe the Frame Relay service. What is the significance of Frame Relay's variable-length frames in terms of types of payloads?

2. What are the primary differences between Frame Relay and X.25? Differentiate between X.25 and Frame Relay in terms of error control.

3. What are the common elements of Frame Relay and X.25? Why is dynamic allocation of bandwidth an important feature of Frame Relay?

4. Describe the switching tree structure.

5. What is a switched virtual circuit? What are the four operational states?

6. What is a permanent virtual circuit? How is it different from a SVC? Why are multiple PVCs per access line an important feature of Frame Relay?

7. What layers of the OSI model are used by Frame Relay?

8. What is the format of the Frame Relay frame?

9. What is the format of the Frame Relay header?

10. How is throughput and capacity controlled in Frame Relay? How is flow control handled in a Frame Relay network?

11. What is the difference between CIR and DE? How do they complement each other?

12. Describe how the CIR process works.

13. What is committed burst rate and excess burst rate? Why are they important to CIR?

14. What are the two mechanisms of congestion control? How do they work?

15. How does FECN and BECN interact with the DE bit?

16. What error-checking mechanism is used by Frame Relay? How does it work?

17. What are the DCE devices required for connection to the DTE devices for Frame Relay access to the network?

18. What is the difference between a DCE and DTE access point? What devices define the DCE and DTE interfaces in Frame Relay?

19. What is an access line? Is this different than a local loop?

20. Describe a Frame Relay WAN switch.

21. Give an example of connectivity of two DTE devices through the Frame Relay network. Include all of the components necessary for communication. What are the DCE

devices that are required for X.25 and Frame Relay? What are the functions of each?

22. What is the function of LMI? What is meant by global significance? Local significance?

23. What is a DLCI? Give an example.

24. What are the different categories of DLCIs? How many DLCI addresses can be created from the two-octet address format?

25. What is the difference between a NNI and an UNI?

26. Produce a simple drawing that shows the relationship between an UNI and a NNI.

27. What are the NNI operations?

28. Discuss the standards that are relative to Frame Relay. What is Recommendation I.122?

29. What are the primary Frame Relay MIBs?

30. List five advantages of the Frame Relay service. Discuss each in detail.

31. List disadvantages of Frame Relay and give the reasons why. What are the transmission speeds supported on Frame Relay?

32. List five applications for the Frame Relay service. Provide supporting reasons why they would be good candidates.

33. Discuss the application that replaces private lines with the Frame Relay service. Give reasons why this is an economical application.

34. What technologies can be utilized in lieu of Frame Relay? Give your reasons why.

35. There are a number of issues and considerations that must be addressed when deploying Frame Relay. Pick one and discuss.

36. What is the difference between a public and private Frame Relay network? Where would they be deployed? In what instance can frame relay be superior to dedicated line service?

37. What is meant by the term *burstiness?*

38. Explain the encapsulation process.

39. What is meant by the term *virtual circuit?*

40. What is latency and why would a network manager care?

ACTIVITIES

1. Research the Frame Relay topic on the Web, DataPro, and other online resources. Develop a five minute presentation to explain the technology, its usage, and its advantages and disadvantages over other technologies.

2. Arrange for an ILEC representative to give a presentation on the Frame Relay technology.

3. Visit a Frame Relay central office installation. Develop a drawing depicting the various components that are part of this network.

4. Develop a comparison of Frame Relay with such competing technologies as ATM, ISDN, and T-1.

5. The following activities can best be conducted in a class-room environment using boarding and brainstorming to identify the information requested. Rank-order the details identified in each category.

- Make a list of users and their respective applications that might benefit from a Frame Relay implementation.
- Make a list of pros and cons for migrating from a private line network to a Frame Relay network.

- Describe how you would connect Frame Relay to a remote customer. A sample network drawing would be required in this activity.
- Role-play as a network system designer and make a list of all the questions that would need to be answered to successfully design a Frame Relay solution. Someone must play the customer in this activity.

CASE STUDY/PROJECT

Overview
Implemented in 1992, Frame Relay is a public network offering that allows customers to send data among multiple locations. In using Frame Relay, organizations do not have to plan, build and maintain their own network paths to each site. For example, consider a large bank; without Frame Relay it would have over 200 dedicated lines, (paths) for corporate use connecting all of the branches to corporate headquarters. This would require a huge investment in staffing a telecommunications department for bill payment, tracking of repairs, and planning capacity for each line. Equipment at the main corporate office in the form of CSUs/DSUs and multiplexers to terminate each of the 200 lines would also be needed. This would require an investment in technical personnel and equipment. When network and device management becomes unmanageable in this one-to-many configuration of a central location headquarters to remote banks, Frame Relay is a possible technology fit. Smaller organizations could also benefit from this technology. The following is such an example.

Situation
East Coast National Bank (ECNB) is located in the Washington, D.C. area with branch banks in several locations. This bank has been created from a consolidation of three small banks which are located in Virginia, Maryland, and North Carolina. Each of these banks transacts its own business activities with the host computer located in Washington, D.C. via a private line network, which consists of a number of 19.2 kbps leased lines. Performance has degraded since the consolidation and the telecommunications manager has been advised to correct the situation.

Details
There are four branch banks in Virginia, four branch banks in Maryland and three branch banks in North Carolina. Each bank offers automated teller terminal access for their customers. Access to the teller machines is via 19.2 kbps dial-up circuits to each branch bank's server. The host

computer that will service this banking network will be located in Arlington, Virginia. Currently there are 19.2 kbps leased-line circuits between the host mainframe located in Arlington and the branch banks. Banking hours are 9 a.m. – 4 p.m., Monday through Friday. Automatic Teller Machine access is open 24 hours a day. Several ILECs, CLECs, and IXCs have contacted ECNB with proposals. The CFO is very conservative.

Requirement

Prepare a list of questions to help determine the speed and number of access circuits for a Frame Relay implementation. The questions should be oriented toward strategic (long-term), tactical (current), and day-to-day operations personnel who are involved in the daily banking operation.

Prepare a proposal for this customer that includes a Frame Relay network design. Include information to address the Automated Teller Machine (ATM) system requirements and the host-to-branch network. Provide all components that are necessary to implement the Frame Relay solution, and the necessary information to program the routers at all locations. Determine the amount of CIR that is necessary for this corporate customer.

Draw a network configuration that depicts the solution. Include both DTE and DCE devices and show connectivity between these devices. Include such information as is necessary to configure the routers in this configuration. IP addresses and interface specifications will enhance the usefulness of this drawing.

Asynchronous Transfer Mode

◼ INTRODUCTION

Asynchronous Transfer Mode (ATM) is a cell-switching and multiplexing technology that combines the benefits of circuit switching with those of packet switching. It provides scalable bandwidth from a few megabits per second to many gigabits per second. Because of its asynchronous nature, ATM is more efficient than synchronous technologies, such as Time Division Multiplexing (TDM). Because ATM is asynchronous, time slots are available on demand with information identifying the source of the transmission contained in the header of each ATM cell.

ATM provides a physical level, transparent solution for corporate communications requirements. The typical speeds of ATM networks are 155 Mbps (OC-3) or 622 Mbps (OC-12). Through the use of adaptation layer services, ATM provides adequate transmission quality across a broad spectrum of user services. ATM can interface and interoperate with various media and transmission types, including coax, twisted-pair, fiber optic media; T-3, FDDI, SONET, Frame Relay, and X.25 networks.

ATM can work hand in hand with other technologies, such as X.25, Frame Relay, and Ethernet to provide an efficient multipurpose physical transmission network. ATM provides a framework for integration of a corporate network, combining LAN, mainframe, voice, and video services onto a single platform, which transmits these services over high-bandwidth circuits, typically at hundreds of megabits per second. Digital lines and fiber optics are used to support ATM. The resulting noise- and error-free communications enable ATM to deliver at amazing transmission rates.

ATM is ideally suited to applications that cannot tolerate much time delay, as well as for transporting LAN, Frame Relay, and IP traffic that are characterized as **bursty.**

OBJECTIVES
Material included in this chapter should enable you to:

- understand ATM technology and its relation to the current broadband market.

- become familiar with the terms and definitions that are relevant in the Enterprise Network environment [ATM Forum, 2000].

- understand the function and interaction of ATM in the Enterprise Network environment. Become familiar with the ATM Layers and the ATM Adaptation Layers (AALs). Identify the various ATM classes of service.

- understand where ATM fits in the different applications of the technology. Identify the characteristics that make ATM a viable solution for an enterprise backbone network.

bursty

- identify components that comprise the ATM networking environments.Become familiar with the ATM frame format and the ATM cell functions.
- understand the Quality of Service (QoS) parameters of ATM and what role Network Management plays in the environment.
- identify ATM network solutions. Look at the impact of capacity and throughput on network traffic.

asynchronous transfer mode (ATM)

ATM is a dedicated-connection switching technology that organizes digital data into fifty-three octet cells or packets and transmits them over a medium using digital signal technology. Individually, a cell is processed asynchronously relative to other related cells and is queued before being multiplexed over the line.

ATM works by initially setting up a connection between two sites during which a virtual circuit is established between them. The virtual circuit corresponds to a specific path determined during signaling and all cells that are sent through the virtual circuit follow the same path.

ATM is based mostly on an ITU-T protocol known as Broadband Integrated Services Digital Network (B-ISDN). It makes use of hardware switching devices to transmit at the Data Link layer of the OSI model.

4.1   WHAT IS ATM?

Asynchronous Transfer Mode (ATM) is a high-speed packet switching service (currently up to 622 million bps) capable of carrying voice, data, fax, real-time video, CD-quality audio, and multimedia images. It was developed as part of the B-ISDN and is intended to be carried on the Synchronous Optical Network (SONET) infrastructure. ATM has been used principally in the telephone company network infrastructure. Costing more than Frame Relay for end users to implement, it is emerging as the mode of choice for large users as a way to switch large files. For example, the airport control centers use ATM to ensure the proper function of the air travel system. ATM is used by the following groups:

- Governments
- Long-distance providers
- Bell telephone companies
- Internet service providers
- Large banking firms
- Frame relay networks
- Large universities

ATM Benefits
Key benefits of the ATM architecture include the following:

- Constant cell length affords faster, predictable delivery times.
- Constant cell length and predictable times allow voice, video, and data to be transported effectively.
- ATM protocols are supported from the LAN to the WAN, and from Network Interface Cards (NICs) to ATM WAN switches, thereby removing the necessity for multiple protocol conversions.

4.2   ATM CHARACTERISTICS

ATM is able to transport a wide range of information transmissions at high speeds by dividing data into equal-size cells and attaching a header to ensure that each cell is routed to its destination. This cell structure is able to transport voice, video, and data equally well. The primary attributes of ATM are:

- Fixed-size packets called cells
- Connection-oriented device
- Packet switching
- High-speed, low-delay transmission
- Speeds of 155 Mbps (OC-3) and 622 Mbps (OC-12)
- Designed for Multimedia applications
- Supported by Telcos
- B-ISDN technology

Because ATM is a switch-based technology, it is easily scalable. As traffic loads increase or the network grows in number, more ATM switches are added to the network. ATM physical links operate over many cable types, including categories 3, 4, and 5 UTP and STP, coaxial cable; and multi-mode and single-mode fiber.

Protocol Architecture

ATM is a streamlined protocol with minimal error and flow control capabilities. This reduces the overhead of processing ATM cells and reduces the number of overhead bits required with each cell.

To achieve high data rates, low delay, and low jitter, ATM technology divides data into small, fixed-size packets called **cells.** Each cell contains exactly fifty-three octets: five octets of header information and forty-eight octets of data. ATM allows the user to specify the **Quality of Service (QoS)** requirements for each communication. When using ATM for audio, the user must specify both low delay and low jitter. In ATM terminology, a **Constant Bit Rate (CBR)** is specified for voice and video. In contrast, applications that use ATM to send data can request an **Available Bit Rate (ABR)** connection.

Two layers of the protocol architecture relate to ATM functions. The ATM layer common to all services provides packet transfer capabilities, and an **ATM Adaptation Layer (AAL)** that is service dependent. The ATM layer defines the transmission of data in fixed-size cells and also defines the use of logical connections. The use of ATM creates the need for an adaptation layer to support information transfer protocols not based on ATM. The AAL maps higher layer information into ATM cells to be transported over an ATM network, then collects information from ATM cells for delivery to higher layers.

ATM uses the concept of **Virtual Channels (VCs)** and **Virtual Paths (VPs)** to accomplish the routing of cells. A VC is a connection between

cells

quality of service (QoS)

constant bit rate (CBR)

available bit rate (ABR)

ATM adaptation layer (AAL)

virtual channels (VCs)
virtual paths (VPs)

Figure 4–1
ATM Cell

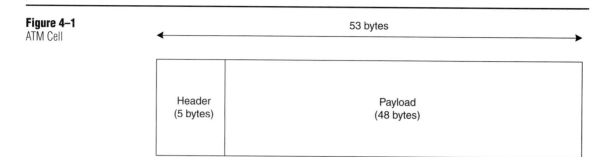

communicating entities. It may consist of several ATM links between Local Exchange Carriers (LECs). All communications occur on the VC, which preserves cell sequence. In contrast, a VP is a group of VCs carried between two points and may consist of several ATM links.

As with other WAN technologies, an ATM network uses switches as the primary building block in the network. A switch has multiple ports, where each port can be connected to another switch or to a computer (DTE). To achieve the highest bit rates, most ATM networks use optical fiber as the interconnection media.

Three ATM characteristics produce its speed:

1. The cells are fixed in size.
2. The cells are switched in hardware in a connection-oriented manner.
3. Switching is performed asynchronously.

Fixed-Size ATM Cells

ATM packages the data it switches into distinct bundles called cells. This is analogous to having the same number of digits for every kind of credit card. These bundles are called fixed-size cells (Figure 4–1). Because each cell has exactly 53 bytes, the ATM knows when each cell begins and ends. Five of the 53 bytes are header information and include bits that identify the type of information contained in the cell—voice, data, or video—so the cell can be prioritized. Small fixed-length cells are well suited to transferring voice and video traffic because such traffic is intolerant of delays that result from having to wait for a large data packet to download, among other activities. Also, cell relay switches can perform much faster than Frame Relay switches by being able to depend upon constant cell length, since more instructions for processing can be included in firmware and hardware. The constant-size cells also lead to a predictable and dependable processing rate and forwarding rate.

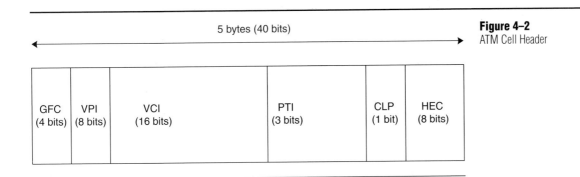

5 bytes (40 bits)

Figure 4–2
ATM Cell Header

| GFC (4 bits) | VPI (8 bits) | VCI (16 bits) | PTI (3 bits) | CLP (1 bit) | HEC (8 bits) |

Voice and video communications, which need constant bit rate transmission, are provided higher priorities with ATM so there is no interruption in voice or picture. Other header information is used for routing, putting the cells in correct sequence, and checking for errors. The other 48 bytes are the "payload," user data such as voice, video, bank records, or other business information. Individual cells can be mixed together and routed to their destination via the telecommunications network.

The heart of the ATM communications process lies in the content and structure of the cell. The ATM cell contains all of the network information for relaying individual cells from node to node over a preestablished ATM connection. The header field of the cell, which includes its first five bytes, has been designed for the addressing and flow control. Its role is only for networking purposes.

ATM Cell Header Format
Fixed-size cells with fixed header sizes were chosen as the cell structure for ATM, primarily because of the advantages produced for the switching equipment by the fixed approach. The ITU-T decided in 1989 that 48 bytes for cell payload and a 5-byte header (Figure 4–2) would comprise the ATM cell. The ATM cell header's main function is to provide each cell with channel and path information. When the cell enters an ATM switch, that switch identifies the virtual circuit connection to which the cell is routed.

The header contains the following fields:

- Generic Flow Control (GFC)
- Virtual Path Identifier (VPI)
- Virtual Channel Identifier (VCI)
- Payload Type Identifier (PTI)
- Cell Loss Priority (CLP)
- Header Error Control (HEC)

Figure 4–3
ATM Channel and Path
Multiplexing

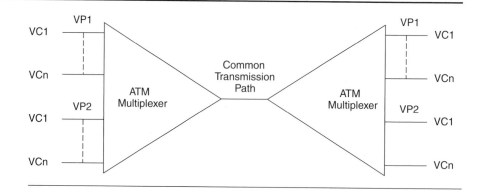

4.3   ATM CELL FUNCTIONS

Generic Flow Control

generic flow control
(GFC)

Generic Flow Control (GFC) is used for local control functions such as reducing short-term overload conditions. The first 4 bits of the first byte controls the flow of traffic across the User-to-Network Interface (UNI) and into the network. The ATM routing address is divided into two parts—**Virtual Path Identifier (VPI)** and **Virtual Channel Identifier (VCI)**. Channels are fed into an ATM multiplexer that identifies the virtual path, then aggregates all of the channels and paths onto the common transmission path through the network. The process is reversed as the paths and channels are demultiplexed. ATM channel and path multiplexing is shown in Figure 4–3.

virtual path identifier
(VPI)
virtual channel identifier
(VCI)

Virtual Path Identifier and Virtual Channel Identifier

The next 24 bits, after the GFC, make up the ATM address. The Virtual Path Identifier (VPI) contains the first part of the ATM routing address to identify the virtual path between users or between the user and the ATM network. It consists of an 8-bit value allowing up to 256 paths from the User-to-Network Interface. A Virtual Path consists of a bundle of virtual channels and is assigned to a Virtual Path Identifier.

The VPI and VCI subfields are used for multiplexing, switching, and demultiplexing a cell through the network. Neither are network addresses; they are explicitly assigned at each segment of a connection when a connection is established, and remain for the duration of the connection. Using the VCI/VPI, the ATM layer can asynchronously interweave (multiplex) cells from multiple connections.

virtual path connection
(VPC)

Multiple VPIs can form a **Virtual Path Connection (VPC)**. VPI operates in a connection-oriented fashion. Virtual Paths are switched by network nodes. VPI numbers are unique to the physical transmission link on which they arrive. VPIs can be point-to-point or multipoint and are typically established as Permanent Virtual Circuits (PVCs).

The Virtual Channel Identifier (VCI) contains the second part of the ATM routing address to identify the virtual channel between the users or between the user and the network. Virtual Channels are the connection-oriented paths on which user data are sent. VCI numbers are unique only to the virtual path on which they arrive. Virtual Channels map to virtual paths and are managed by the switch. VCIs are allocated 16 bits that allow up to 65 K virtual channels.

Payload Type

The **Payload Type (PT)** identifies the type of data in the payload area and could include user, network, or management information. This 3-bit field, which is encoded into one of eight types, indicates whether the contents of the cell are user data or network signaling information. If it is 00, it indicates user information. Types 0 to 3 are reserved at this time for identifying the type of user data. Types 4 and 5 indicate management information and 6 and 7 are reserved for future definition.

payload type (PT)

Cell Loss Priority

The **Cell Loss Priority** (**CLP**) field provides guidance to the network in the event of congestion and indicates whether the cell can be dropped. ATM cells can be lost from time to time due to buffer overflows, physical layer impairments, or errors in the cell routing field. A value of zero means that the cell has a high priority and cannot be dropped. The Header Error Control is used for error detection and for correction of single-bit errors. CLP is part of ATM Quality of Service (QoS) support and is set by the user.

cell loss priority (CLP)

Header Error Control

The final byte of the header field is designed for error control. The value of the **Header Error Control (HEC)** is computed based on the four previous bytes of the header field. The HEC is designed to detect header errors and correct single-bit errors within the header field. This provides protection against the misdelivery of cells to the wrong address. The HEC does not serve as an entire cell check character. Table 4–1 provides a summary of the ATM header fields and their functions.

header error control
(HEC)

4.4   NNI, UNI, PNNI

Two different header schemes exist for ATM cells: **User-to-Network Interface (UNI)** and **Network-to-Network Interface (NNI)**. Both UNI and NNI rely on 5 byte headers. ATM UNI describes premise-level standards, whereas ATM NNI describes network switch-to-switch standards. These

user-to-network interface
(UNI)
network-to-network inter-
face (NNI)

Table 4–1
ATM Cell Header Summary

ATM Header Field	Explanation
Generic Flow Control	Allows multiple devices of various types to gain access to an ATM network through a single access circuit
Virtual Path Identifier	Uniquely identifies the connection between two end nodes and is equivalent to the virtual circuits of Frame Relay.
Virtual Channel Identifier	Uniquely identifies a particular channel of information with the virtual path.
Payload Type	PT identifies whether the cell contains user information or network control information.
Cell Loss Priority	If congestion occurs with the ATM network, it can discard the cells that have been marked by the policing function.
Header Error Control	Ensures that header information contains no errors.

Figure 4–4
NNI / UNI Connectivity

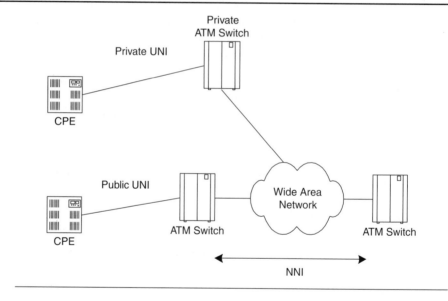

standards provide for easy implementation between different device and network types. The header format of the network node cell is identical to the UNI, except the GFC field has been replaced by an additional 4 bits of VPI, resulting in a 12-bit VPI for NNIs. Figure 4–4 depicts the NNI/UNI network and user environment.

UNI is designed for a connection between an end node and a network switch connection. The two types of UNI are public and private. A private UNI is an interface between an endpoint device and a private network switch. A public UNI is an interface between an endpoint device and a public switched network (WAN).

A NNI is the interface between two ATM switches. As with UNIs, NNI interfaces can be both public and private. NNI interfaces can occur between ATM switches within a LATA or across LATA boundaries.

The **ATM Forum** has proposed that multivendor networks be implemented according to the **Private Network-to-Network Interface (PNNI)** specifications [ATM Forum, 2000]. PNNI defines a protocol that allows the design and implementation of networks utilizing SVCs between any two PNNI-compliant ATM devices.

ATM forum
private network-to-
network interface (PNNI)

PNNI compliance allows the device to establish paths through the network without the entire network topology residing in memory, which is made possible by dividing the entire network into peer groups. Each peer group has one switch that acts as the group leader, gathering network information and communicating with other group leaders.

4.5 SWITCHING IN HARDWARE

ATM is significantly fast because the cells are switched in the hardware. An ATM switch does not have to look up each cell's address in software, as routers must do. Rather, an ATM switch sets up a route through the network when it sees the first cell of a transmission. Specifically, the network sets up a sender-to-receiver Virtual Circuit composed of a sequence of links between ATM switches.

The next time the switches see a cell with the same header router information, they send it down the Virtual Circuit they have already established. The network switches the data along the Virtual Circuit from switch to switch, one cell at a time. The ATM client uses signaling to tell the network that the Virtual Circuit is no longer needed. The network releases the Virtual Circuit created for that call.

The process of setting up a Virtual Circuit, using the same path for each cell, and then tearing down the Virtual Circuit when it is no longer needed makes ATM a connection-oriented service.

4.6 ACCESS CONNECTIVITY

Each ATM cell sent into the network contains addressing information (five octets) that the network utilizes to sequentially forward cells on a virtual connection from the sending location to one or more destinations. A *virtual circuit* is a pathway between two nodes on a switched network, which appears to be a dedicated point-to-point link and is transparent to the subscriber. These virtual connections are commonly called Permanent Virtual Circuits (PVCs) and Switched Virtual Circuits (SVCs).

Permanent Virtual Circuit

The distinction between a PVC and a SVC is similar to the distinction between a voice private line and a dial-up connection. A private line, and likewise a PVC, is set up or provisioned by the carrier. If the subscriber wants to make changes, a service order is placed with the carrier that will effect the change.

A **Permanent Virtual Circuit (PVC)** is a dedicated circuit that has a preassigned path and can have a fixed allocated bandwidth between two designated endpoints. This type of circuit is always up and active once it is created, thus eliminating delays caused by the setup and teardown of the circuit. This circuit must be active at all times for each router to facilitate transmissions and update routing information between the routers.

Switched Virtual Circuit (SVC)

switched virtual circuit (SVC)

A **Switched Virtual Circuit (SVC)** is, in effect, dialed by the subscriber. The subscriber establishes a service agreement with the carrier, and thereafter can set up a virtual circuit to other destinations by signaling.

A SVC is set up and torn down on an as-needed basis. It is a temporary connection that is created by a request for transmission facilities, and is only active as long as the devices are communicating. After communication is complete, the circuit is torn down and all of its resources are returned to the resource pool. The SVC is a dynamically established circuit that is created by signaling software and parameters defined by the device applications, communications equipment, and ATM facilities, and it requires no manual intervention.

4.7 COMPONENTS OF ATM

A discussion about ATM would not be complete without looking at the technological aspects that make up ATM. These include the ATM service layers, ATM adaptation layers, and ATM logical connections [McDysan and Spohn, 1998].

ATM Service Layers

ATM utilizes a four-layer architecture, called the ATM Protocol Reference Model (Figure 4–5), which enables multiple services to function at the same time over a single network.

The ATM layer is application independent: It provides the same frame transmission process regardless of the application. This would be similar to a train carrying boxcars whose contents are irrelevant to the railroad.

The **ATM Adaptation Layer (AAL)** handles different requirements in different applications. The job of the AAL is to build on ATM layer ser-

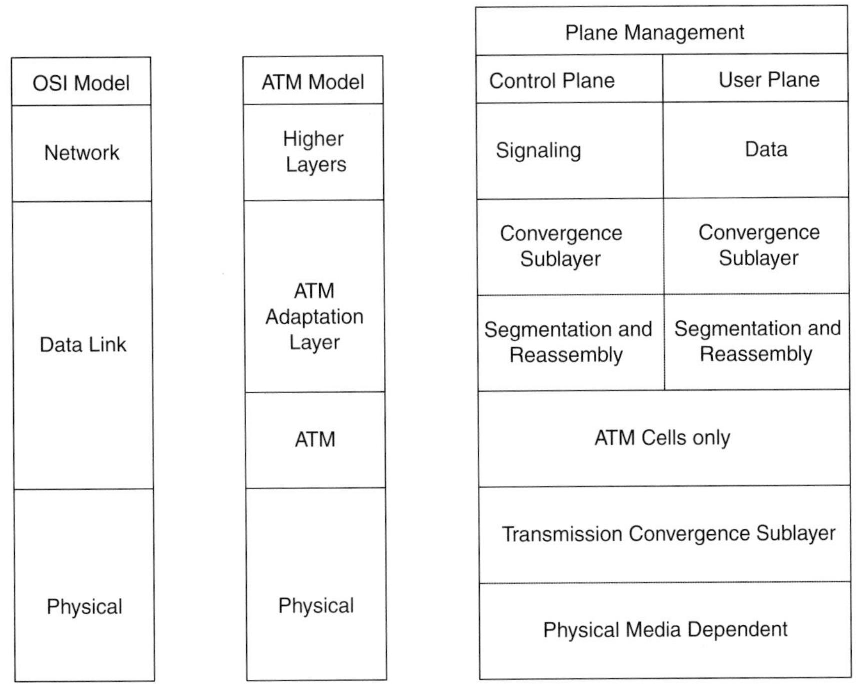

OSI Model	ATM Model
Network	Higher Layers
Data Link	ATM Adaptation Layer
	ATM
Physical	Physical

Plane Management	
Control Plane	User Plane
Signaling	Data
Convergence Sublayer	Convergence Sublayer
Segmentation and Reassembly	Segmentation and Reassembly
ATM Cells only	
Transmission Convergence Sublayer	
Physical Media Dependent	

Figure 4–5
ATM UNI Protocol
Reference Model

vices to provide the specific transmission characteristics that each application needs.

The protocol reference model involves three separate planes:

- **User plane**—Provides for user information transfer, associated controls for flow control, and recovery from error.
- **Control plane**—Performs call control and connection control functions. It also deals with other aspects pertaining to connection establishment and release.
- **Management plane**—Performs management functions related to the system as a whole and provides coordination between all planes. It is divided into two subplanes, plane management and layer management.

Nearly any type of upper layer protocol can be placed with the ATM cell. The four ATM layers are shown in Table 4–2. Two of these layers, the ATM layer and the AAL, will specifically perform ATM functions.

Table 4–2
ATM Layers

Layer	Function
ATM services and applications layer	Establishes the link between the node sending data and the AAL.
ATM adaptation layer (AAL) Contains two sublayers—the convergence and the segmentation and reassembly.	Segments data for ATM cell creation, and defines how data are received from and delivered to upper layers.
ATM Layer	Creates the ATM cell and handles routing and error control.
Physical Layer Contains two sublayers—the transmission convergence sublayer and the physical medium dependent sublayer.	Transforms cells into bits for transport over a physical medium, and contains the electrical and physical interfaces for ATM.

ATM Physical Layer

The ATM physical layer converts the cell stream into transportable bits and handles functions at the physical medium. The electrical and physical interfaces, line speeds, and transmission control are defined at this layer. A goal of the ATM working groups has been to standardize ATM for as many varieties of cable as possible. The ATM physical layer has two sublayers—**Physical Medium (PM)** and **Transmission Convergence (TC)**.

physical medium (PM) transmission convergence (TC)

Physical Medium Sublayer The physical medium sublayer includes only physical-medium-dependent functions. It provides bit transmission capabilities including bit alignment, line coding, and electrical/optical conversion. The functions of bit timing and insertion and extraction of bit timing information are the additional responsibilities. ATM works with many different transport methods and media, which include the following physical interfaces:

- FDDI at 100 Mbps
- SONET at OC-3 (155 Mbps) and OC-12 (622 Mbps)
- DS-1 at 1.544 Mbps
- DS-3 at 44.736 Mbps
- E-1 at 2.048 Mbps and E-3 at 34.368 Mbps
- UTP and STP at 25.6 Mbps

Additionally, work is in progress to develop standards for wireless ATM.

Transmission Convergence Sublayer Transmission frame adaptation functions are responsible for adapting the cell flow according to the payload structure of the transmission system. In the receiving process, this amounts to extracting the cell flow from the received information. The transmission frame may be cell equivalent or Synchronous Digital Hierarchy equivalent. The cell-based interface consists of a continuous stream of fifty-three octet cells.

The sublayer also handles transmission speed variance at the physical interface and on the ATM layer by inserting idle cells into the bit stream. This is necessary because the ATM layer in the switch may be able to process cells faster than the transmission speed of the channel.

The physical layer in an ATM network is defined by other protocols, and is normally implemented at the physical layer using fiber-optic cable. An ATM layer is defined to connect the physical layer to the AAL. The ATM layer generates the five-octet header when passing data to the physical layer. Cells arriving at an endpoint are sent to the AAL. Cells arriving at a switch are routed to the correct bit stream in an ATM multiplexer, and a **Virtual Channel Connection (VCC)** is created to connect two channels together.

virtual channel connection (VCC)

ATM Layer

The ATM layer is responsible for creating the ATM cell. This layer determines the cell structure, how the cell is routed, and error-control techniques, and ensures the quality of service of a circuit. The functions of this layer are performed by the ATM network's devices. These basic devices are the ATM switch and the ATM attached devices.

An ATM switch passes ATM traffic within the network and provides Quality of Service (QoS) for each circuit. A QoS marker in the cell header enables the ATM network to identify traffic types. Each type of traffic has a different tolerance for delay, accuracy, and throughput, and the QoS marker indicates the level of QoS required by the type of data in the cell. If the QoS cannot be maintained for a specific type of transmission, then the request for an ATM virtual circuit is refused. The main responsibility of ATM switches is to ensure that the cells are delivered to the correct receiving node in the same order as they are sent. If a lost cell is detected, then a retransmission is required from the sending node.

An ATM attached device translates data streams into or out of an ATM cell stream. Attached devices are workstations or servers that contain an ATM interface.

Some ATM QoS implementations, such as multimedia, require the ATM layer to establish the service level by having the transmitting node negotiate with the ATM switch. The network's connection management service establishes an approved transmission rate and throughput for a virtual circuit.

4.8   ATM LOGICAL CONNECTIONS

Logical connections in ATM are referred to as **Virtual Channel Connections (VCC)**, which are the basic unit of switching in an ATM network. A VCC is set up between two end users through the network and a variable-rate, full-duplex flow of fixed-size cells is exchanged over a connection. VCCs are also used for user-network exchange of control signaling and network-to-network exchange of network management and routing.

For ATM, a second sublayer of processing deals with the concept of virtual path. A **Virtual Path Connection (VPC)** is a bundle of VCCs that have the same endpoints. Thus, all of the cells flowing over all of the VCCs in a single VPC are switched together.

The virtual path concept was developed in response to a trend in high-speed networking in which the control cost of the network is becoming an increasingly higher proportion of the overall network cost. The virtual path technique helps control the cost of connections sharing common paths through the network by combining them into a single unit. Network management actions can then be applied to a small number of groups of connections instead of a large number of individual connections. Advantages for the use of virtual paths include:

- Simplified network architecture
- Reduced processing and short connection setup time
- Enhanced network services
- Increased network performance and reliability

The endpoints of a VCC may be end users, network entities, or both. In all cases, cells are delivered in the same order in which they are sent.

The ATM adaptation layer connects the ATM layer to higher layers in the protocol stack at an endpoint in the ATM network. Format conversion is made to add data from an incoming ATM cell to a packet used in the higher layer protocol. Outgoing data are taken from a higher layer **Protocol Data Unit (PDU)**, forty-eight octets at a time, and sent to the ATM layer for the addition of an ATM header.

protocol data unit (PDU)

The entire ATM network can be constructed using ATM and AAL switching and multiplexing principles. Four types of AAL information fields were originally recommended by the Consultation Committee on International Telegraphy (CCITT). The CCITT has been replaced by the **International Telecommunications Union-Telev (ITU-T).** Two of these (3 and 4) have now been merged into one, called AAL 3/4. For ATM to support many kinds of services with different traffic characteristics and system requirements, it is necessary to adapt the different classes of applications to the ATM layer. This adaptation function is performed by the AAL, which is service dependent. At this layer, different types of user traffic are adapted to a cell format. It is the AAL that gives the ATM the versatility to carry many different types of services from continuous processes such as voice to the highly bursty messages generated by LANs, all within the same format.

International Telecommunications Union (ITU-T)

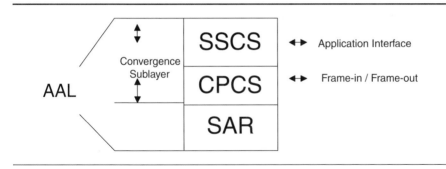

Figure 4–6
AAL Sublayers

The 48-byte information field of the ATM cell contains the user's data. Insertion of the user's data into the ATM cell is accomplished by the upper half of Layer 2 of the OSI model. The PDU, whether an IP datagram or Ethernet frame, must be segmented into a size that will fit into the 48-byte cell payload. At the destination, these segments must be reassembled into the original PDU. This function occurs in the AAL.

AAL Sublayers

The ATM adaptation layer is sub-divided into **Segmentation/Re-assembly (SAR)** and convergence sublayers. The convergence layer is further sub-divided into the **Service Specific Convergence Sub-layer (SSCS)** and the **Common Part Convergence Sub-layer (CPCS)**. Figure 4–6 depicts the relationships between the SAR, CPCS and SSCS sub-layers.

segmentation/re-assembly (SAR)
service specific convergence sub-layer (SSCS)
common part convergence sub-layer (CPCS)

The convergence sublayer provides the functions needed to support specific applications using AAL. Each AAL user attaches at a **Service Access Point (SAP)**, which is simply the address of the application. For ATM to support many kinds of services with different traffic characteristics and systems requirements, it is necessary to adapt the different classes of applications of the ATM cells into suitable high-level PDUs. This function, which is service dependent, is performed by the AAL convergence sublayer.

service access point (SAP)

The SAR sublayer accepts information that is provided by the convergence services sublayer. It is then responsible for packaging the information that is received from this sublayer into cells for transmission and for unpacking the information at the other end. The SAR must pack any SAR headers plus convergence sublayer information into forty-eight octet blocks, which is the size of the information fields. At the receiving end, transmissions are converted from transmitted cells to packets, and then processed at the higher layers of the receiving end.

4.9 AAL SERVICES

ATM adaptation layer protocols are designed to optimize the delivery of a wide variety of possible types of user inputs or traffic. This traffic can vary due to delay sensitivity, cell loss sensitivity, guaranteed bandwidth requirements, and overhead requirements. These AALs include services that provide for the transport and interworking of circuit and packet services such as:

- Voice
- Packet Data (SMDS, IP, Frame Relay)
- Video
- Imaging
- Circuit emulation
- LAN extension and emulation

AAL services assist many of the functions needed to interface a protocol such as Ethernet TCP/IP to ATM cells. It attempts to make the usage of ATM transparent to the user's communication applications. The AALs perform the following functions:

- Segmentation and reassembly. At the transmitting end, data are segmented into 48-byte chunks and reassembled into their original form at the receiving end.
- Sequence numbering. Each cell is given a sequence number to aid reassembly and detect lost or misinserted cells.
- Cyclic Redundancy Check (CRC). Performs error checking.
- Length Identification. Provides an indication to the receiver about the total length of the message, so an appropriate buffer size can be reserved for the message.
- Buffer Allocation Size. Each virtual circuit is given a timed buffer at every switched port. The buffer on the input side provides enough time for incoming packets or frames to be segmented into cells. The buffer on the output side provides enough time to reassemble the cells into the original packets or frames.

ATM handles both connection-oriented traffic, either directly (cell based) or through adaptation layers, and connectionless traffic, through the use of adaptation layers. Connection-oriented systems must connect with their recipient prior to information exchange. It provides high-integrity service over marginal facilities at the cost of a low throughput. Connectionless systems can send to a recipient without prior contact. ATM offers higher throughput than connection-oriented systems if the transmission path is clean.

4.10   ATM SERVICE CLASSES

Because most user's data require more than 48 bytes of information, the AAL divides this information into 48-byte segments that are suitable

Service Class	Class A	Class B	Class C	Class D
Service Type	Voice, real-time video	Packet video	Local ATM traffic	SMDS traffic
AAL Type	Type 1	Type 2	Type 3 / 4 or 5	Type 3 / 4 or 5
Bit-Rate	CBR	VBR	VBR	UBR and ABR
Connection Type	Connection Oriented	Connection Oriented	Connection Oriented	Connectionless

Table 4–3
ATM Services Summary

for packaging into a series of cells to be transmitted between endpoints on the network. AAL services are divided into five information field categories.

ITU-T specifies the use of AAL-1 through AAL-4 and the ATM Forum developed AAL-5. ATM service can be used without AAL if the application does not need AAL support and can work directly with the 48-byte ATM payload. This would require the application to perform its own segmentation and sequencing, as well as integrity checks.

AAL is organized around a concept called service classes. The purpose of this classification is to convert and aggregate different traffic into standard formats to support different user applications. These classes, grouped according to the following service categories, are summarized in Table 4–3. Additionally, there is a Class X which allows for user-defined traffic and timing relationships.

Class A ATM service is connection oriented and maintains a constant bit rate and a timing relationship between the source and the destinations systems. Class A service can be used to replace a standard circuit switched telecommunications facility and is useful for high-bandwidth isochronous applications such as transmission of full-motion video signals. Normal voice conversations are isochronous.

Class B ATM service is also connection oriented. Class B provides a variable bit rate while still maintaining a timing relationship between the source and destination systems. Class B service is intended for audio and video applications such as teleconferencing, where a variable bit rate can be tolerated as long as delays are within defined boundaries.

Class C ATM service is also connection oriented with a variable bit rate, but it maintains no timing relationship between the source and destination systems. Class C service is intended to be used by data transmission applications and provides a service similar to that provided by a X.25 or Frame Relay virtual circuit.

A Class D ATM service is connectionless, supports a variable bit rate, and maintains no timing relationship between the source and destination systems. This service provides a datagram data transmission facility where no error correction or flow control mechanisms are required in the Data Link layer.

Constant Bit Rate

peak cell rate (PCR)

Constant Bit Rate (CBR) is intended for real-time traffic that cannot tolerate variable bit rates or delay. CBR is characterized by a continuously available **Peak Cell Rate (PCR)** for all cells. Typical applications include videoconferencing, video on demand, and interactive voice and video. These applications must have the cells arrive with a constant delay time, constant bit rate, and in the same order as sent. The constant delay time means that the timing between end-to-end users must be maintained. Cell arrival in the same order as sent implies connection orientation. A constant bit rate ensures that the voice or video will be presented to the end user at the proper speed.

A negative side of CBR is that if this guaranteed amount of bandwidth is not required 100 percent of the time, no other applications can use the unused bandwidth.

Class A service provides AAL type-1 support. This AAL handles T-carrier traffic such as DS-0s, DS-1s, and DS-3s. This permits the ATM network to emulate voice and DS-n services. AAL-1 Circuit Emulation Service (CES) is used with a PVC between Private Branch Exchanges (PBXs) to eliminate leased T-1 lines. The adaptation can be accomplished in a PBX ATM card, ATM Data Service Unit (DSU), or an ATM access device.

Variable Bit Rate

variable bit rate (VBR)

Variable Bit Rate (VBR) provides enough bandwidth for bursty traffic such as transaction processing and LAN interconnection, as long as rates do not exceed a specified average. It provides a guaranteed minimum threshold amount of constant bandwidth below which the available bandwidth will not drop. Traffic is of a synchronous nature. This Class B service provides for AAL-2 support.

sustainable cell rate (SCR)
maximum burst size (MBS)

Type 2 AAL (timing-sensitive services) is currently undefined but reserved for data services requiring transfer of timing between endpoints as well as data. These include time-sensitive applications that can tolerate some variation in cell transfer speed, but must be delivered with minimal delay and delay variation. Traffic parameters include Peak Cell Rate (PCR), **Sustainable Cell Rate (SCR)**, and **Maximum Burst Size (MBS)**.

Type 3 AAL (connection-oriented) transfers VBR data between two users over a pre-established connection. The connection is established by network signaling similar to that used by the PSTN. This class of service is intended for large, long-period data transfer, such as file transfers for banking and order processing or backup and provides error detection at the AAL level. Traffic parameters are the same as a Type 2 AAL.

Type 4 AAL (connectionless) provides for the transmission of VBR data without pre-established connections. It is intended for short, bursty transmission as might be generated by LANs.

Available Bit Rate

Available Bit Rate (**ABR**) makes use of available bandwidth and minimizes data loss through congestion notification. Applications include e-mail and file transfers. ABR provides leftover bandwidth when it is not required by the variable bit rate traffic. Conventional frame and packet data exchange, such as transport of TCP/IP, Ethernet, and Frame Relay, is supported at this layer. A primary purpose, however, is to provide a mechanism for controlling traffic flow from LAN-based routers and their associated workstations. This is a Class C service that provides for AAL-3/5 support.

Type 5 AAL *(simple and efficient adaptation layer [SEAL])* offers improved efficiency over type 3. It serves the same purpose and assumes that the higher layer process will provide error recovery. The SEAL format simplifies sublayers of the AAL by packing all 48 bytes of the information field with data. AAL-5 supports connection-oriented variable bit-rate services. It is a substantially leaner AAL compared with AAL-3/4, at the expense of error recovery and built-in retransmission. This trade-off provides a smaller bandwidth overhead, simpler processing requirements, and reduces implementation complexity.

ABR provides flow control mechanisms that support several types of feedback to control the source rate in response to changing ATM layer transfer characteristics. Traffic sources must have the ability to adjust their information rates when the network requires a change. An end system is expected to adapt its traffic in accordance with the feedback from the control mechanisms, which will result in a low **Cell Loss Ratio** (**CLR**) and thus will allow it to obtain a fair share of the available bandwidth.

cell loss ratio (CLR)

Unspecified Bit Rate

Unspecified Bit Rate (**UBR**) makes use of any available bandwidth for routine communications between computers, but does not guarantee when or if data will arrive at their destination. The underlying cell switching service is connectionless. Variation in both delay and throughput can be quite large. Switched Multimegabit Data Service (SMDS) is supported at this layer. This Class D service provides for AAL-4/5 support.

unspecified bit rate (UBR)

Neither the service guarantee nor the guaranteed bandwidth is negotiated by UBR. Cells are sent on a best-effort basis and may be subject to considerable statistical multiplexing. Traffic is expected to be noncontinuous and bursty.

AAL Functions

As stated previously, the ATM Adaptation Layer is primarily responsible for the segmentation and reassembly of data into and out of ATM cell format. AAL services also enhance the quality of service provided by the ATM layer. AAL-5 is ideal for LAN data transport and **LAN Emulation** (**LANE**). AAL-1 will be used for ATM voice traffic. Some implementations

LAN emulation (LANE)

will skip AAL altogether. The five options for AAL are summarized as follows:

- **AAL-1:** Isochronous, Continuous Bit Rate—voice and video
- **AAL-2:** Isochronous, Variable Bit Rate—compressed video
- **AAL-3/4:** Data mode connection oriented/connectionless—bursty, SMDS
- **AAL-5:** Simple data mode—Frame Relay, LANE
- No AAL: ATM layer services are adequate

4.11   CAPACITY AND THROUGHPUT

As end users transmit more and more data into the network, it will eventually reach a saturation level. At this point the network nodes will begin to discard cells to relieve network congestion. Retransmission of this discarded data will impact the end users' systems. Full saturation of a node can occur due to the compounding effect of cells being discarded and then retransmitted. Several options are available to provide for effective congestion control and flow control.

Congestion Control Techniques

Congestion control is required to ensure that quality of service is met for each connection. Reactive congestion control schemes are used in existing networks, and preventive schemes are attempted in ATM networks. These schemes use policing functions to protect the network against congestion by forcing the traffic to conform to the parameters negotiated at the call connection phase.

The loss of a single cell results in the retransmission of thirty-one cells. Cell errors are reduced by digital/optical carrier, and cell insertion is reduced by header error correction. Cell congestion loss occurs at a time when the network is loaded at the maximum and produces additional loading from retransmission, resulting in a negative snowball effect called *throughput collapse*. Different approaches are utilized to counter the congestion issue.

Generic Cell Rate and Leaky Bucket

generic cell rate (GCR)

Two different algorithms utilized to address the congestion issue are generic cell rate and leaky bucket. The **Generic Cell Rate (GCR)** algorithm measures system load based on the arrival time of cells. It occurs at the User-to-Network Interface (UNI) and guarantees that traffic matches the negotiated connection that has been established between the user and the network. For each cell arrival, the GCR algorithm determines whether the cell conforms to the traffic contract and enforces conformance.

Both constant and variable bit rate traffic typically use the **leaky bucket** technique to meter traffic flow. Basically, committed burst size over time is the reference utilized in this algorithm. The "leaking hole in the bucket" applies to the sustained rate at which cells can be accommodated. Both GCR and leaky bucket fix the overall flow of data through a given channel. The only advantage VBR gives over CBR is that two leaky buckets are utilized.

leaky bucket

Available bit rate techniques (discussed earlier) utilize a form of feedback to inform the sender of route congestion, allowing the sender to utilize the full bandwidth until a congestion problem exists. The Frame Relay congestion control mechanism of FECN/BECN is utilized as a method to throttle traffic flow. FECN uses the forward data stream to indicate to the end system that congestion was experienced en route. The receiving end system then informs the original sender via a return message indicating congestion experienced. The sender should then adjust the forward data flow. BECN provides a faster feedback path than FECN by using the en-route switches to sense and respond to traffic congestion, placing the burden on the switching fabric.

4.12   ATM BACKBONE NETWORKS

ATM is available in several different modes. Most of the major IECs and many LECs have announced ATM support. To use the service, customers must access the network by one of several methods. One method is a dedicated access line, which could be a T-1, T-3, or one of the levels of SONET. With this access method, the customer furnishes data in a continuous bit stream. The carrier's ATM switch does the segmenting and desegmenting.

A second access method is via private ATM network. With this access the customer delivers and receives cells, and provides its own segmentation and desegmentation. An ATM network can also be fed from other sources such as Frame Relay or SMDS.

The most common implementation of ATM in LAN environments is as a backbone network. Using ATM as the backbone technology simplifies network management by reducing the complexity of internetworking environments. ATM works well for both small and midsize legacy backbones and large multi-LAN environments.

Large multi-LAN segments consist of LANs interconnected over large distances. Since there is no provision for protocol conversion in ATM, the user must provide Ethernet and Token Ring MAC frame conversion capabilities. These segments must be able to communicate using multiprotocol equipment such as routers and gateways. An issue that must be addressed is redundancy in the event of a network failure. This can be addressed by designing the ATM backbone with redundant

Figure 4–7
ATM Backbone Network

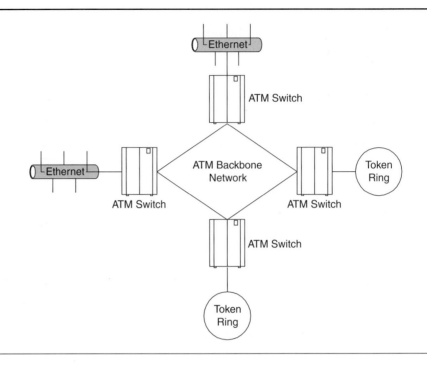

ATM switches and connecting them in a mesh configuration. Figure 4–7 depicts a multinode ATM backbone network that is connecting both Ethernet and Token Ring LANs. Note that there are two paths between each LAN.

4.13   ATM DEVICES

An ATM network consists of an ATM switch and ATM endpoints. An ATM switch is responsible for cell transit through an ATM network. The ATM switch performs the following functions:

- Accepts the incoming cell from an ATM endpoint or another ATM switch
- Reads and updates the cell-header information
- Quickly switches the cell to an output interface toward its destination

An ATM endpoint or end system contains an ATM network interface adapter. ATM endpoints can be workstations, routers, gateways, LAN switches, and video coder/decoders.

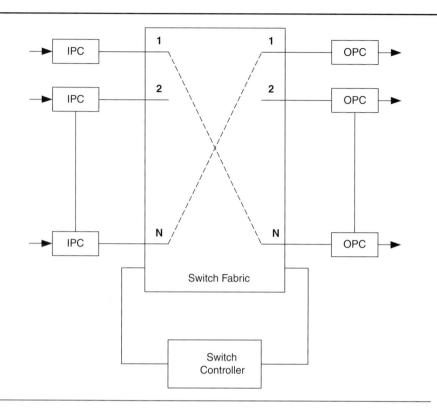

Figure 4–8
General Architecture of an
ATM Switch

ATM Switch

ATM switches are available as **access switches, backbone switches,** and **edge switches.** An access switch is a specialized ATM switch that sits on the end-user premise, providing access into a carrier ATM network. The ATM backbone switch is a specialized switch that sits in the carrier backbone network. The edge switch is an ATM cell switch that sits at the edge of a carrier network, providing access from the end user's environment to the carrier's ATM backbone network. ATM switches can be arranged in full-mesh and hierarchical configurations. The widely used PNNI standard used to connect ATM switches provided by different vendors utilizes a switch-hierarchy configuration.

access switches
backbone switches
edge switches

A general architecture of an ATM switch is depicted in Figure 4–8. Primary components are the Input Port Controller (IPC), Input Ports, Output Port Controller (OPC), output ports, switch fabric, and switch controller. The IPC consists of one table called the VCI table, which maps an input VCI to an output VCI and an output port address. Before cells are released to the switching fabric, the input VCI is replaced by the output VCI and the output port address is appended for self-routing. Each ATM switch has a switch controller that performs different switch management functions including updating tables in the IPCs. The main tasks performed by an ATM switch are VPI/VCI translation and cell transport from input to the concerned output.

4.14 ATM STANDARDS

American National Standards Institute (ANSI)

Since ATM has evolved into a globally implemented technology, a number of standards organizations including the **American National Standards Institute (ANSI),** European Telecommunications Standards Institute (ETSI), and the International Telecommunications Union (ITU-T) have become involved in the standards process. Complementing the standards organizations, several independent groups of vendors, users, and industry experts participate in the evolving standards process. These groups include:

- ATM Forum
- Internet Engineering Task Force (IETF)
- Frame Relay Forum
- Switched Multimegabit Data Service (SMDS) Special Interest Group (SIG)

ATM is based on the efforts of the ITU-T Broadband Integrated Services Digital Network (B-ISDN) ATM Forum standard. It is a high-speed transfer technology for voice, video, and data over public and private networks. The ATM Forum has released work on the following specifications:

- UNI 3.0 and 3.01
- UNI 4.0 and 4.01
- Public-Network Node Interface (P-NNI)
- LAN Emulation (LANE)
- I.361 B-ISDN ATM layer specifications
- I.362 B-ISDN AAL Functional Description
- I.363 B-ISDN AAL Specification
- Q.2110 B-ISDN Adaptation Layer Overview Description
- Q.2110 B-ISDN Adaptation Layer SSCOP
- Q.2130 B-ISDN Signaling AAL—SSCF at UNI
- Q.2931 ITU-T Signaling standard—UNI

Several RFCs that apply to ATM services are included in Appendix C.

4.15   ATM NETWORK MANAGEMENT

Network management issues in ATM LAN and WAN environments include:

- The need to monitor and control all PVCs and SVCs
- The need to manage the network topology so that it is ATM compatible
- The need to monitor the status of each network device

Quality of Service

Certain types of traffic, such as voice and video, are highly time-sensitive. ATM provides a mechanism for negotiating the **Quality of Service (QoS)** at the time of connection. A set of parameters is provided which constitutes a source traffic descriptor that, along with cell delay variation tolerance and a conformance definition, characterizes an ATM connection.

ATM supports QoS guarantees comprising traffic contract, traffic shaping, and traffic policing. A traffic contract specifies an envelope that describes the intended data flow. When an ATM end system connects to an ATM network, it enters a contract with the network, based on QoS parameters. Traffic parameters include the following:

- Peak cell rate
- Sustainable cell rate
- Maximum burst size
- Minimum cell rate
- QoS parameters
    - Cell delay variation
    - Maximum cell transfer delay
    - Cell Loss Ratio

ITU-T I.610 specifies the following ATM **Operations, Administration, and Maintenance (OAM)** components.

operations administration maintenance (OAM)

- Performance Monitoring
- Defect and Failure Detection
- System Protection
- Failure Information
- Fault Isolation

Performance monitoring includes continuous monitoring of throughput load. Defect and failure detection is concerned with malfunction monitoring. System protection is concerned with backup protection to alternate facilities and the changeover to accomplish this task.

alarm indication signal
(AIS)
far end receive failure
(FERF)

Failure information is distributed across management planes and is initiated when a failure is determined and the location of that failure is identified. Link failures are monitored with an alarm function similar to that found in SONET, T-1, and DS-3. These include **Alarm Indication Signal (AIS),** which is used to indicate a failure of an upstream path to a downstream device, and **Far End Receive Failure (FERF),** which is used to alert upstream nodes of a path failure.

The ATM UNI Management Information Base (MIB) offers seven categories of alarm information.

- Physical Layer—media information and alarms
- ATM Layer—VPC VCC information
- Virtual Path Connection (VPC)—status and information
- Virtual Channel Connection (VCC)—status and information
- ATM Layer Statistics—aggregate of ATM information
- Network Prefix—network identification
- Address—local address

This information used in conjunction with the Simple Network Management Protocol (SNMP) is utilized to effectively manage the ATM environment.

4.16   ATM APPLICATIONS

ATM is a cell switching and multiplexing technology that combines the benefits of circuit switching (guaranteed capacity and constant transmission delay) with those of packet switching (flexibility and efficiency for intermittent traffic). It provides scalable bandwidth from a few megabits per second (Mbps) to many gigabits per second (Gbps). Because of its asynchronous nature, ATM is more efficient than synchronous technologies, such as Time Division Multiplexing (TDM).

ATM has the capability of providing flexible solutions in the LAN, MAN, and WAN environments. LAN emulation can be utilized to provide a flexible path to interconnect LANs in a global enterprise network. Two WAN protocols, Frame Relay and SMDS, can be utilized to span large geographic areas and join multiple, separate logical networks. Figure 4–9 depicts an ATM network that provides connectivity for a Frame Relay network. The frames from a Frame Relay network are converted into cells for transport over the ATM network. The process is reversed at the receiving Frame Relay switch.

MANs can be interconnected with telephone carrier facilities that include ATM switches. A MAN is often designed around LAN extender devices. These devices make the network transparent to the operation of the LAN. This environment has some similarities to a FDDI application that will be discussed in Chapter 7.

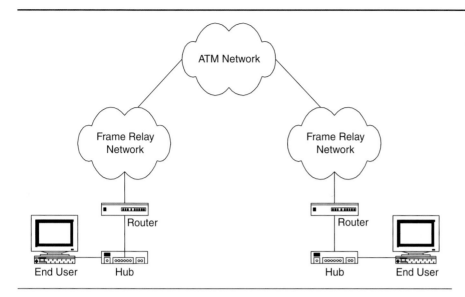

Figure 4–9
ATM / Frame Relay Network

The ATM Forum's **LAN Emulation (LANE)** specification is vital for integrating ATM with non-ATM networks. LANE defines a protocol-dependent method where LAN-attached devices can communicate over an ATM backbone.

Another methodology used to transport legacy traffic over an ATM network is an IETF specification that utilizes an IP-specific MAC layer for ATM. Classical IP over ATM is a simpler implementation than LANE and generates less network overhead. Additional information is available in RFC 1577—Classical IP and ARP over ATM.

Multiprotocol over ATM (MPOA) provides a technique to route traffic over ATM. In MPOA, the network layer protocols use routers to communicate across subnet boundaries, as does LANE, which requires that routers be used to communicate between emulated LANs.

multiprotocol over ATM (MPOA)

ATM has traditionally been considered a LAN and WAN backbone transport method. Efforts are underway to extend these applications to the desktop to relieve network congestion which has been caused by the more demanding software applications of multimedia.

Applications for ATM are as follows:

- Internet Service
  - Intermittent traffic
  - Bursty traffic
  - Small amount of transmit data/large receive data
- Frame relay access
  - Consolidate local frame relay links into ATM backbone switch

- Third-generation ATM backbone network
  - Backbone for ISP
  - Backbone for multistate commercial/governmental operation
- Dedicated access
  - ATM access for ISP
  - Commercial/governmental operations that require large transmission pipes
- Regional dial-up
  - Consolidation of dial-up facilities into a centralized ATM backbone switch

4.17   ADVANTAGES AND DISADVANTAGES

Advantages

One major advantage ATM brings over other alternatives is scalability. No other protocol can operate at such a wide range of speeds, nor can any other alternative be used effectively in a local, metropolitan, and wide area network. This flexibility can help with changing traffic patterns. ATM's classes of service make it effective for voice, video, and data communications, which have widely different requirements.

ATM operates at the physical layer of the OSI model and is thus transparent to payload carried within. Future services will undoubtedly have to use a physical service to be transported and ATM should work with any service.

ATM allocates its resources statistically to provide the most efficient use of corporate resources. Since ATM services provide transparent service to voice, video, and data, it allows a corporation to implement a single backbone communication network.

ATM is a virtual circuit switching technology, which means that these virtual connections isolate the traffic being carried from other traffic on the same physical connection. It also provides for a connection-oriented path through the network, which eliminates the possibility of cells arriving out of order.

ATM provides for a deterministic Quality of Service. This improves throughput for delay-sensitive applications such as voice and video.

Disadvantages

ATM is not a good substitute technology for today's services. It is a technology that enables new applications today, with a future migration path for voice. If network traffic does not include voice or video, then ATM may not be required in the enterprise. It might not be beneficial to combine voice, video, and data into a single network.

The cost of implementing the technology may be too high in comparison with the other alternatives that are available. If transmission speed requirements are not 100 Mbps or greater, then ATM may be overkill. If the current enterprise network is not congested or overloaded, consideration might be given to delaying an ATM implementation.

Premise wiring systems may have to be upgraded to support the higher speeds of ATM. Fiber-optic premise cabling might be required.

4.18   ATM TECHNOLOGY ALTERNATIVES

When looking at alternatives, it is necessary to understand the issues of using frames versus cells, multiplexing techniques, shared versus dedicated, and service guarantees. It is also necessary to determine if ATM is needed as a backbone or as a native service.

ATM competes with other technologies when used as a transport of user traffic, including Frame Relay, SMDS, FDDI, ISDN, and private lines. ATM, however, is better suited for multimedia applications than the other technologies.

When used as a backbone topology, ATM competes with FDDI and SONET. It works in conjunction with virtual networks (VPNs). In the backbone environment, ATM is basically a pipe and is not concerned with the content of the traffic being carried. ATM is often used to carry both Frame Relay and SMDS traffic. In these situations, the frames are converted to cells before passing through the ATM network.

4.19   ATM PRODUCT AND SERVICE PROVIDERS

Because ATM is a mainstream technology, it is available from numerous sources, including local and long-distance carriers. ATM switches are available in both CO configurations and customer network models. Costs vary considerably due to the wide variety of configurations. Most of these switches are scalable, so a user can start with a small system and migrate the network by adding components and functionality.

ATM Providers

ATM service providers include numerous interexchange carriers (IXCs), ILECS, and CLECs. Following are providers listed on the Web.

AT&T	GTE Corporation	SBC Communications, Inc.
BellSouth	Global NAPs, Inc.	2nd Century
Sprint	Gabriel Communications, Inc.	Communications, Inc.

ATM Switch Manufacturers and Equipment Providers

A search of the Web shows that ATM products are available from the following suppliers:

Alcatel Telecom	Digital Equipment Corporation
First Virtual Corporation	FORE Systems, Inc.
General DataComm	Hitachi Telecom Inc.
IBM Corp.	Lucent Technologies
Madge Networks	NEC America
NewBridge Networks Inc.	Nortel
Telematics International	Xylan Corp.

Additional ATM product information is available at the Ohio State University Web site [http://www.cis.ohio-state.edu/Research/research.html].

4.20   ISSUES AND CONSIDERATIONS

Packet size is a key issue in designing a general network technology. Data achieve maximum throughput when packet sizes are large, because the overhead from headers is minimized by carrying a maximum payload with each header. Networks designed to optimize data transport often have packet sizes of 8 k bytes or larger; however, voice cannot use such large packets. Sampling delay and echo cancellation techniques introduced in the voice systems create these throughput problems.

ATM networks must rely on considerable user-supplied information about the traffic profile in order to provide the connection with the desired quality of service. These sources of traffic are easier to describe than others, and herein lies the cost/performance challenge for best bandwidth utilization in an ATM interface.

ATM technology as it stands today faces many of the bottlenecks that any new technology faces, such as high initial cost of the ATM equipment, installation, training, and more. Many analysts believe that the expense of converting the ATM technology will not be justified unless multimedia matures throughout the general public and private sectors. ATM research is also available at www.cis.ohio-state.edu/Research/research.html.

Every layer of ATM adds overhead to cell transmission. In the case of the streaming transmission option that is good for voice transmission, the added overhead is especially problematic. Although this overhead adds to the capabilities of ATM, it makes ATM less attractive compared with switched Ethernet for local transmission.

Pricing ATM appropriately is a problem plaguing many carriers hoping to deliver commercially viable service. While traditional networks priced voice calls based on distance, connection time, and time of day, ATM must also consider the type of service provided. A video call will

clearly cost more than a voice call, but the price will not be directly proportional to the number of bits transmitted. Carriers will need to strike a
balance between call setup, quality of service, bandwidth, call duration,
and time-of-day usage charge.

Issues such as traffic, congestion control, and tariffs will probably be
refined during deployment of these networks.

■ SUMMARY

ATM is a high-performance, high-speed networking technology that utilizes virtual circuit switching and statistical time division multiplexing
(STDM). It provides for the performance guarantees of time division multiplexing and the efficient utilization of resources of statistical time division multiplexing.

ATM provides a framework for integration of an enterprise network,
combining LAN, mainframe, voice, and video services onto a single platform. ATM provides a physical level transparent solution for enterprise
networking needs. Through the use of adaptation layer services, ATM
provides a good transmission quality across a broad spectrum of user services and allows users to implement ATM without any modifications to
their legacy systems.

Both Frame Relay and SMDS provide switched networking solutions—
Frame Relay for WANs and SMDS for MANS. Frame Relay provides global
connectivity, whereas SMDS provides multimegabit connectivity. Both
Frame Relay and SMDS use megabit facilities between switches. ATM ser-
vice can provide switched access between Frame Relay/SMDS nodes.
Frame Relay uses ATM AAL-5 while SMDS uses AAL-4.

ATM also provides a means to construct ATM virtual LANS. This is
accomplished by aggregating the rate by utilizing ATM intelligent hubs.
Connecting a LAN to an intelligent hub makes the LAN into a single user
access method to the hub. ATM takes the multiuser enterprise transmission burden from the traditional LANs and becomes the focus for enterprise connectivity.

Because ATM operates at the physical layer of the OSI model, it is
transparent to the payload carried within. ATM allocates its resources statistically to provide the most efficient use of enterprise resources. Transparency allows the enterprise to implement a single backbone communication network.

The 53-byte ATM cell consists of a 5-byte header and a 48-byte payload. VPI/VCI identifies path information for users while PTI identifies
payload contents and CLP identifies priority. HEC forwards errors, cor-
rects header bit errors, and provides cell delineation and scrambles the
payload to assist in cell delineation.

Point-to-point and point-to-multipoint operations are supported by E.164 addressing. ATM Forum UNI 3.1 Specification defines the Q.2931 signaling. A number of choices exist for ATM interfacing. Public carrier access can be accomplished with DS-1, DS-3 and SONET. Private access can be implemented with 100 Mbps for fiber and CAP for UTP copper.

Key Terms

AAL-1

AAL-2

AAL-3/4

AAL-5

Access Switches

Alarm Indication Signal (AIS)

ATM Adaptation Layer (AAL)

Available Bit Rate (ABR)

American National Standards Institute (ANSI)

Asynchronous Transfer Mode (ATM)

ATM Forum

Backbone Switches

Bursty

Cell

Cell Loss Priority (CLP)

Cell Loss Ratio (CLR)

Common Part Convergence Sub-layer (CPCS)

Constant Bit Rate (CBR)

Edge Switches

Far End Receive Failure (FERF)

Generic Cell Rate (GCR)

Generic Flow Control (GFC)

Header Error Control (HEC)

International Telecommunications Union (ITU-T)

LAN Emulation (LANE)

Leaky Bucket

Maximum Burst Size (MBS)

Multiprotocol over ATM (MPOA)

Network-to-Network Interface (NNI)

Operations, Administration, and Maintenance (OAM)

Payload Type (PT)

Peak Cell Rate (PCR)

Permanet Virtual Circuit (PVC)

Physical Medium (PM)

Private Network-to-Network Interface (PNNI)

Protocol Data Unit (PDU)

Quality of Service (QoS)

Segmentation and Reassembly (SAR)

Service Access Point (SAP)

Service Specific Convergence Sublayer (SSCS)

Sustainable Cell Rate (SCR)

Switched Virtual Circuit (SVC)

Transmission Convergence (TC)

Unspecified Bit Rate (UBR)

User-to-Network Interface (UNI)

Variable Bit Rate (VBR)

Virtual Channel (VC)

Virtual Channel Connection (VCC)

Virtual Channel Identifier (VCI)

Virtual Path (VP)

Virtual Path Connection (VPC)

Virtual Path Identifier (VPI)

1. Provide a list of ATM users.

2. What are the primary characteristics of ATM? What are three characteristics that produce the speed of ATM?

3. What is a concise definition of ATM? What is the foundation for ATM? What is the relationship between cell relay and ATM?

4. How does ATM achieve the high data rates and low delay?

5. What are the two protocol layers that relate to the ATM functions? What functions do each perform?

6. What are three characteristics of ATM that produce its speed?

7. Describe the ATM cell.

8. Describe the ATM cell header. What are the fields?

9. What is the function of generic flow control?

10. What is the purpose of the virtual path identifier and virtual channel identifier? What is the difference between the two?

11. What is the purpose of the payload type?

12. Why does ATM need a cell loss priority field?

13. What are some differences between a NNI and a UNI?

14. ATM cells are switched in hardware. What are the implications for this?

15. What is the difference between a PVC and a SVC?

16. What are the four layers of the ATM architecture?

17. What are the three separate planes of the protocol reference model? How does the OSI model differ from the ATM model?

18. What are the functions of the four ATM layers?

19. What are the two sublayers of the ATM physical layer?

20. What are the different transport methods and media that work on the ATM physical interface?

21. What is the function of the transmission convergence sublayer?

22. What is the responsibility of the ATM layer?

23. What is a VCC? A VPC? What is the relationship between them?

24. What are the two sublayers of the adaptation layer? What are their functions?

25. What are some services of the AALs? What functions do they provide? What is the purpose of AAL protocols?

26. What is an ATM class of service? How many are there?

27. What are the differences between the class-of-service categories?

28. What is constant bit rate? What is the type?

29. What is variable bit rate? What are the types?

30. What is available bit rate?

31. What is the use of unspecified bit rate?

32. Differentiate between CBR, VBR, and ABR in terms of architecture and applications.

33. Provide a summary of the five options of the adaptation layer.

34. How is congestion control handled in ATM? Comment on the various congestion control techniques.

35. How would you use ATM as a backbone network?

36. What are the primary ATM device categories? What device is utilized to connect a legacy device to an ATM switch?

37. What is the function of an ATM switch?

38. What is meant by ATM quality of service? Identify the parameters.

39. What are the components of ATM OAM?

40. What are the major applications for the ATM service? What is LANE? What are the benefits of deploying ATM?

41. What are the advantages and disadvantages of deploying ATM?

42. How do you determine if a user can cost-justify ATM access?

43. Provide an overview of some issues and considerations that are appropriate when deploying ATM. What is an important element of designing an ATM backbone?

ACTIVITIES

1. Research into the various aspects of the ATM technology should preclude these activities. Each of the following topics can be assigned to individuals or groups for classroom discussion. Each of these activities can be enhanced by having a role-play exercise during which one group offers the questions and the other group provides the answers.

- Prepare a list of potential alternatives to ATM that might be proposed by a vendor. Develop a list of pros and cons for each alternative.

- Prepare a list of arguments that would be a plus for a vendor that is proposing an ATM backbone infrastructure.

- Prepare a list of questions to qualify a user for an ATM solution. Provide sample answers for each question.

- Develop a list of issues and considerations for implementing ATM in various industries.

- Research various resources to identify ATM service and equipment providers. Develop a matrix of features and cost.

2. An exercise that can be used as a project is to research RFC1212 in addition to the standards references noted in this chapter. Prepare an overview that describes the contents of these documents. Note that there are a number of RFC references in Appendix C.

3. Schedule a tour of an IXC that utilizes an ATM switch in its network infrastructure.

4. Schedule a presentation by a representative of an ATM supplier. Since costs are difficult for students to obtain, have this representative provide prices for the service.

5. Develop a matrix of a feature comparison between ATM and Frame Relay.

CASE STUDY/PROJECT

Overview

Figure 4–10 provides a representation of a network provider's ATM infrastructure. This ATM provides a backbone infrastructure for a Frame Relay cloud. There are access lines from each of the Frame Relay switches to subscriber locations. Given this infrastructure, develop a solution that would use this network as a backbone transport for your user.

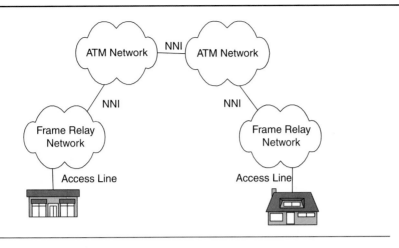

Figure 4–10
Backbone

Using this basic logical backbone configuration, it will be necessary to identify components for an ATM backbone that supports multiple protocol networks such as Frame Relay and SMDS. It will also be necessary to include all routers, hubs, and end-user devices in the design. Refer to Chapter 1 for the required devices.

Details

LongHaul Communications (LHC) is building a broadband network to support a Frame Relay infrastructure. LHC currently has Frame Relay switches installed and operational in these eight northeastern cities:

New York, NY (2)	Arlington, VA (2)
Newark, NJ (1)	Richmond, VA (1)
Washington, DC (1)	Pittsburgh, PA (1)
Baltimore, MD (1)	Philadelphia, PA (1)

Each location has sufficient space and facilities to add an ATM backbone switch.

Requirements

- Research ATM switch competitors and select an ATM switch for installation in each of these eight central offices.
- Provide a Visio or comparable design of each location that would include both the ATM switch and the Frame Relay switches.
- Produce a design of the entire eight-city backbone network, including the Frame Relay switches.
- Develop a plan for implementing these ATM switches.
- Develop a plan for upgrading the backbone based on some percentage of growth.

- Provide information as to the various applications that would be transported and the user interactions that would take place in the transactions.
- Develop a list of questions that must be researched and answered in order to develop the specifications for the ATM network. The questions developed in the Frame Relay project can be used as a stepping stone to ATM.

Synchronous Optical Network

■ INTRODUCTION

Synchronous Optical Network (SONET) is the U.S. (ANSI) standard for synchronous data transmission on optical media. The international equivalent of SONET is **Synchronous Digital Hierarchy (SDH).** Together, they ensure standards so that digital networks can interconnect internationally, and so existing conventional transmission systems can take advantage of optical media through tributary channels.

Bellcore proposed this new networking transmission approach in 1985. It was developed because of a demand by the industry for a new technological approach that addressed high-speed digital transmission over optical fiber. In 1988, ANSI published the first of a series of SONET recommendations that have evolved into the current standards. The SONET standards originated out of a cooperative effort between various standards bodies and private industry. SONET standards not only define transmission rates and network management, but also define optical interfaces.

SONET consists of dual, counterrotating SONET fiber rings that connect and pass through a number of SONET-equipped Central Offices (COs). The ring has multiple paths (diverse) and so can reroute itself (self-heal) when a failure occurs.

SONET provides standards for numerous line rates, up to the maximum of 13.271 gigabits per second (Gbps). SONET defines a base rate of 51.84 Mbps and a set of multiples of the base rate known as **Optical Carrier (OC)** levels. SONET is a standard way to interconnect high-speed communications from multiple vendors. SONET's major attribute is its ability to transport many different digital signals using a standard **Synchronous Transport Signal (STS)** format.

Although the SONET standard defines a technology that can be used to build a high-capacity ring network with multiple data circuits

OBJECTIVES

Material included in this chapter should enable you to:

- understand SONET technology and its relation to the current broadband market. Look at the structure of the STS-1 frame and learn how it is utilized in transporting the synchronous payload.

- become familiar with the terms and definitions that are relevant in the Enterprise Network environment. Look at the relationships between fiber optics and SONET.

- understand the function and interaction of SONET infrastructure in the Enterprise Network environment. Identify Enterprise networking technologies that could benefit from a high-speed backbone.

synchronous optical network (SONET)
synchronous digital hierarchy (SDH)
optical carrier (OC)
synchronous transport signal (STS)

• understand where SONET fits in the different applications of the technology. Look at the alternatives to SONET.

• identify components that make up the SONET networking environments. Identify the functions of add/drop multiplexers, digital cross-connects, and regenerators.

• identify opportunities to effectively utilize the SONET technology. Complete a SONET worksheet for an application.

unidirectional path switching
automatic protection switching
bidirectional line switching

multiplexed across the fibers that constitute the ring, most data networks only use SONET to define framing and encoding of a leased circuit.

5.1   SONET INFRASTRUCTURE

Fiber-optic transmission facilities have expanded at a rapid pace in all segments of telecommunications networks. This widespread implementation of fiber systems motivated the establishing of fiber transmission standards. SONET was originally proposed by Bellcore in 1985 and was created as a standard in 1988. The SONET standard consists of signal rates, multiplexing schemes, network element functions, synchronous network operation, and single-mode optical interface. The ITU-T version of SONET is called Synchronous Digital Hierarchy (SDH). The fundamental building blocks of SDH are called synchronous transport modules which begin at 155 Mbps.

SONET travels in a ring topology, and offers three possible methods of failure recovery: **unidirectional path switching, automatic protection switching,** and **bidirectional line switching.**

Unidirectional Path Switching   Unidirectional path switching involves one fiber-optic ring. The data signal is transmitted in both directions around the ring. The receiving node determines which signal to accept. If there is a break in one path, the signal on the alternate path still reaches the destination node. The data sent along the alternate path warn the receiving node that only one path is open.

Automatic Protection Switching   In automatic protection switching (APS), if a failure is detected at some point on the SONET network, the data are directed to an alternate switching node, and then redirected to the assigned destination. Protection switching operations are initiated for several reasons, including loss of connection and deterioration in the quality of the signal. APS is also useful when performing maintenance on the network or initiating testing procedures.

Bidirectional Line Switching   Bidirectional line switching provides the highest level of redundancy—up to 99 percent. It uses a dual-ring topology, so there are always two paths to a node. The data is sent to both rings, but in opposite directions. If is a break occurs along one path, the data on the second path will still get through.

Synchronous Multiplexing
SONET synchronous multiplexing refers to combining (multiplexing) low-speed digital signals into a high-speed signal format that allows for easy extraction of the low speed constituents from the high speed line.

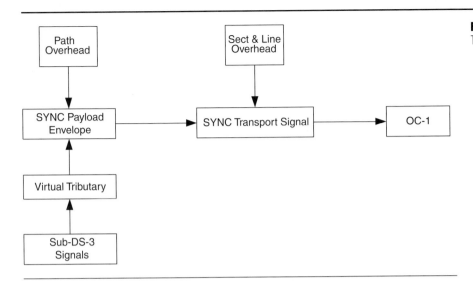

Figure 5–1
Transport Hierarchy

These signals are usually DS-1 (1.544 Mbps), DS-1C (3.152 Mbps), DS-2 (6.312 Mbps), and DS-3 (44.736 Mbps).

SONET's major attribute is its ability to transport many different asynchronous or synchronous digital signals using a Standard Synchronous Transport Signal (STS) format. The **Synchronous Transport Signal Level 1 (STS-1)** is defined as 51.840 Mbps and is formed using a byte interleaved multiplex scheme, which in turn can be readily multiplexed with other STS-1s to form a N * OC-1 hierarchy. Figure 5–1 depicts the SONET transport hierarchy. These components will be explained in detail later in the chapter. The basic building block of SONET is a DS-0 (64 kbps) channel that is multiplexed into a STS-1 transmission frame. STS-1 uses a synchronous frame alignment, which enables DS-1 data frames to be extracted from the STS-1 frames without the frame being disassembled and reassembled. The majority of the STS-1 frame consists of the SONET payload envelope.

Payload signals are mapped into the STS-1, which is then converted to an **Optical Carrier Level 1 (OC-1)** or multiplexed to a higher level STS-N. STS-N signals are electrical and are converted to optical carriers (OC-Ns) for transport over single-mode fiber. Table 5–1 depicts OC-1 through OC-256. The ability to directly access the lower bit rate constituents, combined with the additional overhead capability, allows automated control of the network by the carriers and their customers. This automated control is required to implement enhancements such as time-of-day circuit provisioning and customer-controlled reconfiguration.

synchronous transport signal level 1 (STS-1)

optical carrier level 1 (OC-1)

Table 5–1
SONET Optical
Transmission Rates

OC#	Speed	ITU-T	Media	Utilization
OC1	51.84 Mbps		Optical fiber	Smaller links within internet infrastructure
OC3	155.52 Mbps	STM-1	Optical fiber	Large company backbone, internet backbone
OC12	622.08 Mbps	STM-4	Optical fiber	Internet backbone
OC24	1.244 Gbps	STM-8	Optical fiber	Internet backbone
OC48	2.488 Gbps	STM-16	Optical fiber	Internet backbone
OC192	9.953 Gbps	STM-64	Optical fiber	Backbone
OC256	13.271 Gbps	STM-90	Optical fiber	Backbone

Signal Rates

The existing digital transmission hierarchy consists of DS-0 (64k bps) through DS-4NA (139.264 Mbps) signals. SONET transports these signals (called payloads) in a synchronous fashion over fiber transmission systems. Mapping of these signals is accomplished by a structure called **Virtual Tributaries (VTs).** The DS-3 signal is mapped directly into the STS-1.

virtual tributary (VT)

The SONET signal architecture is formed by a byte interleaved multiplexing scheme. Interleaving is the transmission of pulses from two or more digital sources in a time-division sequence over a single path. Payloads can be readily added or dropped from the carrier by a network element called an **Add/Drop Multiplexer (ADM).** The ADM will be discussed in detail later in the chapter. STS-N signals (electrical) are converted to optical carriers (OC-Ns) for transport over fiber.

add/drop multiplexer
(ADM)

5.2   SONET FRAME

The STS-1 frame structure is a 90-column x 9-row byte matrix, which consists of a separate transport overhead and information payload area. The OC-1 is obtained from the STS-1 after scrambling, which provides uniform ones and zeroes density, allowing clock recovery at the receivers. Figure 5–2 depicts the STS-1 frame structure. The frame is transmitted starting with the byte in the first column of the first row and then followed sequentially by the remaining bytes in the row. The transmission continues with each row following in order. The frames repeat every 125 microseconds (ms), which facilitates easy definition of 64 kbps channels.

The components of a STS-1 are the transport overhead consisting of 27 (3 * 9) bytes and the synchronous payload consisting of 783 (87 * 9) bytes including the path overhead. The overhead plus payload sum equals 810 bytes, which is contained in a 125 ms frame.

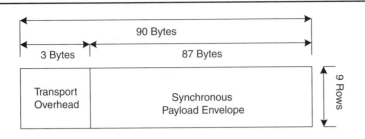

Figure 5–2
STS-1 Frame Structure

SONET Framing

In many ways, SONET framing is the same as T-1 framing. The basic purpose of each is to establish markers with which to identify individual channels. Because of the higher bandwidth of SONET and potential for sophisticated mixed-media services, more overhead is reserved surrounding each frame than the single bit that is reserved every 193rd character in a T-1 frame.

Rather than fitting twenty-four channels per frame delineated by a single framing bit, a single SONET row is delineated by three octets of overhead for control information, followed by eighty-seven octets of payload. Nine of these ninety-octet rows are grouped together to form a SONET superframe. The eighty-seven octets of payload per row in each of the nine rows of the superframe are known as the **Synchronous Payload Envelope (SPE).**

synchronous payload envelope (SPE)

5.3 SONET MULTIPLEXING

Higher rate SONET signals are formed by byte interleaving the N lower level constituents. Byte interleaving and frame alignment are referenced at the STS-3 level (155.52 Mbps). Levels higher than STS-3 (OC-3) are obtained through byte interleaving the lower level constituents, yet maintaining a multiple STS-3 arrangement. Because SONET carries voice, video, image, and data, the basic unit of measurement is referred to as an octet of 8 bits rather than a byte of 8 bits. Byte is usually only reserved for referring to data, and is often synonymous with a character.

The STS-N electrical signal is then scrambled and converted to its optical carrier level N, the OC-N counterpart. The OC-N will have a line rate of exactly N times that of an OC-1. The American National Standard-allowed values of N are 1, 3, 9, 12, 18, 24, 36, 48, 192, and 256. The primary rates carried by current CO multiplexers are OC-3, OC-12, OC-48, and OC-192.

Figure 5–3
STS-1 Frame Structure

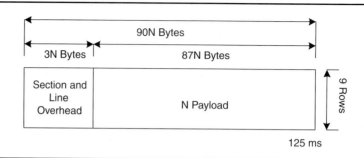

Note that Figure 5–3 is similar to the STS-1 frame depicted in Figure 5–2. Transport overhead is divided into section and line overhead and there are multiple frames (1 – N) that are multiplexed.

SONET Synchronization

stratum-one

The SONET network is synchronous and therefore must conform to the Digital Synchronization Network. The purpose of this synchronization is to prevent data loss by ensuring that all network elements derive timing from a common clock source, traceable to a primary reference **stratum-one** clock source. All SONET network elements must accept timing from primary and backup sources. These timing signals are derived from a stratum three or higher clock. Note that a stratum level refers to the accuracy of a SONET clock and that stratum timing is one of the fundamental elements that makes SONET possible. The Stratum Level one clock is located in Paris, France.

5.4   SONET COMPONENTS

The transport system adopted in SONET utilizes a synchronous digital bit stream comprised of bytes organized into a frame structure in which the user's data are filled. The basic electrical signal of SONET is the STS-1. Its optical counterpart is OC-1. The components of an STS-1 are transport overhead and the synchronous payload envelope. As depicted earlier, the transport overhead consists of section and line components and the synchronous envelope consists of the path overhead component.

transport overhead (TOH)

The first three columns comprise the **Transport Overhead (TOH)**, which contains the section and line layer overhead. The section layer deals with the transport of an STS-N frame across the physical medium.

SONET Component	Function
Line Overhead (LOH)	Processed at all nodes Communicates with higher level components, such as terminals, switches, multiplexers, and digital cross-connects Pointer for frequency justification and data communications channel
Section Overhead (SOH)	Between network elements such as regenerators and terminals Used for framing and performance monitoring
Transport Overhead (TOH)	Alarms Error monitoring Contains section overhead and line overhead
Path Overhead (POH)	End-to-end information Stays with SPE until the final node

Table 5–2
Functions of SONET
Components

The line layer deals with the transport of path layer payload and its overhead. The available overhead capacity enables communications between intelligent controllers at network nodes. This permits an increasing level of software-controlled flexibility to reconfigure the network in response to customer needs. The remaining eight-seven columns and nine rows (783 bytes) of the frame contains the STS-1 information payload, designated the **Synchronous Payload Envelope (SPE)**.

SONET Overhead

A **path** is defined as the area between the point a SONET signal (OC-N) is constructed to the point where it is torn down. A **line** is between any two SONET elements that have protection switching capability. A **section** is between any two SONET elements.

path
line
section

The SONET STS-1 contains an abundance of overhead capacity. The overhead is functionally layered into a section, line and path. **Section Overhead (SOH)** applies to the physical function, **Line Overhead (LOH)** applies to the multiplexing and switching functions, and **Path Overhead (POH)** applies to the end-to-end communications across the network. Table 5–2 provides a summarization of the functions of the SONET components.

section overhead (SOH)
line overhead (LOH)
path overhead (POH)

This division clearly delineates the segregation of processing functions in the network elements. In the formation of the signal, overhead is added to the payload that is then handed off to the next layer. This process continues until the actual OC-N signal is formed at the photonic layer, which has no overhead of its own.

Figure 5–4
SONET Overhead

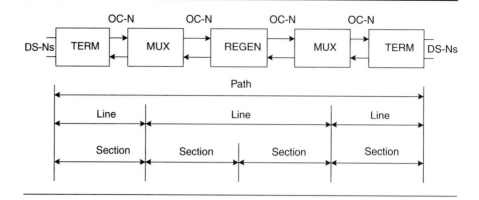

Figure 5–4
SONET Overhead

To summarize, overhead falls into three functional areas:

- POH is relevant to end-to-end communications
- LOH is relevant to the maintenance span
- SOH is relevant to the transceiver functions

Figure 5–4 shows how the line, path, and section elements flow through the terminal, multiplexer and regenerator components. The various DS-N signals flow from the terminals to the multiplexers and pass through the regenerators to other multiplexers and finally back through other terminals and resurface as DS-N signals.

In the figure, the path consists of all components between the terminals. The line exists between terminals and multiplexers and between multiplexers. Sections exist between all elements in the path. Note that the bidirectional signals between each of the sections are OC-N levels.

Transport Overhead

The Transport Overhead (TOH) portion of the STS-1 frame is for alarm monitoring, bit error monitoring, and data communications overhead necessary to ensure the reliable transmission of the SPE between nodes in the synchronous network.

Path Overhead

The Path Overhead (POH) is end-to-end information and remains with the payload until the payload is finally demultiplexed at the end multiplexer. The path overhead bytes are processed by path-terminating equipment at all points of the SONET system. There are four classes of functions at the path overhead:

- Class A—Payload independent functions. Path-terminating equipment verifies a connection to the sending device. It performs a parity check on all bits in the path overhead. It is used to indicate the construction of the STS payload envelope and to inform the network that different types of systems are being used, such as SMDS or FDDI.

- Class B—Mapping dependent functions. Used for VTs to signal the beginning of frames.
- Class C—User-specific overhead functions. This is used by the network provider.
- Class D—Growth and future-use functions.

Line Overhead

The Line Overhead (LOH) is processed at each node. It consists of communications for terminals, switches, multiplexers, and digital cross-connects. The line overhead occupies the bottom six octets of the first three columns in the SONET frame. It is processed by all equipment except for the regenerators. These octets indicate the offset in bytes between the pointer and the first byte of the SPE. This pointer allows the SPE to be allocated anywhere within the SONET envelope, as long as capacity is available. All of the data communications channel octets are used for line communication and are part of a 576 kps message that is used for maintenance control, monitoring, alarms, and so forth.

Section Overhead

The Section Overhead (SOH) is processed at each node and also at each regenerator. It is used for framing and performance monitoring. The first two octets (16 bits) are the framing bits and are provided with all STS-1 through STS-N signals. The bit pattern is always 1111011000101000 in binary or F628 in hexadecimal. The purpose of these octets is to identify the beginning of each STS-1 frame. The receiver initially operates in a search mode and examines bits until the candidate two-octet pattern is detected.

5.5   SYNCHRONOUS PAYLOAD ENVELOPE

The Synchronous Payload Envelope (SPE) is designed to transport a tributary's signal across the synchronous network from end to end. The SPE is assembled and disassembled only once even though it may be transferred from one transport system to another on its route through the network. In most cases, the SPE is assembled at the point of entry to the synchronous network and disassembled at the point of exit from the network.

STS-1 Payload

The payload can be thought of as the revenue-producing traffic that is being transported and routed over the SONET network. Once the payload is assembled into the SPE, it can be routed through the SONET network to its destination. The STS-1 payload has the capacity to transport the circuits listed in Table 5–3.

The STS-1 bit rate is 51.840 Mbps and is calculated as follows:

$$OC\text{-}1 = 90 * 9 * 8 * 8000 = 51.840 \text{ Mbps where } 90 * 9 \text{ is the SONET envelope,}$$
$$8 \text{ is the number of bits, and } 8{,}000 \text{ (125 ms) is the sample speed.}$$

The synchronous transport signal and Optical Carrier (OC) terminology has an additional feature called concatenation, specified with the letter "C." The presence of the suffix denotes a circuit with no inverse multiplexing. An OC-3 circuit normally consists of three OC-1 circuits, each operating at 51.840 Mbps, and is intended to be used as three circuits. An OC-3C (STS-3C) circuit, however, is a single circuit operating at 155.520 Mbps.

The synchronous payload envelope consists of an 87-byte/row times 9-row field containing 783 bytes total. Within the SPE, the first byte of every row is designated as the Path Overhead (POH) byte. This format is presented in Figure 5–5.

Table 5–3
STS-1 Payload Capacity

Capacity	Signal Type	Signal Rate (Mbps)	Voice Circuits	T-1s	DS-3s
28	DS-1	1.544	24	1	-
21	CEPT-1	2.048	30	-	-
14	DS-1C	3.152	48	2	-
7	DS-2	6.312	96	4	-
1	DS-3	44.736	672	28	1

Figure 5–5
Synchronous Payload Envelope

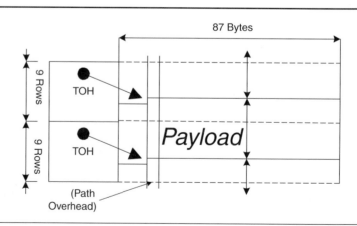

The information payload is contained in a SPE, where the first column (9 bytes) contains the POH, and the remaining 774 bytes (9 * 86) are for data. The POH deals with the transport of services across the network. It consists of bytes for various monitoring and statusing capabilities. The information payload capacity of the STS-1 can carry a clear channel DS-3 signal or a variety of lower rate signals (DS-1, DS-1C, DS-2). The SPE may begin anywhere in the interior of the frame, but typically begins in one frame and ends in the next. The flexible alignment of the SPE within the frame is accomplished by the use of payload pointers.

5.6   SONET POINTERS

SONET uses a concept called *pointers* to deal with the timing variations in a network. The purpose of pointers is to allow the payload to float with the payload. The SPE can occupy more than one frame. The **payload pointer** is an offset value that shows the relative position of the first byte of the payload. Pointers provide a method of flexible and dynamic alignment between a payload and its synchronous transport package, independent of the payload contents. During the transmission across the network, if any variations occur in the timing, the pointer need only be increased or decreased to compensate for the variation.   payload pointer

  The address where the SPE begins is contained in a pointer that resides in the Transport Overhead (TOH). The TOH contains synchronizing bit patterns that correlate the STS-1 to a synchronous time base. Note that the payload within the SPE may be either synchronous or asynchronous. Once the payload is mapped within the SPE, the STS pointer accounts for fluctuations between the SPE and the TOH time base. Figure 5–6 provides an example of this SPE/STS alignment that shows the pointers from the

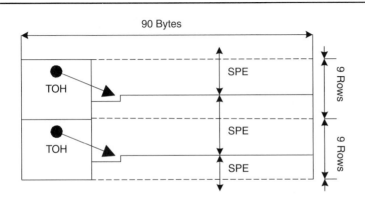

Figure 5–6
SPE/STS Alignment

TOH(N)s to the addresses where the SPE(N)s begin. There is one complete SPE and two partial SPEs shown in this example.

Payload identification is done using a payload pointer (number) to identify the starting byte location of each STS-1 or Virtual Tributary (VT) payload. The pointer scheme is a synchronous multiplex technique with the capacity of accommodating slight frequency variations. The conventional multiplexing techniques that involve positive pulse stuffing accommodate large frequency variations, but do not allow easy access to tributaries. Synchronous multiplexing of tributaries with fixed frame alignment gives easy access to tributaries, but would require additional buffers. The pointer payload mechanism has advantages over the previous two techniques, but adds complexity for pointer processing.

Virtual Tributaries

Unlike the T-1 frame with its twenty-four predefined 8-bit channels, SONET is flexible in its definition of the use of its payload area. It can map DS-0 channels into the payload area just as easily as it can map an entire T-1. The virtual tributaries of SONET are equivalent to circuit-switched transmission services.

The **Virtual Tributary (VT)** is a structure designed for the transport and switching of sub-STS-1 payloads. The VT is constructed to accommodate both asynchronous and synchronous payloads flexibly. The payload is subdivided into partial areas, each carrying these VTs. The full STS line then transports these multiple VTs as a single transmission. At the received end, they are removed and returned to their original protocol state.

There are four sizes of VTs based on their appropriate payload capacities:

- VT-1.5   (DS-1)   1.544 Mbps   +   27 bytes OH   =   1.728 Mbps
- VT-2   (E-1)   2.048 Mbps   +   36 bytes OH   =   2.304 Mbps
- VT-3   (DS-1C)   3.152 Mbps   +   54 bytes OH   =   3.456 Mbps
- VT-6   (DS-2)   6.312 Mbps   +   108 bytes OH =   6.912 Mbps

Figure 5–7 depicts this capacity graphically.

A STS-1 SPE can accommodate up to twenty-eight VT-1.5s for transport through the SONET network.

VT Groups

VT groups

To accommodate mixes of various VTs with a VT-structured SPE, the VT group architecture is used. VTs are packaged into **VT groups.** As depicted in Figure 5–8, a VT group is always 9 rows times 12 columns or 108 bytes. The number of VTs in a VT group depends on the VT size: 4 VT-1.5s, 3 VT-2s, 2 VT-3s, or 1 VT-6 are packaged into a VT group. Mixes within a VT group are not allowed. VT groups are formed by byte-interleaving VTs.

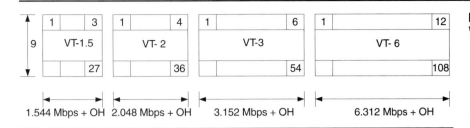

Figure 5–7
Virtual Tributaries

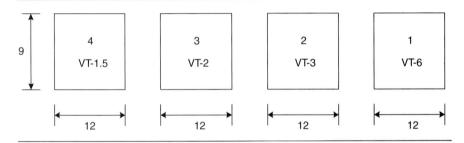

Figure 5–8
VT Groups

Virtual tributaries are byte interleaved, byte-by-byte, row-by-row, to form the VT group. VT-1.5s, VT-2s, VT-3s, and the VT-6 are byte-interleaved into their respective group. As noted previously, the VT Group's size is a constant of 9 rows by 12 columns, or 108 bytes. Seven VT Groups are similarly byte interleaved into the synchronous payload envelope.

VT Modes

There are two VT modes of operation: locked and floating. The locked mode minimizes the interface complexity in DS-0 switching by utilizing a fixed mapping of all VTs within a STS-1 SPE. The locked mode introduces signal processing delay during VT switching operations that does not exist in floating mode. The floating mode minimizes delay by utilizing per-VT pointers to identify the beginning of each VT SPE. During a VT switching function between SONET interfaces, the VT pointer of each switched VT is changed to align with the new STS-1 SPE.

5.7   SONET/SDH END-USER TOPOLOGY

End user devices operating on LANs (e.g., FDDI, Ethernet, Token Ring) and digital transport systems, such as DS-1 and DS-3, are attached through a service adapter to a network. The service adapter may be called

Figure 5–9
SONET Topology

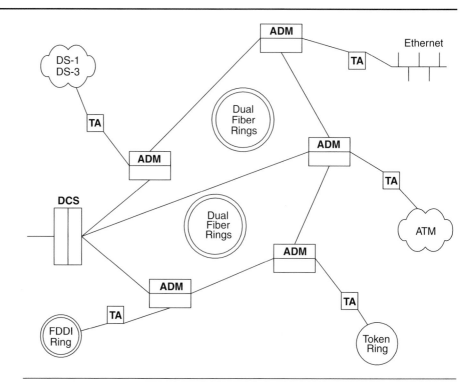

an access terminal, a terminal, or a terminal multiplexer. This device (or concentrator) is responsible for supporting the end-user interface by sending and receiving traffic from LANs, DS-1, DS-3, E-1, and so on. Figure 5–9 shows a typical SONET topology. The user signals are converted into a standard format called the Synchronous Transport Signal (STS). As discussed, STS is an electrical signal (51.840 Mbps). An Add/Drop Multiplexer (ADM) interleaves various input streams onto optical-fiber channels. The digital cross-connect consolidates different types of traffic and makes two way cross-connections between different carrier rates, such as DS-1 and OC-N. The terminal (or service) adapter (TA) provides connectivity between the ADM and various technologies. The backbone network consists of the counter rotating dual fiber-optic rings that provide transparent transport for the various dissimilar technologies.

5.8 SONET EQUIPMENT

The SONET signal format provides capabilities that allow for a highly flexible and intelligent network. Flexibility is due to the signal observability of the low-speed constituents, and intelligence is due to the abundance of allocated signal overhead. These attributes make possible a variety of digital terminals and interfaces to support such an intelligent network. SONET utilizes multiplexers, digital cross-connect systems, digital switch interfaces, digital loop carrier systems, and regenerators.

Short-haul carrier systems converge in COs in both private and common carrier networks for connection to long-haul facilities. If twenty-four-channel digroups (2 * 12 channels) are connected through the office to a single terminating point, no channel bank is required. Instead, the incoming T-1 line is connected to the outgoing line with an express office repeater, and channel banks are needed only at the terminating ends. If fewer than twenty-four channels are needed, back-to-back channel banks or a drop-and-insert multiplexer must be used to access the bit stream for channel cross-connection.

Multiplexers

SONET is a multiplexed transport mechanism that uses a system wide clocking signal to synchronize all activity on the network. The STS multiplexing function accepts multiple electronic signals, converts them into optical signals, and multiplexes them into a single STS.

Besides performing the conversion and multiplexing process, the STS multiplexer is defined as the entry and exit points of a SONET link. The inputs and outputs to the multiplexer can be either end stations or network node access points. The multiplexed STS is sent with a section connection to a regenerator, which extends the length of the link by regenerating optical signals that could have experienced some loss.

Generally, adding and dropping of constituent signals is achieved through back-to-back multiplexer arrangements. The SONET ADM in the add/drop mode is used to add or drop DS-1, DS-1C, DS-2, or DS-3 signals that are packaged in the SONET signal structure, and without back-to-back multiplexers. These multiplexers are undesirable for several reasons:

- Cost of the channel banks
- Added source of potential circuit failure
- Labor cost of making channel cross-connections
- Extra analog-to-digital conversions, which are a source of distortion

Add/Drop Multiplexers

For smaller networks, a drop-and-insert multiplexer, which splits a certain number of channels out of the T-1 bit stream, can be used. A drop-and-insert multiplexer is similar to a channel bank with respect to the channels that are dropped or inserted, and thus, gets its name as an **Add-Drop Multiplexer (ADM).** These are connected as individual channels or as a portion of the T-1 bandwidth to a host computer or similar device. Drop-and-insert capabilities may also be contained in a T-1 multiplexer.

The ADM gets the STS from the regenerator. Additional channels can be multiplexed or demultiplexed from the signal path without having to demultiplex the entire signal first. Add/Drop Multiplexers accomplish this by examining header information to identify individual links and determine if they are headed toward the same destination as the current STS, or if they are to be routed elsewhere. In the latter case, the individual lines can be added to another STS line through another ADM linked to the one in the current path.

Add/Drop Multiplexers are required to convey the DS-N signals as they are, without alteration. They operate bidirectionally, which means they can add or drop DS-1, E-1, or other types of signals from either direction. ADMs support both locked and floating mode signals. They use both electrical and optical interfaces, which are specified in great detail in the ITU-T and ANSI documents.

The SONET ADM may be configured in either terminal mode or add/drop mode. In the terminal mode, the ADM multiplexes between DS-N and OC-N signals. In the add/drop mode, the ADM allows the adding and dropping of DS-N and OC-N payloads without the need for back-to-back multiplexers. The DS-N interface can accommodate either synchronous or asynchronous inputs.

In the terminal configuration in Figure 5–10A, the ADM multiplexes up to N * 28 DS-1s or equivalent signals into an OC-N. The SONET ADM in the add/drop mode (Figure 5–10B) interfaces two full-duplex OC-N signals and one or more full-duplex DS-N signals. Nonpath-terminated incoming payloads are passed through the ADM to be transmitted by the OC-N interface on the other side.

T-1 Multiplexers

Channel banks are somewhat inflexible devices for subdividing the T-1 bit stream. Voice channels and data occupy a full 64 kbps each, though the data channels may operate at lower speeds. Even intelligent channel banks lack the ability to switch and reroute signals. A more versatile device is the T-1 multiplexer, which uses time division multiplexing (TDM) techniques to combine multiple low-speed bit streams into a 1.544 Mbps signal. T-1 multiplexers contain control logic to provide clocking, generate frames, and enable testing.

T-1 multiplexers are more expensive than channel banks; however, multiplexers offer significantly increased functionality over channel banks. A valuable feature of T-1 multiplexers is their network management capability. Some multiplexers have the capability of monitoring multiple points in the network, reporting malfunctions and keeping the network manager supplied with performance and usage information from all of the nodes.

Digital Cross-Connect Systems

digital cross-connect system (DCS)

A **Digital Cross-Connect System (DCS)** is a network element that terminates standard signals and automatically grooms the lower level con-

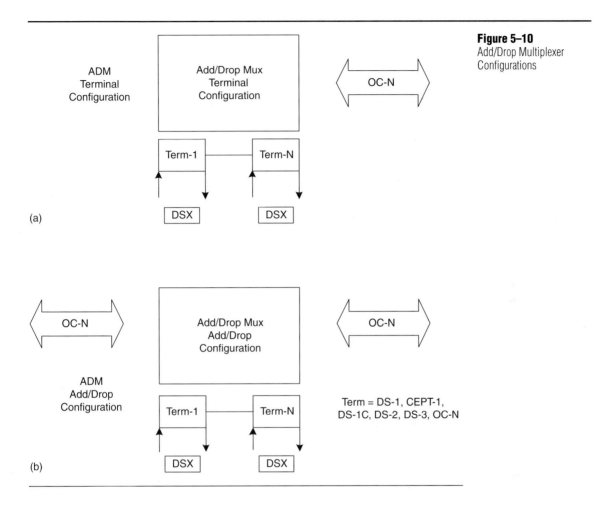

Figure 5–10
Add/Drop Multiplexer
Configurations

stituents of these signals per an electronically alterable map. DCSs are deployed in the network to groom DS-1 signals at both DS-0 and sub DS-0 levels. **Grooming** means combining partially filled input T-1 trunks into fully filled outgoing T-1 trunks (24 DS0s). The DCS is a specialized electronic switch that terminates T-carrier lines with channel banks and routes the individual channel bit streams to the desired output line. A DCS establishes a semipermanent path for the bit stream through the switch. This path remains connected until it is disconnected or changed by programmer command or administrative action.

grooming

The DCS is also tasked with trouble isolation, loopback testing, and diagnostic requirements. It must respond to alarms and failure notifications. The DCS performs switching at the VT level, and the tributaries are accessible without demultiplexing. It can segregate high-bandwidth

Figure 5–11
Digital Cross-Connect
System

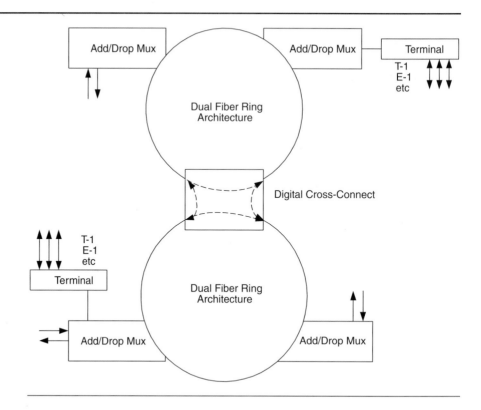

traffic from low-bandwidth traffic and send both out to different ports. Figure 5–11 shows two dual ring architectures that include a DCS with multiple ADMs.

wideband DCS (W-DCS)

The **Wideband DCS (W-DCS)** has the capacity of interfacing SONET high-speed facilities to the electrical interfaces that exist in today's network. The W-DCS cross-connect network provides the capability of interconnecting at a twenty-four channel (DS-1) level. Internally the cross-connects are made in the VT-1.5 format, but are converted to DS-1 and DS-3 interfaces for connections to non-SONET network elements. These interfaces are clear-channel interfaces; thus, whatever format is input on the electrical interface can be output electrically at the distant terminal without changes in format. Figure 5–12 illustrates this multiplexer.

A W-DCS terminates higher rate signals such as DS-3 and OC-Ns and automatically cross-connects DS-1s. It has the basic functionality of a floating VT-1.5. The W-DCS reduces the need for fiber optic terminal equipment in the central office, because the SONET W-DCS may terminate an optical signal.

broadband DCS (B-DCS)

The **Broadband DCS (B-DCS)** interfaces SONET signals and DS-3s, and has the basic functionality of STS-1 and DS-3 cross-connection. The

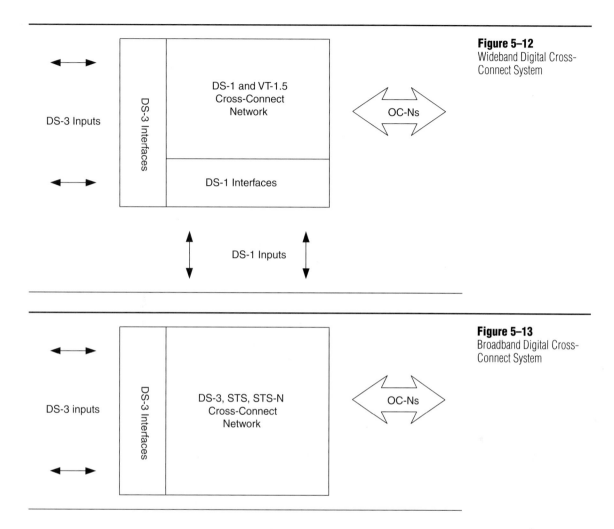

Figure 5–12
Wideband Digital Cross-Connect System

Figure 5–13
Broadband Digital Cross-Connect System

main function of the B-DCS is to provide automated DS-3 and STS-1 level grooming of facilities and cross-connection of DS-3s to larger bandwidth signals (OC-Ns).

The B-DCS (Figure 5–13) provides transparent cross-connections between DS-3 interfaces and between DS-3s and OC-Ns that terminate into the DCS. Transparent (clear channel) implies that any DS-3 at the nominal DS-3 rate can be cross-connected irrespective of the nature and format (synchronous or asynchronous).

Digital Switches

Digital switches have digital interfaces for trunks at the DS-1 rate. An incoming DS-3 is terminated on a multiplexer, demultiplexed into

Figure 5–14
Digital Switch Interface

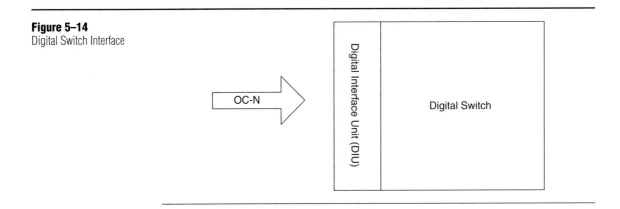

Figure 5–15
SONET Regenerator

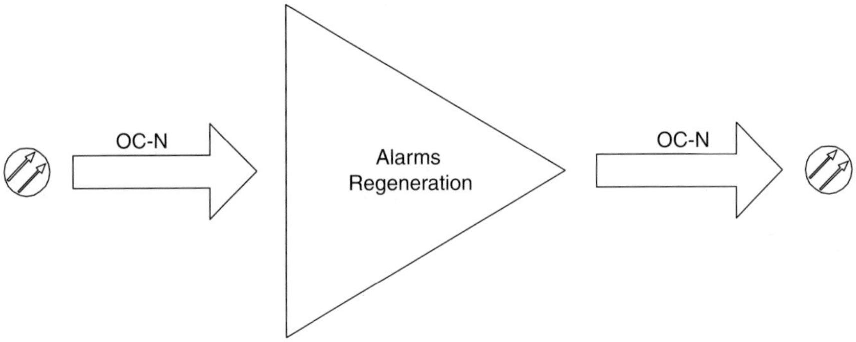

twenty-eight DS-1 signals, and is then terminated on twenty-eight separate switch interfaces. The SONET digital switch interface (Figure 5–14) provides direct termination of the OC-N, thus eliminating stand-alone multiplexers and multiple DS-1 interfaces. Therefore, a SONET digital switch interface will provide direct access to constituent DS0s transported through the network in a terminating OC-N or STS-N. Integrated digital loop carrier generic requirements address the integration of the remote digital terminal (RDT) into a local digital switch at the DS-1 rate.

Regenerators
A regenerator (Figure 5–15) is a network element that receives a digital signal and retransmits it within specified limits of amplitude, shape, and timing. There is an incompatibility between different vendors' regenerators. SONET regenerators comply with standards that make them compatible.

Multi-mode ST Fiber   Single-Mode SC Fiber

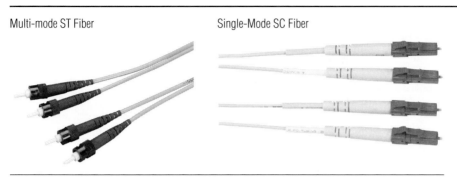

Figure 5–16
Fiber-Optic Cable
(Reproduced with
Permission from Black Box
Corp.)

Regenerator functions include:

- Performance monitoring
- Framing
- Regeneration
- Alarm recognition
- Embedded Operations Channel passing
- STS-1 identification

Optical Media

SONET high-speed communications use single-mode fiber-optic cable and T-carrier communications. The main transport method occurs at the OSI physical layer, which enables other transmission technologies such as FDDI, SMDS, and ATM to operate over SONET.

The two types of fiber-optic cable that are utilized in the SONET environment are produced to single-mode and multi-mode specifications.

- Single mode—LASER utilized in the infrastructure and backbone networks; usually has a yellow jacket
- Multi-mode—Light Emitting Diode (LED) utilized on a customer site; usually has an orange jacket

Connectors come in two versions, namely ST (round) and SC (square). Figure 5–16 shows both ST and SC connectors and single- and multi-mode fiber-optic cable.

Attenuators will adjust the light to ensure proper transmission of the signals.

5.9 SONET STANDARDS

The SONET/SDH topology is based on standards developed by ANSI and the Exchange Carriers Standards Association (ECSA). Bellcore has also been instrumental in the development of these standards.

Table 5–4
SONET Layers and
Functionality

OSI Layer	SONET Function
Layer 4	Path layer responsible for mapping signals into the right channel and for ensuring channel reliability
Layer 3	Line layer responsible for signal switching, monitoring for transmission problems, and error recovery
Layer 2	Section layer responsible for encapsulating data and ensuring that data are sent in the right order
Layer 1	Photonic layer responsible for physical connectivity such as conversion and transport of the optical signals

Four protocol layers are used in SONET. Table 5–4 shows the functions that occur at each of the four layers. The bottom layer 1 is the photonic layer, which corresponds to the physical layer of the OSI model. It handles transportation and conversion of the transported signals. The transmitted electrical signals are changed into optical signals and placed onto the fiber-optic cable, and the received optical signals are changed back to electrical signals. This layer also monitors aspects of the signal transmission, including the optical pulse shape, transmission power levels, and wavelength of the transmitted signal.

Optical Interface Standards
The culminating efforts of standards organizations have resulted in a worldwide standard for optical communications. The major goal of SONET is to standardize fiber-optic network equipment and allow internetworking of optical communications systems from different manufacturers.

The terminology of Optical Carrier (OC) arises because higher data rates associated with the STS standards require optical fiber. The STS standards refer to the electrical signals used in the digital circuit interface, whereas the OC standards refer to the optical signals that propagate across the fiber. The **Fiber Optics Transmission System (FOTS)** is a high-level SONET multiplex that is used to multiplex DS-3 and OC-N signals to higher OC-N levels.

fiber optics transmission system (FOTS)

The SONET optical ANSI interface parameters were concurrently developed with the rates and ANSI formats standard. The ANSI 106 standard addresses the spectral characteristics for 1,310 nanometer (nm) and 1,550 nm systems such as spectral widths and maximum wavelength skew. Other parameters addressed in this standard are the optical logic levels, interface power levels, and pulse shapes for the 1,310 nm and 1,550 nm systems.

SONET standards are designated for single-mode fiber. SONET's use with multi-mode fiber is restricted. Additional SONET RFCs are included in Appendix C.

5.10   QUALITY OF SERVICE

The enhanced network management attributes of SONET are one reason why carriers are migrating to this technology. The OAM procedures are associated with the hierarchical layered design of SONET and ATM. SONET maintenance functions include trouble detection, repair, and restoration. To support these functions, SONET is designed with alarm surveillance operations to detect a problem or a potential problem. Failure states that are monitored by the SONET network elements are:

- Loss of signal
- Loss of frame
- Loss of pointer
- Equipment failures
- Loss of synchronization

The alarm surveillance signals and other OAM signals are conveyed in the SONET headers.

5.11   SONET ADVANTAGES AND DISADVANTAGES

Advantages
SONET provides several attractive features when compared with current technology:

- It is an integrated network standard on which all types of traffic can be transported.
- Fiber-based architecture offers fault tolerance and reliability.
- It offers bandwidths more commensurate with today's fiber-optic systems than the older digital signal hierarchy.
- It merges North American and European hierarchies at higher rates.
- It offers multivendor interoperability over fiber-optic systems.
- It offers centralized end-to-end network management and performance monitoring.
- The SONET standard is based on the optical fiber technology that provides superior performance vis-à-vis the older microwave and cable systems.
- SONET efficiently combines, consolidates, and segregates traffic from different locations through one facility.
- Network costs are lower because SONET permits direct access to any signal level from DS-0 to the top of the hierarchy.
- It provides for new switching standards such as ATM and B-ISDN.

Disadvantages

SONET is not for everyone. It is primarily utilized in carrier networks and as high-speed backbones. As might be expected, SONET is not cheap, and it is also not available everywhere.

5.12   SONET APPLICATIONS

With the advent of SONET, new applications that are impractical because of high-bandwidth requirements have become a reality. Video is a prime market for the use of high bandwidth. As the television industry matures, High Definition Television (HDTV) will require digital capabilities able to support gigabit transmission rates. Another service called video on demand will allow users to program and schedule their television viewing based upon specific schedules or interest and have the programs delivered at specific times.

New applications such as virtual reality will mature, spawning a new industry. Using SONET as a delivery vehicle, users will be able to access programs in a three-dimensional interactive fashion. Virtual reality can be described as a technology that places the user in the middle of the program.

SONET is being used by LECs to deliver clear-channel DS-1 and DS-3 digital services between offices and directly to end users. Private network users also employ SONET over private fiber-optic systems. As bandwidth demand increases, SONET will likely be the physical layer technology that the LECs use to carry services from the customer premises to the local central office [SONET Interoperability Forum (SIF), www.atis.org/atis.sif.sifhom.html, 2000].

SONET is particularly useful for providing:

- Very high-speed data connectivity between distant networks
- Video conferencing between distant sites
- High-quality sound and video reproduction
- High-speed transmission of complex graphics

Mission-critical applications include the following:

Telemarketing	WANs
Broadcast FAX	Customer Service
Electronic Commerce	Videoconferencing
MANs	Telecommuting
Internet Access	Distance Learning
Telemedicine	

Two principal architectures for SONET deployment are:

- Unidirectional path-switched rings (UPSRs), in which all users share transmission capacity around the ring rather than using dedicated segments

- Bidirectional Line-Switched Rings (BLSRs), in which each user's traffic is specifically rerouted in the case of a fiber failure

5.13   SONET ISSUES AND CONSIDERATIONS

SONET availability is currently limited to large metropolitan areas. SONET availability implies that a high-capacity, dual ring, fiber-optic transmission service is available between the customer premises and the carrier's central office.

SONET services cost about 20 percent more than conventional digital service of identical bandwidth. If a subscriber has not identified a mission-critical network transmission requiring fault-tolerant circuits, then the benefits of SONET may not be cost effective.

5.14   SONET TECHNOLOGY ALTERNATIVES

FDDI, wireless, and Cable Modem are possible competitors for high-speed backbone network transmissions. Availability is an issue with these technologies. Several carriers and ILECs (Telcos) offer SONET in the larger metropolitan areas. Many suppliers offer components for implementing high-speed backbone and CO-based networks. These components include add/drop multiplexers, digital cross-connect systems, regenerators, and digital switches.

■ SUMMARY

SONET is utilized in long-distance and local telephone company networks. It was introduced in 1984 by the Bell System and has been approved by the ITU-T for Optical Carrier (OC) speeds.

SONET is a Layer 1 transport service standard used on fiber optic cabling. OSI Layer 1 functions define interfaces to physical media such as copper and fiber-optic cabling. SONET takes data and transports it at high speeds on optical carriers.

SONET links transport data from ATM switches and T-1 and T-3 multiplexers. The differences between T-1 and SONET transmission services lie chiefly in the higher transmission capacity of SONET due to its use of fiber-optic media and the slightly different framing techniques used to channelize this higher transmission capacity.

SONET is (1) a transport technology that provides high availability with self-healing topologies; (2) a multivendor system that allows multivendor connections without conversions between the vendors' systems; (3) a network that uses synchronous operations with powerful multiplexing and demultiplexing capabilities; and (4) a system that provides extensive embedded overhead channels in support of Operations, Administration, and Maintenance (OAM).

SONET WAN service is spreading rapidly via Regional Bell Operating Companies (telcos) and independent telephone companies, because it offers extremely high-speed WAN networking at speeds greater than 1 Gbps, and because it provides the best solution for linking high-speed LANs—including those based on ATM—without a reduction in bandwidth.

Key Terms

Add/Drop Multiplexer (ADM)

Automatic Protection Switching

BiDirectional Line Switching

Broadband DCS (B-DCS)

Digital Cross-Connect System (DCS)

Fiber Optics Transmission System (FOTS)

Grooming

Line

Line Overhead (LOH)

Optical Carrier (OC)

Optical Carrier Level 1 (OC-1)

Path

Path Overhead (POH)

Payload Pointer

Section

Section Overhead (SOH)

Stratum-one

Synchronous Payload Envelope (SPE)

Synchronous Digital Hierarchy (SDH)

Synchronous Optical Network (SONET)

Synchronous Transport Signal (STS)

Synchronous Transport Signal Level 1 (STS-1)

Transport Overhead (TOH)

Undirectional Path Switching

Virtual Tributary (VT)

VT Groups

Wideband DCS (W-DCS)

REVIEW QUESTIONS

1. Develop a succinct definition of SONET.

2. What is SONET and where does it fit in the WAN architecture? What are two principal architectures for the SONET deployment?

3. What is the difference between SONET and Synchronous Digital Hierarchy?

4. What is SONET's major attribute?

5. Describe the topology of SONET. Draw the SONET transport hierarchy.

6. What are the possible methods of failure recovery for SONET?

7. Define SONET synchronous multiplexing.

8. What is the relationship between STS-1 and OC-1? How are these signals different? What are the speeds designated by OC-1, OC-3 and OC-12?

9. What are the different optical carrier designations and speeds? How do these relate to the ITU-T designations?

10. What are the allowable signal rates for SONET? What is the function of the clock and what clock is in the SONET distribution network?

11. How is the signal architecture formed for SONET? The SONET signal format is functionally layered. Name these layers.

12. Describe the SONET frame. Identify the major components and subcomponents.

13. What is the purpose of SONET framing? How does this apply to the frame format?

14. Describe the SONET multiplexing frame format. How does this relate to the signal architecture? What is the purpose of SONET framing?

15. What are the components of the synchronous transport signal 1? Describe these components and give a short description of their functions.

16. What is the function of the transport overhead?

17. What is the function of the path overhead? What are the four classes and their functions?

18. Describe the line overhead functionality.

19. Why is the line overhead important? What is the special bit pattern in the section overhead and why is it necessary?

20. What is the synchronous payload envelope? What is the payload?

21. Give examples of capacities that can be transported by the STS-1 payload.

22. Calculate the value of the STS-1 rate.

23. What is the function of concatenation?

23. SONET overhead is functionally divided into three layers. Identify each and describe the processes of each.

25. What is a SONET pointer and how is it implemented?

26. What is the function of a virtual tributary? Identify the four sizes that are allowable.

27. What are the two modes of VT operation? Describe each.

28. What medium is utilized in the SONET networks? What are the two types? What is the difference between single-mode and multi-mode fiber?

29. Describe a simple SONET end-user topology. Describe the service adapters.

30. SONET utilizes a number of devices in the network. List the most prevalent.

31. What is the function of a multiplexer in the SONET network?

32. Describe the functionality of an Add/Drop multiplexer. How is the add/drop multiplexer utilized in the SONET network?

33. How is a T-1 multiplexer different from an Add/Drop multiplexer?

34. Why is a digital cross-connect system required in the SONET network?

35. What are the functions of a digital cross-connect system?

36. What is the difference between a broadband DCS and a wideband DCS?

37. Why is a digital CO switch useful in the SONET environment?

38. What is a regenerator and where would it be deployed?

39. What are the layers of the SONET model? What are the functions of each layer?

40. Describe the optical interface standards for SONET.

41. What failure states are monitored by the SONET network elements? How does OAM fit in the SONET environment? Identify the network elements that work in the SONET environment.

42. What are advantages for utilizing the SONET network?

43. Identify applications that could benefit from utilizing SONET.

ACTIVITIES

1. Utilize the Web and other online resources to research the SONET technology. Prepare a five-minute overview for an in-class oral presentation.

2. Prepare a list of advantages that can be gained by utilizing the SONET backbone infrastructure for transporting the user's traffic.

3. Develop a list of users and their applications that could benefit from SONET transport.

4. Draw a logical diagram of a SONET infrastructure. Identify all the major components and the functionality of each.

5. Draw a physical diagram of a SONET infrastructure and the user's end points. Identify all of the individual devices and components of this network.

6. Utilize the list of SONET RFCs in Appendix C to prepare an overview of SONET standards.

7. Arrange for a carrier representative to give a presentation on the SONET technology.

CASE STUDY/PROJECT

It is recommended that this case study be assigned to a work group. Because it contains sufficient requirements for a lengthy project, allow an extended time for completion.

The focus of this case study is to complete a worksheet concerning the SONET technology. The questions and requirements are designed so that the student must use additional resources to complete the worksheet. Some recommended sources are the DataPro Communications series in DataPro Information Services, *Emerging Communications Technologies* [Black, 1997], and the Internet.

A graphical network design should be developed in conjunction with the completion of the form. It should be documented with a graphics package such as Visio. This design would take the form of a SONET/SDH deployment.

SONET Case Study Worksheet
- Identify the applications.
- Describe the topology.
- Identify the media requirements.
- Identify the WAN, MAN, and LAN interactions.
- Identify and describe the transport envelopes.
- Identify the bandwidth options.
- Describe how flow control/congestion control is implemented.
- Identify data integrity capabilities.
- Describe the addressing schemes.
- Identify the competition.
- How is the service managed?
- Provide a logical SONET/SDH network drawing.

6

Virtual Private Network

■ INTRODUCTION

A **Virtual Private Network (VPN)** is a private data network that makes use of the public telecommunication infrastructure, maintaining privacy through the use of a tunneling protocol and security procedures. The VPN is so named because an individual user shares communications channels with other users. Switches are placed on these channels to allow an end user to have access to multiple end sites. A VPN can be contrasted with a system of owned or leased lines which can only be used by one company. The connection between sender and receiver acts as if it were completely private, even though it uses a link across a public network to carry information.

The idea of the VPN is to give the company the same capabilities at much lower cost by using the shared public infrastructure rather than a private one. Phone companies have provided secure shared resources for voice messages. A VPN makes it possible to have the same secure sharing of public resources for data. Figure 6–1 depicts a VPN between Raleigh and New Orleans that utilizes the Internet for transport. Access to the Internet cloud occurs at Raleigh and then again at New Orleans, which saves on transport costs.

Virtual networks are economical for large companies that have a considerable amount of on-net calling. Most of the features of a dedicated private network can be provided. Companies today are looking at using a virtual private network for both extranets and wide area intranets.

Using a VPN involves encrypting data before sending the data through the public network, and decrypting it at the receiving end. An additional level of security involves encrypting not only the data but also the originating and receiving network addresses. A VPN combines the advantages of private public networks by allowing a company with

OBJECTIVES
Material included in this chapter should enable you to:

● understand VPN technology and its relation to the current broadband market. Identify the various types of virtual private networks.

● become familiar with the terms and definitions that are relevant in the Enterprise Networking environment. Look at the similarities between VPNs and other network technologies.

● understand the function and interaction of VPN in the Enterprise Network environment. Look at the issue of tunneling and the security techniques that must be deployed in a VPN.

● understand where a VPN fits in the different applications of the technology. Identify these applications and the organizations that can benefit from utilizing a VPN.

virtual private network (VPN)

- identify components that comprise the VPN networking environments. Look at router and firewall components that are part of the VPN.
- identify the benefits, issues, and considerations that must be part of the VPN deployment process.
- become familiar with the various protocols that are part of the VPN infrastructure.
- identify the advantages and disadvantages of implementing a VPN.

multiple sites to have the illusion of a completely private network, and to use a public network to carry traffic between sites.

The voice private network question is greatly affected by virtual networks being offered by the major Interexchange Carriers (IXCs). A VPN operates as if it was composed of voice private circuits; however, it is actually part of the IXC's switched network. AT&T's virtual network is Software Defined Network (SDN), MCI's VPN is V Net, and Sprint's VPN is Virtual Private Network. The VPN architectures of the IXCs are all similar, although the terminology may somewhat differ.

6.1   VPN ENVIRONMENT

The virtual private network allows an individual user to share communications channels with other users. Switches are placed on these channels to allow an end user to have access to multiple end sites. Ideally, users do

Figure 6–1
VPN over the Internet

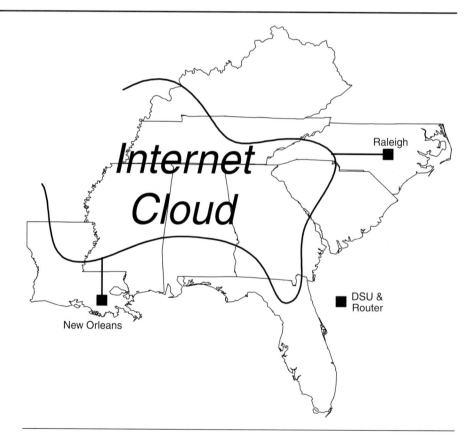

not perceive that they are sharing a network with each other, thus the term *virtual private network*. Users think that they are the only ones on the network, although they are not.

Two fundamental components of the public network make VPNs possible. The process of tunneling enables the virtual part of a VPN and security services keep the VPN data private.

VPN is a relatively new term in the computer/communications industry, but the ideas behind the concept are not new. Public X.25 networks have offered VPN services for years, and switched T-1 services also offer VPN-like features. However, it will become evident that several of the emerging technologies offer more powerful VPNs than these older technologies.

VPNs represent temporary or permanent connections across a public network, such as the Internet, that make use of special **encryption** technology to transmit and receive data meant to be impenetrable to anyone who attempts to monitor and decode the packets transmitted across the public network. The connection between the sender and receiver acts as if it were completely private, even though it uses a link across a public network to carry information. This makes something virtually public act as though it were private; thus, a Virtual Private Network.

Windows NT 4.0 and Windows 95 support a special TCP/IP protocol called **Point-to-Point Tunneling Protocol (PPTP).** PPTP permits a user running Windows 98, for example, to dial into a Windows NT server running the **Remote Access Service (RAS)**, and it supports the equivalent of a private, encrypted, dial-up session across the Internet. Similarly, a VPN could be established permanently across the Internet by leasing dedicated lines to an Internet Service Provider (ISP) at each end of a two-way link, and maintaining ongoing PPTP-based communications across that dedicated link. This means that organizations can use the Internet as a private dial-up service for users with machines running Windows 95 and Windows NT 4.0 or as a way to interconnect multiple LANs across the Internet, one pair of networks at a time.

The VPN uses PPTP or other equivalent protocols to extend the reach of private networks across public ones, easily, economically, and transparently, by utilizing on-demand, dial-up connections across the Internet.

encryption

point-to-point tunneling protocol (PPTP)
remote access service (RAS)

6.2   VIRTUAL NETWORKS OVERVIEW

Some Interexchange Carriers (IECs) offer VPNs to their subscribers, a service that requires support from a CO tandem switch. The typical tandem switch has a nonblocking switching fabric, which primarily provides access to CO trunks. Digital trunks terminate in digital interface frames that couple incoming T-1 or T-3 bit streams directly to the switching network. Peripheral equipment detects signaling, and the central processor sets up a path through the switching network from the incoming time slot to an outgoing time slot which it assigns to an outgoing digital channel. The digital switch acts as a large time-slot interchange device that is transparent to the bit stream in the terminating circuits.

service control point (SCP)

service switching point (SSPs)

A VPN is one that operates as if it is composed of switched private lines, but in reality, is derived by shared use of a carrier's switched facilities. The database for a VPN is contained in a **Service Control Point (SCP)**, that is, a computer connected to tandem switches by 64 kbps data links. The switches in a virtual network are known as **Service Switching Points (SSPs).**

A virtual private network handles calls in three manners:

- Dedicated access line to dedicated access line (DAL)
- Dedicated access line to switched access line (DAS)
- Switched access line to switched access line (SAL).

A DAL-to-DAL call bypasses the LEC's access charges in both the originating and the terminating direction, reducing the cost significantly. The DAL-to-SAL is handled on a VPN the same way it is handled with a customer using conventional DAL service. The access charge is eliminated in the originating, but not the terminating, direction. For SAL-to-SAL calls, the VPN handles calls like regular long-distance calls except for features and restrictions.

Virtual private networks also emulate many features of electronic tandem networks. They offer a full restriction range such as blocking calls to overseas locations, selected area codes, central office codes, or even selected station numbers. If the virtual network is used in conjunction with account codes, calls can be restricted for certain station numbers.

6.3   VPN TYPES

A VPN is customer connectivity deployed on a shared infrastructure with the same policies as a private network. The shared infrastructure can provide a service provider with IP, Frame Relay, ATM backbone, or Internet access. There are three types of VPNs, which align with how businesses and organizations use them, namely—access VPN, intranet VPN, and extranet VPN.

Access VPN

An **access VPN** provides remote access to a corporate intranet or extranet over a shared infrastructure with the same policies as a private network. Access VPNs enable users to access corporate resources when, where, and however they require. Access VPNs encompass analog, dial, ISDN, DSL, mobile IP, and cable technologies to securely connect mobile users, telecommuters, or branch offices.

access VPN

Intranet VPN

An **intranet VPN** links corporate headquarters, remote offices, and branch offices over a shared infrastructure using dedicated connections. Businesses enjoy the same policies as a private network, including security, Quality of Service (QoS), manageability, and reliability.

intranet VPN

Extranet VPN

An **extranet VPN** links customers, suppliers, partners, or communities of interest to a corporate intranet over a shared infrastructure using dedicated connections. Businesses enjoy the same policies as a private network, including security, QoS, manageability, and reliability.

extranet VPN

6.4   VIRTUAL NETWORKS CONNECTIVITY

In VPNs, *virtual* implies that the network is dynamic, with connections set up according to the organizational needs. Unlike the private lines used in traditional virtual private networks, Internet VPNs do not maintain permanent links between the endpoints that comprise the corporate network. Instead, a connection is created between the two sites when needed, and torn down when no longer required. This makes the bandwidth and other network resources available for other uses.

The term virtual also means that the logical structure of the network is formed only from the company's devices, regardless of the physical structure of the underlying network. Devices such as routers and switches that are part of the ISP's network are hidden from the devices and users of the company's private network.

The connections that comprise the company's VPN do not have the same physical characteristics as the hard-wired connections used on the Local Area Network. Hiding the ISP and other infrastructure from the VPN applications are made possible by **tunneling,** one of the two fundamental components that comprise the VPN architecture. The other component is security services.

tunneling

Figure 6–2
Tunneling Structure

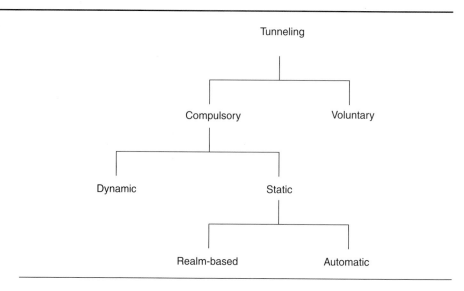

Tunneling

A key component of VPN is tunneling, which is a vehicle for encapsulating packets inside a protocol that is understood at the entry and exit points of a given network. These entry and exit points are defined as tunnel interfaces. The tunnel itself is similar to a hardware interface, but is configured in software.

voluntary tunnel
compulsory tunnel

End-user software capabilities and ISP support has resulted in the division of tunnels into two classes: voluntary and compulsory. **Voluntary tunnels** are created at the request of the user for a specific use. **Compulsory tunnels** are created automatically without any action from the user, and without allowing the user any choice in the matter. Within the compulsory category are two subclasses: static and dynamic. The static tunnels can be subdivided again, into realm-based and automatic classes. Figure 6–2 illustrates the tunneling structure.

When using a voluntary tunnel, the end user can simultaneously open a secure tunnel through the Internet and access other Internet hosts via basic TCP/IPs without tunneling. The client-side endpoint of a voluntary tunnel resides on the user's computer. Voluntary tunnels are often used to provide privacy and data integrity for intranet traffic being sent over the Internet.

Because compulsory tunnels are created without the user's consent, they may be transparent to the end user. The client-side endpoint of a compulsory tunnel typically resides on a RAS. All traffic originating from the end user's computer is forwarded over the PPTP tunnel by the RAS. Access to other services outside the intranet would be controlled by the network administrators. PPTP enables multiple connections to be carried over a single tunnel.

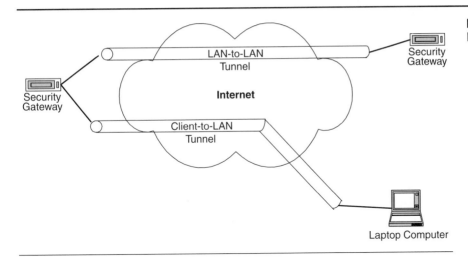

Figure 6–3
LAN and Client VPN Tunnels

Because a compulsory tunnel has predetermined endpoints and the user cannot access other parts of the Internet, these tunnels offer better access control than voluntary tunnels. If corporate policy states that employees cannot access the public Internet, then a compulsory tunnel would keep them out of the public Internet, while still allowing them to use the Internet to access the VPN.

Static compulsory tunnels typically require dedicated equipment or manual configuration. These dedicated, automatic tunnels might require the user to call a special telephone number to make the connection. On the other hand, in realm-based, or manual, tunneling schemes, the RAS examines a portion of the user's name, called a realm, to decide where to tunnel the traffic associated with that user.

A more flexible approach would be to dynamically choose the tunnel destination on a per-user basis when the user connects to the RAS. These dynamic tunnels can be set up in PPTP.

Tunneling has two definitions: one is a LAN term and the other is an Internet term. As a LAN term, tunneling means to temporarily change the destination of a packet in order to traverse one or more routers that are incapable of routing to the real destination. This text will use the Internet term, which says that tunneling is used to provide a secure, temporary path over the Internet. As an example, a telecommuter might dial into an ISP, which would recognize the request for a high-priority, point-to-point tunnel across the Internet to a corporate gateway. The tunnel would be established, effectively making its way through other, lower-priority Internet traffic. Figure 6–3 illustrates a tunneling network example with tunneling between security gateways and between a security gateway and a client workstation.

serial line internet protocol (SLIP)
point-to-point protocol (PPP)

Tunneling that involves three types of protocols is depicted in Table 6–1. These include (1) the passenger protocol, (2) the encapsulating protocol, and (3) the carrier protocol. The passenger protocol is that which is being encapsulated. In a dial-up scenario, this protocol could be **Serial Line Internet Protocol (SLIP), Point-to-Point Protocol (PPP),** or text dialog. The encapsulating protocol is used to create, maintain, and tear down the tunnel. The carrier protocol is used to carry the encapsulated protocol. IP is the first carrier protocol used by the L2F Protocol because of its robust routing capabilities, ubiquitous support across different media, and deployment within the Internet.

No dependency exists between the L2F Protocol and IP. In subsequent releases of the L2F functionality, Frame Relay, X.25, and ATM virtual circuits could be used as a direct Layer 2 carrier protocol for the tunnel.

authentication header (AH)
encapsulating security payload (ESP)

To create a tunnel, the source end encapsulates its packets in IP packets for transit across the public network. For VPNs, the encapsulation may include encrypting the original packet and adding a new IP header to the packet. At the receiving end, the gateway removes the IP header and decrypts the packet if necessary, forwarding the original packet to its destination. Figure 6–4 depicts a packet prepared for tunneling. The illustration shows the original packet and the elements that have been added for tunneling. The new IP header includes an **Authentication Header (AH)** and an **Encapsulating Security Payload (ESP).**

Table 6–1
Tunneling Packet Format

Passenger Protocol	Encapsulating Protocol	Carrier Protocol
SLIP PPP Text dialog	L2F	IP UDP

Figure 6–4
A Packet Encapsulated for Tunneling

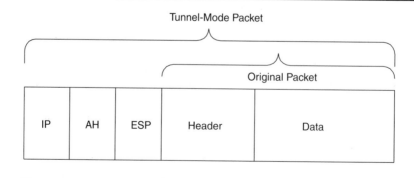

Security Services

Equally important to a VPN's use is the issue of privacy or security. In its most basic use, the *private* in VPN means that a tunnel between two users on a VPN appears as a private link, even when running over shared media. For business use, especially for LAN-LAN links, private has to mean security that is free from prying eyes and tampering. VPNs need to provide four critical functions to ensure the security of data. These functions are as follows:

- **Authentication**—Ensuring that the data is coming from the claim.   authentication
- Access control—Restricting unauthorized users from gaining admission to the network.
- Confidentiality—Preventing anyone from reading or copying as the data travels across the network.
- Security—Ensuring that no one tampers with the data as the data travel the network.

Although tunnels ease the transmission of the data across the network, authenticating users and maintaining the integrity of the data depends on cryptographic procedures such as digital signatures and encryption. These procedures use shared secrets called keys, which have to be managed and distributed with care.

Implementation of security is implemented at the lower levels of the OSI model, the Data Link and Network layers. Deploying security services at these OSI layers makes much of the security services transparent to the user. Implementation of security at these levels can take two forms, which affect the user's responsibility for securing personal data. Security can be implemented either for end-to-end communications, such as communication between two computers, or between other network components, such as firewalls and routers. The latter case refers to node-to-node security and is depicted in Figure 6–5.

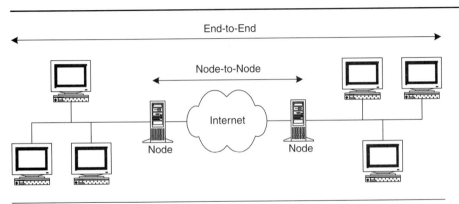

Figure 6–5
End-to-End versus Node-to-Node Security

Using security on a node-to-node basis can make the security services more transparent to the end users and relieve them of some of the heavy-duty computational requirements such as encryption. Node-to-node security, however, requires that the networks behind the node be trusted networks, which means that they are secure against other attacks of unauthorized use. End-to-end security, because it involves each host, is inherently more sound than node-to-node security. End-to-end security, however, increases the complexity for the end user, and can be more challenging to manage.

6.5   VPN CONNECTIVITY AND DESIGN

Figure 6–6 depicts a simple VPN network. The VPN server for each network is connected through the Internet or public network via tunneling. The remote workstation is also connected to the VPN server through the Internet or public network via the tunneling technique.

Basically, a virtual circuit connection is set up on a network between a sender and a receiver in which both the route for the session and bandwidth is allocated dynamically. VPNs can be established between two or more LANs or between remote users and a LAN.

Access is dedicated (on-net) or switched (off-net). Calls placed over the network are rated in three categories: (1) on-net to on-net, (2) on-net to off-net, and (3) off-net to off-net. The on-net portion of the calls does not incur LEC access charges, and thus reduces the cost of the call.

Internet VPN

Rather than depend on the traditional telcos for dedicated leased lines or Frame Relay's PVCs, another option is to use an Internet-based VPN. This would use the open, distributed infrastructure of the Internet to transmit

Figure 6–6
VPN Network with
Tunneling

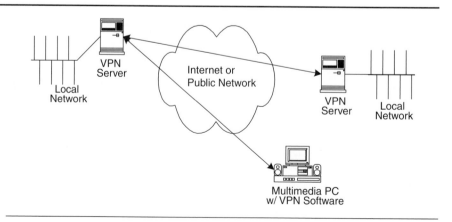

data between corporate sites. Companies using an Internet VPN would set up connections to the local connection points, or Points-Of-Presence (POP) of their Internet Service Provider (ISP) and let the ISP ensure that the data is transmitted to the appropriate destinations via the Internet. The rest of the connectivity details for the network and Internet infrastructure would be the ISP's responsibility. The link created to support a given communications session between sites is dynamically established, reducing the load on the network. Permanent links are not part of the VPN's structure. The bandwidth required for a session is not allocated until required and is released for other uses when a session is completed.

Because the Internet is a public network with open transmission of data, Internet VPNs include the provision of encrypting data passed between VPN sites, which protects data against eavesdropping and tampering by unauthorized parties. Issues still outstanding for Internet VPNs are guaranteed performance and security. Additional information concerning the Internet VPN can be found in RFC 2764—A Framework for IP-Based Virtual Private Networks.

Reasons to Implement an Internet VPN

Whether a VPN is being built from scratch or the traditional VPN is being converted to one using the Internet, certain factors may indicate the use of Internet-based VPNs. These factors include cost savings, flexibility, scalability, reduced technical support, and reduced equipment requirements.

Cost Savings   Often the cost savings alone makes it worthwhile to adopt Internet VPNs in the business. Conceivably, the organization's network cost can be reduced by outsourcing the entire VPN operation to a service provider. Service providers include technical support, help-desk services, and security audits as part of their support package.

Flexibility   In an Internet-based VPN, not only can T-1 and T-3 lines be used between the various locations and the ISP, but many other connection types can be used to connect smaller offices and mobile workers to the ISP, and therefore, to the corporate VPN. The only restriction is the media that the ISP supports.

Scalability   Because VPNs use the same media and underlying technologies as the Internet, they are able to offer businesses two dimensions of scalability that are otherwise difficult to achieve. These are geographic scalability and bandwidth scalability. As for geographic scalability, an Internet VPN, offices, teams, telecommuters, and mobile workers can become part of a VPN wherever the ISP offers a point of presence (POP). This scalability can be dynamic; for example, a remote field office can be easily linked to a local POP within a matter of minutes and just as easily removed from the VPN when the remote office closes up shop. Regarding bandwidth scalability, ISPs charge by usage, so fees for a little-used T-1 are less than the charges for a highly used T-1. ISPs, therefore, can quickly offer a choice of bandwidth according to the needs of the user. Links are not hard-wired between each site, so it is not necessary to upgrade the equipment at every site to support changes at one location.

Reduced Technical Support   VPNs can reduce the demand for technical support resources. A considerable amount of this reduction stems from standardization on one type of connection (IP) from mobile users to an ISP's POP and standardized security requirements. Internal technical support requirements are reduced because the service providers take over many of the support tasks for the network.

Reduced Equipment Requirements   Offering a single solution for enterprise networking, dial-in access, and Internet access, Internet VPNs require less equipment. Rather than maintain separate modem pools, terminal adapters, and remote access servers, a business can set up its Customer Premises Equipment (CPE) for a single medium, such as a T-3 line, with the rest of the connection types handled by the ISP. WAN connection and maintenance can be reduced by replacing modem and DSU pools and multiple frame relay circuits with a single wide area link that carries remote user, LAN-to-LAN, and Internet traffic at the same time.

6.6   VPN HARDWARE AND SOFTWARE

After the user has a connection to the Internet or the public access network, the important network devices for the VPN are the ones that control access to the protected LAN from remote and external sources. The external source might be another of the corporate LANS, a mobile worker with a laptop, or a corporate partner. VPN access devices should be able to handle all of these situations; however, not all are equally adept at handling the different connectivity situations.

VPN hardware and software can be placed at various locations in the network. These include security gateways, policy servers, and certificate authority holders for preventing unauthorized intrusions. This section will look at firewalls and routers, which are implemented in the VPN to provide access control.

Firewalls and Routers

firewalls
Firewalls have long been used to protect LANs from other parts of an IP internetwork by controlling access to resources on the basis of packet type, application type, and IP address. Figure 6–7 shows the locations of

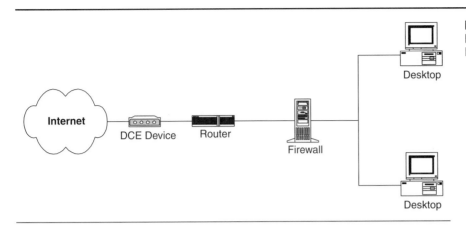

Figure 6–7
Firewall and Router
Locations

routers and firewalls in the VPN. These devices can be placed anywhere between the DCE device and the LAN.

A firewall is a device acting as a network filter to restrict access to a private network from the outside, implementing access controls based on the contents of the packages of data that are transmitted between two parties or devices on the network. There are three main classes of firewalls: packet filters, application and circuit gateways (proxies), and stateful multi-layer inspection firewalls.

Packet Filtering Firewalls   Packet filtering firewalls were the first generation of firewalls. Packet filters track the source and destination address of IP packets, permitting packets to pass through the firewall based on rules set by the network manager. Two advantages of packet filter firewalls are that they are fairly easy to implement and they are transparent to the end users. However, they can be difficult to configure properly, particularly if a large number of rules have to be generated to handle a wide variety of applications and users.

Packet filtering often requires no separate firewall, because it is often included in most TCP/IP routers at no extra charge; however, it is not the best firewall security that can be implemented. One of its deficiencies is that filters are based on IP addresses that can be forged, and not authenticated user identification.

Packet filters can be used as part of the organization's VPN because they can limit the traffic that passes through a tunnel to another network, based on the protocol and direction of traffic. For example, it is possible to configure a packet filter firewall to disallow File Transfer Protocol (FTP) traffic between two networks, while allowing Hypertext Transfer Protocol (HTTP) and Simple Mail Transfer Protocol (SMTP) traffic between the two, further refining the granularity of the control on protected traffic between sites.

proxy server

Application and Circuit Gateways   Application and circuit gateways enable users to utilize a **proxy server** to communicate with secure systems, hiding valuable data and servers from potential attackers. The proxy accepts a connection from the other side and, if the connection is permitted, makes a second connection to the destination host on the other side. The client attempting the connection is never directly connected to the destination. Because proxies can act on different types of traffic or packets from different applications, a proxy firewall (server) is usually designed to use proxy agents. In this case, an agent is programmed to handle one specific type of transfer, such as FTP or TCP traffic. The more types of traffic that must pass through the proxy, the more proxy agents need to be loaded and running on the machine.

Circuit proxies focus on the TCP/IP layers, using the network IP connection as a proxy. Circuit proxies are more secure than packet filters because computers on the external network never gain information about internal network IP addresses or ports. A circuit proxy is typically installed between the company's network router and the public network (that is, the Internet), communicating with the public network on behalf of the company's network. Real network addresses can be hidden because only the addresses of the proxy are transmitted to the public network.

Circuit proxies are slower than packet filters because they must reconstruct the IP header to each packet to its correct destination. Also, circuit proxies are not transparent to the end user, because they require modified client software.

Stateful Multilayer Inspection   A firewall technique called Stateful Multi-layer Inspection (SMLI) was invented to make security tighter while making it easier and less expensive to use without slowing performance. SMLI is the foundation of a new generation of firewall products that can be applied across different kinds of protocol boundaries, with an abundance of easy-to-use features and advanced functions.

SMLI is similar to an application proxy in the sense that all levels of the OSI model are examined. Instead of using a proxy, SMLI uses traffic-screen algorithms optimized for high-throughput data parsing. With SMLI, each packet is examined and compared against known states of friendly packets.

One of the advantages to SMLI is that the firewall closes all TCP ports and then dynamically opens ports when connections require them. Stateful inspection firewalls also provide features such as TCP sequence-number randomization and User Datagram Protocol (UDP) filtering. These firewalls, however, have to be supplemented with proxies in order to support other important functions such as authentication.

Many firewall vendors include a tunnel capability in their products. Like routers, firewalls have to process all IP traffic. Because of all the processing performed by firewalls, they are ill-suited for tunneling on large networks with large amounts of traffic. Combining tunneling and encryption with firewalls is probably best used on small networks with low volumes of traffic. Also, like routers, firewalls can be a single point of failure for the VPN.

Routers

A **router** is an intelligent device that connects like and unlike LANs and   router
WANs. Routers are protocol sensitive, and typically support multiple protocols. Routers usually operate at Layer 3 of the OSI model and are responsible for making decisions about which of several paths network traffic will follow. To do this, a routing protocol to gain information about the network, and algorithms to choose the best route, are based on several criteria known as routing metrics.

Because routers examine and process every packet that leaves the LAN, it is natural to include packet encryption on routers. Vendors of router-based VPN services usually offer two types of products, either add-on software or an additional feature of a coprocessor-based encryption engine. However, adding the encryption tasks to the same box as the router increases the risks of losing access to the VPN if the router has a failure.

Many of the requirements for an encrypting router are the same as those for firewalls. Encrypting routers are appropriate for VPNs if they have the following features:

- Supports both transport mode and tunnel mode **IP Security**   IP security (IPSec)
  (IPSec).
- Restricts access by operations personnel to keys.
- Supports a cryptographic key length that best matches the security needs.
- Includes separate network connections for encrypted and unencrypted traffic.
- Supports the default IPSec cryptographic algorithms.
- Supports automatic rekeying at regular periods or for each new connection.
- Logs failures when processing headers and issues alerts for repeated disallowed activities.

Because routers are generally designed to investigate packets at Layer 3 of the OSI model and not authenticate users, it will probably be necessary to add an authentication server in addition to the encrypting router to create a secure VPN.

Another possible VPN solution is to use special hardware that is designed for the task of tunneling and encryption. These devices usually

Figure 6–8
Router and Firewall Device
(Reproduced with
permission from Black Box
Corp.)

Rack-Mounted Router   Desktop Firewall

operate as encrypting bridges that are typically placed between the network routers and the WAN links. Although most of these hardware devices are designed for LAN-to-LAN configurations, some products also support client-to-LAN tunneling. Figure 6–8 provides examples of a firewall and router that can be utilized in a VPN.

VPN Software

To provide virtual private networking capabilities using the Internet as an enterprise network backbone, specialized tunneling protocols needed to be developed that could establish private, secure channels between connected systems. Four protocols were originally suggested as VPN solutions. Three are designed to work at the OSI Layer 2:

- Microsoft's Point-To-Point Tunneling protocol (PPTP).
- Cisco's Layer 2 Forwarding (L2F) Protocol
- A combination of the two called Layer 2 Tunneling Protocol (L2TP)

The only VPN protocol for OSI Layer 3 is **IPSec,** which has been developed by the Internet Engineering Task Force (IETF) over the past few years.

An effort is underway to have the IETF propose a unification of the two rival standards, to be called Layer 2 Tunneling Protocol (L2TP). One shortcoming of the proposed specification is that it does not deal with security issues such as authentication. Figure 6–9 depicts the use of tunneling protocols to build a virtual private network. Various tunneling protocols are utilized to pass traffic across the VPN, a process that is transparent to the end users.

Point to Point Tunneling Protocol   The Point-to-Point Tunneling Protocol (PPTP) was first created by a group of companies calling themselves the PPTP Forum. The group consisted of 3Com, Ascend Communications, Microsoft, ECI Telematics, and U.S. Robotics. The basic idea behind PPTP was to split up the functions of remote access in such a way that individuals and corporations could take advantage of the Internet's infrastructure to

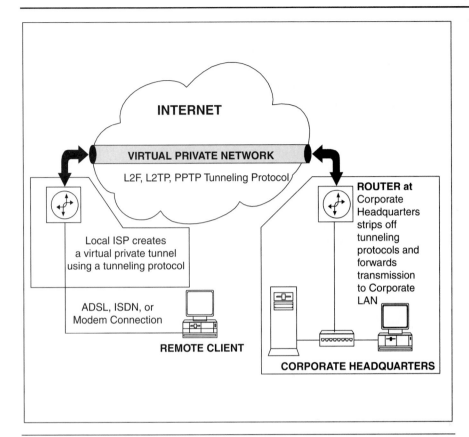

Figure 6–9
Utilizing Tunneling
Protocols

provide secure connectivity between remote clients and private networks. Remote users would dial into the local number of the ISP and then securely tunnel into their corporate network.

The most commonly used protocol for dial-up access to the Internet is the Point-to-Point Protocol (PPP). The first connection is to the remote access host using PPP. A second connection is then made over the PPP connection to a PPTP server on a private LAN. A typical example of a PPTP installation is a client computer running PPP and PPTP to access an ISP. Figure 6–10 illustrates this connectivity. PPTP permits a user running Windows 95 to dial into a Windows NT Server running the Remote Access Service (RAS), and it supports the equivalent of a private, encrypted dial-up session across the Internet. Similarly, a VPN could be established permanently across the Internet by leasing dedicated lines to an ISP at each end of a two-way link, and maintaining ongoing PPTP-based communications across that dedicated link.

Figure 6–10
PPTP Environment

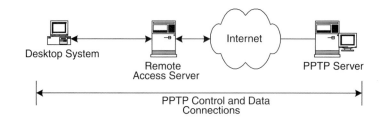

Desktop System

Remote
Access Server

Internet

PPTP Server

PPTP Control and Data
Connections

Since it is based on PPP, PPTP is well suited to handling multiprotocol network traffic, particularly IP, IPX, and NetBEUI protocols. PPTP's design also makes it easier to outsource some of the support tasks to an ISP. By using RADIUS proxy servers, an ISP can authenticate dial-in users for corporate customers and create secure PPTP tunnels from the ISP's network access servers to the corporate PPTP servers. These PPTP servers then remove the PPTP encapsulation and forward the network packets to the appropriate destination on the private network. RADIUS is a popular remote authentication dial-in user service security system.

Layer 2 Tunneling Protocol   The Layer 2 Tunneling Protocol (L2TP) was created as the successor to two tunneling protocols, PPTP and L2F. Like PPTP, L2F was designed as a tunneling protocol, using its own definition of an encapsulation header for transmitting packets at Layer 2. One major difference between PPTP and L2F is that the L2F tunneling is not dependent on IP, enabling it to work with other physical media.

There are two levels of authentication of the user. Level one is by the ISP prior to setting up the tunnel and level two is when the connection is set up at the corporate gateway.

L2TP defines its own tunneling protocol, based on the work of L2F. Work has continued on defining L2TP transport over a variety of packetized media such as X.25, Frame Relay, and ATM. Although many initial implementations of L2TP focus on using UDP on IP networks, it is possible to set up a L2TP system without using IP as a tunnel protocol entirely. A network using ATM or Frame Relay can also be deployed for L2TP tunnels. Figure 6–11 depicts the protocols involved in a L2TP connection between an access concentrator and a network server.

Because L2TP is a Layer 2 protocol, it offers users the same flexibility as PPTP for handling protocols other than IP, such as IPX and NetBEUI. Also, because it uses PPP for dial-up links, L2TP includes authentication mechanisms within PPP. The components of a L2TP system are essentially the same as those for PPTP: point-to-point protocols, tunnels, and authentication systems.

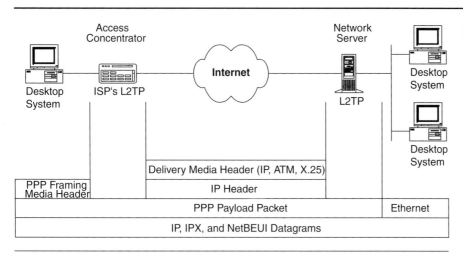

Figure 6-11
L2TP Connection Protocols

L2TP should be considered the next-generation VPN protocol, particularly for dial-in VPNs. Many vendors already have plans to supplant PPTP-based products with L2TP products. L2TP offers a number of advantages of PPTP, particularly for handling multiple sessions over a single tunnel as well as assigning QoS parameters of different tunnels to the same site. In addition, L2TP's capability to run over technologies like X.25, Frame Relay, and ATM, while handling multiple network layer protocols in addition to IP, affords users and ISPs a considerable amount of flexibility in designing VPNs. L2TP also provides stronger security for the corporate data, because it uses IPSec's Encapsulating Security Payload (ESP) for encrypting packets, even over a PPP link between the end-user and the ISP.

Layer 2 Forwarding Protocol   Layer 2 Forwarding (L2F) Protocol is a tunneling protocol developed by Cisco Systems, Inc. The key management requirements of service that are provided by Cisco's L2F implementation are as follows:

layer 2 forwarding (L2F) protocol

- Neither the remote end system nor its corporate hosts should require any special software to use this service in a secure manner.
- Authentication as provided by dial-up PPP and the various authentication protocols, as well as support for smart cards and one-time passwords; the authentication will be manageable by the user independently of the ISP.
- Addressing will be as manageable as dedicated dial-up solutions; the address will be assigned by the remote user's respective corporation, and not by the ISP
- Authorization will be managed by the corporation's remote users, as it would in a direct dial-up solution.
- Accounting will be performed by both the ISP and the user.

Figure 6–12
PPP Connection for Virtual
Dial-up Topology

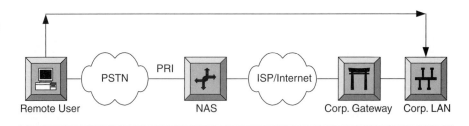

These requirements are primarily achieved based on the functionality provided by tunneling the remote user directly to the corporate location using the L2F protocol. In the case of PPP, all link control protocol and network control protocol negotiations take place at the remote user's corporate location. PPP is allowed to flow from the remote user and terminate at the corporate gateway. Figure 6–12 illustrates this process.

IP Security   Internet Protocol Security (IPSec) is a standard created to add security to TCP/IP networking. It is a collection of security measures that address data privacy, integrity, authentication, and key management, in addition to tunneling. The IPSec system includes a considerable amount of flexibility in authentication and encryption algorithms, allowing it to meet the demands of both current and future networking situations.

The original TCP/IP did not include any inherent security features. To address the issue of providing packet level security in IP, the IETF has been working on the IPSec protocols within their IP Security Working Group. The first protocols comprising IPSec, for authenticating and encrypting IP datagrams, were published by the IETF as RFCs 1825 to 1829 in 1995.

These protocols set out the basics of the IPSec architecture, which includes two different headers designed for IP packets. The IP Authentication Header (AH) is for authentication and the other, the Encapsulating Security Payload (ESP), is for encryption purposes. In addition, Security Association (SA), which is a security agreement on key management, is an important concept in IPSec.

The AH and ESP protocols can be applied either to authenticate and/or decrypt just the packet's payload or the entire IP header, including the IP addresses of the source and destination. The greatest degree of security is provided by applying authentication and encryption in tunnel mode.

To enable secure communications between two parties, a system for exchanging keys is required. An IPSec Security Association is created between VPN sites to exchange keys and any pertinent details on the cryptographic algorithms that will be used for a session. Although manual exchanges of security associations and keys are possible for a small number of VPN sites, IPSec includes an involved, but workable framework for automatic key management called Internet Key Exchange (IKE) or ISAKMP/Oakley.

IPSec software can reside in stationary hosts, mobile clients, or security gateways. Only security gateways are needed if LAN-to-LAN tunnels are to be created. Mobile workers would require IPSec client software if they wanted to connect to a VPN site.

6.7 VPN IMPLEMENTATION

Carrier Implementation

Carriers implement many VPNs within a single physical network. A corporate-wide ATM WAN provided over a public ATM service network is one example of a VPN. Shared network resources are assigned in fair proportion to the bandwidth required by subscribers, and users receive specific QoS guarantees.

In a VPN, a single access circuit from the site to the network is usually sufficient, because multiple virtual circuits and paths can be provided via multiple users from a site to their destination. For example, each virtual circuit or path can be allocated a peak rate equal to the access circuit, but have a sum of average rates that is less than the access circuit.

Figure 6–13 illustrates this concept by showing how users A, B, and C at location one all have a single physical circuit into their premise ATM

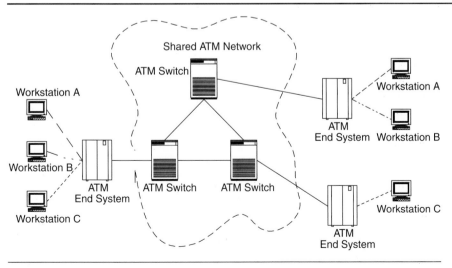

Figure 6–13
Virtual Circuits over a Shared ATM Network

device. This ATM device converts these inputs to three ATM virtual circuits and then transmits them over a single physical link between network ATM switches to the destination-based ATM switch (ATM End System). These individual user virtual circuits are logically switched across the ATM network to the destination premises device, where they are delivered to the physical access circuit of the end user. Note that while this single circuit provides good circuit aggregation, it can also be a single point of failure for all users accessing the network, so alternative facilities may be desirable for backup.

The availability of a public switched ATM service, with its bandwidth-on-demand sharing capabilities, will likely lure users away from private point-to-point networks and onto public networks. They attain the increased reliability of very large public network platforms and cost savings versus private line operations as a result of the economies of scale of the larger backbone infrastructure with its high availability and resiliency to failure through built-in alternate routing.

Virtual Dial-up Service

The major elements of a virtual dial-up service include authentication and security, authorization, address allocation, and accounting. For the virtual dial-up service, the ISP pursues authentication to the extent required to discover the user's apparent identity and the desired corporate gateway. No password interaction is performed at this point. As soon as the corporate gateway is determined, a connection is initiated with the authentication information gathered by the ISP. The corporate gateway completes the authentication by either accepting or rejecting the connection.

In a virtual dial-up service, the burden of providing detailed authorization based on policy statements is given directly to the remote user's corporation. By allowing end-to-end connectivity between remote users and their corporate gateway, all authorization can be performed as if the remote users are dialed into the corporate location directly.

For the virtual dial-up service, the corporate gateway can exist behind the corporate firewall and allocate addresses that are internal.

Because virtual dial-up is an access service, accounting for connection attempts is of significant interest. The corporate gateway can reject new connections based on the authentication information gathered by the ISP, with corresponding logging. Because the corporate gateway can decline a connection based on the ISP information, accounting can easily draw a distinction between a series of failed connection attempts and of brief successful connections.

Dial-up Components

network access server
(NAS)

A dial-up service can be implemented to access the VPN infrastructure. The four components of this scenario are the remote user, the **Network Access Server (NAS),** the Internet Service Provider (ISP), and the corporate gateway.

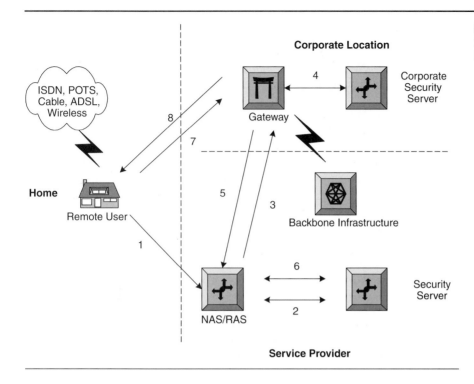

Figure 6–14
Dial-Up VPN
Communications

Remote users access the corporate LAN as if they were dialed directly into the corporate gateway, although their physical dial-up is through the ISP NAS. Figure 6–14 provides a step-by-step illustration of this process.

To illustrate how the virtual dial-up works, the following eight steps describe what might happen when a remote user initiates access.

1. Remote user initiates a PPP connection and the NAS accepts the call.
2. The NAS identifies the remote user.
3. The NAS initiates a L2F tunnel to the desired corporate gateway.
4. The corporate gateway authenticates the remote user and accepts or declines the tunnel.
5. The corporate gateway confirms acceptance of the call and the L2F tunnel.
6. The NAS logs the acceptance and traffic (optional).
7. The corporate gateway exchanges PPP negotiations with the remote user. The IP address can be assigned by the corporate gateway at this point.
8. End-to-end data is tunneled from the remote user to the corporate gateway.

6.8 VPN STANDARDS

Currently there are no standards outlining the software and hardware components on a VPN. Every vendor that provides a VPN service performs best when supported by its own hardware platforms and software applications.

Microsoft, 3Com, and several other companies have proposed the standard protocol, PPTP, and Microsoft has built the protocol into its Windows NT server. VPN software such as Microsoft's PPTP support, as well as security software, would usually be installed on a company's firewall server. Cisco utilizes Network Access Servers (NASs) and routers with its Internetwork Operating System (IOS) software to provide a virtual dial-up functionality. This functionality is based on the L2F Protocol IETF draft RFC. The PPTP is the most recent of the remote access protocols defined by the IETF in RFC 1171. PPTP provides on-demand, multiprotocol support for VPNs. PPTP is an extension of PPP that requires two connections to be made. A number of IETF working groups are currently addressing issues relating to VPNs. These efforts can be viewed on the IETF Website at www.ietf.org/html.charters. Appendix C also contains a listing of RFCs that are related to the VPN environment.

6.9 VPN APPLICATIONS

Virtual networks are economical for large companies that have a considerable amount of on-net calling. Most of the features of a dedicated private network can be provided. Locations can call each other with an abbreviated dialing plan, calls can be restricted from selected areas or country codes, and other dialing privileges can be applied based on trunk group or location. Special billing arrangements are provided. Call detail furnished online or on magnetic media enables the company to analyze long-distance costs.

VPN applications can be considered for the following situations:

- Long-distance usage in excess of 200,000 minutes a month
- Multiple locations, at least one of which is large enough to justify one or more T-1 access lines
- The needs for tie line services, such as extension number dialing, without the volume to justify a dedicated tie line network
- Remote users such as Clients who are dialing ISDN/PSTN from either home or a remote location
- Internet Service Providers (ISP) access
- Corporate Gateway that provides services to remote user
- Mobile workers that utilize dial-in connections to ISPs

6.10   VPN ISSUES AND CONSIDERATIONS

Although there is proven demand for Internet-based VPNs, the market is still in its relative infancy, as protocols and devices continue towards standardization. Internet VPNs offer more flexibility and scalability than other alternatives. There is also a reduction in the requirements for both technical support and communications equipment (VPN Consortium, www.vpnc.org/, 2000).

To summarize, a virtual network should be considered by any company that has the following characteristics:

- Long-distance usage in excess of 200,000 minutes a month
- Multiple locations, at least one of which is large enough to justify one or more T-1 access lines
- The need for tie line services, such as extension number dialing, without the volume to justify a dedicated tie-line network.

Corporations have concerns when utilizing the Internet to conduct business. These include the following issues that can impact both the design and deployment of an Internet VPN.

Reliability	Bottlenecks
Congestion	IP address issues
Throughput	Performance
Security	Multiprotocol support
Interoperability	Integrated Solutions

Security Issues

Security problems can be solved when using a VPN. Securing tunnels for private communications between corporate sites will do little if employee passwords are openly available or if other holes are in the security of the network. VPN-related security management—of keys and user rights—must be integrated into the rest of the organization's security policies. As noted earlier, several areas of security that are implemented in VPNs are authentication, encryption, integrity, and content filtering.

Authentication   Before users can connect to a site, they must first authenticate themselves; that is, they must be able to prove they are who they say they are. Passwords offer only weak forms of authentication. VPNs use stronger forms.

Encryption   Before IP packets are sent, they are first encrypted so that if someone intercepts an IP packet, that person will not be able to read its contents.

Integrity   The danger exists that someone will change IP packets in transit. Integrity checks ensure that packets have not been destroyed, changed, or reordered in their passage from the sender to the receiver.

Content Filtering   Some companies check transmission for improper content such as viruses.

6.11   VPN ADVANTAGES AND DISADVANTAGES

Advantages

One advantage of a private network is the reduction in long-distance charges that result from bypassing local carrier access charges for both the originating and terminating end of a call.

Virtual networks are economical for large corporations that have a considerable amount of on-net calling. Most of the features of the dedicated private network can be provided. Special billing arrangements can be provided, and call detail can be furnished either on-line or on CD-ROM.

Because point-to-point links are not part of the Internet VPN, a company does not have to support the same media and speeds at each site, which reduces equipment and support costs.

Dial-up PPTP use produces two advantages:

- It is not necessary to install several modems on a Remote Access Server (RAS) so that the users can dial directly into the server; instead, they can dial into any ISP. This saves money on both hardware and systems management.
- Remote users can usually access the RAS server by making only a phone call, no matter where in the world they might be located. The user needs only local access to an ISP. Distance from the RAS server, therefore, does not matter, which saves money on long-distance telephone charges.

VPN extends the reach of private networks across public ones, both easily and transparently. Today, VPN is used for more on-demand, dial-up connections, but increasingly, dedicated PPTP connections are also being used to connect LANs across the Internet.

VPNs provide a considerable advantage over fully meshed private point-to-point systems. The use of a VPN normally requires one dedicated local loop to the VPN switch. Private line networks, however, require one circuit for each switch in the meshed network. Figure 6–15

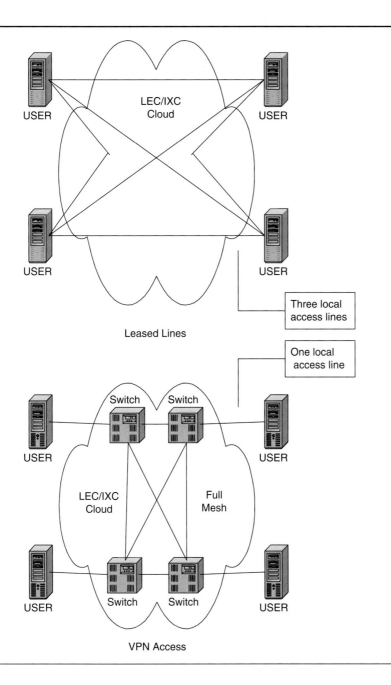

Figure 6–15
Leased Lines vs VPN
Access

LEC/IXC
Cloud

USER

USER

USER

USER

Three local
access lines

Leased Lines

One local
access line

Switch   Switch

USER

USER

LEC/IXC
Cloud

Full
Mesh

USER

USER

Switch   Switch

VPN Access

depicts a network with four sites that are connected by a fully meshed private line access and the same network that is served by a VPN. Note the difference in requirements of connections to the end locations. A leased line configuration requires three local access lines at each location and the VPN access requires only one local access facility.

An Internet VPN supports secure connections for mobile workers by virtue of the numerous dial-in connections that ISPs typically offer their clients.

The advantages of public VPN ATM network services are:

- Reduced access line charges
- Capability to satisfy high-peak bandwidth demands
- Cost impacts proportional to usage
- Enhanced availability and reliability, fault tolerance, security
- Access to public network engineers and support
- Provider network management, intelligent routing, and bridging
- Bandwidth management
- Integration and standardization
- Order entry and processing

Disadvantages

Using VPNs involves some important disadvantages as well. VPNs have not solved the issues of reliability, congestion, and throughput, which are all problems for corporate networks. Internet VPNs, specifically, still have problems with guaranteed performance and security. There is currently a restriction on the amount of media supported on VPN networks. Encryption is fundamental to maintaining privacy and integrity on the VPN. The management of secret keys can be problematic as the number of corresponding parties grows. Someone in the organization must also manage the infrastructure for distributing digital certificates. Disadvantages of utilizing VPN over ATM network services include less user control and less predictable peak capacity.

6.12   VPN TECHNOLOGY ALTERNATIVES

Competition for a VPN is primarily from leased or private line services. Most of the competition in the VPN arena revolves around the numerous products and services that are being offered.

VPN Product and Service Providers

There are a number of commercial VPN providers for both products and services. Suppliers are constantly entering this market. Trade magazines and periodicals are an excellent source for current participants in the private network field. A partial list identified during this research included the following companies and their respective Websites:

ANS	www.ans.net/
AT&T WorldNet VPN Services	www.att.com/worldnet/
Concentric Network	www.concentric.com/index.shtml
GRIC Communications	www.gric.com/
GTE Internetworking	www.gte.net
IPass, Inc.	www.ipass.com
Netcom	www.netcom.com/
NetworkMCI	www.networkmci.com/
TCG CERFnet	www.cerfnet.net
UUNET	www.uunet.net/

There are also a number of vendors that offer hardware and software products and VPN solutions. Online systems such as DataPro and Black Box Corp. are examples of information sources that are available for VPN product research. A partial list identified during the research included the following companies and their respective Websites:

3Com	www.3com.com
Ascend Communications, Inc.	www.ascend.com/
Cisco Systems, Inc.	www.cisco.com/
Intel Corp.	www.intel.com/network/
IBM	www.software.ibm.com/ enetwork/technology/
Lucent	www.lucent.com/ins
Microsoft Corp.	www.microsoft.com/
NEC Systems Laboratory, Inc.	www.socks5.nec.com
Nortel Networks	www.nortelnetworks.com/ products/
Novell Inc.	www.novell.com/ bordermanager/
Storage Technology Corp.	www.network.com/
VPNet Technologies, Inc.	www.vpnet.com/

■ SUMMARY

Virtual Private Network (VPN) technology allows a company with multiple sites to have a private network, but use a public network as a carrier. Although the company can use the public network as a link between its sites, VPN technology restricts traffic to packets that can travel only between the company's sites. Even if an outsider accidentally receives a copy of a packet, VPN security technology prevents the understanding of its contents.

To build a VPN, a company buys a special hardware and software system for each of its sites. The system is placed between the company's private, internal networks and the public network. Each of the systems must be configured with the addresses of the company's other VPN systems. The software will then exchange packets only with the VPN systems at the company's other sites. To guarantee privacy, VPN encrypts each packet before transmission.

Three types of firewalls are utilized in a VPN. These include packet filters, proxies, and stateful inspection systems. Each of these systems differ in the security they provide, their difficulty in configuration, and their performance.

In addition to configuring the VPN system at each site, a network manager must also configure routing at the site. When a computer at one site sends a packet to a computer at another site, the packet is routed to the local VPN system. The VPN system examines the destination, encrypts the packet, and sends the result across the public network to the VPN system at the destination site. When the packet arrives, the receiving VPN system verifies that it came from a valid peer, decrypts the contents, and forwards the packet to its destination.

Basically, a VPN combines the advantages of private and public networks by allowing a company with multiple sites to have the illusion of a completely private network and by using a public network to carry traffic between sites.

Key Terms

Access VPN

Authentication

Authentication Header (AH)

Compulsory Tunnel

Encapsulating Security Payload (ESP)

Encryption

Extranet VPN

Firewalls

Intranet VPN

IP Security (IPSec)

Layer 2 Forwarding Protocol (L2F)

Network Access Server (NAS)

Point-to-Point Protocol (PPP)

Point-to-Point Tunneling Protocol (PPTP)

Proxy Server

Remote Access Service (RAS)

Router

Service Control Point (SCP)

Service Switching Point (SSP)

Serial Line Internet Protocol (SLIP)

Tunneling

Virtual Private Network (VPN)

Voluntary Tunnel

1. Provide a general overview of a VPN.

2. Name several vendors of VPNs. Why would a company want to implement a VPN?

3. What are the two fundamental components of a VPN?

4. What are two advantages of dial-up VPNs accessing Remote Access Servers that are used to connect to ISPs?

5. A VPN handles calls in three manners. Identify them and give a brief description of each.

6. What are the three types of VPNs? How are they utilized?

7. Two fundamental concepts comprise the VPN architecture. Provide a brief description of each.

8. Define tunneling. How is it used in the VPN environment?

9. What is the difference between voluntary and compulsory tunnels?

10. What are the three types of protocols utilized in tunneling? How are they utilized?

11. Draw a simple sketch of tunnels in the Internet environment.

12. What is the encapsulation process? What is the format of an IP packet that has been encapsulated?

13. What are the four security functions that are necessary for securing data across the VPN?

14. Where in the OSI Model is security implemented?

15. How would an Internet VPN be implemented? What part would an ISP and POP play in this connectivity?

16. Provide several reasons why anyone would want to implement an Internet VPN.

17. What are the hardware and software components that might be found in a VPN?

18. What is a firewall? What are the three main classes of firewalls? How are they different?

19. What is a router? Give a general description of the part that a router plays in the VPN environment.

20. How does the functionality of a router differ from a firewall? What is the difference between a firewall and a gateway server?

21. Why would an encrypting router be beneficial in the VPN environment?

22. Four protocols were originally suggested for the use in VPNs. List them and give a brief description of each.

23. Describe the Point-to-Point Tunneling protocol (PPTP). What is its purpose? Contrast PPP with PPTP.

24. Describe the Layer 2 Tunneling Protocol (L2TP).

25. What is the difference between PPTP and L2F protocols?

26. Describe the L2F protocol.

27. Describe the IPSec protocol.

28. How are VPNs implemented in the carrier environment?

29. Describe the elements of a virtual dial-up service.

30. What are the components of a dial-up service?

31. Describe the process for initiating dial-up access.

32. Discuss the current standards for VPNs. What is the significance of RFC 1171?

33. Provide a list of VPN applications. Give an example for each application.

34. Installing a VPN in the organization involves a number of issues. Discuss several in the context of a particular industry.

35. What is meant by authentication? What is the difference between authentication and authorization?

36. Where would a proxy server be deployed? What is its purpose? Where is a network access server utilized in a VPN?

37. Where does an ISP fit in the VPN environment?

38. What concerns do organizations have about using the Internet to conduct business?

39. What does the function of encryption accomplish?

40. Compare the advantages and disadvantages of utilizing VPNs.

ACTIVITIES

Each of these research projects should require a paper and/or an oral presentation.

1. Utilize the Web and online resources to research the VPN topic. Identify the applications that can benefit from the implementation of a VPN. Utilize the brainstorming approach and have access to a whiteboard or flipchart for their suggestions. These applications should be rank-ordered according to their contributions to the corporate goals specified and also to the difficulty and cost of implementation.

2. Research the network management aspects of virtual networks. Look at encryption methods and the security issues associated with implementing a VPN.

3. Research the encapsulation and tunneling process of a VPN. Show how they work together to make the VPN secure. Produce a drawing that utilizes tunneling.

4. Research the middle-ware software that is utilized in the VPN environment. Provide a matrix of features and functionality.

5. Develop a matrix of the various tunneling protocols showing advantages and disadvantages of each.

6. Research the VPN environment. Where are VPNs being implemented? What have been the difficulties associated with these implementations?

7. Research the various security and encryption methods available for public implementation.

CASE STUDY/PROJECT

The first two case studies involve the conversion of a private line network WAN configuration into a VPN that utilizes a public ATM backbone. The third case study converts the private line network into a Frame Relay network that provides transmission capabilities over the ATM VPN. Try using a graphics package such as VISIO 2000 to depict both of these configurations.

Remember that a VPN is defined as a partition of a shared public network which is utilized by multiple users but appears to be a private network to the users. A corporate-wide WAN over public ATM infrastructure facilities is an example of a VPN. Shared resources are assigned as the need dictates. In a VPN, a single access circuit from the customer location to the network is usually sufficient.

Case Study 1: Private Line Network

The initial private line network included circuits from the host site to a number of remote locations that contained server devices and the associated LANs. Each of these private line circuits required a data service unit at each termination. Because private lines are distance sensitive, the monthly cost to the service providers was considerable.

The details for the WAN are as follows:

Headquarters host location:
- Ten 56 kbps private lines with 56 kbps DSUs
- IBM host computer system with a FEP and a gateway.

Remote locations:
- There are ten remote locations in the southeast United States, each in a different LATA.
- Each remote location utilizes a 56 kbps DSU to access the host location.
- There is a server at each location that is connected to a Token Ring LAN.
- There are a number of workstations and printers at each location.

Requirements:
- Prepare a graphic depicting this private line WAN network.

Case Study 2: Private Line Network over an ATM VPN

Requirements:
- Utilize the design and graphic that was developed in the case study 1. Upgrade this design from a private line WAN network to an ATM VPN.
- Add the shared ATM infrastructure to the design.
- Add the ATM tunneling function to the design.
- Add the ATM end system capability at the host and remote locations.
- Replace the private lines with virtual circuits.

Case Study 3: Frame Relay Network over an ATM VPN

- Utilize the information on Frame Relay networks from Chapter 3 to convert the virtual circuits to access a premises Frame Relay switch. Utilize the graphic created in Case Study 1 and 2.
- Provide the Frame Relay requirements.
- Provide the information that is necessary to program the routers/gateway.

7

Fiber Distributed Data Interface

■ INTRODUCTION

The **Fiber Distributed Data Interface (FDDI)** standard, like 802.5 Token Ring Networks, uses a ring topology at the physical layer, token passing at the **Media Access Control (MAC)** layer, and 802.2 at the **Logical Link Control (LLC)** layer. FDDI was derived from the 802.5 technology; however, it is compatible due to the extensive modifications developed between the two technologies. FDDI was designed to run at 100 Mbps, whereas Token Ring is designed to run at 4 bps or 16 Mbps.

FDDI networks are wired as a physical ring, connected by a number of **concentrators** that serve as central connection points for user access. FDDI uses two physical rings operating in different directions to avoid cable problems. Figure 7–1 illustrates this counterrotational 100 Mbps fiber ring. This allows for cable-break resolutions to be accomplished by beaconing and network reconfiguration.

FDDI supplies not only a great deal of bandwidth, but also a high degree of reliability and security while adhering to standards-based protocols not associated with a particular vendor. Much of FDDI's reliability comes from the fiber itself, which is immune to both Electromagnetic Interference (EMI) and Radio Frequency Interference (RFI).

FDDI was designed to be a MAN technology that could span distances typically found in urban areas. A FDDI ring has a maximum circumference of 200 km. Although the MAN market did not emerge as a major force in networking, the wide distance span of FDDI made it perfect as a backbone network within large site networks to link individual LANs. Two versions of the FDDI technology currently exist: FDDI-I and FDDI-II.

OBJECTIVES
Material included in this chapter should enable you to:

- understand FDDI technology and identify where it fits in the current broadband networking environment.

- become familiar with the terms and definitions that are relevant in the Enterprise Networking environment.

- understand the function and interaction of FDDI in the Enterprise Network environment. Look at the dual counterrotating ring topology and see how it interfaces with the various networks.

- understand where FDDI fits in the different applications of the technology.

- identify components that make up the FDDI networking environments.

fiber distributed data interface (FDDI)
media access control (MAC)
logical link control (LLC)
concentrators

Figure 7–1
FDDI Counterrotating
100 Mbps Fiber Ring

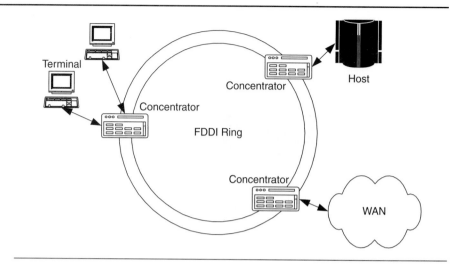

- become familiar with the contents of the various frame formats and learn how they are used in the FDDI technology.

- look at the various standards associated with the FDDI technology. See how they are compatible with the other broadband technologies.

Table 7–1
FDDI Parameters and
Values

FDDI Parameter	Value
Optical Source	Infrared LED or Laser
Wavelength	1,300 nm
Energy Requirements	−20 db minimum
Fiber Types	62.5/125 nm in multi-mode
Maximum Nodes	1,000
Maximum Total Length	100 km
Distance Between Stations	2 km—multi-mode/40 km single mode
Power Budget Between Nodes	11

7.1 FDDI TECHNOLOGY

copper distributed data
interface (CDDI)
token

The FDDI standard defines a ring-structured network that uses a token-passing form of medium access control. The FDDI standard specifies the use of full-duplex, point-to-point, fiber-optic physical links to interconnect DTE devices. A variation of the FDDI standard named the **Copper Distributed Data Interface (CDDI)** has been adopted for twisted-pair copper wire. A FDDI topology operates at a data rate of 100 Mbps. A special data unit called the **token** circulates around the ring, and a station can transmit frames only when it has possession of the token. Table 7–1 provides details on the values associated with the various FDDI parameters.

Table 7-2
FDDI Device Types

Device Type	Definition	Connectivity
Dual Attachment Station (DAS)	Has two pairs of PHY and PMD entities and one or more MAC entities; participates in the trunk dual ring	DAS, DAC
Dual Attachment Concentrator (DAC)	A DAS with additional PHY and PMD entities beyond those required for attachment to the dual ring; the additional entities permit attachment of additional stations that are logically part of the ring, but are physically isolated from the trunk ring	DAS, DAC, SAC, SAS
Single Attachment Station (SAS)	Has one each PHY, PMD, and MAC entities, and therefore cannot be attached into the trunk ring, but must be attached by a concentrator	DAC, SAC
Single Attachment Concentrator (SAC)	A SAS with additional PHY and PMD entities beyond those required for attachment to a concentrator; the additional entities permit attachment of additional stations in a tree-structured fashion	DAC, SAC, SAS

7.2 FDDI TOPOLOGY

In FDDI terminology, a station is an addressable network component capable of generating and receiving frames. Each instance of a **Physical Layer (PHY)** Protocol sublayer entity and a **Physical Layer Medium Dependent (PMD)** sublayer entity within a station is called a port. PHY and PMD will be explained in detail later in the chapter. A station can implement one or more ports. Each port is attached to the transmission medium through a **Medium Interface Connector (MIC).**

The FDDI standard specifies two classes of stations. The **Dual Attachment Station (DAS)** connects to both rings, and the **Single Attachment Station (SAS)** connects only to the primary ring through a wiring concentrator. In addition, concentrators, which are FDDI devices that can support non-FDDI stations, can be **Single (SAC)** or **Dual Attachment (DAC)**. Table 7-2 shows the types of devices that can exist on the FDDI ring. The dual ring appearance of DAS permits double the throughput, as the station can send on both rings simultaneously although the second ring is reserved for backup.

Figure 7-2 shows the basic connectivity for single attachment stations, dual attachment stations, and dual attachment concentrators. A dual attached station is connected to both rings and a single attached station is connected to only one ring.

physical layer (PHY)
physical layer medium dependent (PMD)

medium interface connector (MIC)
dual attachment station (DAS)
single attachment station (SAS)
single attachment concentrators (SAC)
dual attachment concentrators (DAC)

Figure 7–2
FDDI Basic Components

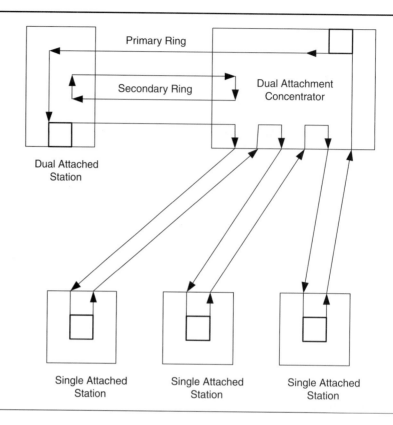

Another configuration, called dual homing, can provide additional redundancy and help guarantee operation of the network. If the application is of a critical nature, some customer devices such as routers, gateways, or mainframe hosts may need a fault-tolerant capability, In this arrangement, the devices are connected to two different concentrators on the FDDI ring. Figure 7–3 shows this connectivity.

Three types of networks are supported by FDDI: backbone networks, front-end networks, and back-end networks. Following is a brief explanation of each.

- Backbone networks—These are connections between the main nodes on the network. Backbone local networks are used to provide a high-capacity LAN data link that can be used to interconnect other, lower-capacity LANs.
- Front-end networks—These are networks connecting workstations through a concentrator to a host computer over the FDDI network. The need for high-speed local networks has arisen from the increased use of image and graphics processing devices in the desktop computing environment.

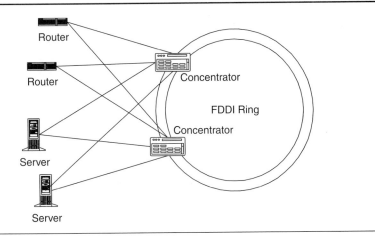

Figure 7–3
Dual Attached/Dual Homed
Connectivity

- Back-end network—These allow connections from a host computer to peripherals such as high-speed disks and printers to be connected over FDDI to replace parallel bus connections. A back-end local network typically has a few devices to be connected, which will be close together.

FDDI uses the token-passing channel access method while using dual counterrotating rings for redundancy. FDDI transmits at 100 Mbps and can include up to 1,000 nodes over a distance of 200 km (120 miles). The maximum cable segment allowed without repeaters is 2 km. The 200 km distance can be attained by connecting 100 such segments. Like Token Ring, FDDI uses token passing; however, a FDDI network is wired as a physical ring and not as a star. It does not utilize hubs: rather, devices are generally connected directly to each other utilizing a concentrator device.

Since FDDI is organized as a token passing ring, it allows for every advantage of Token Ring such as increased throughput saturation between 90 and 95 percent. FDDI does not, however, require stations to relinquish the token after sending a frame. Thus, an entire transmission package, such as a large file, can be sent without waiting for other stations. This way, data can be transmitted more quickly around the network; and, after the computer has finished sending its data, it can immediately pass the token along. It does not need to wait for confirmation of receipt.

Unlike Token Ring, FDDI supports the capability to assign a priority level to a particular station or type of data. A DTE device, such as a server, can be given a higher priority than other computers, which means that video or other time-sensitive data can be given a higher priority than normal data traffic.

Table 7–3
Comparison of FDDI and
Token Ring

Feature	FDDI	Token Ring
Transmission Medium	Optical fiber Shielded twisted pair (STP) Unshielded twisted pair (UTP)	Shielded twisted pair (STP) Unshielded twisted pair (UTP)
Data Rate	100 Mbps	4 or 16 Mbps
Signaling Rate	125 Mbaud	8 or 32 Mbaud
Signal Encoding	4B/5B (optical fiber) MLT (twisted pair)	Differential Manchester
Maximum Frame Size	4,500 bytes	4,500 bytes (4 Mbps) 18,000 bytes (16 Mbps)
Clocking	Distributed	Centralized
Token Release	Release after transmit	Release after receive
Capacity Allocation	Timed token rotation (TTR)	Priority and reservation bits
Reliability Specification	Yes	No

FDDI operates much like 802.5 Token Ring with the following important exceptions:

- Multiple frames can circulate simultaneously on the FDDI network
- FDDI has a dual-ring configuration, where Token Ring does not.
- FDDI supports a 5-bit encoding system, whereas Token Ring uses Manchester coding.
- The FDDI protocol includes a reliability specification whereas Token Ring does not.
- Clocking on Token Ring is centralized, while in FDDI each station has a stable autonomous clock that synchronizes to incoming data.

Table 7–3 summarizes the comparisons between FDDI and Token Ring.

As mentioned, FDDI consists of two physical rings operating in different directions. More than one advantage is gained by ring redundancy. If the primary ring were to develop a fault, transmission would switch to the secondary ring and the transmission would continue to the destination in the opposite direction. If the facility were to develop a fault in both the primary and secondary rings, the nodes adjacent to the break would immediately begin receiving on their primary rings and sending on their secondary rings. In this manner, the integrity of the ring is preserved. It is also possible for both rings to be used simultaneously. In this mode, the transmission speed increases to 200 Mbps. Figure 7–4 provides a view of a FDDI topology connecting a host computer with multiple LANs and a gateway to the WAN. Please note that all figures in this chapter that depict fiber

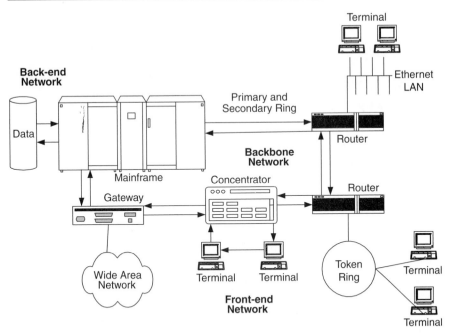

Figure 7–4
FDDI Topology

connectivity assume that there is a transmit and a receive element for each fiber segment.

7.3   FDDI PHYSICAL SPECIFICATIONS

The FDDI physical layer specification details the encoding scheme and transmission medium characteristics for both optical fiber and twisted-pair media. It also includes considerations of timing jitter and station and FDDI network configurations.

Physical implementation at each node connected to the FDDI ring is accomplished using a dual port transceiver that functions as the interface converter between the fiber light source and the electrical signals. Each transceiver contains ports for each fiber ring and the media (either fiber or coaxial cable) to the DTE devices and nodes. The transceiver performs the automatic reconnection process in the event of a fault in the fiber link. Note that coax is not needed if the transceiver is implemented on the Network Interface Card (NIC). Figure 7–5 depicts a fiber transceiver and a fiber extender. Note the transmit (TR) and receive (RX) connectors for the fiber cable.

Figure 7–5
Fiber Transceiver and
Extender (Reproduced with
permission from Black Box
Corp.)

Fiber Transceiver (ST connectors)

Fiber Extender

Table 7–4
FDDI Physical Layer
Medium

Transmission Medium	Multi-mode Optical Fiber	Twisted Pair
Data Rate (Mbps)	100	100
Signaling Technique	4B/5B NRZI	MLT-3
Maximum Number of Repeaters	100	100
Maximum Length between Repeaters	2 km	100 m

The FDDI standard specifies a ring topology operating at 100 Mbps. Two media are included (Table 7–4). The optical fiber uses the 4B/5B **Non-Return-To-Zero Alternate Mark Inversion (NRZI)** encoding scheme [Fiber Optics Association, www.std.com/fotec/foa.htm 2000]. Two twisted-pair media are specified, which include a 100-ohm Category 5 UTP and 150-ohm STP. MLT-3 encoding is used for both twisted-pair media. The wavelength specified for data transmission is 1,300 nm.

non-return-to-zero
alternate mark inversion
(NRZI)

The original specification lists the use of multi-mode fiber transmission. A mode is a ray of light that enters the fiber at a particular angle. Although today's long-distance networks primarily rely on single-mode fiber, that technology generally requires the use of lasers as the light source rather than the cheaper and less powerful light-emitting diodes (LEDs), which are adequate for FDDI requirements.

modal dispersion

Multi-mode fiber allows multiple modes of light to propagate through the fiber. Because these modes of light enter the fiber at different angles, they will arrive at the end of the fiber at different times. This characteristic is known as **modal dispersion**. Modal dispersion limits the bandwidth and distances that can be accomplished using multi-mode fibers.

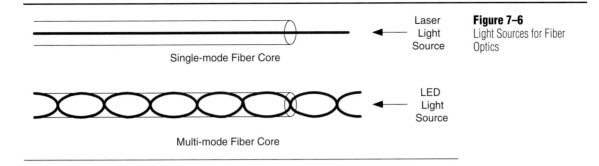

Figure 7–6
Light Sources for Fiber Optics

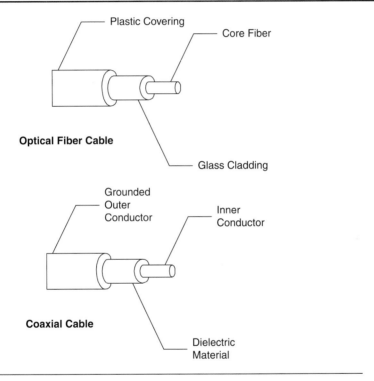

Figure 7–7
Optical Fiber and Coaxial Cable Construction

Single-mode fiber allows only one mode of light to propagate through the fiber core. Because only a single-mode is used, modal dispersion is not present with single-mode fiber. Figure 7–6 illustrates this difference.

Dimensions of the fiber cable are specified in terms of the diameter of the core of the fiber and the outer diameter of the cladding layer that surrounds the core. The combination specified in the standard is 62.5/125 nm. Smaller diameters offer higher potential bandwidths, but also higher connector loss. Figure 7–7 shows the construction of both

Figure 7–8
Fiber Optic and Coaxial
Cable (Reproduced with
permission from Black Box
Corp.)

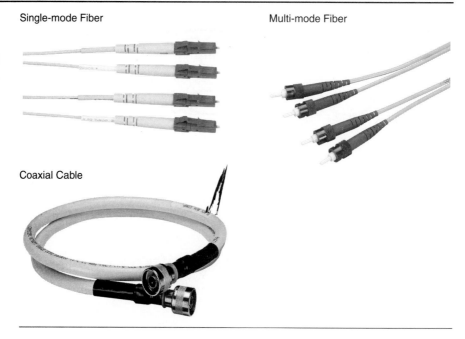

Single-mode Fiber

Multi-mode Fiber

Coaxial Cable

coaxial cable and optical fiber media. The material composition of the two layers can be any of the following:

- Glass cladding and glass core
- Plastic cladding and glass core
- Plastic cladding and plastic core

Because single-mode fiber allows only one mode of light to propagate through the fiber, modal dispersion is not present. The core of the fiber is manufactured substantially smaller relative to multi-mode fiber. In addition, the index of refraction of the core is further reduced, thus increasing the critical angle or decreasing the angle at which the light ray must penetrate the core with respect to its central axis. The intent is to permit light rays to propagate in one mode only: through the central axis of the fiber. Examples of single-mode and multi-mode optical fiber are shown in Figure 7–8.

Recently, an additional fiber medium specification has been added, as described in Table 7–5. The single-mode fiber specification can be used to configure much longer links between repeaters. The low-cost fiber specification provides lower-cost optical fiber connections for lengths up to 500 m. The main savings are achieved by relaxing some of the specifications for the optical transceivers.

Specification	Multi-mode Fiber	Single-mode Fiber	Low-cost Fiber
Light Source	LED	Laser	LED
Wavelength	1,300 nm	1,300 nm	1,300 nm
Cable Size (core/cladding)	2 km	40–60 km	500 m
Maximum distance Between Repeaters	62.5/125 nm	8–10/125 nm	62.5125 nm

Table 7–5
FDDI Optical Fiber Medium Alternatives

Port Type	Definition	Device Type
A	Connects to the incoming primary ring and the outgoing secondary ring of a dual ring	DAS, DAC
B	Connects to the outgoing primary ring and the incoming secondary ring of a dual ring	DAS, DAC
M (master)	Connects a concentrator to a SAS, DAS, or another concentrator	DAC, SAC
S (slave)	Connects a SAS or SAC to a concentrator	SAS, SAC

Table 7–6
FDDI Port Types

7.4 FDDI PORT TYPES

The FDDI standard specifies connection rules to ensure against the construction of illegal topologies. These rules are expressed in terms of allowable connections between port types. Port types include A ports, B ports, M ports, and S ports. The A port connects to the incoming primary ring and the outgoing secondary ring of a dual ring. The B port connects to the outgoing primary ring and the incoming secondary ring of a dual ring. The M(aster) port connects a concentrator to a SAS, DAS, or another concentrator. The S(lave) port connects a concentrator to a single attached station, dual attached station, or another concentrator. Table 7–5 summarizes the FDDI port types.

In addition to the connections allowed in Table 7–6, configurations include dual homing and optical bypass.

As discussed, the common implementation of FDDI includes DACs and DASs in the dual ring. SACs and SASs exist in the tree elements below the dual ring. This arrangement provides for redundancy for the dual-ring stations only. The FDDI standard allows redundant

Figure 7–9
FDDI DAS Ring Attachment

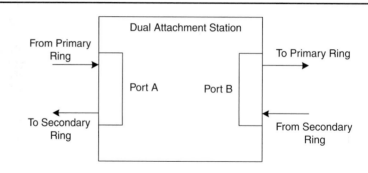

paths in tree topologies and in the dual ring, by using DACs and DASs in the tree portion. This concept is called dual homing. Figure 7–9 illustrates the dual ports and dual ring connectivity of a DAS.

An additional degree of reliability can be provided by the use of an optical bypass switch, which can be installed in any DAC or DAS. The switch bypasses the station's receiver and transmitter connections so that the optical signal from the preceding station is passed directly to the next station.

7.5 RING CONNECTIVITY

Two addressing modes are allowed in the FDDI network. One mode uses a 15-bit address and the other uses a 46-bit address. The FDDI standard does not stipulate the exact format of the addresses.

FDDI employs two rings, so that if one ring malfunctions, data can reach its destination on the other ring. DTE devices are connected to the rings in three different modes: Class A, Class B, and Class C.

> **Class A**—Attachment to both rings in either redundant failure prevention mode or a bandwidth-doubling mode that uses both rings for transmission rather than redundancy. These nodes consist of network equipment such as hubs. They have the ability to reconfigure the ring architecture to use a single ring in the event of a network failure.

> **Class B**—Attachment to only one ring. These nodes connect to the FDDI network through Class A devices. They can be devices such as servers and workstations.

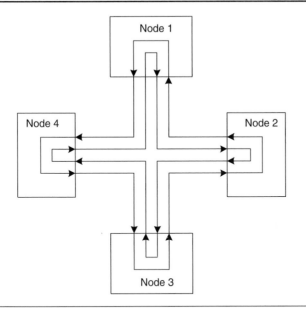

Figure 7–10
FDDI Ring with
Counterrotating Fiber Rings

Class C—Implementation of the ring within a concentrator, similar to a Token Ring Multistation Access Unit (MAU), with DTE attached in a star configuration to the concentrator. This mode provides the greatest flexibility and ease of troubleshooting.

Utilizing a Class A configuration, the following process is initiated: (1) Incoming light signals are routed through the transceiver input on the outer ring, (2) converted to electrical signals, (3) sent to the NIC for processing, (4) converted to light, and (5) sent on around the outer ring.

Simultaneously, the light signals are also directed to the inner (redundant) ring and passed on to the next node. If a fault should occur in the outer ring, the transceiver takes its light source from the inner ring. If the fault should effect both rings, the transceiver sets its input to the outer ring and the input to the inner ring. All other nodes must detect the changes and implement an appropriate process. Figures 7–10 and 7–11 depict this fault process. The ring is operating without a fault in Figure 7–10. In Figure 7–11, the fault occurs between nodes 3 and 4. Nodes 3 and 4 initiate the healing process with a wrap in the opposite direction on the ring. Remember there are separate transmit and receive elements for each fiber segment.

Figure 7–11
FDDI Ring with a WRAP

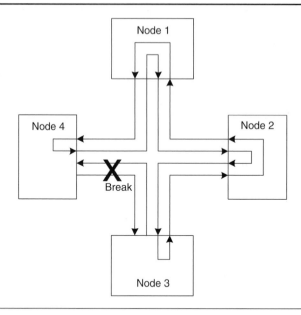

7.6   FDDI ARCHITECTURE

FDDI architecture can be divided into three areas: Protocol architecture, ring scheduling, and ring operation. The FDDI architecture consists of four operational layers that map into Layers 1 and 2 of the OSI model. These are the Physical Medium Dependent (PMD), Physical Layer Protocol (PHY), Media Access Control (MAC), and Station Management (SMT).

FDDI Protocol
The FDDI protocol is described by the following documentation components:

- Physical Medium Dependent (PMD) sublayer
- Media Interface Connector (MIC)
- Physical layer (PHY)
- Media Access Control (MAC)
- Station Management (SMT)
- Protocol Data Units (PDU)

Physical Medium Dependent Sublayer   The Physical Medium Dependent (PMD) sublayer corresponds to the OSI Model Physical Layer 1. The PMD requirements include definitions for the fibers, transceivers, optical power, and the Media Interface Connectors.

The FDDI Physical layer is divided into two sublayers—PMD, the lower sublayer, and PHY, the upper sublayer. PHY provides information on the connection between the physical and MAC sublayer of the OSI Data Link Layer 2.

Sublayer 1 (PMD) defines the fiber-optic interface. The PMD interface is responsible for defining the transmit-receive signals, providing proper power levels, and specifying cables and connections. The physical layer protocol is intended to be medium independent. It defines symbols, coding and decoding techniques, clocking requirements, the states of the lines, and data framing conventions. The PMD requirements include definitions for the fibers themselves, transceivers, optical power, and the MIC. It specifies a 62.5/125 micron multi-mode graded-index fiber operating in the 1,300 m wavelength at 100 Mbps.

Physical Layer   Sublayer 2, the Physical Layer (PHY) Protocol, specifies the type of data encoding used, how control information is generated and detected, and receiver characteristics.

Media Access Control Layer   The second layer, Media Access Control (MAC), defines the data frame format and the token passing protocol, describing how access is gained to the backbone ring. The MAC corresponds to the OIS Data Link Layer 2. FDDI frame formats are defined in the FDDI standard using the term symbol to refer to a group of 4 bits. Symbols are encoded in a way that allows both data and nondata values to be represented. This layer provides the procedures for frame formatting, error checking, token handling, and managing the data link addressing.

The FDDI MAC sublayer uses a timed Token Ring access protocol that governs the way in which a MAC sublayer entity gains access to the ring to transmit data. The basic FDDI MAC protocol is fundamentally the same as IEEE 802.5. There are, however, two key differences:

- In FDDI, a station waiting for a token seizes the token by aborting the token transmission as soon as the token frame is recognized.
- In FDDI, a station that has been transmitting data frames releases a new token as soon as it completes data frame transmission.

The protocol implemented by the FDDI MAC sublayer performs the following functions in supplying its services:

- Ring initialization
- Providing fair and deterministic access to the transmission medium
- Address recognition and address filtering
- Generation and verification of FCS fields

- Frame transmission and reception
- Frame repeating
- Removal of frames from the ring

station management
(SMT)

Station Management Layer The final layer addresses **Station Management (SMT)** and handles connection management and fault detection functions of the FDDI ring. Additional responsibilities include ring initialization, ring monitoring, and fault recovery.

Station Management provides information on the supervisory functions for station control and monitoring. It is the function of SMT to detect cable faults and redirect data as required. This layer provides procedures for managing the station attached to the FDDI and provides for node configuration, ring initialization, error statistics, error detection and recovery, and connection management. SMT is unique to FDDI and does not exist in the OSI reference model.

Connection management is concerned with the insertion of stations onto the ring and removal of stations from the ring. This involves establishing or terminating a physical link between adjacent ports and the connection of ports to MAC entities. Connection management can be considered as comprising three subcomponents: entity coordination management, physical connection management, and configuration management.

Entity coordination management is responsible for the media interface to the FDDI ring, including the coordination of the activity of all the ports and the optional optical bypass switch associated with that station.

Physical connection management provides for managing the point-to-point physical links between adjacent PHY/PMD pairs. This includes initializing the link and testing the quality of the link.

Configuration management provides for configuring PHY and MAC entities within a node. It is concerned with the internal organization of the station entities and may be thought of as controlling a configuration switch that implements the desired interconnections.

protocol data units
(PDUs)

Protocol Data Unit FDDI uses **Protocol Data Units (PDUs)** in the same way as Token Ring. The information field data packets can range from 128 bytes to 4,500 bytes, and the overall frame length is limited to a maximum of 9,000 bytes.

Table 7–7 illustrates the FDDI protocol stack. The SMT Layer consists of the MAC sublayer, PHY layer, and PMD sublayer. Optical drivers and receivers provide the interface between the PMD and the physical fiber cable.

Ring Operation

FDDI is similar to the Token Ring access method because it uses token passing for network communications. It differs from standard Token Ring in that it uses a timed token access method. FDDI ring operation includes

Model Layers	Layer Function	
Upper Layers	Upper Layer	**Table 7–7** FDDI Protocol Stack
Application through Network	Functionality	
Logical Link Control (LLC) Sublayer	LLC functions	
Media Access Control (MAC) Sublayer	Packet Interpretation	
	Token passing	
Physical Layer (PHY)	Medium dependent	
PMD Sublayer	Optical drivers/receivers	

connection establishment, ring initialization, steady-state operation, and ring maintenance. Timers are used to regulate these operations.

A FDDI ring consists of stations connected in a series by medium segments that form a closed-loop. Data are transmitted serially as a bit stream from one attached station to its downstream neighbor. Each station, in turn, regenerates and repeats each bit stream, passing it to the next station. If the node possessing the token does need to transmit, it can send as many frames as desired for a fixed amount of time, called the **Target Token Rotation Time (TTRT).** Because FDDI uses a timed token method, it is possible for several frames from several nodes to be on the network at a given time, providing high-capacity communications.

target token rotation time (TTRT)

Fundamental concepts of FDDI include the use of a **Timed Token Protocol (TTP)**. The TTP defines the rules for acquiring access to the ring. It guarantees that the token appears at a station within twice the TTRT. The MAC standard specifies the rules by which the TTRT is negotiated by the attached stations.

timed token protocol (TTP)

FDDI ring operation includes connection establishment, ring initialization, steady-state operation, and ring maintenance. Timers are used to regulate these operations. Each station on the FDDI ring uses three timers to regulate its operation. These timers are administered locally by the individual stations, and include the following:

- Token Rotation Timer—Times the duration of operations in a station.
- Token Holding Timer—Controls the length of time that a station can initiate asynchronous frames.
- Valid Transmission Timer—Times the period between valid transmission on the ring.

Once a node transmits a frame, the frame goes to the next node on the network ring. Each node determines if the frame is intended for it, and each node checks the frame for errors. When the frame arrives back at the originating node, the frame is read to determine if it was received by the target node. The frame is also checked for errors. If an error is detected, the frame is retransmitted; however, if no errors are found, then the frame is removed from the ring by the originating node.

A transmitting station is responsible for removing all frames that it sends. This process leaves frame fragments on the ring, which are removed by the next transmitting station or by the cumulative action of repeat filters in many stations.

Timing Jitter   In a ring LAN, an important concern involves the ability to keep all of the repeaters around the ring synchronized. Each repeater recovers clocking information from incoming signals by means imposed by the FDDI (4B/5B) encoding scheme. As data circulate around the ring, each repeater receives the data and recovers the clocking. This clocking recovery method results in timing errors that are called **timing jitter.**

The approach taken by FDDI for dealing with timing jitter is to use a distributed clocking scheme with elastic buffers. Each repeater uses its own autonomous clock to transmit bits from its MAC layer onto the ring. For repeating incoming data, a buffer is imposed between the receiver and the transmitter. Data are clocked into the buffer at the clock rate recovered from the incoming stream but are clocked out of the buffer at the station's own clock rate.

timing jitter

FDDI Ring Monitoring

The responsibility for monitoring the functioning of the Token Ring algorithm is distributed among all stations on the ring. Each station monitors the ring for invalid conditions requiring ring initialization. Three processes are involved in error detection and correction: claim token process, initialization process, and beacon process.

FDDI nodes monitor the network for two types of error conditions: long periods of no activity and long periods when the token is not present. In the first instance, the token is presumed to be lost; in the second instance, a node is assumed to be transmitting continuously. If either error condition is present, the node that detects the error sends a stream of specialized frames called claim frames.

Claim Process   In performing the claim token procedure, a station bids for the right to initialize the ring. The station begins the claim token procedure by issuing a continuous stream of control frames. The claim frames contain a proposed TTRT value. The first node stops transmitting, and the next node on the ring compares its proposed TTRT value with the value sent by the previous node. After the comparison, it sends the lower of the TTRT values in its claim frames to the next node. By the time the last node is reached, the smallest TTRT value has been selected. At this point the ring is initialized by transmitting the token and the new TTRT value to each node, until the last node is reached.

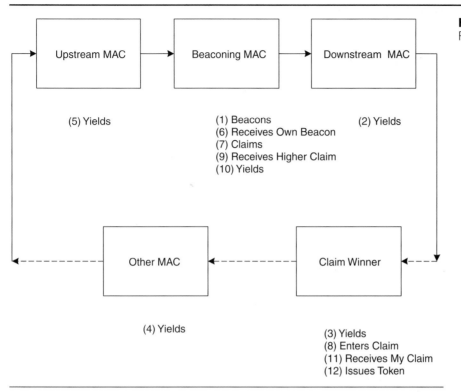

Figure 7–12
Ring Initialization

Initialization Process   The station that has won the claim token process is responsible for initializing the ring. All the stations on the ring recognize the initialization process as a result of having seen one or more claim frames. The initializing station issues a nonrestricted token. On the first circulation of the token, it may not be captured. No frames are transmitted until the token has passed once all the way around the ring. Rather, each station uses the appearance of the token for transition from an initialization state to an operational state, and to reset its Token Rotation Timer (TRT).

Beaconing Process   The **beacon** frame is used to isolate a serious ring failure such as a break in the ring. Upon entering the beacon process, a station continuously transmits beacon frames. A station always yields to a beacon frame received from an upstream station. If the logical break persists, then the beacon frames of the station that are immediately downstream from the break will normally be propagated. If a station in the beacon process receives its own beacon frames, it assumes that the ring has been restored, and it initiates the claim token process. Figure 7–12 provides a graphical overview of a normal FDDI ring initialization.

beacon

Ring Management

Ring management receives status information from media access control and from connection management. Services provided by ring management include stuck beacon detection, resolution of problems through the trace process, and detection of duplicate addresses. A beacon is a MAC frame used to isolate a serious ring failure such as a break in the ring. A stuck beacon indicates that a station is locked into sending continuous beacon frames. The trace function uses PHY signaling of symbol streams to recover from a stuck beacon condition. The result of the directed beacon is to localize the fault to the beaconing MAC and its nearest upstream neighbor. If two or more MAC entities have the same address, the ring cannot function properly. Duplicate address detection is performed during ring initialization and consists of monitoring the ring for conditions that indicate that duplicate addresses are present.

7.7 FDDI FRAME AND TOKEN FORMAT

The FDDI frame (PDU) and Token frame format is shown in Figure 7–13. The standard defines the contents of this format in terms of symbols, with each data symbol corresponding to four data bits. Symbols are used because data are encoded in 4-bit chunks at the physical layer.

Figure 7–13
FDDI Frame and Token Format

(a) FDDI MAC Frame

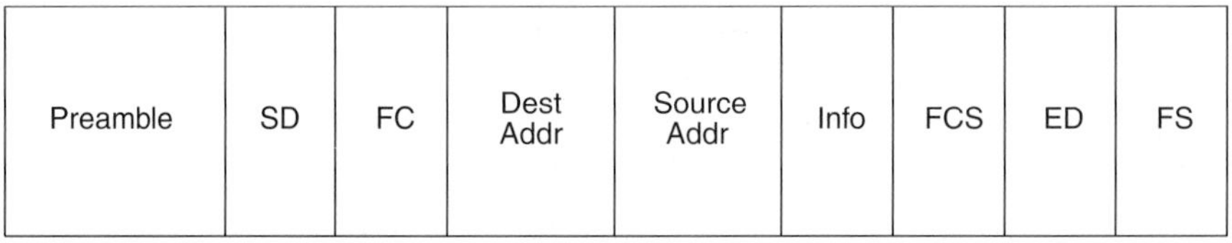

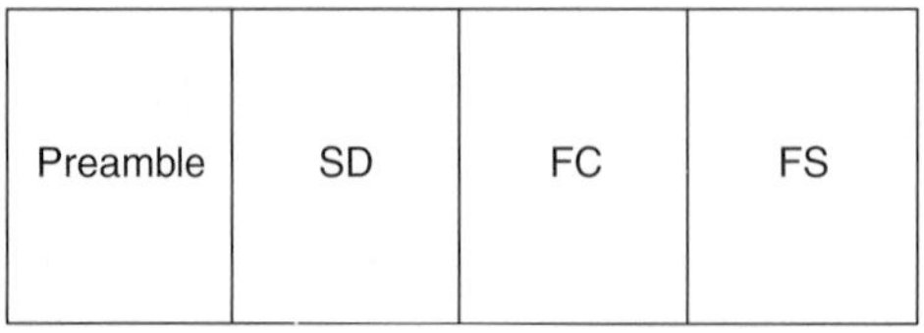

| Preamble | SD | FC | Dest Addr | Source Addr | Info | FCS | ED | FS |

(b) FDDI Token Format

| Preamble | SD | FC | FS |

The preamble field precedes each frame and is used for synchronization procedures. The Starting Delimiter (SD) field is used to uniquely identify the beginning of the frame. The Frame Control (FC) field defines the type of frame, such as a token, data frame, or administrative frame. The next two fields contain the destination address and source address. They may be either 16 or 48 bits and they contain the address of the sender and the receiver of the frame. The information (Info) field carries user information and headers from upper layers. The **Frame Check Sequence (FCS)** field is a 32-bit field that is used to calculate a Cyclic Redundancy Check (CRC) on the frame. The End Delimiter (ED) field is used to uniquely identify the ending of the frame. The Frame Status (FS) field indicates if the addressed station has recognized its address, if the frame was copied successfully by the destination, or if an error occurred in the detection operations at the station.

frame check sequence (FCS)

A token frame consists of the following fields: preamble, Starting Delimiter (SD), Frame Control (FC), and Ending Delimiter (ED). The frame control field contains the bit format of 10000000 or 11000000 to indicate that this is a token.

FDDI Packets

Two types of packets can be sent by FDDI: synchronous and asynchronous. Synchronous communications are used for time-sensitive transmissions requiring continuous transmission, such as voice, video, and multimedia traffic. Asynchronous communications are used for normal data traffic that is not time-sensitive. On a given network, the TTRT equals the total time needed for a node's synchronous transmissions plus the time it takes the largest frame to travel around the ring.

All stations having a synchronous allocation are guaranteed an opportunity to transmit synchronous frames; however, a station sends asynchronous frames only if time permits. Asynchronous frames can optionally be subdivided using levels of priority that are then used to further prioritize the sending of asynchronous traffic.

7.8 FDDI STANDARDS

The FDDI standard allows up to 500 stations to be connected to the ring in a priority system. High-priority stations can access the ring for longer periods of time.

ANSI has developed the FDDI standard (X3T9.5) to support users who require the flexibility, reliability, and speed provided by fiber-optic technology. A dual counterrotating ring structure is used for the high-speed tokenpassing network. The standard specifies a 100 Mbps transmission signaling rate with up to 2 km between stations. This is an improvement over the 10 Mbps for Ethernet and 4 Mbps and 16 Mbps for Token

FDDI-II

Ring, but the development of Fast Ethernet at 100 Mbps has hampered the utilization of FDDI. FDDI is defined in IEEE 802.8, ISO 9314, and ISO 8802.8. There are a number of RFCs listed in Appendix C that provide information concerning FDDI.

FDDI-II is a newer standard that provides additional capability beyond that offered by FDDI. FDDI-II provides the ability to handle circuit switched and packet switched traffic.

7.9   FDDI-I VERSUS FDDI-II

Despite the speed of FDDI, it still delivers information in packet form. Because packets vary in size, information tends to arrive at a station at an uneven rate. Although this may be acceptable for many applications, it does not work well for multimedia applications. To provide for continuous video and sound, information must arrive at a constant rate.

Because of the restrictions on the original version of FDDI, a new specification is under development that addresses those limitations. This new specification defines extensions to FDDI and is known as FDDI-II or twisted-pair FDDI (TPFDDI). This new specification allows data to be transmitted at the FDDI of 100 Mbps over copper twisted-pair wires. It uses a modified token passing technique for accessing the ring and will additionally support voice and video traffic. FDDI-II is not compatible with the original FDDI, however.

A FDDI-II network can operate in either basic or hybrid mode. This means that FDDI-II provides the ability to handle both circuit switched (hybrid) and packet switched (basic) traffic. It also provides the ability for a constant data rate connection between two stations, which is not possible with FDDI. Constant data rate connections are required for voice and video applications.

When a FDDI-II network is operating in basic mode, the entire 100 Mbps transmission capacity is used for packet switching services. When the network is operating in hybrid mode, the transmission capacity is split between a packet-data channel and several wideband channels.

The FDDI-II ring is controlled by one of the stations, called the cycle master. The cycle master maintains a rigid structure of cycles on the ring. Within each cycle a certain bandwidth is reserved for circuit switched voice and data traffic. This guarantees bandwidth for established connections and ensures adequate delay performance. The remaining bandwidth with the cycle is available for packet data use.

The voice and video transport capability of FDDI-II is possible because of its interworking with the Integrated Voice Data (IVD) LAN standard defined in IEEE 802.9.

It is now possible to connect a large campus or corporate complex into a single LAN or to easily connect smaller departmental LANS. FDDI-II will provide the functionality where there is a requirement to support applications such as full-motion color video, high-resolution graphics and photographs, traditional voice, and data concurrently. It is also designed to provide native-mode throughput on a LAN-to-LAN connection, as well as packetized isochronous services.

The FDDI-II isochronous transmission mechanisms impose a 125 ms frame structure of the time. The frame structure is used to divide the total transmission capacity into a number of discrete channels by allocating regularly repeating time slots to users that require them. These discrete channels are used to provide a virtual circuit between pairs of communicating stations.

FDDI-II represents a refinement of the shared access technology used in LANs that allows the LAN to be used for applications better suited for circuit switching.

7.10   FDDI APPLICATIONS

The primary application of FDDI is in the backbone for many LANs. This high-speed backbone is used to connect a group of lower speed LANs, such as Token Rings and Ethernets. Recent developments in the local exchange carriers' areas have indicated that FDDI has the robustness and reliability to be delivered in the metropolitan area.

FDDI, however, is a technology match for a limited number of applications. These include the following:

- Mission-critical requirements for fault tolerance
- Requirements where multiple LANs need connectivity in a metropolitan area
- A need for very high bandwidth
- Isochronous, high-bandwidth data traffic requirements such as interactive voice and video
- Situations where distances between stations are in excess of 100 meters
- Locations where EMI and RFI problems exist

7.11   FDDI ADVANTAGES AND DISADVANTAGES

Advantages

As a general practice, FDDI LANs are installed in organizations to serve as backbone networks. The FDDI LAN is used to interconnect lower speed LANs such as 802.3, Ethernet, and 802.5 networks. Utilization of FDDI in a MAN can produce several advantages:

- An industry standard
- Large numbers of users
- 100 Mbps bandwidth in each direction
- High reliability due to redundant configurations
- High immunity to EMI
- Efficient connectivity availability
- Multimedia support due to the high bandwidth
- Very long transmission distances in the ring and between workstations

Disadvantages

FDDI has not fared well in the market for LANs because of its high cost. Network Interface Cards cost between $2,000 and $5,000, making them far too expensive for desktop PCs. FDDI operates at 100 Mbps and is not scalable. FDDI was popular as a backbone; however, its single 100 Mbps speed makes it unattractive in today's environment. Other issues that impact the installation of FDDI are as follows:

- High cost of installation and hardware
- Copper interface compatibility issues
- Speed conversions for embedded devices
- No active monitor
- Migration to the 100 Mbps Ethernet standards

7.12   FDDI TECHNOLOGY ALTERNATIVES

Since FDDI functions as a Token Ring LAN topology, a limited number of topologies serve as alternatives in the WAN environment. FDDI actually works best as a LAN extender. Services that provide a backbone support would be the best alternatives. FDDI was designed to do the following:

- Support higher data rates at up to 100 Mbps
- Work in a deterministic token passing environment, presenting collisions
- Support attachment of up to 1,000 addressable nodes
- Extend the distance of the ring up to 200 km
- Provide robustness after equipment and cable failures
- Use optical fiber as the preferred medium to overcome the physical limitations of electrical and mechanical interference
- Allow spacing of up to 2 km between repeaters.

Competitive technologies to FDDI include Asynchronous Transfer Mode (ATM), Synchronous Optical Network (SONET), and Ethernet 100Base-TX.

FDDI versus ATM

Asynchronous Transfer Mode transport is a possible competitor for high-speed local transport. Multimedia applications may not work well on a shared media such as a ring. Indeed, fast cell switching will probably support voice and video better than a shared ring technology such as FDDI.

FDDI versus SONET

SONET is a fiber-optic carrier technology, whereas FDDI is basically a fast Token Ring technology. FDDI does not have the speed or capacity to operate in the fiber backbone environment.

The relatively high cost of fiber-optic transmission media for FDDI has kept it from dominating the PC and workstation LAN market. The ANSI X3T9.5 committee continues to work on an alternative to lower the cost. The alternative is to attain 100 Mbps over both STP and UTP copper wire.

FDDI versus Ethernet 100Base-FX

Ethernet 100Base-FX is a 100 Mbps LAN technology. It is based on FDDI's physical layer standard for optical fiber transmission. This means that 100Base-FX can benefit from the existence of mature chips at the physical layer while avoiding the complexity and cost of Token Ring operation.

Product and Service Providers

Because FDDI requires a fiber-optic infrastructure, service providers are limited to carriers, ILECs, and large CLECs. Fiber-optic products such as coaxial cable, single-mode and multi-mode fiber cable, fiber extenders, transceivers, and other hardware devices, are available from the following sources.

Cabletron Systems	www.ctron.com/
DSI	www.dsiinc.com/
BayNetworks	www.nortelnetworks.com/
3Com	www.3com.com/
Madge	www.madge.com/

■ SUMMARY

The Fiber Distributed Data Interface (FDDI) standard defines a multiple-access form of data link that uses a ring-structured network topology. Stations are connected to one another using point-to-point, fiber-optic cable segments to form a ring, and stations pass frames from one station to the next station in the ring so that eventually all stations receive all frames that have been transmitted.

The FDDI standard defines architectural models for various types of stations, including dual attachment stations, dual attachment concentrators, single attachment stations, and single attachment concentrators. Stations are connected using full-duplex transmission segments which allow dual ring and tree structures to be formed. The dual ring formed by dual-attachment stations and concentrators allow the ring structure to be dynamically reconfigured to recover from station and link failures.

With the FDDI access control method, a token is passed from one DTE device to the next around a physical ring. When a DTE device receives the token, it is allowed to transmit MAC frames for a specified time. After a station completes transmitting frames, it transmits a token granting the next DTE device on the ring the right to transmit frames. Frames from multiple devices can all be circulating at any time around the ring.

The FDDI standard defines a fiber-optic transmission medium that supports baseband transmission at a 100 Mbps data rate. Some implementations of the standard (CDDI) use twisted-pair cable.

FDDI has served the industry well as a high-speed backbone and LAN extender for Metropolitan Area Networks, but its 100 Mbps speed limits its effectiveness for a multiple-user, multimedia enterprise network.

Key Terms

Beacon

Concentrators

Copper Distributed Data Interface (CDDI)

Dual Attachment Concentrators (DAC)

Dual Attachment Station (DAS)

Fiber Distributed Data Interface (FDDI)

FDDI-II

Logical Link Control (LLC)

Media Access Control (MAC)

Medium Interface Connector (MIC)

Model Dispersion

Non-Return-to-Zero Alternate Mark Inversion (NRZI)

Physical Layer (PHY)

Physical Layer Medium Dependent (PMD)

Protocol Data Unit (PDU)

Single Attachment Concentrators (SAC)

Single Attachment Station (SAS)

Station Management (SMT)

Target Token Rotation Time (TTRT)

Timed Token Protocol (TTP)

Timing Jitter

Token

1. Define FDDI.

2. What is the function of the token in the FDDI technology?

3. What is the difference between FDDI and CDDI?

4. What are the two classes of stations in FDDI? What are their functions?

5. What are the classes of concentrators available in the FDDI architecture? How are they deployed?

6. What are the three types of networks supported by FDDI? Give a brief description of each.

7. What are the transmission speeds of FDDI?

8. What are the distances that can be spanned by FDDI? How is this accomplished?

9. How is FDDI different from Token Ring? How is it similar?

10. What is the media supported by FDDI? Describe the attributes of each.

11. What is the composition of fiber media? What are the different classifications of fiber cable?

12. Provide a comparison between single-mode and multi-mode fiber-optic cable.

13. What are the different ports allowable in the FDDI topology? What are their functions?

14. What process takes place when one of the FDDI rings fails?

15. What are the three areas of the FDDI architecture?

16. Define the PMD and PHY components of FDDI.

17. What is the function of the MIC component?

18. Provide an overview of the MAC layer?

19. What is the difference in functionality in the MAC layer between FDDI and 802.5?

20. What are the functions supplied by the FDDI MAC sublayer?

21. What are the responsibilities of the station management layer?

22. What is the protocol data unit in the FDDI topology?

23. What are the differences between the OSI model and the FDDI protocol stack?

24. Define the timed token protocol and explain its functionality.

25. What is the TTRT and how is it utilized?

26. What are the timers that each FDDI station utilizes? How are they used?

27. What is timing jitter?

28. What are the processes involved with error detection and correction?

29. Explain the claim process.

30. How does the station use the claim process in the initialization process?

31. What is beaconing and why is it important?

32. What are the services provided by ring management?

33. What is the format of the FDDI frame and token frame?

34. What is the difference between a synchronous packet and asynchronous packet?

35. Provide a general overview of FDDI-II.

36. How is FDDI different from FDDI-II?

37. What are the primary applications for FDDI?

38. What are the advantages incurred when using the FDDI topology?

39. Why would you not want to implement FDDI?

40. Why was FDDI designed? List a number of features.

41. Why does FDDI have a bleak future?

ACTIVITIES

1. Utilize DataPro, textbooks, and the Web to develop a comparison spreadsheet between FDDI-I and FDDI-II. Look at the applications supported and the various speed requirements.

2. Research the various sources to identify pros and cons for the utilization of FDDI in the Enterprise Network environment. Rank order these issues according to importance.

3. Identify alternatives to FDDI in the multimedia environment. Identify the features of each alternative. Create a spreadsheet that depicts these features and functions.

4. Draw a diagram that steps through the beaconing process.

5. Draw a diagram that steps through a ring break and ring wrap.

6. Identify different applications that can utilize the FDDI topology. Why are these applications candidates for FDDI?

7. Develop and present a five minute overview of the technology. Include Powerpoint slides or transparencies in the presentation.

8. Arrange for a carrier representative to give a FDDI presentation.

CASE STUDY/PROJECT

The FDDI technology is often deployed to connect multiple LANs in a metropolitan area. The limiting factor is the location of the FDDI backbone that can provide access to the various LANs.

Create a system design for a Metropolitan Area Network (MAN) using FDDI that provides connectivity of two LANs. Show all of the hardware components that make up this design. Show the flow of data traffic on the rings. Information that has been presented in Chapters 1 and 2 can be utilized to develop the LAN designs. Information presented in this chapter can be used to link these two LANs.

Details

The Greater Atlanta metropolitan area covers a large geographical area. ATL Corp. has a corporate location in Alpharetta and a warehouse facility at the Hartsfield Airport location. There are 30 airline miles between these two locations. Private line facilities have been utilized to provide network communications between ATL Corp. and the airport. The amount of network traffic has increased considerably and the 56 kbps circuits are no longer sufficient for this traffic. Routers and 56 kbps DSUs are currently being utilized for this connectivity.

There is a 802.3 LAN at both of these locations. The local carrier representative has proposed linking these two LANs with the FDDI technology. Carrier end offices are available close to each of these corporate locations. This FDDI access can provide a backbone access of 100 Mbps to ATL Corp.

Requirements

Develop the FDDI infrastructure that will be utilized to support the Atlanta metro area. Show the concentrators that will be required at both of the end offices. Show the connectivity between the end offices and the fiber bridges that will be located at the ATL Corp. locations. Show 802.3 LAN components at each location that will participate in this network.

Optional Requirements

In reality, there are multiple FDDI rings traversing the Greater Atlanta metropolitan area. Show how the FDDI backbone would look if three FDDI rings would be required to traverse the distance between the two ATL Corp. locations. Note that this network is transparent to the user; however, there are implications for the introduction of multiple FDDI rings in the infrastructure.

PART THREE

CHAPTER 8
Digital Subscriber Line

Chapter 8 begins with a description of POTS and its relationship with the DSL technologies. The components discussed include the Digital Subscriber Line Access Multiplexer (DLAM) and splitters that are utilized in the network. In-depth presentations follow on Asymmetrical DSL (ADSL), High-bit-rate DSL (HDSL), Symmetrical DSL (SDSL), Very-high-data-rate DSL (VDSL), and Rate Adaptive DSL (RADSL). The DSL sections are followed by information concerning cable television systems and their components. The chapter concludes with a discussion on DSL and cable modem standards, issues, and considerations.

CHAPTER 9
Integrated Services Digital Network/Braodband ISDN

Chapter 9 discusses ISDN, both broadband and narrowband ISDN, as well as the ISDN architecture, connectivity, out-of-band signaling, and both Basic Rate (BRI) and Primary Rate (PRI) interfaces. A technical discussion involves the functional groups for user access arrangement, including terminal adapters, network terminations, and ISDN reference points. Other topics include ISDN devices, wiring configurations, the ISDN frame, and ISDN standards. The chapter concludes with an overview of ISDN advantages, benefits, issues, considerations, and disadvantages.

CHAPTER 10
Switched Multimegabit Data Service/ Metropolitan Area Network

Chapter 10 begins with an overview of metropolitan area networks and the Switched Multimegabit Data Service (SMDS), along with service characteristics. A detailed explanation concerns SMDS connectivity, including in-depth information on the Distributed Queue Dual Bus (DQDB) operation. A section is devoted to explaining the Sustained Information Rate (SIR) and the various access classes and addressing. Included in the chapter are details concerning the cell structure and the various standards associated with the service. The chapter concludes with SMDS applications, issues, considerations, advantages, and disadvantages.

CHAPTER 11
Wireless/Personal Communications Service

Chapter 11 begins with an overview of mobile services history and wireless transmission. It continues with a discussion of the number of frequency bands and the variety of possible wireless network configurations. These primary categories of wireless networks include local area networks, extended LANs, and mobile computing. An in-depth analysis is directed at wireless LAN, radio, and microwave technologies. These include both satellite and terrestrial transmission systems. Cellular telephony is another topic, which includes information on pagers, mobile telephones, personal digital assistants, and the various network components of the mobile telephone service. A separate section provides an in-depth study of the Personal Communications Service (PCS). Information concerns both Frequency Division Multiple Access (FDMA) and Code Division Multiple Access (CDMA) transmission techniques. Wireless data technologies, including packet data networks, cellular telephone services, and cellular digital packet data, are important topics, as are wireless standards and associated issues and considerations. The chapter concludes with a significant number of advantages and disadvantages for each of the different wireless technologies.

CHAPTER 12
Fibre Channel

Chapter 12 looks at Fibre Channel and reviews its underlying technology, emphasizing the medium that is utilized to carry the signals. The chapter describes the Fibre Channel features in relation to the hardware devices utilized, and presents technical aspects of the Fibre Channel topology, which includes the concept of ports and fabrics, the Fibre Channel protocol, and the Fibre Channel classes of service. The discussion also involves the concept of a high-availability architecture that includes storage pools and computer clusters. Other topics include components of Fibre Channel such as hardware, software, cable, and converters, and the integration of these elements into the Fibre Channel system. The chapter also compares Fibre Channel, Gigabit Ethernet, and ATM. The chapter concludes with a section on applications, advantages, and disadvantages.

CHAPTER 13
Internet/Intranet/Extranet

Chapter 13 begins with a historical perspective of the Internet and the World Wide Web. A section describes the various search tools employed in the network. A detailed explanation includes the Internet components, access and transport across the network, Internet Service Providers (ISPs), and Network Service Providers (NSPs). Other topics are the Internet standards and IP addressing. Complete sections for both intranets and extranets present information about the technologies, topologies, and applications deployed. Discussions also involve Electronic Data Interchange (EDI) and Computer Telephony Integration (CTI) applications. The chapter concludes with issues, considerations, and network competition.

Digital Subscriber Line

■ INTRODUCTION

Digital Subscriber Line (DSL) is a technology for bringing high-bandwidth information to homes and small businesses over ordinary copper telephone lines. A number of DSL suppliers have projected a substantial increase in DSL installations during 2001. This demonstrates that the communications industry is committed to deploying and expanding DSL availability around the world. The ADSL Forum (www.adsl.com/) is one of many communications organizations that are actively promoting the DSL technology.

xDSL refers to different variations of DSL, such as ADSL, HDSL, RADSL, and SDSL. Assuming the user's home or small business is close enough to the central office that offers DSL service, it is possible to receive data at rates of up to 6.1 megabits (millions of bits) per second (of a theoretical 8.448 Mbps). This will enable continuous transmission of motion video, audio, and even 3-D effects. More typically, individual connections will provide from 1.544 Mbps to 512 kbps downstream and about 128 kbps upstream.

A DSL can carry both data and voice signals, and the data part of the line is continuously connected. DSL installations began in 1998 and will continue at a greatly increased pace in a number of communities in the United States and elsewhere. Compaq, Intel, and Microsoft, working with telephone companies, have developed a standard and easier-to-install form of DSL called **G.lite** which is expected to accelerate deployment. Within a few years, DSL is expected to replace ISDN in many areas and to compete with the cable modem to bring multimedia and 3-D to homes and small businesses.

DSL requires a xDSL terminating device at each end of the cable pair which accepts a data stream and overlays a high-speed analog signal. The three modulating techniques currently in use for xDSL divide the signal frequency range into three basic elements to carry POTS and the upstream and downstream high-bandwidth signals.

OBJECTIVES

Material included in this chapter should enable you to:

- understand Digital Subscriber Line (DSL) technology and its relation to the current broadband market and become familiar with the differences between Asymmetrical DSL, High-bit-rate DSL, Rate Adaptive DSL, and Very-high-data-rate DSL.

- become familiar with the DSL terms and definitions that are relevant in the Enterprise Networking environment. See how DSL relates to the current voice and digital technologies.

- understand the function and interaction of DSL in the Enterprise Network environment. Look at how splitters and DSL modems fit in the environment.

- understand where DSL fits in the different applications of the technology. Look at how current communication technologies can be applied to DSL.

Digital Subscriber Line (DSL)
G.lite

- identify components that comprise the DSL networking environments. Look at the operation of splitters and DSLAMs.
- identify applications of DSL and cable modem. Identify opportunities to implement DSL and cable modem systems. List the advantages and disadvantages of each technology.
- become familiar with the standards that relate to DSL and cable modem technologies, and the modulation techniques associated with the DSL technology.

DSL technology provides both symmetric and asymmetric bandwidth configurations to support one-way and two-way high bandwidth requirements. Traditional POTS is a symmetric application, as the bandwidth requirements are the same in both directions. Asymmetric applications are those in which bandwidth needs are higher in one direction than the other. Web applications and database accesses usually have a small amount of data in the request and a large amount of data in the response.

DSL is fundamentally another name for ISDN-BRI, which operates with two 64 kbps circuit switched channels and one 16 kbps packet switched and signaling channel. DSL refers to the various arrangements in which advanced modulating techniques are imposed onto the channel in order to derive higher throughput in one or both directions.

Unlike ISDN, which integrates voice and data onto a single channel, DSL uses narrowband filters called splitters to split the voice channel to then carry voice and data on separate lines. This method eliminates interference between the two signals as well as interference to the data signal from household appliances. Because voice and data are using separate lines, an Internet connection can be maintained at the same time a voice communication is in progress over the same channel.

8.1 POTS AND DIGITAL SUBSCRIBER LINE

Traditional phone service, sometimes called "Plain Old Telephone Service" or POTS, connects the home or small business to a telephone company office over copper wires that are wound around each other and called twisted pair. Traditional phone service was created to let the user exchange voice information with other phone users. The type of signal used for this kind of transmission is called an analog signal. An input device such as a phone set takes an analog signal and converts it into an electrical equivalent in terms of volume (signal amplitude) and pitch (frequency of wave change). Since the telephone company's signaling is already set up for this analog wave transmission, it is easier for it to use that mechanism as the way to get information back and forth between your telephone and the telephone company.

This fact explains why the computer needs to have a modem: It can demodulate the analog signal and turn its values into the string of 0 and 1 values called digital information. DSL services are dedicated, point-to-point, public network access over twisted-pair copper wire on the local loop. The network provides connectivity over the "last mile" between a network service provider's central office and the customer's location.

Digital Subscriber Line Benefits
The benefits of the DSL technologies are superior to those currently available, because they lend themselves to accessing the Internet, which is be-

coming pervasive in today's society. These benefits include high-speed access, instant connectivity, ease of use, reliable service, and security.

8.2   DSL TECHNOLOGY

DSL technology uses advanced modulation technologies on existing telecommunications networks (Figure 8–1) for high-speed networking between a subscriber and a telco [Goralski, 1998]. DSL supports the transmission of data, voice, and video communications, including multimedia applications. The Telecommunications Act of 1996 was influential in the development of DSL, because the act encourages telecommunications and cable TV providers to develop interactive communications over existing telephone networks.

Figure 8–2 depicts a simple diagram of DSL connectivity over common telephone facilities. The two pairs (four wires) that are provided to the user's location are split into one pair for data and one pair for voice.

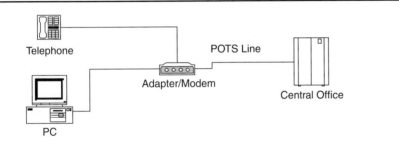

Figure 8–1
DSL Adapter (modem)

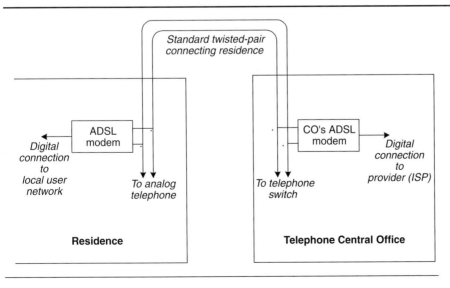

Figure 8–2
ADSL Modems Connected to Existing Local Loop Wiring

Because analog transmission uses only a small portion of the available amount of information that could be transmitted over copper wires, the maximum amount of data that can be received using ordinary modems is about 56 kbps.

The ability of the computer to receive information is constrained by the fact that the telephone company filters information that arrives as digital data, puts it into analog form for the telephone line, and requires the modem to change it back into digital. In other words, the analog transmission between the home or business and the phone company is a bandwidth bottleneck.

The technology of DSL assumes digital data does not require a change into analog form and back. Digital data are transmitted to the computer directly as digital data, which allows the telco to use a much wider bandwidth for transmitting it to the user. It is also possible for the signal to be separated so that some of the bandwidth is used to transmit an analog signal to allow the telephone and computer to be used on the same line simultaneously. A big advantage of DSL is the utilization of the existing copper facilities for the service. DSL is a digital technology that works over copper wire that is already in place in most residences and businesses for telephone services. DSL can carry both voice and data signals at the same time, in both directions, as well as signaling data used for call information and customer data.

To use DSL, the user must install an intelligent adapter in the DTE device (PC, server) that is connected to the DSL network. The adapter can be a card similar in appearance to a modem, but it must be fully digital. Two wires are connected between the adapter and the telco facilities. Communication over the copper wire is simplex, which means that one pair is for outgoing transmissions and the other pair is for incoming transmissions.

8.3 DIGITAL SUBSCRIBER LINE ACCESS MULTIPLEXER AND SPLITTER TECHNOLOGY

Several new technical terms associated with the DSL technology are DSLAM, splitter, splitterless, symmetrical, and asymmetrical. These terms identify both hardware components and techniques.

digital subscriber line access multiplexer (DSLAM)

A **Digital Subscriber Line Access Multiplexer (DSLAM)** concentrates traffic in ADSL implementations through Time Division Multiplexing (TDM) at the Central Office (CO) or remote line shelf. The DSLAM combines streams of bits in the channels coming upstream from the user and splits up a large bit stream coming downstream from the IP or ATM network. The DSLAM splits this bit stream based on channels, as a multiplexer always does. Figure 8–3 depicts the basic architecture of a DSL access multiplexer. The DSLAM resides in the CO along with the PSTN

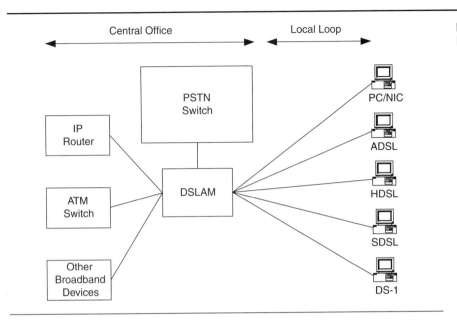

Figure 8–3
DSLAM Architecture

switch. The DSLAM is connected to both the CO network devices and to the remote user sites. These remote sites can be personal computers with Network Interface Cards (NIC), ADSL service with DSL modems for splitting voice and data, HDSL service, SDSL service for voice and data, and DS-1 Remote Access Multiplexer (RAM) service.

The DSLAM Backbone

To interconnect multiple DSL users to a high-speed backbone network, the telephone company uses a DSLAM. The network shown in Figure 8–4 consists of two DSLAM devices which provide the backbone connectivity. Typically, the DSLAM connects to an Asynchronous Transfer Mode (ATM) network that can aggregate data transmission at gigabit data rates. At the other end of each transmission, a DSLAM demultiplexes the signals and forwards them to the appropriate individual DSL connections. The DSLAM has the capability of providing both IP routing and ATM switching.

The terms **splitter** and **splitterless** are used to designate where the splitting of the analog and digital signals are accomplished and what device is necessary to accomplish this activity. To use DSL, an intelligent adapter must be installed at the user site. The adapter card can be similar in appearance to a modem, but it must be a digital device because digital signals are transmitted instead of analog signals. Two pairs of wires from the network are connected to the adapter. One pair is for outgoing transmissions and the other pair is for incoming transmissions. The splitter

splitter
splitterless

Figure 8–4
DSLAM Voice and Data
Switching

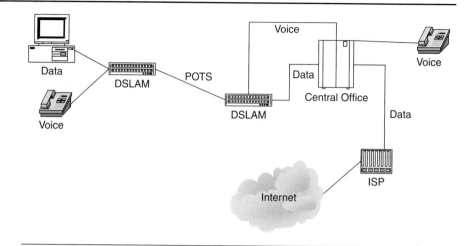

Figure 8–5
DSL Modem Splitters

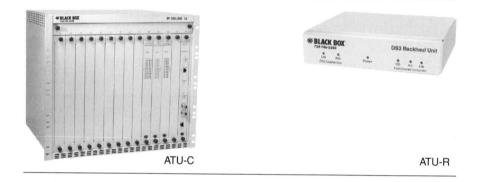

ATU-C   ATU-R

ADSL transceiver unit—
central office (ATU-C)
ADSL transceiver unit—
remote (ATU-R)

splits the incoming bit stream into voice and data. Typically the
LEC/CLEC would install this at the customer's site. Figure 8–5 shows ex-
amples of splitter devices that are available. The **ADSL Transceiver Unit—
Central Office (ATU-C)** splitter is located at the central office and the
ADSL Transceiver Unit—Remote (ATU-R) splitter is located at the
user's location.

The alternative to the splitter is an arrangement that is splitterless.
This xDSL device splits the bit stream into voice and data; however, the
LEC/CLEC does not have to go to the customer's premise and install it,
because the function resides at the central office. Most DSL technologies
require that a signal splitter be installed at a home or business, requiring
the expense of a phone company visit and installation. With a splitterless
DSL, however, it is possible to manage the splitting remotely from the
central office.

A newer version of ADSL, known as Universal Asymmetric Digital Subscriber Line (UADSL), and ADSL Lite, DSL Lite, and G.Lite, are attempts to standardize ADSL deployment across all carriers, since early implementations had so many speed and transmission differences. It operates using a 1.544 Mbps downstream and 384 kbps upstream speed combination.

The various DSL services are based upon whether the transmission is **symmetrical** or **asymmetrical.** The transmission technique is symmetrical if the transmission speed is the same in both directions. The transmission technique is asymmetrical if the transmission speed is different in each direction.

symmetrical transmission asymmetrical transmission

Traditional POTS is a symmetric application. Asymmetric applications are those in which bandwidth needs are much higher in one direction than the other. For example, Web browsing requires very little bandwidth from the user, since the data transmitted are small. The response, however, for graphics and/or data could be extremely large.

8.4 ASYMMETRICAL DIGITAL SUBSCRIBER LINE

An alternative digital local loop transmission technology is **Asymmetrical Digital Subscriber Line (ADSL).** Unlike ISDN, ADSL works along with POTS for traditional voice services. Actually, ADSL works over POTS, at higher frequencies, on the same pair of copper wires that currently carries voice transmission. Unlike using a modem on a voice line, ADSL does not interfere with the voice services. ADSL is the form of DSL that will become most familiar to home and small-business users.

asymmetrical digital sub-scriber line (ADSL)

ADSL Reference Model
To understand the structure of ADSL, it is essential that the various terms and components be described. There are two major parts of this structure—the central office network and the customer premises network. In ADSL terminology, the major reference elements are the ADSL Transceiver Unit—Central Office (ATU-C) and the ADSL Transceiver Unit—Remote (ATU-R). Two other components provide the functionality for separating the voice signals from the data signals. These include the *High-Pass Filter (HPF)* and the *Low-Pass Filter (LPF).* The signal from the local loop is split into a voice and data component. The LPF passes the voice signals to the PSTN/POTS and the HPF passes the data signals to the broadband network/CPE network. Figure 8–6 provides a general reference model that includes these components.

ADSL Architecture
ADSL is called asymmetric because most of its two-way or duplex bandwidth is devoted to the downstream direction, sending data to the user.

Figure 8–6
ADSL Reference Model

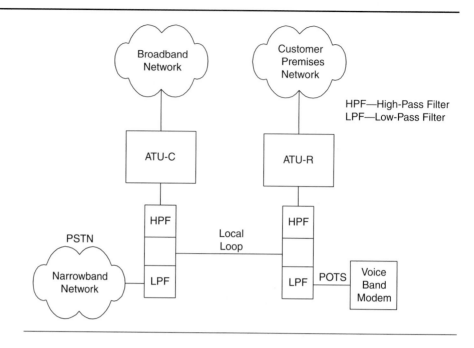

HPF—High-Pass Filter
LPF—Low-Pass Filter

Asymmetric in ADSL refers to the service's differing upstream and downstream bandwidths. The bandwidths and associated distance limitations from the carrier's CO are the two issues most associated with an ADSL implementation. Only a small portion of bandwidth is available for upstream or user-interaction messages. Most Internet, and especially graphics or multimedia intensive Web data, however, need lots of downstream bandwidth, but user requests and responses are small and require little upstream bandwidth. Using ADSL, up to 6.1 Mbps of data can be sent downstream and up to 640 kbps can be sent upstream. The high downstream bandwidth means that the standard telephone line will be able to bring motion video, audio, and 3-D images to your computer or hooked-in TV set. In addition, a small portion of the downstream bandwidth can be devoted to voice rather than data, and it is possible to hold phone conversations without requiring a separate line. Figure 8–7 depicts an ADSL configuration. Note that the splitter at the user location is used to separate the voice and data transmissions, and the same function occurs at the CO through a DSLAM.

ADSL Design
ADSL was originally targeted at the expected need for video on demand and related services. Since the introduction of the ADSL technology, however, the demand for high-speed access to the Internet has become the

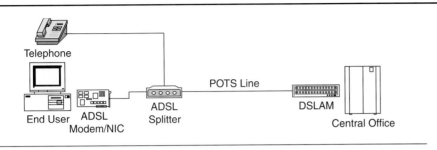

Figure 8–7
ADSL Configuration

service of choice. Typically, the user requires a higher capacity for downstream rather than upstream transmission. Most user transmissions are in the form of short messages, whereas the responses, especially Web traffic, include large amounts of data. The data often contain images and video transmissions. Thus, ADSL provides a good fit for the Internet environment.

ADSL is used over the local analog loop. This DSL technology is asymmetrical because it offers a higher download rate than upload data rate. Data rates may be from 1.5 Mbps to 9 Mbps over the entire 18,000 ft of a local loop in download mode from CO to the local loop. Data uploaded from the end of the local loop to the CO have a useful bandwidth between 16 kbps to 640 kbps. If the local loop is limited to 12,000 ft, upload data rates can be up to 1.544 Mbps.

ADSL is an attractive alternative from a carrier perspective because it does not require carriers to upgrade existing switching technology. Separate ADSL units, about the size of a modem, are deployed at subscriber sites and at the central office, where voice frequencies are stripped off and passed to existing voice switching equipment. Data frequencies are also separated at this time and forwarded to an Internet Service provider (ISP). The ADSL units provide an Ethernet 10BaseT (10 Mbps) interface for data that may be connected to the 10BaseT interface in the user's terminal, to a shared 10BaseT hub, or to other LAN devices.

ADSL uses Frequency Division Multiplexing (FDM) to exploit the 1-MHz capacity of twisted-pair cable. POTS occupies the lowest frequencies from 0 kHz to 4 kHz, upstream data use from about 25 kHz to 200 kHz, and downstream data use from about 250 kHz to 1.1 MHz. There are three elements of the ADSL strategy:

- Reserve the lowest 25 kHz for voice (POTS).
- Use FDM to allocate two bandwidths, a smaller upstream and a larger downstream.
- Use FDM within the upstream and downstream bandwidths.

Echo cancellation can be utilized instead of FDM to allocate the two bandwidths. When employed, the entire frequency bandwidth for the upstream channel overlaps the lower portion of the downstream channel. There are two advantages to using echo cancellation instead of creating two distinct frequency bandwidths for upstream and downstream traffic.

- The higher the frequency, the greater the attenuation. With echo cancellation, more of the downstream bandwidth is in the "good" part of the spectrum.
- The echo cancellation technique is more flexible for changing upstream capacity. The upstream channel can be extended upward without running into the downstream; instead, the area of overlap is extended.

The disadvantage is the need for echo cancellation functions on both ends of the circuit.

Modulation Technologies

discrete multitone (DMT)
carrierless amplitude
phase modulation (CAP)
Multiple Virtual Line
(MVL)

Several modulation technologies are used by various kinds of DSLs, although these are being standardized by the International Telecommunications Union (ITU). Different DSL modem makers are using either **Discrete MultiTone (DMT)** technology or **Carrierless Amplitude Phase (CAP) modulation.** A third technology, known as **Multiple Virtual Line (MVL),** is another possibility. CAP and DMT are currently the two competing standards for how ADSL units manage bandwidth for data transmission. MVL is a Paradyne© product designed for the mass market.

quadrature amplitude
modulation (QAM)

Carrierless Amplitude and Phase   CAP treats the frequency range as a single channel and uses a technique similar to **Quadrature Amplitude Modulation (QAM)** to build constellations and avoid interference. CAP is a de facto standard, developed by AT&T Paradyne, and deployed in many trial ADSL units. CAP is a proprietary digital modulation technique that is the official ANSI standard for ADSL.

Discrete MultiTone (DMT)   To accommodate differences in local loop characteristics, ADSL is adaptive. When ADSL modems are powered on, they probe the line between them to find its characteristics, and are then set up to communicate using techniques that are optimal for the line. The underlying technology for ADSL is Discrete MultiTone (DMT), which uses multiple carrier signals at different frequencies, sending some of the bits on each channel. DMT uses a combination of FDM and inverse multiplexing techniques to accomplish this. The available bandwidth (upstream or downstream) is divided into several 4 kHz subchannels.

On initialization, the DMT modem sends out test signals on each subchannel to determine the signal-to-noise ratio. The modem then assigns more bits to channels with better signal transmission qualities and less bits to channels with poorer transmission qualities. Each subchannel can

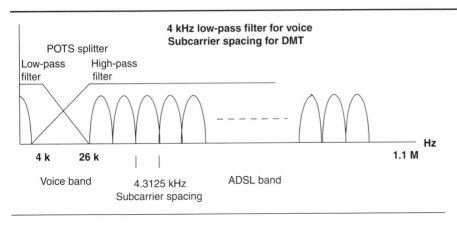

Figure 8–8
Discrete Multitone

carry a data rate of from 0 to 60 kbps. Present ADSL/DMT designs utilize 256 downstream subchannels. Current implementations operate at 1.5 Mbps to 9 Mbps, depending on circuit distance and quality.

If a particular frequency has a high signal-to-noise ratio, ADSL selects a modulation scheme that encodes many bits per baud; if the quality on a given frequency is low, ADSL selects a modulation scheme that encodes fewer bit per baud. The result of adaptation is a robust technology that can adapt to various line conditions automatically.

Figure 8–8 provides a graphical representation of the DMT line-encoding technology. The spectrum from 0 to 4 kHz, voice band, is designated for POTS and the spectrum from 26 kHz to 1.1 MHz is designated for data.

DMT has been approved as an ADSL standard (ANSI Standard T1.413) by the ANSI T1E1.4 working group.

2B1Q (4-PAM)

It will be useful to look at the relationship between the 2B1Q line coding and ISDN. **2B1Q** represents a signal type that has 2 bits per baud, arranged as one quaternary (four level) Pulse Amplitude Modulation (PAM) scheme. It basically transmits data at twice the frequency of the signal. It is utilized in HDSL, SDSL, and ISDN BRI.

2B1Q

The 2B1Q line encoding was intended for the use of **ISDN DSL.** The 2B1Q line coding was seen as a major enhancement over the original T-1 line coding, which was called bipolar Alternate Mark Inversion (bipolar AMI). 2B1Q line coding was intended to deliver ISDN BRI speeds of 144 kbps through local loops up to 18,000 feet. This was done on only one pair of wires, and basically, provided 144 kbps full-duplex transmission in each direction using the same frequency range. The familiar 2B1Q code for ISDN is a member of a family of PAM line codes. Figure 8–9 illustrates the frequency ranges of BiPolar AMI, 2B1Q, and CAP.

ISDN DSL

Figure 8–9
Bipolar AMI, 2B1Q, and
CAP Frequency Ranges

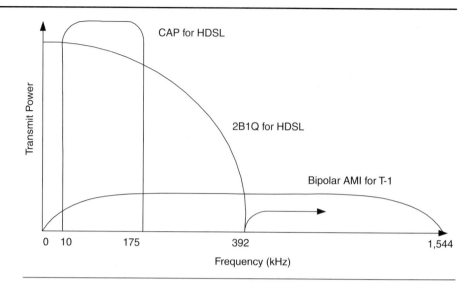

Figure 8–10
ADSL PPP Frame Format

7E Flag	FF	03	PPP Protocol ID	PPP Protocol Data Unit (payload)	FCS CRC-16	7E Flag

Why is 2B1Q important to the xDSL technology? It turns out that 2B1Q is not sophisticated enough to achieve multimegabit speeds over long distances. Although 2B1Q requires less bandwidth than bipolar AMI, 2B1Q still uses the frequency range that analog voice would normally use on a purely analog local loop. Note in Figure 8–9 that CAP HDSL uses much less of the available spectrum on twisted-pair loops than either bipolar AMI or 2B1Q. Also, CAP, as with any passband modulation method which dictates the range of frequencies that can pass through a filter without being attenuated, holds out a chance to preserve the 300 Hz to 3000 Hz passband for analog voice service on the same wires.

ADSL Frame

An example frame format on an ADSL line is shown in Figure 8–10. It is based on the HDLC data link format. The frame format is used to transport Point-to-Point Protocol (PPP) data packets. PPP is a basic lower level networking protocol whose payload can be used to encapsulate higher level protocols. The frame begins with the standard HDLC flag followed by a PPP address code field. Two bytes of protocol ID identify the payload type and possible protocol that has been encapsulated in it. The frame check sequence field uses CRC-16 for error detection and the frame ends with another flag. There are similar frame formats to support other protocol services using ADSL lines.

8.5 HIGH-BIT-RATE DSL

The earliest variation of DSL to be widely used has been **High-bit-rate DSL (HDSL)** which is used for wideband digital transmission within a corporate site and between the telephone company and a customer. HDSL was developed in the late 1980s by BellCore to provide a more cost-effective means of delivering a T-1 data rate. The standard T-1 circuit uses AMI coding, which occupies a bandwidth of about 1.5 MHz. Because such high frequencies are involved, the attenuation characteristics limit the use of T-1 to a distance of about 1 km between repeaters. Therefore, for many subscriber lines, one or more repeaters are required, which adds to the installation and maintenance expense.

high-bit-rate DSL (HDSL)

The main characteristic of HDSL is that it is symmetrical, which means that an equal amount of bandwidth is available in both directions. For this reason, the maximum data rate is lower than for ADSL. HDSL transceivers can reliably transmit a 2.048 Mbps data signal over two non-loaded, 24 gauge, unconditioned twisted-pair loops, utilizing the 2B1Q coding scheme, at a distance of up to 4.2 km without the need for repeaters. Eliminating the need for repeater equipment significantly simplifies the labor and engineering effort to provision the service.

There are two disadvantages of using HDSL versus ADSL. HDSL has a short-distance limitation on local loops and the current version of HDSL requires two independent twisted pairs of wire.

There are, however, several advantages of utilizing HDSL over ADSL:

- HDSL is tolerant of local loop modifications made for the telephone system. HDSL can be used on a loop that includes a telephone bridge tap.
- Because its bit rate is compatible with a T-1 circuit, moving data between a T-1 circuit and HDSL is straightforward.
- HDSL has an ability to tolerate failure gracefully. The technology is designed so that if one of the two twisted pairs fails, the modems do not fail completely, but instead continue to operate at one half the maximum bit rate.

Unlike ADSL, which uses a single twisted pair, note that HDSL requires two independent twisted pairs. To overcome the wiring disadvantage, a variant known as HDSL2 has been proposed that runs over two wires.

Figure 8–11
HDSL Configuration

Figure 8–11 depicts a HDSL configuration. Note the requirements for CO and Remote HDSL transceiver units.

8.6 SYMMETRICAL DSL

Although HDSL is attractive for replacing existing T-1 circuits, it is not suitable for residential subscribers because it requires two twisted-pair wires, whereas the typical residential subscriber has a single twisted-pair wire access.

symmetrical digital subscriber line (SDSL)

A single-pair or **Symmetrical Digital Subscriber Line (SDSL)** operates on a single copper pair as opposed to the two-pair requirements of HDSL. SDSL allows easy implementation of applications that require symmetrical data rates on a single local loop while maintaining the existing POTS on the same loop. Because only one pair is needed, the capacity of the entire local loop infrastructure is greatly magnified. This is an advantage to the local providers because it allows deployment of new capacities at a lower capital expenditure. SDSL speeds are T-1 or E-1, with a maximum distance between the subscriber and the telco end office of approximately 3 km (10,000 ft.) over 24 gauge UTP. As with HDSL, 2B1Q coding and echo cancellation are used to achieve full-duplex transmission over a single pair of wires.

SDSL has data rates of 768 kbps in both directions, which is preferred by the business community. Web server applications are better served with a SDSL-type link. SDSL usually does not include simultaneous POTS, which is not normally a problem for the business environment.

8.7 VERY HIGH-DATA-RATE DSL

very high-data-rate digital subscriber line (VDSL)
fiber to the curb (FTTC)
fiber to the neighborhood (FTTN)

Very High-Data-Rate Digital Subscriber Line (VDSL) provides very high bandwidth asymmetrically to users with broadband access requirements over a **Fiber To The Curb (FTTC)** and **Fiber To The Neighborhood (FTTN)** network. VDSL provides data rates of 12.9 Mbps to 52.8 Mbps downstream and 1.5 Mbps to 2.3 Mbps upstream. With the higher speeds, the maximum distance between the user and the telco end office is much shorter, approximately 1.35 km (4,500 ft) over 24 gauge UTP. VDSL's range is relatively short at 300 to 1,800 meters (980 to 5,900 ft), which limits its practicality for wide area networks. VDSL requires a fiber-optic feed and ATM to function.

VDSL functions similarly to **Rate Adaptive Digital Subscriber Line (RADSL)** because rates can also be dynamically allocated. It is also similar to ADSL because it creates multiple channels over UTP wires, and allows transmission of voice at the same time as data. VDSL does not use echo cancellation, but provides separate bandwidths for different services. These allocations are as follows:

rate adaptive digital subscriber line (RADSL)

- POTS: 0 – 4 kHz
- ISDN: 4 – 80 kHz
- Upstream: 300 – 700 kHz
- Downstream: GT 1 MHz

Although high data rates can be achieved over twisted copper pairs, VDSL cannot be used on existing wiring between the telephone CO and subscribers because the distances are too long. To overcome this restriction, VDSL requires intermediate concentration points with optical fiber connecting the concentration points back to the CO. In VDSL terminology, a concentration point is called an Optical Network Unit (ONU). Because versions of VDSL with lower data rates run over longer distances of copper, they do not require concentration points to be as close to the subscriber. This means that lower data rates require fewer concentration points to cover a given geographic area.

Fiber In The Loop (FITL) facilities include **Fiber To The Curb (FTTC), Fiber To The Home (FTTH), Fiber To The Neighborhood (FTTN)** and **Hybrid Fiber Coaxial (HFC)**. FTTC is deployed to the curb, with a copper facility from the curb to the home. The distance between the curb and the user should be less than 1,000 ft. FTTH is deployed past the curb, all the way to the user's home. FTTN extends from the central office to a neighborhood node. Copper is run from the node to the home. The distance from the node to the home is 3,000 ft. HFC is a network that includes a FTTN node with a coaxial cable to individual homes. Coaxial cable can be run from the FTTN node to the homes in either a star or bus arrangement.

fiber in the loop (FITL)
fiber to the home (FTTH)
hybrid fiber coaxial (HFC)

8.8   RATE ADAPTIVE DSL

Rate Adaptive Digital Subscriber Line (RADSL) adapts the transmission speed depending on the length of the local loop and the quality of the lines being used for transport. Otherwise, it is similar in distance and speed to ADSL. Originally developed for on-demand movie transmissions, RADSL applies ADSL technology, but enables the transmission rate to vary depending on whether the communication is data, multimedia, or voice. RADSL is well suited to applications in which more data flow in one direction than in the other. RADSL is advantageous where there is a lower bandwidth demand and also where the line quality is less than needed for full bandwidth implementations. There are two ways that the transmission rate can be established. One is for the telco to set a specific rate per each subscriber line, based on the anticipated use of the line. Another is for the telco to dynamically adjust the rate to the demand on the line. One popular combination is 6.1 Mbps downstream and 640 Kbps upstream for specific distances over certain quality wire. RADSL techniques are now included in the ADSL T1.413 Issue 2 standards, which is now an ANSI standard.

8.9 DSL SERVICE ALTERNATIVES

There are numerous DSL versions, all with their own speeds and distance limitations, and the technology is continually improving. No single DSL choice is clearly superior to others; rather, it is a question of matching the transmission speed with the user's application. All DSL versions simply provide a digital pipe from the CO to the user. Once the data reaches the CO, the carrier must then pass it to the data network (i.e., the Internet) using the B Channels in the ISDN or some other means. Table 8–1 provides a comparison of the xDSL alternatives. (NOTE: u-upstream d-downstream)

DSL Service Limitations

The actual DSL transmission rate is determined by a number of factors, including the type of DSL service used, the condition of the cable in the network, the distance to the closest DSL demarcation point, and the bus speed in the user's device. DSL modems follow the data rate multiples established by North American and European standards.

Table 8–1
DSL Service Alternatives

Specification	ADSL	HDSL	SDSL	VDSL	RADSL
Bits/second	1.5 – 9 Mbps d/s	1.544 or	1.544 or	13 – 52 Mbps d/s	6.1 Mbps d/s
	16 – 640 kbps u/s	2.048 Mbps	2.048 Mbps	1.5 – 2.3 Mbps u/s	640 kbps u/s
Mode	Asymmetric	Symmetric	Symmetric	Asymmetric	Asymmetric
Copper Pairs	1	2	1	1	1
Range	3.7 – 5.5 km	3.7 km	3.0 km	1.4 km	3.7 – 5.5 km
Signaling	Analog	Digital	Digital	Digital	Analog
Line Code	CAP/DMT	2B1Q	2B1Q	DMT	CAP/DMT
Frequency	1 – 5 MHz	196 kHz	196 kHz	10 MHz	1 – 5 MHz
Bits/cycle	Varies	4	4	Varies	Varies

Data Rate (Mbps)	Wire Gauge (AWG)	Distance (Feet)	Distance (Km)	Wire Size (mm)
1.5/2	24	18,000	5.5	0.5
1.5/2	26	15,000	4.6	0.4
6.1	24	12,000	3.7	0.5
6.1	26	9,000	2.7	0.4

Table 8–2
DSL Data Rates/Distance/Wire Gauge Comparisons

Because DSL is distance sensitive, service is not always available from the ILEC/CLEC. The maximum distance from user to the telco demarcation without a repeater is 5.5 km (3.4 mi). As distance decreases toward the local CO, the data rate increases. Another factor is the gauge of the copper wire. The heavier 24 gauge wire carries the same data rate farther than 26 gauge wire. If the customers are beyond the 5.5 km range, they may still be able to have DSL if your company has extended the local loop with optical fiber cable.

Some versions of DSL cannot operate on existing wiring between the telco CO and the subscriber. Additional fiber facilities must be deployed for the VDSL service.

Before considering a DSL design, it is essential to check with the LEC/CLEC to verify that facilities are available or planned for the various locations being considered. Table 8–2 presents a chart of DSL data rates when compared with distance and cable wire gauge.

8.10   ANOTHER TECHNOLOGY OPTION

Thus far, the technologies that have been examined deliver digital information over the twisted-pair wiring, which forms the local loop of the analog telephone system. This section examines the use of an alternative wiring scheme that can deliver even higher bit rates.

The primary motivation for considering alternatives to the telephone local loop arises from inherent limitations. The chief problem lies in the electrical characteristics of twisted-pair wiring. Although technologies such as DSL can achieve much higher bit rates than dial-up modems, the wiring places an upper bound on how fast data can be transferred. Also, the lack of shielding makes the wiring susceptible to interference, which can substantially degrade performance for some subscribers.

In an effort to overcome the limitations of twisted-pair wiring, researchers have investigated both wireless and wired technologies for use in the local loop. One alternative technology stands out as particularly attractive because it offers higher speeds than telephone wiring, is less susceptible to electromagnetic interference, and does not require a completely new infrastructure. This technology is the Cable Television (CATV) system.

Cable Television System

A CATV system has almost all the facilities needed to send digital information downstream at high speed. The medium consists of coaxial cable, which has high capacity and is immune to electromagnetic interference, and the cable system hardware which uses broadband signaling to deliver multiple TV channels simultaneously. More important, because cable systems are designed to carry many more signals than are currently available, the hardware has unused bandwidth that can be used to send data.

Most CATV systems were built for one-way broadcast transmission, downstream from the cable head end to the subscriber's residences. To provide the necessary upstream bandwidth, cable providers have two basic options:

- Provide upstream bandwidth over POTS while providing downstream bandwidth in a coordinated fashion over the installed cable plant. This architecture does not deliver simultaneous voice capability as in ADSL architecture.
- Modify cable architecture to support simultaneous upstream and downstream transmission. Current implementations of such an architecture provide up to 30 Mbps downstream and 768 kbps upstream.

Providing upstream bandwidth is only one of the architectural obstacles that cable providers must overcome, however. Whereas phone carriers provide voice services via circuit switching, cable companies provide cable service via a shared media architecture in which an entire neighborhood may be served by the same shared coaxial cable. Therefore, although 30 Mbps downstream bandwidth may sound impressive, a subscriber needs to know the number of users sharing that bandwidth.

8.11   CABLE MODEM

cable modem

Cable television providers are beginning to move into the data communications business, primarily to provide Internet access. A subscriber can realize a speed of 10 Mbps and greater with a **cable modem** attached to a cable TV line. The cable modem technology allows simultaneous transmission of television and Internet signals, so one person could be surfing the Internet while others are viewing a television program.

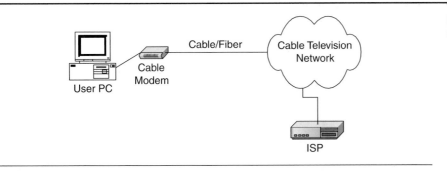

Figure 8–12
Cable Modem

A cable modem is usually a card that fits in a PC and is attached to the CATV coaxial cable. Figure 8–12 depicts the use of a cable modem in a connection to an ISP.

Even though speeds of 10 Mbps are available, this capacity is shared with many other users, so there will be a competition for bandwidth with neighbors in the geographic area. In many areas, the CATV lines may need upgrading to provide two-way communication. Until this upgrade occurs, users must have a separate telephone line for upstream communication. The primary motivation for considering alternatives to the telephone local loop arises from the electrical characteristics of twisted-pair wiring. Although technologies like ADSL can achieve much higher bit rates than dial-up modems, the wiring places an upper bound on how fast data can be transferred. The lack of shielding on UTP makes the wiring susceptible to interference, which can substantially degrade performance for subscribers.

Most cable modems operate by taking some of the bandwidth available on a CATV system away from television signals and dedicating it instead to data traffic. On some older analog cable television networks, data signals are modulated onto sine waves such as occurs with traditional modems, and placed on some of the existing analog system channels. The transmission speed of these modems is much higher than that of conventional modems, because they are not limited by the 3,000 Hz bandwidth of the telephone line. Newer digital CATV systems, with digital cable converter boxes, actually send the television signals and the user's data on separate digital channels on the system. These digital cable modems are not truly modems, but more like Data Service Units (DSUs).

Cable networks can also dedicate channels for voice transmission, offering an alternative to the local telephone company for telephone service. Cable television networks offer one possibility of combining voice, broadcast video, and data on a single connection to the subscriber. The ability to send high-quality video in both directions, as in videoconferencing, requires more bandwidth than today's CATV systems provide to cable modems.

8.12   CABLE MODEM TECHNOLOGY

A CATV system has almost all the facilities needed to send digital information downstream at high speed. The media consists of coaxial cable, which has a high capacity and is immune to *Electromagnetic Interference (EMI)*. EMI occurs when one device leaks so much energy that it affects the operation of another device. The cable system hardware uses broadband signaling to deliver multiple television channels simultaneously and also has sufficient bandwidth for growth.

The cable modem communicates using upstream and downstream channels that are already allocated by the cable service. The upstream channel is used to transmit the outgoing signal over a contiguous range of channels that are carrying data, sound, and TV downstream signals. The downstream channel is used to receive signals and is also blended in with other data, sound, and TV downstream signals. Multiplexing of downstream signals is accomplished via a form of TDM. The cable company uses one frequency for a set of subscribers (a neighborhood). The subscriber's cable modem listens to the assigned frequency for incoming packets and verifies that the traffic is for that user. This system resembles the operation of a local area network instead of a point-to-point network.

Like ADSL, cable modems are designed to provide higher data rates downstream than upstream. The data rate for upstream delivery can be as high as 1.5 Mbps to 2.0 Mbps. However, because the data from multiple subscribers must be multiplexed into a 6 Mhz bandwidth, the throughput rate declines when many users transmit data simultaneously.

coaxial cable (coax)

Cable modems will be provided by cable companies and will connect to standard RG-59 **coaxial cable (coax)** for the network connection while offering a 10BaseT Ethernet connection for user's local data access.

The two technologies for carrying cable system transmissions have already been mentioned, namely Hybrid Fiber Coaxial (HFC) and Fiber To The Curb (FTTC). The HFC system uses a combination of optical fibers and coaxial cables, with fiber used for the central facilities and coax used for the connections to the individual subscribers. Fiber optics are used for the portion of the network that requires the highest bandwidth and coax is used where lower capacities are required.

FTTC is similar to HFC as it uses optical fiber for high-capacity facilities. Fiber-optic facilities are run close to the end subscriber and then copper is used for the feeder circuits. FTTC differs from HFC in that it uses two media (coax and UTP) for each subscriber to provide for interactive video and voice; thus, the cable company must run an additional wire to each residence to use HFC. The first circuit uses existing coaxial cable to deliver interactive video and the second circuit uses twisted pair, which can be used to carry voice.

head end

Figure 8–13 shows a typical cable network implementation. The cable company **head end** includes a number of components to support the

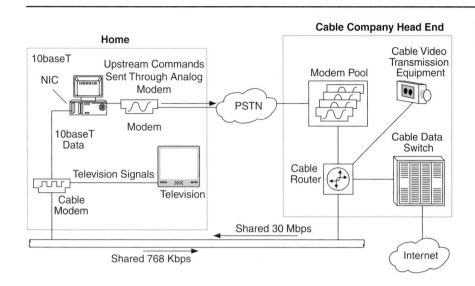

Cable Company Head End

Figure 8–13
Cable Modem Environment

service. These include the cable data switch that provides access to the Internet, the cable video transmission equipment, the modem pool, and the cable router. The residential configuration includes both a cable modem and a modem for connection to the PSTN. A number of users share the cable facility.

8.13   STANDARDS

RFC2662 provides definitions of managed objects for ADSL lines. It contains a standard Management Information Base (MIB) for ADSL lines based on the ADSL Forum standard data model. The standard describes the ADSL Transceiver Unit—Central (ATU-C) and ADSL Transceiver Unit—Remote (ATU-R) components of the ADSL line. Three interface types are defined in this document. As depicted in Figure 8–14, they include a physical, fast channel, and an interleaved channel interface. For each ADSL line, a physical interface always exists.

DSL Standards
A coalition of vendors formed the ADSL Forum in 1994 to promote the concept and facilitate development of ADSL system architecture, protocols, and interfaces. The following activities have impacted the development of ADSL:

- The American National Standards Institute (ANSI) approved an ADSL standard at rates up to 6.1 Mbps in ANSI Standard T1.413.

Figure 8–14
ADSL Interface Types

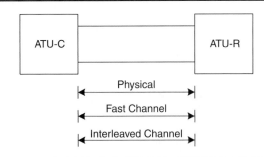

- The European Telecommunications Standards Institute (ETSI) provided an Annex to T1.413 reflecting European requirements.
- The ATM Forum and the Data and Video Council (DAVIC) have both recognized ADSL as a physical layer transmission protocol for unshielded twisted-pair media.
- The ADSL Forum, working in conjunction with the ATM Forum and ANSI, wrote Technical Report TR-002, which specifies how to interwork ATM traffic over ADSL modem links.
- The ATM Forum Residential Broadband Working Group is working on its version of an ADSL/ATM standard that will work with either xDSL or cable modem technologies.

8.14   DSL ISSUES AND CONSIDERATIONS

Issues and considerations that must be addressed before embarking into the DSL environment include those in the network, with equipment, and in the service category. Network issues include items that affect the entire operation and functioning of an xDSL network [DSL Forum, www.adsl.com/dsl.forum.html, 2000], including:

- How are DSL links to be tested, repaired, and managed?
- How should premise issues about splitters and wiring be addressed?
- How should varying DSL bit rates be billed over time and among subscribers?
- How should loading coils and Digital Loop Carrier loops be addressed?
- What technology will be utilized on the service side of the DSLAM?

Equipment issues include items that affect the various product packages in which DSL equipment will be available, including:

- What types of customer equipment interfaces should DSL device support?

- How should CPE be packaged to ensure customer acceptance?
- How much will all this equipment cost?
- Will there be end-to-end turnkey DSL systems?
- Should different DSL technologies be compatible with each other?
- What will be the best technology interface for DSL DSLAM access?

Issues in the service category include items that affect the various service packages that DSL systems will deliver, including:

- Can a subscriber change the voice provider and keep the same DSL service provider?
- Will DSL deployment be as slow as ISDN deployment?
- Will DSL services negatively shift bottlenecks to elsewhere in the network?
- How should DSL services be priced?
- What impact will DSL have on the revenue streams of other services?

Because DSL technologies are designed to run over the local loop, some additional considerations of local loop characteristics are necessary. When the telephone system was digitized in the 1960s and 1970s, the local loop was left as an analog circuit. T-1 service was initiated that used repeaters to clean up a local loop and render it capable of passing digital signals at up to 1.544 Mbps. T-1 service introduced the use of remote terminals to extend the length of a T-1 line; unfortunately, DSL signals stop at remote terminals. The following problems may be encountered on the local loop:

- Carriers may need to recondition or replace some lines within the 18,000 foot distance limitation in order to provide ADSL services.
- ADSL equipment cannot work through bridge taps and loading coils that carriers have installed over the years to boost voice signals.
- It has been a common practice to use different gauges of wire on local loops. Mismatched gauges cause impedance mismatches that produce reflected signals.

Any local loop and T-1 problems must be resolved before DSL can be implemented.

8.15 CABLE MODEM ISSUES AND CONSIDERATIONS

Cable companies, like voice-service carriers, must either develop their own Internet access services or buy these services from an ISP in order to provide transparent Internet access to their subscribers. Although technologies such as FTTC or HFC can deliver digital services to most subscribers, they do not handle all circumstances. The primary problems arise in remote and rural areas. Telephone service for remote farms and villages is normally provided by twisted pair facilities. Rural areas are also the least likely to have cable television service.

8.16   TECHNOLOGY ALTERNATIVES

The xDSL technologies basically provide an access to the public network services. ATM is a networking and transport technology that can be offered as a service over xDSL access, where the xDSL line provides the user access at a lower cost per Mbps. This combination is very attractive to service providers, as it leverages the existing copper plant and offers ATM services over it without the requirement to replace existing copper with fiber-optic cabling. Also, ATM offers virtual connections at specified bandwidths with a guaranteed Quality of Service (QoS).

Technologies that compete for the customer base with DSL are as follows:

- Cable modem
- ISDN
- Frame Relay
- POTS—leased line
- Wireless
- ATM
- SMDS

Product and Service Providers

DSL products and services are available from several manufacturers, suppliers, IXCs, ILECS, and CLECs. The Web is a good place to start product and service research.

DSL Network Service Providers

A representative sample of vendors is presented along with the respective Web URLs.

Alltel	www.altel.com/
Ameritech	www.ameritech.com/
AT&T	www.att.com/dsl/
BellSouth Telecommunications	//fast1.corp.bellsouth.net/adsl/
GTE	//dsl.gte.net
Interspeed	www.interspeed.com/
MCI-Worldcom	www.wcom.com/main.phtml
Pacific Bell	www.pacbell.com/Products_Services/
SBC Technology Resources	www.sbc.com/

SouthWestern Bell	www.swbell.com/Products_Services/
Sprint	www.sprint.com/
UUNet Technologies	www.uu.net/
Verizon Communications	verizon.com/home.html/

DSL Product Suppliers and Manufacturers

Many competing companies provide DSL components. A partial list is included as a starting point for additional research.

3Com	www.3com.com/
ADC Telecommunications	www.adc.com/
Cabletron Systems	www.ctron.com/
Cisco Systems	www.cisco.com/
Copper Mountain	www.coppermountain.com/
Ericsson	www.ericsson.com/
Metalink, Ltd.	www.metalink.co.il/
Netspeed	www.netspeed.com/
Nokia	www.nokia.com/main.html/
Orckit Communications, Ltd	www.orckit.com/
PairGain	www.pairgain.com/index.asp
Paradyne	www.paradyne.com/
Pulsecom	www.pulse.com/
Redback	www.redbacknetworks.com/
Westell Technologies, Inc.	www.westell.com/

8.17   DSL ADVANTAGES AND DISADVANTAGES

Generic advantages that xDSL technologies offer for the service providers include the following:

- xDSL goes in only when a subscriber requests service. Initial costs are much lower than other competing technologies.
- There is no change to central office switch software. In most cases, a splitter carries normal analog voice traffic into the switch.
- xDSL can be used for residential users, small office/home office (SOHO) users, and large organizations alike. The service is basically the same except for streaming video services.
- The various xDSL versions can interface with a number of different premise arrangements. Individual set-top boxes and PCs are supported.
- Most important, xDSL is not a future technology – it is available now!

The subscriber implementing the DSL and cable modem service has some pros and cons to consider. These attributes should help in deciding whether DSL is the proper service for implementation.

DSL Advantages

- DSL operates over ordinary two-wire phone lines (speeds up to 6 Mbps).
- This service is always connected to the Internet like a dedicated line to the home or business.
- There is no waiting for connection or busies.
- DSL enables high bandwidth and push applications.
- DSL is faster than Analog, ISDN, or wireless access.
- DSL provides a secure remote access over a dedicated copper phone line.
- DSL is cost effective to implement, requiring no expensive engineering or infrastructure build-outs, because the infrastructure is in place.
- DSL utilizes existing copper wiring infrastructure, and so DSL is deployable today.
- The service is the same for large and small customers alike.
- The service utilizes standardized premises products such as splitters and PC cards.
- DSL access to other broadband services such as ATM.
- DSL allows rapid file transfer and faster downloads.
- DSL allows constant access to E-mail and shared files.

DSL Disadvantages

- The service is distance sensitive.
- DSL is not available everywhere.
- Splitters require in-house installation.

Cable Modem Advantages

- A cable modem allows high-speed access.
- The service may be available where DSL is not.

Cable Modem Disadvantages

- Cable modem is not available everywhere.
- Subscriber density can impact throughput.
- The service may require an additional line for upstream commands.

8.18   DSL AND CABLE MODEM APPLICATIONS

DSL is a modem technology that uses existing, twisted-pair telephone lines to transport high-bandwidth data, such as multimedia and video, to service subscribers. Any subscriber that has copper access is a candidate for the DSL technology. Faster access for both the residential user and the commercial customer is the primary advantage and selling point. Cable modem technology, although a different technology, allows for similar applications.

A typical application may involve a remote user requesting a large file. The request could be sent at 64 kbps and the file could be returned at T-1 speeds.

Applications for the DSL market can be categorized into (1) general data access for general business applications, (2) home video applications, (3) shared Internet access, and (4) video on demand.

Data Access
- Personal shopping—video catalogs
- Telecommuting—work at home, corporate LANs
- Library research—personal and business
- Education—distance learning
- File downloads—remote LAN access
- Computer Telephone Integration (CTI)
- World Wide Web (WWW) access

Video Applications
- Interactive games—recreation
- Educational programming—distance learning
- Movies—remote CD-ROMS

Shared Internet Access
- Surfing
- Research
- E-commerce

Video on Demand
ADSL comes in two varieties of video conferencing: ADSL-1 and ADSL-3. ADSL-1 provides one MPEG-1 channel on subscriber loops as long as 18,000 ft. ADSL-3 provides four 1.5 Mbps MPEG-1 or one MPEG-2 HDTV signal on loops as long as 9,000 ft. ADSL is not a long-term solution for video on demand. ADSL-1 allows only one video signal at a time and ADSL-3 only works on conventional television.

■ SUMMARY

Since the advent of the Internet, new and faster technologies are needed to deliver data to individual businesses and residences. Telcos have developed solutions that utilize the local loop to provide such a service. Digital Subscriber Line (DSL) services are dedicated, point-to-point, public network access over twisted-pair copper wire on the local loop between a service provider and the user location. The services of choice for development are ADSL and VDSL.

ADSL provides high-speed digital communication over the existing twisted-pair wiring used for analog telephone service. ADSL is attractive because it allows the telephone service to operate on the line simultaneously with the data traffic. An ADSL circuit connects an ADSL modem on each end of the line, creating three information channels: a high-speed downstream channel, a medium-speed duplex channel, and a basic telephone service channel.

VDSL transmits high-speed data over short segments of twisted-pair copper telephone lines, with a range of speeds depending on the line length. VDSL is one of the enabling technologies for Fiber To The Neighborhood (FTTN). VDSL is still in the beta stage of development.

Cable companies have also investigated technologies to deliver digital information to subscribers. One chief obstacle to overcome is that cable is designed to carry information in only one direction. The cable company must replace significant components of their infrastructure to carry traffic upstream and downstream simultaneously.

Where this service is offered, a cable modem is used to attach to cable data services. This type of modem is usually a card that fits into a computer expansion slot and is connected to the coaxial cable used for the CATV system. The cable modem communicates using upstream and downstream frequencies that are already allocated by the cable service.

Whether the user implements a DSL or cable modem technology, or none at all, depends upon the availability of the facilities in any particular geographic area.

Key Terms

2B1Q

ADSL Transceiver Unit–Central Office (ATU-C)

ADSL Transceiver Unit–Remote (ATU-R)

Asymmetrical Digital Subscriber Line (ADSL)

Asymmetrical Transmission

Cable Modem

Carrierless Amplitude Phase (CAP) Modulation

Coaxial Cable (Coax)

Digital Subscriber Line (DSL)

Digital Subscriber Line Access Multiplexer (DSLAM)

Discrete MultiTone (DMT)

Fiber In The Loop (FITL)

Fiber To The Curb (FTTC)

Fiber To The Home (FTTH)

Fiber To The Neighborhood (FTTN)

G.lite

Head End

High-bit-rate DSL (HDSL)

Hybrid Fiber Coaxial (HFC)

ISDN DSL

Multiple Virtual Line (MVL)
Quadrature Amplitude Modulation (QAM)
Rate Adaptive Digital Subscriber Line (RADSL)
Splitter

Splitterless
Symmetrical Digital Subscriber Line (SDSL)
Symmetrical Transmission
Very High-Data-Rate DSL (VDSL)

REVIEW QUESTIONS

1. Identify the numerous different versions of DSL that are on the market today.

2. How does the traditional POTS work? Why is this relevant to DSL?

3. What impact did the Telecommunications Act of 1996 have on DSL?

4. Does DSL require any digital-to-analog transformations? Why or why not?

5. What is the function of a DSLAM? Where is it implemented?

6. What is a splitter? Where is it utilized?

7. What is the difference between splitter and splitterless DSL services?

8. What is the difference between asymmetrical and symmetrical DSL services?

9. What is ADSL and what performance characteristics does it offer subscribers?

10. Describe the ADSL architecture.

11. Describe the ADSL design. How does ADSL manage bandwidth for data transmission?

12. What are the different modulation technologies implemented in DSL?

13. Describe carrierless amplitude and phase.

14. Describe discrete multitone.

15. What is 2B1Q? Where is it utilized?

16. Describe the ADSL frame.

17. Describe the HDSL service.

18. What are the differences between ADSL and HDSL?

19. Describe SDSL.

20. Describe VDSL.

21. How is SDSL different from VDSL?

22. Describe RADSL. How is it similar to ADSL?

23. What is the difference between FTTC, FTTH, FTTN, and HFC?

24. What are the limitations to DSL services?

25. What is an option to DSL? Why is this viable?

26. Provide a brief description of CATV systems.

27. What is a cable modem? What are some of the limitations facing the widespread use of cable modems?

28. How does a cable modem fit in the CATV system?

29. How does the cable modem provide upstream bandwidth to the user?

30. What are the two technologies for carrying cable system transmissions? Describe each system.

31. Identify the three major categories of issues and considerations that should be addressed before embarking into the DSL environment and provide examples for each.

32. What issues must be addressed in the DSL local loop? What are some additional provisions the carriers must make to implement xDSL services?

33. What issues must be addressed when implementing cable modems?

34. Identify advantages for implementing the DSL technology. Why might carriers be especially interested in deploying DSL services?

35. Compare the advantages in question 34 to the disadvantages for implementing DSL.

36. What applications would benefit from the DSL technology? What applications would benefit from the cable modem technology? How would they differ from those same applications using the DSL technology?

37. Define asymmetric and compare with symmetric. Why does it matter which technique is utilized?

38. What is the function of a head end?

39. What is the difference in the physical wiring requirements between DSL and cable modems? How is distance affected by these requirements?

40. Which standards organizations are working on the DSL technologies? What does the IEEE 802.14 standard specify?

41. What equipment must the customer install to utilize xDSL and cable modem?

ACTIVITIES

1. Research the DSL topic on the Web and in other resources and prepare a list of potential alternatives to xDSL that might be proposed by a vendor. Show the advantages and disadvantages for each alternative.

2. Develop an application for xDSL. Show how the DSL technology would enhance the application.

3. Compare cable modem utilization to xDSL utilization. Provide pros and cons for each technology. Show how the applications developed in activity 2 would be impacted when DSL or cable modem is the selected technology.

4. Compare and contrast the different versions of DSL services. Create a spreadsheet with features and options.

5. Research on the Web and in other resources for DSL and cable modem products. This would include splitters and DSLAMs. Develop a spreadsheet showing features and options.

6. Invite a carrier representative to give a presentation that would include service prices.

7. Arrange a visit to a carrier location that supports the DSL or cable modem technology.

CASE STUDY/PROJECT

This case study involves the research and development of an application that would utilize either the DSL or cable modem technology. The project consists of researching the various resources to develop an analysis of the current DSL and cable modem environment. Both of these activities can be group oriented with the requirements of a formal presentation. The research project will consume the most time. Both can qualify for a semester activity.

    1.   Select one of the applications presented in this chapter and develop a network design for either a cable modem or xDSL implementation. Use DataPro for Telecommunications, and the library or the Internet for your research.

The design should include the flow of the data in the application. The analysis should include advantages and disadvantages that may arise from the design. Identify the product and service providers of the requirements, including all costs involved and identified. Consider maintenance and system backup of the application, as well.

   2.   Prepare a comprehensive comparison matrix of the various products and/or services available in the DSL/Cable Modem marketplace. Include features, cost/benefits, limitations, and other information that might be relevant for a telecommunications decision. Prepare a written report and a formal presentation that includes slides. Prepare a list of vital questions to ask in order to obtain the requirements to develop a DSL or cable modem application.

9

Integrated Services Digital Network/ Broadband ISDN

■ INTRODUCTION

One of the first efforts to provide large-scale digital services to subscribers was launched by telephone companies under the name of **Integrated Services Digital Network (ISDN).** ISDN is a set of ITU standards for digital transmission that provides digitized voice and data to subscribers over conventional local loop wiring. ISDN uses the same type of twisted-pair copper wiring as the analog telephone system. ISDN is integrated with the telephone network, so an ISDN subscriber can make a call to a regular telephone number.

Business users who install ISDN adapters (in place of their modems) can see highly graphic Web pages arriving very quickly (up to 128 kbps). ISDN requires adapters at both ends of the transmission, so the access provider must also provide an ISDN adapter. ISDN is generally available in most urban areas in the United States and Europe.

ISDN can best be visualized as digital channels of two types. One type, the bearer (B) channel, carries 64 kbps of digital data. The other type, the delta (D) channel, carries 16 kbps of data and is used for signaling. These two channel types are packaged into two types of access services. The **Basic Rate Access (BRA)** provides two 64 kbps B channels and one 16 kbps D channel. The other type of access is called **Primary Rate Access (PRA).** Primary access provides twenty-three 64 kbps B channels for carrying data and one 64 kbps D channel for signaling.

In addition to providing faster call setup and offering flexible call handling features, ISDN can support the transmission of voice, data, video, and images. The channels can be used separately to handle multiple

OBJECTIVES
Material included in this chapter should enable you to:

- understand ISDN and B-ISDN technology and its relation to the current broadband market. Look at ISDN and identify how it is different from embedded POTS. Look at the differences between BRI and PRI.

- become familiar with the ISDN terms and definitions that are relevant in the Enterprise Networking environment. Identify the various transmission speeds that are allowable with ISDN and B-ISDN.

- understand the function and interaction of the ISDN infrastructure in the Enterprise Network environment. Identify the components that are utilized to implement the ISDN technology. Look at the function of the NT-1 and TA devices.

integrated services digital network (ISDN)
basic rate access (BRA)
primary rate access (PRA)

- understand where ISDN fits in the different applications of the technology. Look at how ISDN can be used to displace POTS and leased line service. Identify the wiring considerations for an ISDN deployment.

- identify the interfaces that comprise the ISDN standard for networking environments. Become familiar with the Q.921, Q.931, and I.420 standards.

- identify alternatives to ISDN. Look at the benefits that can be derived with ISDN. Identify the advantages and disadvantages of the ISDN service.

broadband integrated services digital network (B-ISDN)

applications simultaneously (e.g., fax and voice) or the channels can be used in combination to support a single application at higher speed (e.g., videoconference or Web surfing).

Standards organizations are now developing a **Broadband Integrated Services Digital Network (B-ISDN)** to address the need of new bandwidth-intensive services. Whereas ISDN applies mainly to the narrowband telephony world, the B-ISDN network will use fiber-optic transmission systems and the new multiplexing techniques employed in Asynchronous Transfer Mode (ATM).

9.1 ISDN BASICS

Integrated Services Digital Network (ISDN) was developed in the 1970s by Bellcore and marketed by the seven Bell Operating Companies (BOCs). It was standarized by the ITU-T in 1984. Its purpose was to provide a set of standardized interfaces and signaling protocols for delivering voice and data over a standard telephone line.

In a traditional connection, data originating from a terminal or host computer must be transformed to an analog signal through the use of a modem. The analog signal is then transported to another end user whose modem reconverts the analog signal to digital data for computer and terminal use. With ISDN, the signals originating from digital sources remain digital throughout the network. Terminals and computers connect directly to an ISDN digital line. Even the analog voice signal is digitized by the ISDN telephone prior to being placed on the network. Figure 9–1 provides a visual comparison between traditional and ISDN services.

Rapid advances in computer and communication technologies have resulted in the increasing merger of these two fields. The lines have blurred among computing, switching, and digital transmission equipment, and the same digital techniques are being used for data, voice, and

Figure 9–1
(a) Traditional and (b) ISDN Data Connection to the PSTN

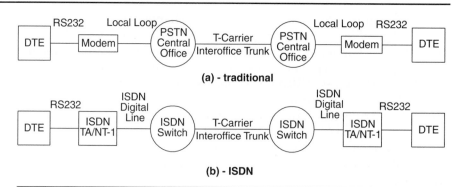

(a) - traditional

(b) - ISDN

image transmission. Emerging and evolving technologies, coupled with increasing demands for efficient and timely collection, processing, and dissemination of information, are leading to the development of integrated systems that transmit and process all types of data.

ISDN services include both packet mode services and circuit mode services. Packet mode ISDN services provide features that are similar to those offered by a X.25 Packet Switched Data Network. Circuit mode services can be used as replacements for conventional telecommunications circuits. As a dial-up service, ISDN is fully switched. Every ISDN station has a telephone number, which means that any ISDN station can call any other station. There is no requirement to set up virtual circuits in advance.

The key service offered by ISDN since its inception has continued to be voice, although many other services have been added. ISDN data transmission services allow users to connect their computers to any other computer across the globe. A new communications service that has become widespread is videotex, which is interactive access to a remote database. Another popular ISDN service is teletex, which is a form of electronic mail for home and business. The goal of ISDN is to integrate these services and make them as commonplace as the telephone.

ISDN Design Criteria

ISDN was designed to provide an end-to-end all-digital transmission methodology regardless of the source—whether data, voice, video, or any mixture of the three. Error rates are reduced over analog data transmission methodologies due to end-to-end control by using network specifications instead of link-to-link error specifications.

ISDN provides a greater capacity because it is compatible with, and will use, the capabilities of SONET and SMDS. ISDN also supports standard bit rates for the various levels of service offered, which makes it compatible with Frame Relay and ATM services. It also provides compatibility with the OSI model and uses the services provided at the upper OSI layers.

The CCITT has defined two major categories of ISDN capabilities— Broadband (B-ISDN) and Narrowband (ISDN).

Broadband ISDN   B-ISDN represents a probable future direction of the telephone industry and will require that the conventional copper-wire local loops that now connect subscriber premises be replaced with optical fiber cables. B-ISDN services are built on top of a physical layer specification called SONET. B-ISDN will use the new multiplexing techniques employed in ATM. B-ISDN services are provided using ATM transmission technology at the OSI Data Link Layer 2 level.

B-ISDN uses packet switching techniques rather than circuit switching methods. It supports variable bit rates, including bursty data (bandwidth on demand). The user is billed for the actual data sent and not the time of the connection.

B-ISDN defines both interactive services (two-way transmission) and distribution services (one-way transmission). B-ISDN has the potential of providing extremely high bandwidth WAN connectivity. Services are intended to support all types of data, including text, documents, graphics, and full-motion video. B-ISDN is explained later in more detail.

Narrowband ISDN   Narrowband ISDN is available today as a copper-based technology, and is supported by standards at the CCITT (international) and ANSI (national) levels. The ISDN access path is a digital pipe to the customer's premise. Regardless of the provisioned service and the equipment type, the connection between the customer and the CO is a digital transmission facility.

basic rate interface (BRI)
primary rate interface
(PRI)

Basic Rate Interface (BRI) and **Primary Rate Interface (PRI)** are the two standard interfaces or access arrangements defined in narrowband ISDN. The capability is based upon a single pair of copper wires to a user's premise. The digital pipe between the provider's CO and the ISDN user carries several communications channels. The capacity of the pipe and the number of channels carried may vary from user to user.

9.2   ISDN ARCHITECTURE

Most major Private Branch Exchange (PBX) vendors offer the ISDN interfaces that are used in business facilities today. Existing two-wire facilities were designed to accommodate analog voice signals, which limit the bandwidth capacity. ISDN signals are digital and require a greater bandwidth than the 3.1 kHz (300 to 3,400 Hz) available over the standard switched voice-grade lines. Unfortunately, the loading coils used in the subscriber loop that have been designed to minimize signal losses within this frequency range have an abrupt increase in signal loss above 3,400 Hz. This loss plays havoc with phase and amplitude characteristics on the ISDN signal.

To overcome these restrictions, the local telephone loops have been upgraded to accommodate the ISDN technology. These upgrades include:

- Higher grade twisted-pair cable installation
- Removal of loading coils
- Installation of digital echo cancellers
- Addition of ISDN-compatible amplifiers and repeaters

ISDN supports a new physical connector for users, a digital subscriber loop, and modifications to all CO equipment. A common physical interface has been defined to provide a DCE/DTE type connection. This same

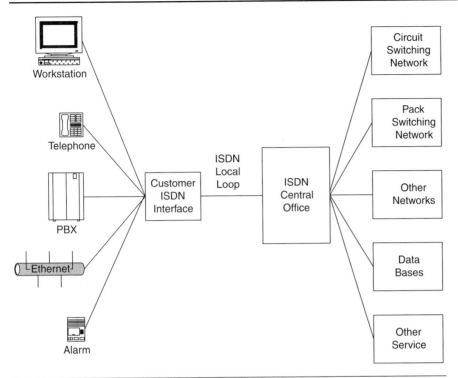

Figure 9–2
ISDN Environment

interface would be usable for telephone, computer terminal, and video terminal devices. Provision has been made for high-speed interfaces to digital PBXs and to Local Area Networks.

The digital CO provides for connectivity of the numerous ISDN subscriber loop signals to the digital network. In addition to providing access to the circuit switched network, the CO provides subscriber access to dedicated lines, packet switched networks, and time-sharing transaction-oriented computer services. The transmission structures are constructed from the following types of channels:

- 4 kHz analog telephone channel
- 64 kbps digital PCM channel for voice or data
- 8 or 16 kbps digital channel
- 16 or 64 kbps digital channel for out-of-band signaling
- 64 kbps digital channel for internal ISDN signaling
- 384, 1,536, or 1,920 kbps digital channel (H channel)

Figure 9–2 provides an illustration of the ISDN architecture [Kessler & Southwick, 1997]. The user has access to ISDN by means of a local interface

Figure 9–3
Separation of User and
Signaling Information

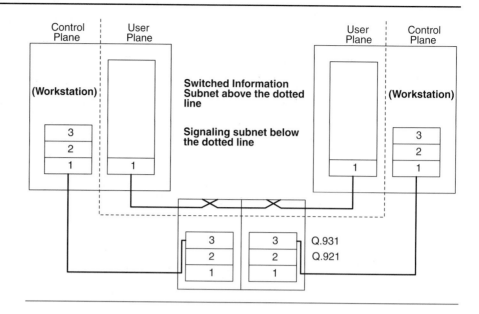

to a digital pipe of a certain bit rate. Pipes of various sizes are available to satisfy differing needs. ISDN can be utilized to provide access to workstations, ISDN telephones, PBX systems, LANs, and alarm systems. These are all accessible through the common equipment that is located in the ISDN central office.

9.3 ISDN CONNECTIVITY

A distinguishing feature of ISDN is that user information and signaling are kept logically separate. The user information lies in the user plane and the signaling lies in the control plane. ISDN is composed of two planes: a switched information plane, and a signaling plane. The signaling protocols are layered in accordance with the OSI reference model (Appendix A).

Q.931
Q.921

     The network layer call control protocol specified in ITU-T standard **Q.931** defines the call control messages and procedures. The link layer protocol defined in **Q.921** also makes sure that the call control messages are reliably passed, without errors, between the terminal and the call control process in the local CO switch. These components are illustrated in Figure 9–3.

I.420
terminal adapter (TA)

     Terminals used in conjunction with the BRI must conform to the **I.420** S/T interface specifications. The diagram depicted in Figure 9–4 shows three of the possible **Terminal Adapter (TA)** configurations, with the ITU-T recommendation numbers pertaining to them. These include

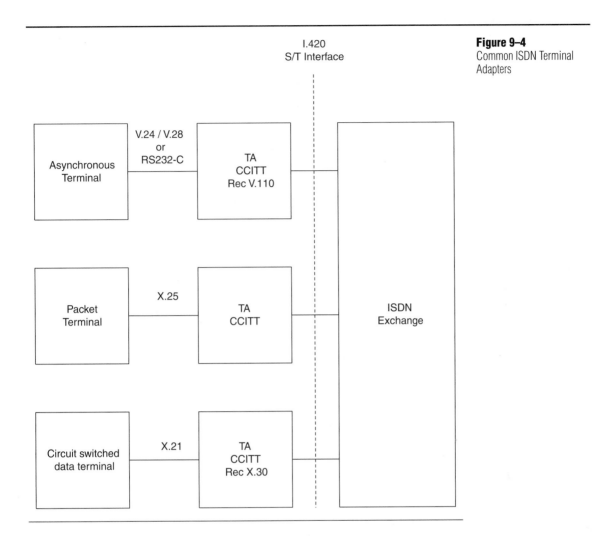

I.420
S/T Interface

Figure 9–4
Common ISDN Terminal
Adapters

an asynchronous terminal, a packet terminal, and a circuit switched data terminal.

Link Access Procedure-D (LAPD) is utilized on the D channel. This standard is based on High-level Data Link Control (HDLC) modified to meet ISDN requirements. All transmission on the D channel is in the form of LAPD frames that are exchanged between the subscriber equipment and an ISDN switching element.

link access procedure-D (LAPD)

The D channel can also be used to provide packet switching services to the subscriber. In this case, the X.25 level 3 protocol is used, and the packets are transmitted in LAPD frames. The X.25 level 3 protocol is used to establish virtual circuits on the D channel to other users and to exchange packetized data.

Figure 9–5
ISDN Interoffice Signaling

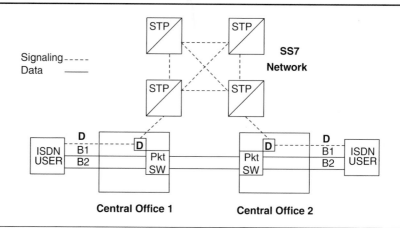

With ISDN, control and signaling information uses out-of-band signaling. Out-of-band signaling provides a noticeable improvement in connection time as well as in overall throughput. ISDN signaling is carried between switches in the network according to the standard known as *Signaling System 7 (SS7)*. SS7 information includes the call's origin and destination phone number, and instructions for the network handling the call. In most cases today, SS7 is used only between switches in the network. If a call is being placed to an ISDN subscriber, the SS7 signaling information is translated to the appropriate D channel format before transmission to the end user. The signaling technology is often used to assist in call routing. Figure 9–5 shows the ISDN/SS7 interoffice signaling diagram. Note that B channels can be packet switched, circuit switched, or one of each.

9.4   ISDN FEATURES

It is important to not confuse ISDN with some kind of new circuit architecture that the telephone company has implemented. ISDN is identified by its service characteristics. These are capabilities that allow new product offerings to become usable. ISDN runs over the same physical network as POTS. There are some new components to handle the advanced features offered, but the network infrastructure remains the same. Once the PCM pulse stream is encapsulated into a DS-1 frame, the user cannot tell the difference.

The following six fundamental features are typically associated with ISDN:

- End-to-End digital connectivity
- Access and Service integration

- Family of Standard Interfaces
- Out-of-Band Signaling
- Common Channel Signaling
- Customer Control

End-to-End Connectivity   Most of the network is digital beyond the local office, although even the local loop is digital. ISDN provides an all-digital network, which avoids the time delay and error possibilities associated with multiple analog/digital conversions in the network. The all-digital network eliminates the requirements for modems on the customer's premise; however, a DCE device such as a NT-1 is required.

Access and Service Integration   For the single-line user, ISDN offers increased capabilities over the same single line, which provides better utilization of existing facilities. The same pair of wires that historically provided a customer premise with either analog voice capabilities or an analog path for carrying data can be used in ISDN to provide both voice and data simultaneously to a user. For data, both circuit and packet switching capabilities are supported, and several applications may be active concurrently on the single line.

Standard Interfaces   ISDN also utilizes standard jacks, thereby allowing a variety of terminal types to be connected to one. Because of the standard jacks, the limited number of CPE interfaces, and the economies of scale that can be achieved through standardized interfaces, ISDN offers the possibility of reduced equipment costs.

Out-of-Band Signaling   ISDN makes more bandwidth available to users by separating all signaling from the information channels. Rather than the "bit robbing" that has been used historically for signaling in telephone networks, ISDN provides a separate signaling channel between the user's premise and the local office, and utilizes the **Common Channel Signaling 7 (CCS7)** Network for interoffice signaling.

common channel signaling 7 (CCS7)

The term "64 clear channel capability" is often used in association with ISDN. Networks have typically robbed 1 kbps on a 64 kbps information channel to provide for signaling and repeater synchronization. Repeater synchronization is maintained by disallowing the all-zero octet, a *Bipolar 8 Zero Substitution (B8ZS)*, string of eight consecutive zeros in the digital bit stream. "Clear channel" or "unrestricted" refers to a user information channel that has no network restrictions on the content of information being carried, and carries no signaling.

Where out-of-band signaling is available, bit robbing of the 64 kbps bit stream is not required for signaling, because the signaling is not passed in the information channel.

Common Channel Signaling   All interoffice and internetwork signaling in ISDN occurs via the Common Channel Signaling (CCS) Network. This network has access to various information databases and intelligent nodes.

Customer Control   Because of the separate signaling channel to the user's premise, ISDN is also characterized by a greater degree of customer control than has been previously possible. Through the use of the signaling channel, customers can reconfigure their networks, change services, and vary available options.

ISDN Circuit Modes   ISDN networking capabilities are divided into bearer services, teleservices, and supplementary services. The bearer services include both circuit mode and packet mode options. The circuit mode options are shown in Table 9–1. Packet mode bearer services include virtual call and permanent virtual call circuits, which are modeled after X.25 switched and permanent virtual circuits. The channel column shows the name of the ISDN communications channel used to deliver the service.

    The teleservices provide for 3.1 kHz speech communications. They also include telex for interactive text communications, fax, and videotext, which provides for retrieval of digital mailbox information. Supplementary services are primarily available for voice communications. These include caller ID and conference calling.

Table 9–1
ISDN Circuit Modes

Information Rate	Channel	Applications
64 kbps	B (bearer)	8 kHz general purpose communications
64 kbps	B	8 kHz digitized speech
64 kbps	B	3.1 kHz audio
64 kbps	B	8 kHz alternate transfer of speech
16 or 64 kbps	D (data)	8 kHz signaling, packet switching, and credit card verification
384 kbps	H-0 (6 B channels)	8 kHz video and PBX link Fast Fax Computer imaging High-speed data LAN internetworking
1.472 Mbps	H-10 (23 64 kbps channels)	Video conferencing LAN-to-LAN communications Computer imaging High-speed data applications
1.536 Mbps	H-11 (H10 + 1 64 kbps D channel)	Same as H10
1.984 Mbps	H-12 (30 64 kbps B channels)	Throughput of 1.920 Mbps
155 Mbps	H-4X	High-speed data, voice, and video

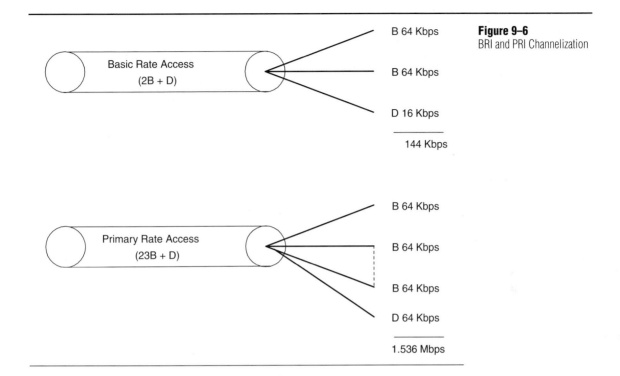

Figure 9–6
BRI and PRI Channelization

9.5 PRIMARY RATE AND BASIC RATE

PRI, with its greater bandwidth, provides small, midsize, and large businesses and telecommunications service companies with sufficient service, whereas BRI is suited more for very small businesses and home use. ISDN lines cannot transmit data or video to analog lines. ISDN lines can only communicate with other ISDN-equipped devices. Both BRI and PRI ISDN can communicate with each other. BRI and PRI users can communicate with POTS for voice communications.

ISDN is based on 64 kbps digitized channels. One channel is dedicated as the control or *Data (D) channel* and the others are information carriers or *Bearer (B) channels*. The PRI employs the use of one or more T-1 formatted links (may also be fiber) and carries twenty-three B and one D channel. BRI service is carried over a two-wire phone line and contains two B channels and one D channel. The D channel operates at 16 kbps and is used for signaling. Figure 9–6 depicts BRI and PRI channelization. Table 9–2 shows the basic and primary ISDN channel capacities and structure.

Information about the calling party can also be passed over to support services such as Caller ID, Automatic Number Identification (ANI),

ISDN Channel Types	Component and Speed
Basic Rate Access	2 B + D 2(64) + 16 = 144 kbps
Primary Rate Access—US & Japan	23 B + D 23(64) + 64 = 1544 kbps
Europe	30 B + D 30(64) + 64 = 2048 kbps

or Dialed Number Identification Service (DNIS). The ISDN D channel passes control information more efficiently than other methods of signaling and is frequently deployed in call centers as a result.

Note that the D channel accesses the SS7 network for the purpose of setting up and tearing down circuits for long-distance calls. It uses an out-of-band configuration.

Basic Rate Access

Two 64 kbps B channels and one 16 kbps D channel form the "2B+D" Basic Rate Interface (BRI) ISDN. Each of the two B channels provides 64 kbps, which a customer may use for either voice or data applications. The data may be either circuit switched or packet switched. A B channel is a 64 kbps clear channel, and the entire channel is switched by the CO to a single network interface.

A single D channel provides 16 kbps and is responsible for all user network signaling. Additionally, the D channel provides the capability to support multiple applications of low speed (up to 9.6 kbps) packet data. The primary function of a D channel is for ISDN call setup and teardown, for starting and stopping a communications session. The D channel is also used for packet switching and low-speed telemetry at times when no signaling is waiting. Figure 9–7 provides a view of basic rate access.

BRI has grown in popularity with the desire for increased speeds to surf the World Wide Web (WWW). BRI is not suitable for some network applications, such as a large file transfer and graphics applications, unless channels are bonded. For example, one BRI line with two 64 kbps channels can be bonded to achieve a 128 kbps connection. A user, therefore, has 144 kbps of usable bandwidth that is bonded from the three channels (64 + 64 + 16). Note that framing synchronization and other overhead bits bring the total bit-rate on a basic access link to 192 kbps.

The basic service is intended to meet the service requirements of most individual users. It allows for the simultaneous use of voice and several data applications, such as packet switched access, facsimile, and teletex. These services can be accessed through a multifunction terminal or several separate terminals, but a single interface is provided.

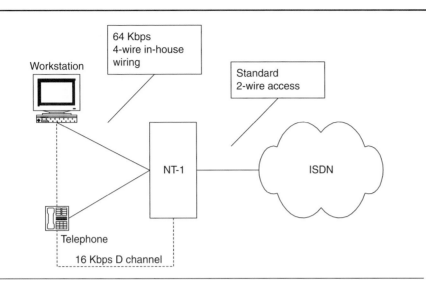

Figure 9–7
BRI Configuration

Primary Rate Access

The Primary Rate Interface (PRI) consists of twenty-three B channels and one 64 kbps D channel in the United States, or thirty B channels and one D channel in Europe. PRI is based on the T-1 technology with two pairs of wires to a user's premise. Thus, a Primary Rate user on a T-1 line can have up to 1.544 Mbps service or up to 2.048 Mbps service on an E-1 line. The actual usable capacity of PRI is 1.536 Mbps. PRI uses the Q.931 messaging protocol over the D channel for call setup and teardown. The Q.931 protocol describes what makes up the signaling packet and defines the message type and content.

The Primary Rate channels are carried on a T-1 line (in the United States, Canada, and Japan) or an E-1 line (in other countries) and are typically used by midsize to large enterprises. The twenty-three (or thirty) B channels can be used flexibly and reassigned when necessary to meet special needs such as videoconference and call-center applications. The Primary Rate user is hooked up directly to a CO by means of a multiplexer or a PBX, and the aggregate of 24 channels is called a **trunk.** A multiplexer   trunk
is typically used when the PRI ISDN provides a LAN-to-LAN connection, or for ISP access. A PBX is used for video conferencing and for telephone call centers that maintain databases of customer telephone numbers linked to customer records. Figure 9–8 depicts a LAN connection over PRI to a telco facility.

The PRI B channels are similar to those supported by BRI; they are 64 kbps clear channels that may be used for voice, circuit switched data, or

Figure 9–8
ISDN PRI Connectivity

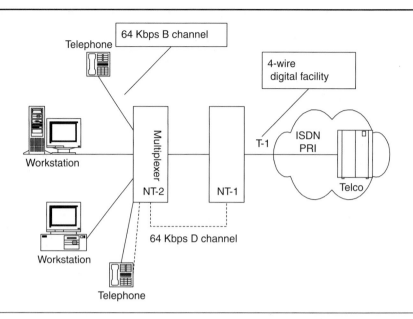

packet switched data. The PRI D channel also provides 64 kbps and carries user network signaling. The D channel on one 23 + D can be used to control a number of twenty-four B PRI connections.

A customer with high data rate requirements may be provided with more than one physical interface. A single D channel on one of the interfaces suffices for all signaling purposes, and all other interfaces may consist solely of B channels.

European Primary Access

The European ISDN PRI has twenty-nine 64 kbps bearer channels and one 64 kbps D channel (29B + 1D), which interfaces easily into E-1 systems. The E-1 system has thirty data channels, a framing channel, and a synchronization channel. The total data rate for all twenty-nine 64 kbps ISDN B channels is 1,856 kbps.

Bearer Services

ISDN B channels are sometimes referred to as Bearer Channels. The B channels are the vehicles for providing Bearer Servers, which are basic subscription type services, to a user's CPE. Bearer Services provide the capability for carrying user information over an ISDN connection, and are used by the ISDN and CPE. Each bearer service is based on the lower three layers of the OSI reference model (Physical, Data Link, and Network). Different carriers may take a variety of approaches in implementing these services.

The B channel can be used to carry digital data, PCM-encoded digital voice, or a mixture of lower rate traffic, including digital data and digitized voice encoded at a fraction of 64 kbps. Four types of connections can be set up over a B channel:

- Circuit-switched Connection is equivalent to the switched digital service available today. The user places a call and a circuit switched connection is established with another network user.
- Packet-switched Connection is the user connected to a packet switching node, and data are exchanged with other users via the X.25 technology.
- Semipermanent Connection is a connection to another user set up by prior arrangement and not requiring a call establishment protocol. It is equivalent to a leased line.
- Frame Mode is the user connected to a Frame Relay node, and data are exchanged with other users via Link Access Procedure—Frame (LAPF).

The B or bearer channels can also be bonded together to provide 128 kbps. The bonding ability of ISDN makes the service more flexible for data communications purposes. Bonding works on both PRI and BRI circuits and produces the benefit known as "bandwidth on demand." The 64 kbps size of the bearer channels was chosen to accommodate a 64 kbps PCM voice channel.

It is important to understand that each of the B channels can be used independently. There might be a data link to a computer on one B channel, and voice communications on the other.

If a user requires a higher rate than 64 kbps, this can be provided using a technique called multirate ISDN. Multiple B channels are synchronized, and their combined data rate is made available to the user as if it were a single channel.

D Channels

A data link (D) channel is primarily used for control purposes, including the establishment, management, and termination of services. A D channel can also be used for packet switching and low-speed telemetry at times when no signaling is waiting.

When packet-switching service is provided internal to ISDN, it can also be accessed on the D channel. For D channel access, ISDN provides a semipermanent connection to a packet switching node with ISDN. The user employs the X.25 level 3 protocol as is done in the case of a B channel virtual call. Here, the level 3 protocol is carried by the LAPD frames. Because the D channel is also used for control signaling, some means is needed to distinguish between X.25 packet traffic and ISDN control traffic. This is accomplished by the means of a link layer addressing scheme.

H Channels

For user services requiring additional bandwidth for high-speed transmission, H channels are used. H channels carry multiple B channels at rates of 384 kbps, 1,536 kbps, and 1,920 kbps. They are used for audio, video, and graphics conferencing; business television; and other high-speed applications.

9.6   FUNCTIONAL GROUPS

Functional groups are sets of functions that may be needed in ISDN user access arrangements. Specific functions in a functional group may be performed in one or more pieces of equipment. These functions occur in the following components.

- Terminal Equipment (TE)
  - Terminal Equipment Type 1 (TE-1)
  - Terminal Equipment Type 2 (TE-2)
- Terminal Adapter (TA)
- Network Termination 1 (NT-1)
- Network Termination 2 (NT-2)
- Line Termination (LT)
- Exchange Termination (ET)

network termination (NT)

NT-1
NT-2

Network Termination (NT) is the general term used to denote the group of functions used to terminate the network connection. There are two NT designations—**NT-1** and **NT-2**. NT-1 is used to terminate the local digital loop and provide such functions as maintenance, performance monitoring, timing, power transfer, channel multiplexing, and multiple device access contention resolution at the network interface. NT-2 includes the functions normally associated with digital PBXs and LANs. The functions include multiplexing and protocol handling, switching, concentration, maintenance, and interface termination. The NT-2 could control hundreds of terminals (TE-1) behind it.

terminal equipment (TE)

TE-1

TE-2

The **Terminal Equipment (TE)** grouping includes digital telephones, data terminal equipment, and integrated workstations. The functions include protocol handling, maintenance, interface functions, and connections to other equipment. **TE-1** is the subset of TEs that can directly connect to the NT-1 or NT-2 at the S or T reference point. This is the ISDN terminal. The **TE-2** is the subset of TEs that cannot directly connect to a NT-1 or NT-2. All existing user equipment fits this definition. They require a Terminal Adapter to interface the ISDN network correctly. Everything from standard 2500 telephone sets to IBM 3270 terminal equipment is considered TE-2.

The Terminal Adapter (TA) functions to adapt TE-2 equipment to properly comply with the ISDN user-to-network interface recommendations. It is basically a device that connects an analog or digital source to the four-wire S interface. In the United States, the terminal adapter is owned by the subscriber and must be purchased before connection to an ISDN line is possible with non-ISDN equipment. An example of a TA is a card in a PC that allows a telephone and the PC to work on a basic access line.

9.7 REFERENCE POINTS

In ISDN, the architecture on the user's premises is specified functionally by groups separated by reference points. This separation permits development of interface standards at each reference point, thereby organizing the standards' work and providing guidance to equipment providers.

The reference points are the conceptual points dividing functional groups [Beyda, 2000]. In specific access arrangements, a given reference point may correspond to a physical interface between pieces of equipment. A reference point, when physically realized by an interface, requires the specification of at least two interface points. It is important to note that reference points are conceptual and do not necessarily indicate an actual physical interface. The reference points are designated as R, S, T, S/T, U, and V. Figure 9–9 depicts these reference points.

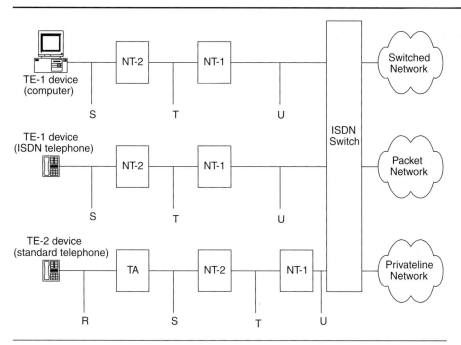

Figure 9–9
Configuration Depicting Reference Points

R is the reference point between a non-ISDN Terminal (TE-2) and a Terminal Adapter (TA). It includes several non-ISDN standard and non-standard interfaces. When physically implemented, this is typically a connector such as a RJ11, RS232-C, or V.35. The R interface is not defined in the CCITT ISDN recommendation because it is not an ISDN compatible interface. Terminal adapters are available to allow numerous types of existing equipment to be connected to the ISDN via its R interface. Basically, this means that the interface is employed to connect to a non-ISDN telephone, but provides limited ISDN service features to the telephone set.

S is the reference point between a TE-1 or TA and a NT-2. This is a CCITT standard. If no NT-2 is present, the S and T interfaces are identical, which is why it is convenient to speak of these together as one, as in the S/T interface.

T is the reference point between a NT-2 and a NT-1. This is a CCITT standard. The actual interface specifications are exactly the same as the S reference point.

S/T is the reference point between a TE-1 and a NT-1, in the absence of a NT-2. This is a CCITT standard. The S/T interface separates the transmit and receive signals onto different wire pairs, provides full ISDN services, and is used to connect computer equipment and telephones to the WAN.

U is the reference point between a NT-1 and a LT. The reference point is of particular interest in the United States, because of the ANSI U interface standard for the U.S. Equipment located on a user's premise, considered Customer Premises Equipment (CPE). Thus, everything described in the CCITT ISDN customer access to the services reference model that is located at the user's location is owned by the user. All is well, except that when the functions of the NT-1 were originally defined, it was intended that the NT-1 be part of the network and owned by the telephone company. The NT-1 was to be designed and manufactured to work with the specific CO that was installed in the area. Since the NT-1 is considered CPE here, a new reference point had to be defined. It is called the U interface. The U interface is that point at a user location where the local loop coming from the CO is connected to the NT-1. Basically, it provides full-duplex communications over twisted-pair cable, and is used for a single connection device.

line termination (LT)
exchange termination (ET)

V is the reference point between the **Line Termination (LT)** and the **Exchange Termination (ET)** in the CO. The LT is typically the location of the line card in the CO, and is the network counterpart of the NT-1. This is not a CCITT standard, but is particular to a switch manufacturer.

9.8 WIRING CONFIGURATIONS

U interface

The **U interface** is a standard two-wire connection to the service provider, as would occur with a standard analog voice line. However, because it uses pulse modulation, it is called a digital local loop or digital subscriber line.

Since loading coils must be removed from the telco facilities for the ISDN service, signal attenuation becomes quite large. This can amount to -40 decibels at the 3 mi limit for ISDN facilities. This large attenuation necessitates a special signal transmission technique. ISDN data are not transmitted by a binary level signal, such as RS232. A four-level two-binary state protocol is used instead. The protocol, as discussed in Chapter 8, is 2B1Q. This protocol reduces the required bandwidth and counteracts the distortion and attenuation on the lines.

Three different Basic Rate Interface wiring configurations defined to connect TE-1s and TAs to NTs are:

1. Point-to-Point
2. Short passive bus
3. Extended passive bus

The Primary Rate Interface access line is electrically and physically identical to T-1 service. The major characteristics are:

1. Line rate: 1.544 Mbps
2. Line code: Bipolar with 8 Zero Substitution
3. Framing: Extended Superframe

ISDN Devices

End-to-end digital connectivity is a big advantage for some users. With an all-digital connectivity, there is no conversion from or to analog signaling, and thus no modem is required. Another important issue of connectivity is service transparency.

Service transparency means that because the line is all digital, any digitized data source can be multiplexed into it transparently. TDM requires that all inputs must be digital data, so if video is digitized, it can be combined and sent along with a PCM-encoded voice transmission and ISDN will not know the difference. ISDN offers TDM channels directly to the user. Figure 9–10 shows the components of ISDN connectivity between several LANs and workstation sites.

Special equipment and phones with ISDN interfaces are required to implement the ISDN rules and protocols. These are referred to as Network Type 1 (NT-1) and Terminal Adapter (TA) interfaces. For businesses, this equipment would most often be implemented in the PBX or ACD.

The NT-1 provides functions related to the physical and electrical termination of the local loop between the network and the user premise. NT-1 devices (Figure 9–11) are required for both BRI and PRI service. A TA allows existing non-ISDN terminals to operate on ISDN lines by providing conversion of the user/ISDN interface. Table 9–3 provides a

Figure 9–10
ISDN Configuration

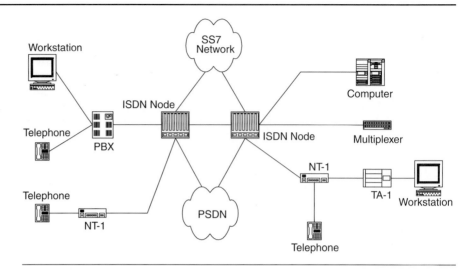

Table 9–3
ISDN Devices

Equipment at the Central Office	Description
LT	Local termination
ET	Exchange termination
PH	Packet handler
Equipment at the subscriber's location	**Description**
NT-1	Network Termination 1—properly terminates the twisted-pair line and gives correct timing, power feed, and error statistics.
NT-2	Network Termination 2—provides distribution switching, multiplexing, and concentrating. This may be used in a PBX or LAN server.
TE-1	Terminal Equipment 1—is fully ISDN compatible
TE-2	Terminal Equipment 2—is non-ISDN compatible
TA	Terminal Adapter—is used to connect TE-2 equipment to ISDN

summary of the devices that can be utilized to connect ISDN service at the CO and the subscriber's location.

A LAN connection to an ISDN WAN is accomplished by using an ISDN interface module in a router, server, or FRAD. When a router is used, it can connect multiple LAN segments to the ISDN WAN, and can minimize network congestion.

ISDN NT-1   ISDN Modem

Figure 9–11
NT-1 and ISDN Modem

When you connect to ISDN without using a leased line, the provider is likely to offer circuit mode services, packet mode services, or both. Circuit mode means that the communications facility lasts for the duration of the communications session, and is used exclusively by the two connected devices until it is terminated. Circuit mode is most commonly used in voice transmissions. Packet mode, used for data, means that several facility paths can be used during a communications session, and each connected device is assigned an address and sequence number at the start of the session to ensure that data arrive at the correct destination. The advantage of packet mode is that it makes maximum use of the available bandwidth.

9.9   ISDN FRAME

ISDN uses a frame format called Link Access Procedure-D (LAPD) channel, which is similar in structure to the LAPB format of X.25. Figure 9–12 depicts the ISDN LAPD frame format.

Link Access Procedure-D is a synchronous data link protocol based on ISO High-level Data Link Control (HDLC). LAPD offers two types of service:

- Acknowledged Information Transfer
- Unacknowledged Information Transfer

The ISDN frame fields are as follows:

- Flag—indicates the start of the frame
- Address—Contains the address of the endpoint node or nodes
- Control—Contains transmission control information which includes the identification of the circuit in use and the type of packet being sent

Figure 9–12
ISDN LAPD Frame

Flag	Address	Control	Data	FCS	Flag

service access point
identifier (SAPI)
terminal endpoint identi-
fier (TEI)

- Data—Contains the packet header and payload data that are transported within ISDN
- FCS—Used to provide error checking
- Flag—Indicates the end of the frame

The protocol is independent of a transmission bit rate and requires a full duplex, bit transparent, synchronous channel. It has a frame format similar to HDLC, but the LAPD provides for octets in the address field. This is necessary for multiplexing multiple sessions and user stations onto the BRI channel.

The address field contains several control bits, a **Service Access Point Identifier (SAPI),** and a **Terminal Endpoint Identifier (TEI).** The SAPI and TEI fields are known collectively as the data link connection identifier.

The SAPI identifies the entity where the data link layer services are provided to Layer 3. The TEI identifies either a single terminal or multiple terminals that are operating on the BRI channel. The TEI is assigned automatically by a separate assignment procedure. A TEI value of all ones identifies a broadcast connection.

9.10   BROADBAND ISDN OVERVIEW

When ISDN was first defined in the mid-1980s, most carriers did not have a widespread deployment of optical fiber. Because ISDN was principally designed for a copper-based environment, no plans had been developed to take advantage of high-speed, high-bandwidth optical fiber for speeds greater than 1.544 Mbps.

At the beginning of the 1990s, the CCITT (ITU-T) began the process of defining a strategy for broadband ISDN to address the new application needs. Broadband ISDN (B-ISDN) using SONET's networking foundation is being designed to take advantage of optical fiber and the gigabit per second speeds it offers. Initial B-ISDN specifications call for bit rates of 155 Mbps (OC-3) and greater using either ATM employed by cell relay technology or a Synchronous Transfer Mode (STM) of operation.

Broadband ISDN is a fiber-based technology characterized by a high bandwidth and is used in such applications as High-Definition Television (HDTV). Ultrahigh-speed computer links to fulfill high-speed data delivery requirements are being used to provide services between LANs, MANs, and WANs. Whereas ISDN applies mainly to the narrowband communications world, the B-ISDN network will use fiber-optic transmission systems capable of transmitting at speeds ranging from 150 Mbps to 2.5 Gbps.

B-ISDN will provide subscribers with higher throughput, better performance, and standard interfaces, because of its relationship to SONET and other emerging technologies designed for the high-speed movement of voice and data. B-ISDN standards are also being developed to address trouble indicators, fault locators, and performance monitors. These will be designed to interface to a variety of customer equipment and to B-ISDN central office switches to provide network operational viability.

B-ISDN will additionally provide a number of new services to its user base. The CCITT has defined two classes of service called interactive services and distribution services. Interactive services are designed as two-way information exchanges such as conversational services, message services, and retrieval services. Distribution services primarily handle information movement in one direction from the service provider to the B-ISDN subscriber. Distribution services include broadcast services and cyclical services (videotext).

The protocol architecture for B-ISDN introduces some new elements not found in the ISDN architecture. For B-ISDN, it is assumed that the transfer of information across the User Network Interface (UNI) will use ATM. The use of ATM for B-ISDN implies that it will be a packet-based network at the interface and in terms of its internal switching. Although the recommendations state that B-ISDN will support circuit mode applications, this will be done over a packet-based transport mechanism.

The set of specifications detail requirements for ISDN networks to handle transmission speeds of 100 Mbps, 155 Mbps, and 600 Mbps over 1 km cable segments utilizing repeater interfaces to extend network distances. In composing this specification, the authors were required to meet N-ISDN interface specifications as well as broadband network needs. A standard N-ISDN terminal or network interface and a Broadband Terminal Interface (BTI) are serviced by the Subscriber's Premises Network (SPN). The SPN multiplexes the incoming data and transfers them to the broadband node, called a Broadband Network Termination (BNT), which in turn codes the information into packets used by the B-ISDN.

Data transfers to and from the B-ISDN may be asymmetric. That is, access on and off the B-ISDN may occur at different rates depending on system requirements. Subscriber's Premises Networks (SPNs) can be a PBX or a LAN configuration. The BNT interface with supporting software is capable of recognizing protocols and frame formats of the data supplied to the interfacing network. This means that ISDN, which began as an evolution from the circuit switching telephone network, will transform itself into a packet switching network as it takes on broadband services.

9.11   ISDN STANDARDS

One objective of ISDN is to allow for international data exchange. This requires interfaces between national and regional providers of such services. The first mission of the ISDN program was to define the functions and characteristics of the network and to establish implementation standards. In 1984, the ITU produced the first of these standards. These "I" series of standards are identified as I.100 through I.700:

- I.100 Series—General concepts
- I.200 Series—Service capabilities
- I.300 Series—Network aspects
- I.400 Series—User Network interfaces
- I.500 Series—Internetwork interfaces
- I.600 Series—Maintenance principles
- I.700 Series—B-ISDN equipment aspects

Additional standards information is available in the following documentation:

- CCITT I.430 and I.431—Basic and Primary Rate physical layer interfaces
- RFC 1618—PPP over ISDN
- RFC 1356—Multiprotocol Interconnect on X.25 and ISDN in the Packet Mode
- RFC 2127—The ISDN Management Information Base (MIB) using SMIv2

Table 9–4 depicts the OSI model and the related ISDN model.

Table 9–4
OSI Model and ISDN Layers

OSI Layer	ISDN Function
Transport	Responsible for the reliability of communications
Network	Responsible for physical and logical routes for data transmission
	Q.931 call control/I.451
	X.25 packet level
Data link	Responsible for point-to-point connectivity, signal formatting, and error detection
	LAPD (Q.921)/I.441
Physical	Responsible for physical connectivity
	I.430 basic interface
	I.431 primary interface

National ISDN

ISDN implementation in the United States has been facilitated by the introduction of the **National ISDN** standards [National ISDN Council, www.nationalisdncouncil.com, 2000]. These are Bellcore standards designed to ensure the consistent deployment of ISDN in a subscriber's private network, regardless of the carrier or equipment manufacturer. National ISDN represents an industry-wide effort to establish consensus on standard technical specifications and implementation agreements. With implementation of National ISDN, subscribers will be able to call anywhere National ISDN is available.

National ISDN

9.12 ISDN ADVANTAGES, BENEFITS, AND DISADVANTAGES

ISDN Benefits

The benefits of ISDN include the following:

- Provides efficient multiplexed access to the public network
- Has the capability to support integrated voice and data
- Has a robust signaling channel, which is important to network management
- Provides a standard open system interface that is internationally defined
- Is offered by most carriers as a circuit-switched service
- Is a dial-up service designed for occasional connections
- Provides digital voice transmission
- Provides for office automation of routing and document access
- Allows for video telephone service
- Transports graphic images for security applications
- Has a layered protocol structure compatible with the OSI model
- Offers communications channels in multiples of 64 kbps
- Has switched and non-switched connection services
- Provides Broadband ISDN capabilities of 155 Mbps and higher
- Provides true 64 kbps clear channel data rate on each B channel
- Provides end-to-end digital connectivity allowing service transparency

ISDN Advantages

For constant use, services that use dedicated connections such as Frame Relay are much more economical. As a dial-up service, ISDN is fully switched, and therefore every ISDN station has a telephone number. Any ISDN station can call any other station, just like any other telephone call. There are no virtual circuits to set up before use. ISDN is integrated with the telephone network, so an ISDN user can call a regular telephone number. On the surface, ISDN can appear relatively expensive compared to the other two PSDN technologies, Frame Relay and ATM. The user pays for the ISDN line's capacity even if it is not being used during a dial-up connection. When positioned for the right application, however, the customer can save a lot of money. Additional advantages include the following:

- ISDN provides the ability to merge different telecommunications services onto the same access line, potentially providing lower costs than separate access for each service. It is possible to switch between voice, data, and video services, or use all of them simultaneously. There is a standard signaling arrangement for all services.
- ISDN PRI trunks can be provided over self-healing networks, reducing the vulnerability to outage.
- Transmission performance is enhanced. PRI Digital trunks can be operated with no loss and imperceptible noise. By contrast, the loss and noise of analog trunks increase with the distance from the central office.
- PRI trunks can be added up to the twenty-three trunk capacity with no wiring work or hardware additions on the customers' premises.
- ISDN network intelligence allows subscribers to instruct the network to respond differently to time of day, day of week, or identification of the calling party. Incoming calls can be treated with the same kind of discrimination the callers would receive if they arrived in person.

ISDN Disadvantages

There are still several impediments to implementing ISDN. These include the following:

- ISDN is still not available everywhere.
- The ordering process for ISDN is complex and cumbersome. Depending on what combinations of voice, video, or data traffic that a user wishes to transmit over ISDN, twenty or more ordering codes are possible.
- ISDN has insufficient bandwidth for many multimedia applications.
- During a power failure, ISDN will not work because it relies on TA and NT devices which require local subscriber electrical power.

9.13   ISDN APPLICATIONS

ISDN applications development is more complex and detailed than that associated with existing services. ISDN offers the integration of voice and data services in a single network. Historically, users required separate access to separate networks for different services. ISDN provides integrated access to a single network for a variety of services, including voice, data, and image [North American ISDN User's Forum, www.niuf.nist.gov, 2000]. Applications are constructed on building blocks of:

- Customer communication needs for both voice and data
- Intelligent CPE
- Highly functional software
- Computer connectivity for multiple database access
- Advanced Intelligent Network (AIN) services
- ISDN-based services
- Multiple vendor participation and coordination
- Internet access
- Digital voice transmission

An application that can be fulfilled by ISDN is desktop videoconferencing, which operates on ISDN's digital channels. This can be utilized effectively in the telecommuting environment. As more incentives develop for workers to spend part of the workweek at home, ISDN can be an economical method of retaining a link to the office at a reasonable transport speed.

Applications that are conducive to dial-up data access such as ISDN are:

- Disaster Recovery
- Video Conferencing
- Large image and data file transfers
- Remote LAN access
- Internet Connectivity
- Caller ID/CTI
- LAN-to-LAN connectivity
- Home office access
- Office automation
- Off-site backup and disaster recovery for business computer systems
- Security
- PBX connection to the local telco
- Concurrent transfer of voice and video

An endless number of features can be provided to the subscriber with ISDN. The extensive signaling capabilities provided in ISDN allow new features to be added later simply by modifying the software in the existing CO switches.

9.14 ISDN TECHNOLOGY ALTERNATIVES

Services and technologies that can be utilized in lieu of ISDN include the traditional leased line or POTS, Frame Relay, and DSL. ISDN competes with private line when a digital end-to-end facility is desired. Private line might be the only alternative if the newer technologies are not available. Private line has some speed advantages over ISDN.

ISDN does not really compete with Frame Relay, but it can be utilized as a fall-back facility for Frame Relay. A number of products, such as routers, have ISDN and other serial interfaces. Frame Relay requires an access line and PVCs, where ISDN is a dial-up method.

ISDN can be utilized instead of DSL; however, ADSL has an advantage if asymmetrical transmission is required. ISDN is available today in more locations than DSL technologies.

ISDN Customer Premises Equipment

Some specialized manufacturers offer ISDN CPE devices [Access Technologies Forum [ACTET], www.via-isdn.org/index.htm, 2000]. These devices include NT-1s, ISDN modems, and communication devices that have ISDN components integrated in the chassis. Because this market is so dynamic, it is essential that research be conducted before selecting the ISDN devices.

ACC Congo	www.acc.com/
Adtran	www.adtran.com/
Arescom	www.arescom.com/
Diamond Multimedia Systems, Inc.	www.diamondmm.com/
Gandalf Systems Corp.	www.gandalf.com/
Intel	www.intel.com/
Link Technology	www.linktechnology.com/
Lucent Technologies	www.lucent.com/micro/isdn/
Tone Commander	www.tonecommander.com/
Trillium	www.trillium.com/
US Robotics, Inc.	www.usr.com/

9.15   ISSUES AND CONSIDERATIONS

Whether ISDN is available locally depends on the services offered by the local telco, and on whether the telecommunications equipment is upgraded for ISDN. When checking on availability of ISDN services, determine which protocol is used by the provider. The most commonly used protocols are National ISDN-1 and National ISDN-2. It is necessary to know which protocol is in use in order to configure the CPE devices.

ISDN cabling can be twisted-pair copper wire or fiber-optic cable. Fiber-optic cable is preferred because it provides the best connectivity and high-speed options, particularly PRI and B-ISDN. If twisted pair is utilized, look at the distance limitations, cable quality, and noise reduction devices that might be utilized in the physical plant.

ISDN is cost effective if used as a dial-up, nondedicated link. Most ISPs charge per hour for ISDN connections. Dedicated service can cost five to fifty times as much as a nondedicated ISDN line. To use ISDN service, two factors must be present:

1. The CO must provide this service.
2. An ISDN-compatible device must be used in the user's office.

One such device is an ISDN adapter card that can be added to a computer; another is an external interface or Terminal Adapter (TA). TAs are generally priced from $250 to $600 each. With a TA, a user can connect an analog and a digital source. The digital source communicates with another ISDN service and the analog source can be used for voice communications to any POTS phone.

■ SUMMARY

ISDN is a network that provides end-to-end digital connectivity to support a wide range of services, including voice and data, to which users have access by a limited set of standard multipurpose customer interfaces. ISDN is an enhancement to the telephone local loop that will allow both voice and data to be carried over the same twisted pair. It is a fully digital network, where all devices and applications present themselves in digital form. This means that the requirements for analog-to-digital and digital-to-analog conversions are eliminated.

ISDN supports a variety of applications, including both switched and nonswitched connections. Switched connections in ISDN include both circuit-switched and packet-switched connections and their concatenations.

The ISDN concept was based on the idea that the functions of transmission and switching could be integrated to form an Integrated Digital Network. To summarize, ISDN:

- Is an architecture rather than a service.
- Provides end-to-end digital connectivity.
- Provides simultaneous transmission of voice and data over a single common access line.
- Supports a wide range of digital telecommunications services.
- Provides user selectability, which allows customer control of services and features.
- Uses a small family of standard multipurpose customer interfaces.

When telephone companies first defined ISDN, 64 kbps seemed fast compared with dial-up modems. As the years passed, however, dial-up modems improved, and alternative technologies were invented to provide higher data rates across the local loop at lower cost. Consequently, ISDN is now an expensive alternative that offers little bandwidth.

An upgrade to ISDN, Broadband ISDN (B-ISDN), has been defined, but has not yet been widely implemented. Broadband ISDN has three services:

- A full-duplex circuit operating at 155.52 Mbps
- A full-duplex circuit operating at 622.08 Mbps
- An asymmetrical circuit with two simplex channels, one operating at 155.52 Mbps and the other at 622.08 Mbps

Narrowband ISDN and Broadband ISDN features can be summarized as follows:

Broadband ISDN	**Narrowband ISDN**
Uses high-speed facilities	Uses medium-speed facilities
Packet switching	Circuit switching
Virtual channels	Fixed channels
Variable bit rates	Fixed bit rates
High throughput	Less throughput

Key Terms

Basic Rate Access (BRA)

Basic Rate Interface (BRI)

Broadband Integrated Services Digital Network (B-ISDN)

Common Channel Signaling 7 (CCS7)

Exchange Termination (ET)

I.420

Integrated Services Digital Network (ISDN)

Link Access Procedure-D (LAPD)

Line Termination (LT)

National ISDN

Network Termination (NT)

NT-1

NT-2

Primary Rate Access (PRA)

Primary Rate Interface (PRI)

Q.921

Q.931

Service Access Point Identifier (SAPI)

TE-1

TE-2

Terminal Adapter (TA)

Terminal Endpoint Identifier (TEI)

Terminal Equipment (TE)

Trunk

U Interface

1. When was ISDN developed? Provide organization names and dates.

2. How does ISDN differ from traditional PSTN connectivity? Is ISDN Dial-up or Dedicated?

3. What was the design criteria for ISDN?

4. What are the two major categories of ISDN? What are their major differences?

5. What had to be done to the network to accommodate ISDN?

6. What transmission channels were constructed for ISDN?

7. What is I.420 and what are its possible configurations?

8. What are the implications for separate user and signaling information channels?

9. What is out-of-band signaling? What is SS7?

10. What are the six fundamental features that are associated with ISDN?

11. Discuss the division of networking capabilities of ISDN. What are the various channel rates and channel designations?

12. What is the difference between PRI and BRI?

13. Identify two applications that are a good fit for PRI. Identify two applications that are a good fit for BRI.

14. Where would you deploy basic rate and primary rate services?

15. What is the difference between U.S. and European BRI services?

16. What are bearer services?

17. What is the use for the D channel?

18. What is an H channel? How is it used?

19. There are a number of functional groups for user access arrangements. Where do these functions occur?

20. What is the function of a terminal adapter?

21. There are a number of reference points on the user's premises. What are the designations for these points? What is the importance of the U interface?

22. What is 2B1Q? Why is it used?

23. What are the three different BRI wiring configurations? What are the PRI access line characteristics?

24. What are the devices that are used to connect a LAN to an ISDN WAN? What equipment is used at the central office? What is the function of a NT-1?

25. What is the format of the ISDN frame? What are the various fields that are contained in the frame?

26. What are the contents of the address field? How are these fields utilized?

27. Provide a general definition of Broadband ISDN. What is B-ISDN and what switching and transmission architectures does it require?

28. What would be the major utilization and applications for B-ISDN?

29. What are the transmission speeds of B-ISDN?

30. Describe the OSI model layers for ISDN.

31. What is National ISDN?

32. List a number of benefits for utilizing ISDN. What advantages does ISDN present over other services?

33. Compare the advantages to the disadvantages of ISDN deployment.

34. What are the building blocks for ISDN applications?

35. List applications that are conducive to dial-up ISDN.

36. Discuss some issues and considerations that need to be addressed when deploying ISDN.

37. Compare and contrast Narrowband ISDN with Broadband ISDN.

38. What is B8ZS and what is its function?

39. What is the difference between LAPB and LAPD?

40. What is the difference between Q.921 and Q.931?

41. What is the difference between I.430 and I.431?

ACTIVITIES

1. Prepare a list of potential alternatives to ISDN that might be proposed by vendors. Provide advantages and disadvantages for each alternative.

2. Develop an application that would effectively utilize ISDN. Show how ISDN would be a better solution than the other alternatives mentioned in activity 1.

3. Identify competing products and services that are utilized in an ISDN installation. Use DataPro for Telecommunications, the Internet, Black Box, and other similar catalogs to obtain this information. Create a matrix with multiple products that are utilized for the same function.

4. Draw a simple example of an ISDN solution. Show all of the DCE and DTE components and give a brief explanation of the function of each component.

5. Identify how you can use PRI capabilities to satisfy BRI customer requirements. Remember that PRI is basically T-1 (1.544 Mbps).

6. Identify customers that might benefit from ISDN. Several sites on the Internet provide this information. Describe the various applications that would utilize the ISDN service.

7. Develop questions that may be asked to uncover needs for ISDN. These questions should concern strategic management (long term), tactical management (near term), and operations network management (daily) personnel.

8. Develop a cost/benefit analysis of leased line, Frame Relay, and ISDN. Prices for these are usually available in the various tariffs. Prices are also available on the various CLEC, IXC, and ILEC Websites.

9. Utilize the list of RFCs in Appendix C to develop a matrix of ISDN standards. Use the IETF home page for additional references.

CASE STUDY/PROJECT

ISDN is currently being implemented as a redundant network access in the event that primary network technology fails. Routers that support both ISDN and Frame Relay access are available from a number of sources.

This case study/project consists of an existing network that does not have a fall-back access and the requirement to design the alternative access using ISDN. The ISDN would be accessible when the primary network failed. Show how this could be accomplished both dynamically and manually.

Details

FoodStuff, Inc. is a large grocery store chain that utilizes Frame Relay to access a centrally located mainframe computer system. Their headquarters are located in Atlanta, Ga. and the stores are located in fifty locations in twenty-five cities. The store configuration consists of a 56 kbps DSU, a router, and a server that supports the local network. The host site consists of a mainframe, a FEP, and a gateway. There are two T-1s with T-1 DSUs connected between the gateway and the local carrier's Frame Relay switch. The local carriers have ISDN capabilities at the local central offices.

Requirement

Develop an ISDN backup capability for FoodStuff, Inc. Show all components and connectivity to accomplish this task. Discuss the process that would be required to dynamically and/or manually activate this redundancy. Provide a graphical design using VISIO or a like product. Price out this ISDN access.

Alternative Requirement

Identify other alternatives to ISDN for redundancy and backup. Develop a cost/benefit analysis of the other alternatives.

Switched Multimegabit Data Service/Metropolitan Area Network

OBJECTIVES
Material included in this chapter should enable you to:

- understand the SMDS/MAN technology and its relation to the current broadband market. Look at how SMDS and the MAN technologies relate to each other.

- become familiar with the terms and definitions of SMDS and how they are relevant in the Enterprise Networking environment.

- understand the function and interaction of SMDS/MAN in the Enterprise Network environment. Look at how Distributed Queue Dual Bus (DQDB) relates to SMDS. Identify the common functions of the DQDB layer.

■ INTRODUCTION

This chapter will delve into the particulars of the Metropolitan Area Network (MAN) environment, and into the aspects of the Switched Multimegabit Data Service (SMDS). This chapter begins with an overview of Metropolitan Area Networks, information on SMDS, and the particulars of Distributed Queue Dual Bus (DQDB).

Switched Multimegabit Data Service (SMDS) is a high-speed connectionless packet-switching service offered by several major telephone companies. SMDS is designed to offer efficient and economical connectivity between LAN and WAN systems over the Public Switched Telephone Network (PSTN). SMDS was first demonstrated in 1990 as a telecommunications-based system to link FDDI networks into a MAN. SMDS is a cell-based data transmission technology that is capable of speeds up to 155 Mbps over T-carrier facilities. SMDS cells are handled through SMDS switches, which are joined by high-speed DS-1, ISDN, and SONET links. Its connectionless transport system is intended to reduce overhead by leaving error checking to higher layers of the OSI model.

SMDS is very much like standard POTS. Once the user establishes the called party's telephone number, that SMDS number need only precede packets addressed to the called party. This type of service makes it unnecessary to add remote ends in a private network, making it ideal for subscribers. SMDS provides a large bandwidth at a bargain price, with

switched multimegabit data service (SMDS)

• identify components that comprise the SMDS/MAN networking environments. Look at the operation of the DQDB architecture. Understand the role of the subscriber network interface (SNI). Understand the operation of the dual ring architecture of DQDB.

• understand the issue of sustained information rate (SIR) and the various classes of services supported.

• identify applications of SMDS. Look at the competing technologies that might replace SMDS. Understand the advantages of quality of service (QoS) provided by SMDS.

• look at the advantages and disadvantages of SMDS. Show how the benefits of utilizing SMDS can justify utilizing it in the Enterprise Network.

IEEE 802.6

the ease of a telephone, which makes it successful and ideally suited for users looking for a wide-bandwidth digital connection.

SMDS primarily works with MANs and supports the ability to interconnect geographically dispersed multiple LANs operating at speeds up to 16 Mbps and is considered an on-demand high-bandwidth service, as well as a protocol-independent service that allows bandwidth to be dynamically allocated.

10.1   METROPOLITAN AREA NETWORKS

As the name suggests, a *Metropolitan Area Network (MAN)* occupies a middle ground between LANs and WANs. Interest in MANs has come about as a result of a recognition that the traditional point-to-point and switched network techniques used in WANs may be inadequate for the growing needs of organizations. The requirement is now for both private and public networks to provide high capacity at low costs over a large area. The high-speed shared-medium approach of the LANs standards provides a number of benefits that can be realized on a metropolitan scale.

A MAN is optimized for a larger geographic area than a LAN, ranging from several blocks of buildings to entire cities. Networks using this technology can be interconnected with bridges or routers to provide service over even larger geographic areas. One approach to metropolitan networks has emerged after many years of research into the various alternatives, which has been standardized by the IEEE 802 committee as IEEE 802.6.

The **IEEE 802.6** DQDB standard is based on the use of a shared transmission medium and defines standards for the Medium Access Control (MAC) and the Physical layer of the IEEE/ISO/ANSI LAN architecture. The IEEE 802.6 standard allows the use of IEEE 802.2 type-1 connectionless service in the Logical Link Control (LLC) sublayer in the same manner as the MAC sublayer standards for LANs.

As with LANs, metropolitan networks can also depend on communications channels of moderate-to-high data rates. Their error rates and delay, however, may be slightly higher than those experienced on a LAN. A MAN might be owned and operated by a single organization, but will primarily be used by many individuals and organizations. MANs might also be owned and operated as public utilities. They will often provide a means for internetworking with local networks. Figure 10–1 is an example of an implementation of various MANs that interface with a number of different technologies.

Where LANs are typically used only to support data traffic, the 802.6 MAN is intended for the support of both data and voice traffic. MANs cover greater distances at higher data rates than LANs, although there is some overlap in geographical coverage.

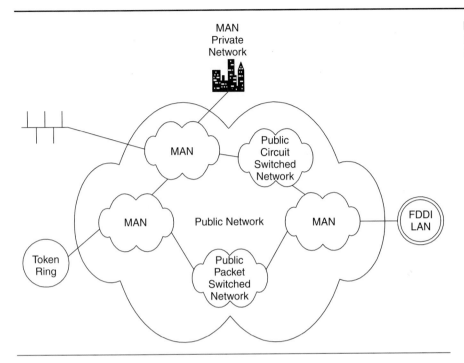

Figure 10–1
Public and Private Use of
MANs

A useful way to view the 802.6 MAN standard is that it is an adaptation of the features of both LANs and wide-area Asynchronous Transfer Mode (ATM) and is well suited to the MAN requirements. Some of the key characteristics of the 802.6 MAN are as follows:

- High Speed
- Shared medium
- Dual bus
- Fixed-length packets
- Logical Link Control (LLC) support
- Addressing scheme

10.2   SMDS OVERVIEW

The IEEE 802.6 DQDB technology is intended for the relatively short-distance communications required to support data transmission within the metropolitan area of a large city. The DQDB transmission is now being used in a telecommunications service offering Switched Multimegabit Data Service (SMDS) [Black, 1997].

SMDS is a high-speed connectionless data service that is deployed by Local Exchange Carriers (LECs) for linking applications within a metropolitan area. The SMDS concept was developed by BellCore, which uses portions of the IEEE 802.6 MAN protocol. SMDS is a service and not a protocol, so it can use any transport mechanism. As an example, ATM provides a connectionless class of service options that is capable of supporting SMDS. Its purpose is to ease the geographic limitations that exist with low-speed WANs. SMDS is designed to span across LANs, MANs, and WANs and carry data at speeds comparable with the speed of LANs.

The original Bellcore standard specified how Customer Premises Equipment (CPE) could access a SMDS switch located at a CO using twisted-pair wiring at the T-1 rate or optical fiber at the T-3 rate. The scope of SMDS has evolved since its original design to support WANs. It now includes higher rate definitions and in the future may operate at the OC-3 (155 Mbps) SONET rate.

The major objective of SMDS is to provide data with any-to-any connectivity that flows from telephone and fax service. The major difference between SMDS and other services such as Frame Relay and ATM is that permanent virtual circuits are not required. Wideband data connectivity is provided at relatively inexpensive prices. Most LECs charge a flat rate price per month, with no additional cost for distance or usage.

SMDS speeds range from 1.544 Mbps to 44.736 Mbps. Access to SMDS is generally via a SMDS-compliant router over T-1, fractional T-3, or T-3 access circuits. For example, the LEC will usually install a twisted-pair (four wire) T1 circuit for the lower end user, or a fiber loop for T-3 access. Variations can include the use of a fiber ring into the customer's premise where a T-1 or T-3 access can be provided. If fiber is not available to the door, the LECs might utilize Fiber-To-The Curb (FTTC) and deliver a metallic coaxial link into the customer location at T-3 speeds.

Using SMDS is very much like using a telephone. Once the user establishes the called party's SMDS telephone number, that number need only precede packets addressed to the called party. This type of service makes it unnecessary to add remote ends in a private network, making it ideal for subscribers. A large bandwidth at a bargain price, with the ease of connection of a telephone, makes SMDS successful and ideally suited for users looking for a wide-bandwidth digital connection worldwide.

SMDS can be viewed as a high-speed protocol or data transfer methodology, and will use either ATM or Frame Relay type packets. For supporting Frame Relay, SMDS breaks the data stream into packets that can be as large as 9,188 bytes in length. ATM type cells will remain at 53 bytes. SMDS cells are routed to their destination down any available path, and do not necessarily follow the same path during the trip. The cells are reassembled into their proper order at the final destination, as happens in any packet switched environment.

SMDS supports the most common protocols including IPX, IP, SNA, and Appletalk, plus providing multicast addressing. SMDS partners who

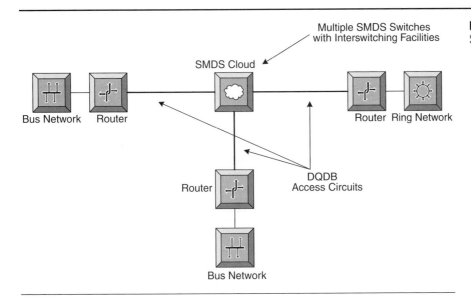

Figure 10–2
SMDS Architecture

wish to communicate must subscribe to the service; however, the service is not available from many independent telephone companies and some ex-Bell companies. It is similar to telephone service, because with a connectionless service like SMDS you can reach any device that will accept the transmission, but the service is far from ubiquitous.

SMDS maintains a database of addresses that are validated to communicate with receiving stations. This ensures that receiving sites get messages only from valid points of origin, which improves security. Addresses are maintained and thus provide a relationship between groups of users. If you are not in that user group, transmission will be denied.

Figure 10–2 depicts a typical SMDS implementation. The SMDS switches are located in LEC central offices, with fiber-optic access circuits operating in a dual bus arrangement to the subscribers. Multiple subscribers are bridged across the dual bus arrangement.

10.3   SERVICE CHARACTERISTICS

SMDS services are unique with each LEC. Although Bellcore developed the service characteristics, pricing and service features were developed by each LEC. Instead of voice traffic, SMDS is designed to carry data. More important, SMDS is optimized to operate at high speeds. For example, header information in packets can require a significant amount of the available bandwidth. To minimize header overhead, SMDS uses a small header and allows each packet to contain up to 9,188 bytes (octets) of data. SMDS also defines a special hardware interface used to connect computers to the network. The special interface makes it possible to deliver data as fast as a computer can handle it. SMDS often operates at speeds similar to Frame Relay.

As SMDS has developed, it has been made compatible with Broadband ISDN (B-ISDN) to provide extremely fast transport of SMDS over long distances, once B-ISDN is widely available. SMDS cells are handled through SMDS switches, which are joined by high-speed DS-1, ISDN, and SONET links. SMDS is a connectionless transport system intended to reduce overhead by leaving error checking to intelligent end devices such as switches and routers.

SMDS is designed to operate over a dual-bus fiber-optic network using the 802.6 protocol, which is also known as Distributed Queue Dual Bus (DQDB). The protocol was developed to be a robust, low-latency protocol. DQDB will be discussed in detail later in this chapter.

The backbone switching system consists of a high-speed packet switch that is located in the wire center or central office. The protocol places a limit on the amount of sustained information that the subscriber can send across the **Subscriber Network Interface (SNI)** to the switching system. SMDS uses the term **Sustained Information Rate (SIR)** as the maximum guaranteed rate at which data can flow across the network. SIR is similar to Committed Information Rate (CIR), which is utilized by Frame Relay. SMDS supports five classes of service with SIRs running from 4 Mbps to 34 Mbps. As with Frame Relay and ATM, the subscriber can send bursts of data up to the speed of the access circuit, which is T-1 or T-3, but the access class limits the average rate of information that can be sent. The switch contains a credit manager that tracks the amount of data sent across the network and compares it with the amount permitted by the access class. If the subscriber has a credit, the traffic is accepted; if not, the traffic is rejected.

Figure 10–3 shows a typical SMDS topology. SMDS uses the DQDB protocol at the SMDS SNI. It defines the procedures for a user's CPE to interface with a SMDS network (Figure 10–4). The DQDB architecture in the public network will be presented later in the chapter. The SNI can also be referred to as the SMDS interface protocol (SIP) or an "access DQDB." The SNI operates between the CPE and the SMDS switching system. The SMDS switch is a high-speed packet switch that supports operations within a LATA.

Multiple SMDS switches can be connected via an Interswitching System Interface (ISSI). The Intercarrier Interface (ICI) connects two LECs through an IXC. The ICI provides inter-LATA service and is also called Exchange Access SMDS. ICI is not used within a LATA.

SMDS has a three-tier architecture. The switching system consists of SMDS-compatible high-speed packet switches. The delivery system is

subscriber network interface (SNI)
sustained information rate (SIR)

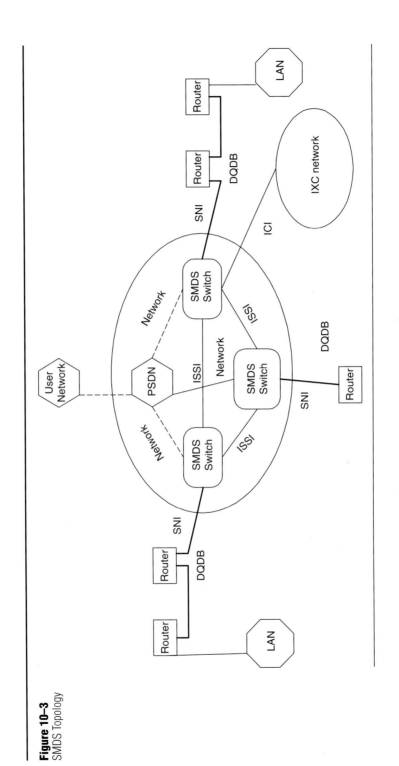

Figure 10-3
SMDS Topology

Figure 10–4
SMDS Subscriber Network
Interface

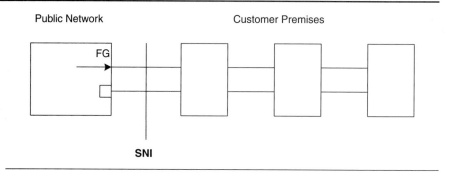

made up of subscriber network interfaces, and the access control system enables subscribers to connect to the switching system. The Layer 3 Protocol Data Unit (PDU) accepts packets up to 9,188 octets in length. These are segmented into a Layer 2 PDU, which uses 53 octet cells. The cells cross the network to the destination, where they are reassembled. Since the protocol is connectionless, the cells do not follow a predefined path through the network. This allows a user in one site to dial up a link to a user in another site without establishing a virtual circuit. Traffic always passes on LEC facilities, but traffic belonging to one subscriber never passes through another subscriber's network.

The SMDS cell structure is similar to ATM, except the cells contain an address in the header, which is necessary in a connectionless system. The address fields in the SMDS header use a ten-digit number following the North American numbering plan.

10.4 SMDS CONNECTIVITY

The difference between SMDS and T-1 service is that a user must lease a T-1 circuit from point-to-point, all the way between the two locations. However, with SMDS the user leases a circuit to the nearest carrier's wire center or CO at both ends, and the carrier handles the transmission in between using normal, shared communication facilities. Depending on the complexity of the network, this usually results in less expense for SMDS, when compared with a private line configuration.

For example, if a company needed to connect each of its six remote sites to a corporate headquarters at 1.544 Mbps (T-1), it could lease six T-1 lines. These would include one from each site to the headquarters, or it could lease one T-1 SMDS line from each remote site to the nearest central office, and then a DS-3 SMDS line from the headquarter's location to its nearest central office. There will be a reduction of circuits, mileage cost,

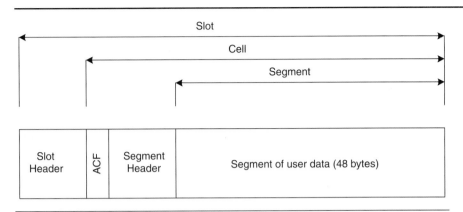

Figure 10–5
DQDB Slot and Segment
Structure

and DSU cost by utilizing SMDS. (Figure 10–12 in the applications section will address such a configuration.)

10.5 DISTRIBUTED QUEUE DUAL BUS (DQDB) OVERVIEW

SMDS networks conform to IEEE 802.6 and use a protocol called **Distributed Queue Dual Bus (DQDB).** DQDB was codeveloped by Telecom Australia, the University of Western Australia, and QPSX Communications Limited. It was designed to provide a basis for initial broadband interconnection of metropolitan areas, and also give a possible migration path to B-ISDN. SMDS became available as a public data communications service in the United States in 1991.

distributed queue dual bus (DQDB)

The DQDB protocol uses two slotted buses of bit rates up to 155 Mbps to transport segments of information between communicating broadband services. Segments are 48 byte frames of user information. Figure 10–5 shows the segment structure of DQDB. The segment header consists of 4 bytes and the Access Control Field (ACF) consists of 1 byte. Two unidirectional high-speed buses run out from master and slave frame generators at opposite ends of the ribbon topology. Each of the devices (nodes) connected to the network are connected to both buses to send and receive data.

Data blocks to be carried by DQDB are formatted in the standard manner of frame header, the user data block, and the frame trailer. The frame header contains the address of the originating and destination nodes. The user data block is the data frame to be carried that may be up to 9,188 bytes in length (192 segments), and the trailer includes the frame check sequence. Data blocks must be broken down into individual segments and then forwarded as slots for transmission. If necessary, the last segment is filled with padding.

DQDB Services

Three types of services have been specified in the service definition for the DQDB standard. These include:

- Connectionless data service—Provides support for LLC type-1 connectionless data communication. The connectionless media access supports the transport of frames up to a length of 9,188 octets. Transmission is in the form of fixed-length 53 octet segments. Accordingly, the service must include a segmentation and reassembly function.
- Connection-oriented data service—Provides support for a form of permanent virtual circuit. A connection is established between the source and destination, but there is no guaranteed arrival rate. As with connectionless service, segmentation and reassembly are required. The control signaling required to establish, maintain, and clear a connection are outside the scope of the current 802.6 standard.
- Isochronous service—Provides transmission with a guaranteed regular arrival rate. The *isochronous* service is suitable for sound and video transmission. The term isochronous refers to the use of regularly repeating time slots. The control signaling required to establish, maintain, and clear a connection is outside the scope of the current 802.6 standard.

Only the connectionless data service and the isochronous service have been fully defined. There may be no further action on supporting additional functions for DQDB.

Because of the emergence of Asynchronous Transfer Mode (ATM) as a universal network technology for the transport of all types of voice, video, and data information in local, metropolitan, and wide area networks, the MAN technologies that use DQDB are already being rendered obsolete.

DQDB Layer

The DQDB layer can be viewed as being organized into three sublayers—common functions, arbitrated functions, and convergence functions.

Common Functions   The common functions module deals with the relay of slots in the two directions and provides a common platform for asynchronous and isochronous services. In addition to the basic transmission and reception of slots, the common functions module is responsible for head-of-bus, configuration control, and MID page allocation functions.

The head-of-bus function is performed only by the one or two nodes designated as the head of bus. It includes generating and transmitting slots. The configuration control function is involved in the initialization of the subnetwork and its reconfiguration after a failure. The MID page allocation function participates in a distributed protocol with all nodes on the subnetwork to control the allocation of message ID values to nodes. The message ID is used in segmentation and reassembly.

Arbitrated Functions   The arbitrated functions are responsible for medium access control. The two functions correspond to the two kinds of slots carried on a bus. *Prearbitrated (PA)* slots are used to carry isochronous data. *Queued Arbitrated (QA)* slots are used to carry bursty data transmissions.

Convergence Functions   The DQDB layer is intended to provide a range of services. For each service, a convergence function is needed to map the data stream of the DQDB user into the fifty-three octet transmission scheme of the DQDB layer. The concept is the same as that of the ATM Adaptation Layer (AAL) for ATM. The MAC convergence function adapts the connectionless MAC service to the QA function. The isochronous convergence function adapts an isochronous octet-based service to the guaranteed-bandwidth octet-based service of the PA function.

DQDB Architecture

The DQDB standard assumes that stations are interconnected using a fiber-optic transmission medium and support data rates in the range of 1.544 Mbps to 155 Mbps. On a DQDB network, stations are interconnected, using fiber-optic cable, to form two buses, each of which carries data in the opposite direction. Data are carried from station to station on each of the buses in the form of fifty-three octet cells called slots. The data units carried in DQDB slots are similar to the cells used in ATM.

Operation of the DQDB subnetwork is controlled by a 125 ms clock. The timing interval provides for isochronous services and reflects the 8 kHz public networking frequency required by voice services. If the DQDB subnetwork is supporting certain isochronous services and is connected to a public network, it may be required that the timing be derived from the public network. The alternative timing source is a node within the DQDB subnetwork that would be designed for that purpose.

Transmission on each bus consists of a steady stream of fixed-size slots with a length of fifty-three octets. Some stations on the network must generate empty slots that flow from station to station through the network. Each station then places data into empty slots passing by. Another station on the network removes slots from the network. Each station of the network has access to both buses and is able to send and receive data using the slots flowing on both of them. On a given bus, a station transmits only to stations downstream of it on that bus. By using both buses, a station can send data to and receive data from any station on the network. The DQDB standard supports the use of two different network topologies: an open bus topology

Figure 10–6
MAN Bus Architecture

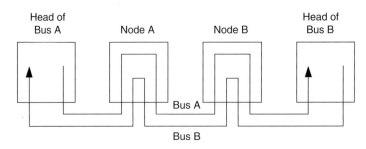

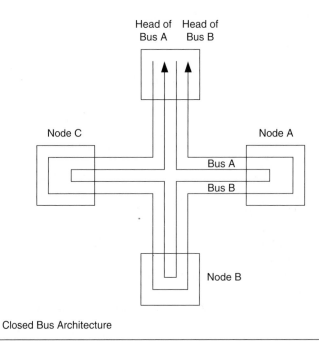

Open Bus Architecture

Closed Bus Architecture

and a looped (closed) bus topology. The difference in the two topologies is the method of generation and removal of slots.

Open Bus Topology   With the open bus topology, the slots on unidirectional bus A (Figure 10–6) are generated by the first node and removed by the last node. Slots for unidirectional bus B are generated by the last node and removed by the first node. Reconfiguration after a break in an open bus topology results in two separate networks.

Looped (closed) Bus Topology   In the looped (closed) bus topology, slots for both buses are generated and removed by the same node. In the event of a break in the transmission medium, a single network results from the reconfiguration process, and all the stations are still able to communicate with one another after the break. The looped bus topology resembles a dual ring topology. Figure 10–6 depicts both the DQDB open and closed bus architectures.

DQDB Operation

The DQDB protocol can be described in terms of a distributed collection of first in, first out (FIFO) queues. At each node, a queue is formed for each bus. For each request read in a passing slot, the node inserts one item in the queue. When the node itself issues a request, it adds an item to the queue for itself. When its own item is at the top of the queue, the node may transmit in the next free QA slot. A node may only have one item for itself in each queue at any time.

The DQDB protocol is based on the use of counters. These counters operate on each bus to determine if a slot is available for use. A DQDB node examines and reserves slots on one bus in order to use the slots on the other bus. Each time a node detects that a slot has been reserved by an upstream node, it increments a counter. As slots that are not busy pass by on the other bus, a counter is decremented. Request counters are used for outstanding requests and waiting counters are used when segments are ready for transmission.

The 802.6 architecture, depicted in Figure 10–7, is similar to FDDI in that it utilizes two counterrotating rings and permits full-duplex communications between any of the nodes. Each node has two attachments to each bus. One attachment reads bus slots and the other writes bus slots. Buses are managed by a head end, which generates slots for use by the downstream nodes. The network is self-healing if there is a break in the fiber.

A frame generator at the master station emits 125 ms frames, which is the length of a T-1 frame. Each frame can contain a fixed number of fixed-length time slots. The number of slots depends on the bit rate of the bus, which depends on the transmission medium. In a freestanding configuration, the network contains its own synchronization. When connected to a public network, it must derive its synchronization from the public network.

Although the dual ring architecture of DQDB appears to be a pair of rings, it is a logically-looped dual bus with each node appearing on each bus. Figure 10–8 shows this logical representation. Two unidirectional high-speed buses run out of the master and slave generators at opposite ends of the ribbon technology. Each of the nodes connected to the network is connected to both buses to send and receive data. The role of the frame generators is to structure the bit stream carried along the buses into

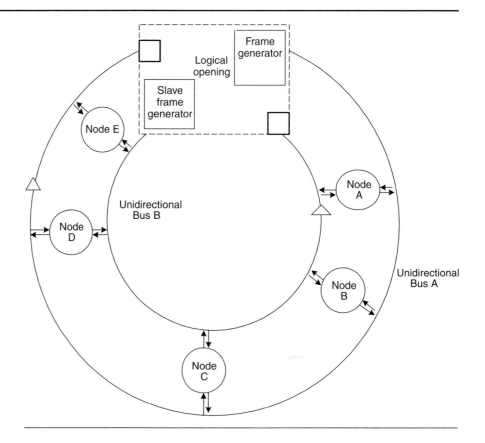

53 byte slots. These slots are filled by nodes wishing to send user information and are then carried downstream along the bus. The relevant receiving node reads information out of the slot being sent to it, but does not delete the slot contents. The slot remains on the bus, traveling further downstream until it falls off the end.

The SMDS interface or DQDB consists of two shared fiber-optic media. The standard, however, does permit the use of coaxial cable. Twisted-pair wire does not have enough bandwidth for the backbone network, but it can be utilized for the access network. Both cables are attached to the CPE at one end, and the carrier switch at the other. Utilizing separate one-way buses eliminates the possibility of collisions. DQDB transmission on each bus is divided into time slots. Any attached device can access the bus as needed, until the bus is saturated with traffic. Slot access is managed by distributing time slots between devices so that no single device can have 100 percent of the allocated slot time. As many as 512 devices can be attached to a SMDS bus, which can extend for up to 160 km (100 miles).

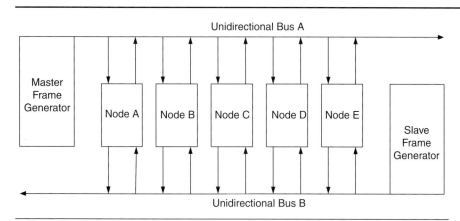

Figure 10–8
Logical Bus Structure of
DQDB

Isochronous services such as digitized voice and video must be transmitted across the network with minimum delay. As mentioned, DQDB has two classes of service, Prearbitrated (PA) and Queued Arbitrated (QA). PA service is utilized for voice and video, which require a fixed amount of bandwidth. QA provides service on demand for bursty applications such as data transmission. The standard defines two separate access control methods for the two types of slots: PA access control and QA access control.

Prearbitrated Access Control
Prearbitrated (PA) Access Control is used to carry isochronous data in PA slots. The head end function generates PA slots at appropriate intervals to provide channels of defined bandwidth. PA slots carry a **Virtual Channel Identifier (VCI)** that identifies a particular channel. An isochronous connection transmits by placing its data in the PA slots associated with a particular virtual channel. If an isochronous connection does not require the entire capacity of the virtual channel, the virtual channel might be shared by multiple isochronous connections. Each connection uses a specified portion of all the slots allocated to a channel.

prearbitrated (PA) access control

virtual channel identifier (VCI)

Queued Arbitrated Access Control
Queued Arbitrated (QA) Access Control is used to implement the connectionless data service using QA slots. Access to QA slots operates using a distributed reservation system. When a node has data to transmit, it sends out a reservation request on each bus. Data are always transmitted downstream, but reservation requests are sent upstream.

DQDB uses an ingenious method of allocating QA access to the bus. Each node is equipped with a request counter and a packet countdown. When a node has a packet ready for transmission, it transmits a request

queued arbitrated (QA) access control

Figure 10–9
MAN Access Process

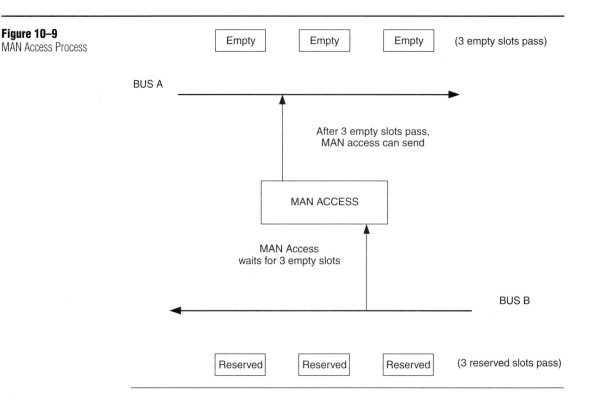

upstream to the head end. As the request passes the upstream nodes, each node increments its request counter by one, so each node knows at any time what its place in queue is based on the number in its request counter. As empty slots flow downstream, each node decrements its request counter, knowing that these slots will be filled by downstream nodes with unfilled requests. Each node has a record of how many slots were requested and how many were filled by passing slots intended for downstream stations. If a node's request counter is set at n, when the nth slot goes by, its counter has reached zero, so it can send on the next empty slot.

With this method of allocating bandwidth, upstream locations have a better chance to seize empty slots than downstream locations. To avoid this problem, the protocol forces stations to pass empty slots, depending on their physical location. This scheduling method provides a high degree of efficiency. Slots are never wasted while a station has traffic to send, and no station can monopolize the network.

The process depicted in Figure 10–9 is as follows:

- A station sends a reservation request by setting bits in the first octet of an empty slot that passes by.

- Each node tracks reservation requests it has received from nodes downstream of that node.
- The reservation requests tell a sending station when it is that station's turn to transmit.
- Each node keeps a reservation request counter for each bus.
- A node uses this counter to track reservation requests that flow by while it is waiting to transmit.
- To implement different priority levels, reservation request counts would have to be kept separately for each priority level.

This creates a deterministic approach to the dual bus very similar to a token passing arrangement. To prevent any one node from saturating the network, provisions are in place to periodically send empty queue slots onto the network so that this deterministic MAN can be achieved.

10.6   SUSTAINED INFORMATION RATE AND ACCESS CLASSES

The SMDS traffic management operations are founded on the Sustained Information Rate (SIR). As mentioned, this is similar to the Committed Information Rate (CIR) that is found in the Frame Relay technology. SIR is founded on access classes that are provided for DS-3 circuits. Each access class identifies the different traffic characteristics for varying applications. The access class places a limit on the amount of SIR that the CPE can send across the SNI to the SMDS switch.

SMDS Access Classes
SMDS defines a connectionless data transfer service that can be used to carry traffic at different data rates. Each SMDS user is assigned an access class that determines the maximum rate of data transmission that can be used over an extended time interval. Access classes and their sustained data rates that are commonly implemented in SMDS networks are:

- Access Class 1—4 Mbps
- Access Class 2—10 Mbps
- Access Class 3—16 Mbps
- Access Class 4—25 Mbps
- Access Class 5—34 Mbps

The access classes do not provide a means of using a portion of a DS-3 circuit. When a CPE sends data across the DS-3 circuit, it is allowed to use the maximum access rate, but the CIR places a limit on the duration of the burst. The access class also places a limit on the average rate of information transferred across the interface.

Figure 10–10
Read and Write Operations
in the Access Unit

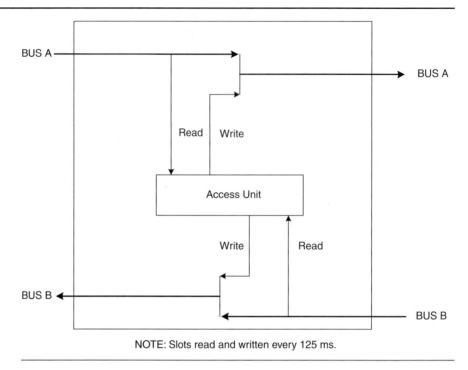

NOTE: Slots read and written every 125 ms.

SMDS Access Unit

Figure 10–10 presents a general view of the MAN access unit residing in a node. The access unit is used to read and write into the DQDB slots. The DQDB signals are read and written every 125 ms (1 second/8,000). The value of 8,000 reflects the conventional sampling rate for a voice channel.

The nodes write traffic to, and read traffic from, the slots on the bus. They gain access to the network by writing into the slots. The read function is placed in front of the write function on each bus, which allows a node to read slots independently of writing into them. Once the slot has been written into, it is read by the downstream nodes to determine if the payload in the slot is to be copied at that node and passed to an upper layer protocol or ignored.

10.7 SMDS ADDRESSING

Each physical interface to a SMDS network is assigned one or more ten-digit addresses. The subscriber to the SMDS service can then assign different addresses, as desired, to the different systems or other devices using the SMDS service for data transmission.

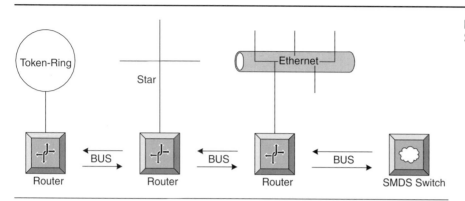

Figure 10–11
SMDS WAN

Group addressing is available with SMDS that allows a device to send data up to 128 different destinations in a single transmission. SMDS group addressing provides a function similar to multicasting on a LAN.

10.8　SMDS WAN

The SMDS bus is typically over T-carrier lines; however, the speed of transmission over these lines is slightly lower than full capacity, because some of the bandwidth is allocated for control and signaling. When a T-3 line is used with DS-3 access rates, SMDS divides the line into different service classes that transmit at a combination of speeds: 4 Mbps, 10 Mbps, 16 Mbps, 25 Mbps, and 34 Mbps. SMDS, which primarily supports data transmission, converts frames received from LANs into cells. The exceptions are routing and network function frames that do not have to be converted, and which are processed by the SMDS **Data Exchange Interface (DXI).** DXI uses frames instead of cells, and these are in the High-Level Data Link Control (HDLC) format, which is similar to the X.25 LAPB and ISDN LAPD formats. Figure 10–11 depicts a logical representation of a SMDS WAN network.

data exchange interface (DXI)

10.9　SMDS INTERFACE PROTOCOL

The **SMDS Interface Protocol (SIP)** is used for communications between CPE and CO equipment. SIP provides connectionless service across the SNI, which allows the CPE to access the SMDS network. SIP is based on the IEE 802.6 DQDB standard for cell relay across MANs. DQDB was designed for compatibility with current transmission standards, and it is

SMDS interface protocol (SIP)

Figure 10–12
SMDS Cell Format

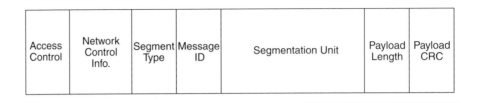

compatible with emerging standards for B-ISDN. The SMDS interface protocol consists of SIP Levels 3, 2, and 1.

- SIP Level 3 (L-3) operates at the MAC sublayer of the OSI data link layer.
- SIP Level 2 (L-2) also operates at the MAC sublayer.
- SIP Level 1 (L-1) operates at the OSI physical layer.

10.10   CELL STRUCTURE

A SMDS cell has a fixed-length format at 53 bytes and consists of a header, a segmentation unit, and a trailer. Figure 10–12 depicts the SMDS format layout.

The header consists of the following elements:

- Access control—Contains information indicating whether the cell was sent by the CPE, such as a router, or by a SMDS switch at the carrier location.
- Network control—Contains information such as whether the cell is carrying control information or data.
- Segment type—Indicates whether the cell contains the beginning, middle, or end of a message sequence or the complete message sequence within one cell.
- Message ID—Contains a unique number assigned to all cells in a message sequence to show that those cells must be interpreted together.

The segmentation unit in a cell contains the cell payload, which is the user data transported over the SMDS network. The cell trailer consists of two fields: Payload length and Payload Cyclical Redundancy Check (CRC). The Payload Length indicates how much of the Segmentation Unit is payload and how much is empty. If there is no payload, the Payload Length field consists of zeros. The Payload CRC enables the recipient node to verify that the following fields are received containing the same information as was sent: Segment Type, Message ID, Segmentation Unit, and Payload

OSI Model	SMDS Model
Layer 3—Network Layer	Responsible for physical and logical routes, and for constructing the cell payload
Layer 2—Data Link Layer	Responsible for cell construction and point-to-point connectivity
Layer 1—Physical Layer	Responsible for physical connectivity

Table 10–1
SMDS Mapping to the OSI Model

Length. The CRC is a value representing the total length of each field added into one sum.

10.11 QUALITY OF SERVICE OPERATIONS

SMDS defines performance and QoS goals to be performed by the SMDS network on behalf of a CPE. These operations occur across the entire network within one LATA from one SNI to another SNI. These operations are divided into three groups: availability, accuracy, and delay.

Availability deals with the ratio of the actual service time provided to the scheduled service time. The accuracy objectives describe operations dealing with lost, misdelivered, duplicated, and missequenced traffic. The delay objective category addresses delay in L2-PDUs and L3-PDUs (SIP layers).

10.12 STANDARDS

SMDS incorporates layers that correspond to the physical, data link, and network layers of the OSI model (Table 10–1). At the physical layer it uses the IEEE 802.6 standard for MAN communications, and at the data link layer it employs LLC sub layer communications. The network layer consists of the communications paths used to transmit data.

The SMDS specifications are published by Bellcore and are included in documents numbered TR-TSV-00073 through 00075, 001059 through 001062, and 001237 through 001240. The standard defines technology intended for a network spanning a range of up to several hundred kilometers. Networks using this technology can be interconnected with bridges or routers to provide service over even larger geographic areas.

RFCs for SMDS include the following documents:

- RFC1694 Definitions of Managed Objects for SMDS Interfaces Using SMIv2
- RFC1209 Transmission of IP datagrams over the SMDS Service

Table 10–2
SMDS and 802.6 Layers

SIP Layers	MAN 802.6 Layers
L3_Protocol Data Unit	Initial MAC PDU (IMPDU)
L2_Protocol Data Unit	Derived MAC PDU (DMPDU)
	Segment
	Slot
L1_Protocol Data Unit	Physical Layer Convergence Procedure (PLCP)

Table 10–3
IEEE 802.6 PDU Formats

PDU	Contents	Sublayer	Description
Slot	Contains a QA segment, a PA segment, or isochronous data	Common functions	Basic unit of data transfer
QA segment	Contained in a slot Carries a DMPDU	Queued arbitrated Functions	Used to carry a portion of a MAC service data unit or other SDU
PA segment	Contained in a slot	Prearbitrated functions	Used to carry isochronous service octets
Initial MAC PDU	Carried in a sequence of DMPDUs	MAC convergence function	Contains a MAC service data unit
Derived MAC PDU	Carries a portion of an IMPDU contained in a QA segment	MAC convergence function	A sequence of DMPDUs carries a single IMPDU

The 802.6 standard defines a medium access control technology named DQDB. The IEEE 802.6 DQDB standard is based on the use of shared transmission medium and defines standards for the MAC sublayer and the physical layer of the ISO LAN architecture. The 802.6 standard allows the use of the IEEE 802.2 Type 1 connectionless data service in the LLC sublayer in the same manner as the MAC sublayer standards for LANs. The technology defined by the DQDB standard can also be used to implement WAN data links that can span very long distances.

The 802.6 DQDB standard supports the use of the same 48-bit or 16-bit MAC sublayer station addresses used on LANs. The standard also includes an alternative 60-bit station address specification.

Table 10–2 contrasts the SMDS Interface Protocol (SIP) layers with the MAN 802.6 layers. Note that the SIP layers do not divide Layer 2 into three sublayers as is done in MAN; however, the same functions are still performed.

Table 10–3 lists the five PDU types and indicates their relationship to each other and the DQDB sublayer that is responsible for generating and reading each type.

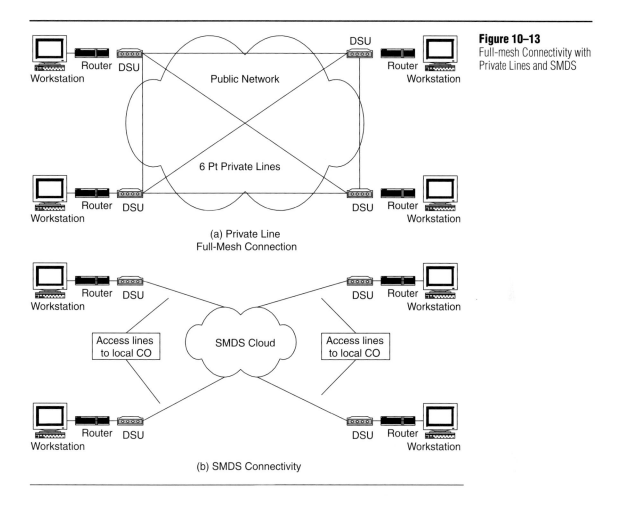

Figure 10–13
Full-mesh Connectivity with Private Lines and SMDS

(a) Private Line
Full-Mesh Connection

(b) SMDS Connectivity

10.13 MAN/SMDS APPLICATIONS

The primary market for MANs is the customer that has high-capacity transmission needs in the metropolitan area. The primary application for SMDS is LAN interconnection across a WAN or a MAN, although it can be used for any data transport. A MAN is intended to provide the required capacity at lower cost and greater efficiency than can be obtained from an equivalent service from the local telephone company. It is attractive for companies with multiple sites that are interconnected with T-1 facilities.

As an example, Figure 10–13 illustrates a full-mesh network utilizing SMDS and the same network utilizing private line service. To provide connectivity between these sites requires $(n * (n - 1))/2$ connecting circuits and DCE devices. To provide the same connectivity requires only n SMDS circuits. If SMDS switches are located in the same wire center or local CO as the terminal points, the mileage charges typical of T-1 circuits are eliminated. Both Frame Relay and ATM offer the same type of connectivity, but they require PVCs for each pair of nodes, which are usually much more expensive.

Configuration (a) depicts a point-to-point private line configuration where all of the sites are connected to each other. This full-mesh configuration requires five private line facilities and the associated data service units (DSUs) for each end of the circuits. Configuration (b) depicts the same customer with SMDS service. DQDB access circuits are required from each site to the local CO. The SMDS network takes care of the switching between all of the sites.

SMDS is being offered by telephone companies in the United States and several public carriers in Europe as a service for high-speed data applications that require bursts of high-bandwidth transmissions for applications such as file transfer, CAD/CAM, and imaging.

SMDS was developed to provide high-speed MAN data connectivity, in order to enable:

- High-speed links for regional networks
- Fast access to electronic catalogs and library holdings
- Transmission of large-image files, such as medical X-rays
- Transmission of architectural drawings
- Transmission of CAD/CAM graphics

Targeted applications are:

- Seamless LANs and WANs
- Graphics, CAD/CAM, and X-rays
- Large database transfers
- LAN interconnections
- Medium quality video
- Data transfer
- Focus on high-bandwidth data applications

New SMDS/MAN services may include the following:

- Video conferencing
- Interactive TV services
- Multimedia
- Broadband ISDN services and interfaces
- Voice
- Medical imaging
- Tele-radiology
- Tele-learning

The current use of SMDS in the industry includes the following applications:

- LAN-to-LAN in a metropolitan area
- Intra- and intercompany document transfer and sharing

- Collaborative processing and development
- Host-to-Host transfers direct access storage device (DASD) mirroring
- Disaster recovery planning for remote hot site

10.14   ISSUES AND CONSIDERATIONS

Several speeds are defined in the 802.6 standard. Speeds depend on the medium used. With coaxial cable, the speed is 45 Mbps; the speed over fiber-optic cable is 155 Mbps.

SMDS is losing the WAN services battle to Frame Relay. The installation base and new installations of SMDS are falling far short of the same projections for Frame Relay. A limited number of service providers in the United States now offer the SMDS. The service has an edge over Frame Relay when the application is for a many-to-many (fully meshed) network.

MAN providers typically offer lower prices than the telephone companies, and offer diverse routing, and backup in emergency situations. They also claim to offer quicker installation and better service than the telephone companies.

SMDS per-cell overhead and formatting of the payload field are identical to ATM's AAL 3/4. The first cell in a SMDS packet contains 44 bytes of information, including the source/destination address and other fields. Much of the flexibility and many of the features of SMDS derive from this header. Packets may be up to 9,188 octets long, slightly longer than the maximum in Frame Relay. This flexibility reduces efficiency, with the best limited to about 80 percent maximum efficiency for very large packets.

If there is a need to connect to other networks via a public service, or require an additional level of multiplexing, then 802.6 / SMDS is probably a good choice, either as a stand-alone service or carried over an ATM-based service.

10.15   SMDS ADVANTAGES AND DISADVANTAGES

Advantages
In addition to high-speed network communications and compatibility with B-ISDN, T-carrier, and ATM technologies, SMDS also provides strong security options for users. Access by nodes can be limited to groups or individual addresses, and private networks can be set up for highly sensitive information. Customers can be billed on the basis of SMDS usage.

Generally, features include:

- High-speed, low delay, connectionless data transport
- Provision for bandwidth on demand

- Robustness in routing
- Many-to-many full-mesh connectivity
- Multicasting and group addressing
- Multiple protocol support
- Network management capability
- Call blocking, validation, and screening for security
- Scalability for network growth

No special equipment is required to implement a SMDS network, although address screening is a feature that must be managed by the user.

Disadvantages

The disadvantage of SMDS is that it is not as universally available as X.25, Frame Relay, and ISDN. SMDS does not operate in the multimedia arena, because it is designed to transport only data and not voice or video.

10.16   SMDS TECHNOLOGY ALTERNATIVES

Competition to SMDS can come from a number of technologies. These include standard POTS and leased lines, Frame Relay, ATM, DSL, and older X.25 networks.

SMDS has some of the same management requirements as Frame Relay. These include SIR for SMDS and CIR for Frame Relay. PVCs must be managed for Frame Relay; however, screening tables must be managed for SMDS. SMDS is most effective when there is a many-to-many network connectivity. Frame Relay is superior when there is a one-to-many connectivity. Frame Relay is available in more locations than SMDS.

SMDS is far superior to leased lines due to the additional requirements for DCE devices at each termination for leased lines. Leased lines are also price sensitive to distance. Leased lines have an advantage over all services in that they are available almost everywhere. There is very little management required for a leased line network. A leased line is usually more secure than connectionless services.

SMDS is not in the same league as ATM. ATM can be utilized as a service and as a backbone network. ATM can be utilized to transport signals at a much higher rate than SMDS. ATM is not available everywhere, which is the same limitation as SMDS.

SMDS is a different animal than DSL. There are multiple flavors of DSL, which provides for a greater selection of alternatives. SMDS is not speed sensitive to upstream and downstream traffic. Today, both SMDS and DSL suffer from availability. DSL requires special splitters and special CO equipment for implementation. SMDS requires only a DSU at the user's location, which is commonly available.

SMDS is superior to X.25-based networks. X.25 is slower than SMDS and is being phased out for the more technologically advanced offerings. X.25 is a point-to-point service that has a topology similar to leased lines. It is a connection-oriented technology and is therefore more secure than SMDS. SMDS is cheaper than X.25 when many circuits are required between locations. This is the same argument used when comparing SMDS with Frame Relay.

Product and Service Providers

The following companies offer SMDS/MAN service, but this list is subject to change at any time. The URLs are subject to change due to the number of mergers taking place.

Ameritech	www.ameritech.com/
Bell Atlantic	www.bellatlantic.com/largebiz/smds.htm/
BellSouth	www.bellsouth.com/
GTE	www.gte.com/
MCI Worldcom	www.wcom/co/
NYNEX	www.nynex.com/
Pacific Bell	www.pacbell.com/
SBC Communications	www.sbc.com/
Sprint	www.sprint.com/
US West	www.uswest.com/

■ SUMMARY

Switched Multimegabit Digital Service (SMDS) is a public high-speed connectionless digital data transmission service originally developed by AT&T and the regional Bell operating companies to support high-speed public local area networking. Created in 1989 and based on Bellcore standards, SMDS was designed to provide a transport service for LANs, host computers, and WANs.

In SMDS, the common carrier provides the user with access points for both sender and receiver. SMDS is broken down into transmission packets. The common carrier provides high-speed switching equipment that routes these packets to their destination address. SMDS speeds range from 1.544 Mbps (T-1) through 44.736 Mbps (T-3). SMDS can be used for high-speed data transmissions such as the connection of LANs across WANs.

A MAN spans a larger geographic area than that of a LAN. A typical MAN might support an area large enough to encompass a large city. MAN technology has much in common with LAN technology. As with local networks, MANs can also depend on communications channels of moderate-to-high data rates. Error rates and delay may be slightly higher than might be obtained on a LAN.

The Distributed Queue Dual Bus (DQDB) protocol defines a technology that can be used to provide LAN-like services over a wide geographic area. The type of network that the DQDB protocol operates over is generally referred to as a MAN. The DQDB protocol is relatively independent of the underlying physical transmission medium and can operate at speeds between 44 Mbps and 155 Mbps. DQDB provides a service that is similar to that defined by IEEE 802.6 LLC.

The DQDB protocol provides a powerful service for both constant bit rate (CBR) and variable bit rate (VBR) applications. Through the use of prearbitrated and queued arbitrated services, DQDB can support multimedia traffic. The IEEE 802.6 MAN standard is based on the DQDB protocol.

Although the DQDB technology was originally intended for relatively short-distance communications, DQDB transmission is also used to provide SMDS WAN data links.

Because of the emergence of Asynchronous Transfer Mode (ATM) as a universal network technology for carrying all types of voice, video, and data information on LANs, MANs, and WANs, technologies are already becoming obsolete.

KEY TERMS

Data Exchange Interface (DXI)

Distributed Queue Dual Bus (DQDB)

IEEE 802.6

Prearbitrated (PA) Access Control

Queued Arbitrated (QA) Access Control

SMDS Interface Protocol (SIP)

Subscriber Network Interface (SNI)

Sustained Information Rate (SIR)

Switched Multimegabit Data Service (SMDS)

Virtual Channel Identifier (VCI)

1. Develop a general definition for a metropolitan area network.

2. Where would a MAN be deployed? What type of traffic would be carried on a MAN?

3. What are the key characteristics of a MAN?

4. Define SMDS. What is the major objective of SMDS service? How is SMDS different from a MAN?

5. What are the transport speeds of SMDS? What are the three tiers of the service?

6. How is SMDS similar to POTS?

7. What protocols are supported by SMDS? What is the difference between connectionless and connection-oriented services?

8. How does addressing work in the SMDS environment?

9. What are the service characteristics of SMDS?

10. Describe the Distributed Queue Dual Bus arrangement of 802.6.

11. What is the mechanism for guaranteeing the rate of service for SMDS?

12. What is the SNI? Why is it important?

13. What is the ISSI?

14. How is SMDS connected to the user's premise?

15. Define the DQDB protocol. How is DQDB similar to FDDI? How is it different?

16. What is the slot, cell, and segment structure of DQDB?

17. What are the three types of DQDB services? How are they utilized? What are the two classes of service for DQDB? How do they differ?

18. What are the three sublayers of the DQDB layer?

19. How is DQDB similar in operation to voice services?

20. What are the two topologies of DQDB? How are they different?

21. How do the queues work with the counters in DQDB? What is a slot? Where does it appear?

22. What is the function of prearbitrated access control?

23. What is the function of queued arbitrated access control?

24. What is the difference between CIR and SIR?

25. What are the SMDS access classes? Identify their sustained data rates.

26. What is a SMDS access unit?

27. What is significant about SMDS addressing?

28. What is a SMDS WAN?

29. What is the cell structure of SMDS? How are the cells identified in SMDS? What is the structure? How does the SMDS cell differ from the ATM cell?

30. What are the three operations of SMDS QoS?

31. How does the OSI model compare with the SMDS model?

32. How does the SIP model compare with the MAN 802.6 model?

33. What are the five 802.6 PDU types?

34. What is the primary market for MANs?

35. Give an example of an application for SMDS. Describe how it would be superior to the alternative design.

36. Give examples of issues and considerations that must be made when implementing MANs and SMDS networks.

37. What are advantages and disadvantages for implementing MANs and SMDS networks?

38. What technology is threatening the life of SMDS? Why?

39. What is DXI and what is its function?

40. Define the HDLC protocol and format. What is the difference between LAPB and LAPD?

ACTIVITIES

1. Develop a list of users who can benefit from the SMDS capabilities. Indicate what applications would be carried for each of these customers.

2. Use the Internet, DataPro, and other texts to identify availability of SMDS from the various carriers.

3. Develop a matrix of the standards involved with the SMDS technology. Appendix C has a list of RFCs, which is a good starting point for the exercise. Another good place to start is the IETF home page.

4. Develop a comparison matrix of the competing technologies. These can include leased line, Frame Relay, ATM, DSL, and SMDS. Show such items of information such as cost, features, speeds, availability, equipment requirements, and application niche.

5. Identify applications that would benefit from using a SMDS platform. For each suggestion state why SMDS would be a good technology to support the application. Use the brainstorming and boarding method in the classroom environment to discuss these arguments.

CASE STUDY/PROJECT

Identify a user who might benefit from the SMDS capability. Identify the current requirements and develop a solution utilizing SMDS. A good application would include one that currently utilizes a full-mesh network of leased lines. This can include a pricing exercise to show that SMDS will save money over a leased line network.

Details
The county library system in Dade County, Florida utilizes leased lines to communicate between the various libraries. It utilizes a bank of terminals at each library to locate references throughout the county library system. The speed of these circuits is 19.2 kbps. The main library is located in downtown Miami. The remaining ten branches are located throughout the county. The traffic volume that is being generated is causing a slow response at all of the library branches. The county commissioners have decided to upgrade the county library's network. A representative of a local carrier has proposed a SMDS solution. The local COs are equipped to handle SMDS traffic.

Requirements
You are the network designer for the county that is assigned to the library project. Design a SMDS network that will provide 56 kbps access to all of the county libraries. Provide the necessary components to install this system. Price out the service. Prepare a graphic.

Alternative Requirement
Since you are the network designer for the county, you are well versed in networking technologies. You realize that there are other alternatives to SMDS that might fit this library application.

Your research has indicated that the local COs also provide access for both ADSL and Frame Relay technologies.

Develop these two alternatives for the library system. Show the advantages and disadvantages for all three technologies. Include a cost/benefit analysis for the technologies. Prepare a graphical network design utilizing VISIO or another graphics package. (Note: This is probably a semester project.)

Wireless/Personal Communications Service

▬ INTRODUCTION

Wireless has been defined as "without wires"! In the past, communication devices and end users were fixed in one location. Wireless communications have broken that barrier by allowing communication to and from a mobile communications device. Wireless, however, requires an infrastructure of wires in conjunction with the wireless system. It is becoming increasingly popular to interconnect computer equipment and to transmit data over wireless links rather than over conventional telecommunications circuits. Wireless transmission can be used to implement either long-distance WAN links or short-distance LAN links. There must be a way of interconnecting physically disjointed LANs and other groups of mobile users with stationary servers and resources. Networks that include both wired and wireless components are called hybrid networks. The ongoing development of a wireless network will allow its users to communicate with anyone, anywhere, and at any time.

As alternatives to cable, there are several wireless media available for transmitting network packets: radio waves, microwave, and infrared signals. All of these technologies transmit a signal through the air or the atmosphere. This characteristic makes them a good alternative in situations when it is difficult or impossible to use cable. It also has an important limitation in that these signals can experience problems due to interference from other signals using the same media, and from sun spots, ionospheric changes, and other atmospheric disturbances.

Wireless technologies continue to play an increasing role in all types of networks. Since 1990, the number of wireless options has been increasing,

OBJECTIVES

Material included in this chapter should enable you to:

- understand the wireless technology and its impact on the current broadband networking environment.

- become familiar with the wireless terms and definitions that are relevant in the Enterprise Networking environment. Understand the history of mobile services and how they impact current wireless services.

- understand the function and interaction of wireless transmission in the Enterprise Network environment. Look at the technical aspects of wireless transmission and identify the various wireless network configurations that are possible. Become familiar with wireless networks that are associated with LANs, extended LANs, and mobile computing.

- understand where wireless fits in the different applications of the technology. Look at wireless LANs, radio, mobile telephony, microwave, and satellite transmission capabilities. Identify the differences between a packet data network, the cellular telephone network, and cellular digital packet data technology.

- identify components that comprise the wireless networking environments. Look at the cell structure and the various frequencies that are allocated to the wireless networking environment.

- look at the differences between AMPS, GSM, FDMA, TDMA, CDMA, EDGE, and GPRS. Become familiar with WAP and 3G environments.

- become familiar with the various standards that are associated with the wireless networking environment.

- identify wireless issues and considerations that are important when deciding to utilize a wireless networking solution. Look at the advantages and disadvantages for each of the different technologies.

advanced mobile phone service (AMPS)

while the cost of these technologies continues to decrease. As wireless networking becomes more affordable, demand will increase, and economies of scale will come into play. Most experts anticipate that wireless networking of all kinds will become more prevalent in the future.

Deployment of wireless communications in North America and Europe has been astronomical. Wireless communications technology has evolved from simple first-generation analog products designed for business use to second-generation digital wireless telecommunications systems for residential and business environments. For 2001, the next generation of wireless information networks is emerging. Complete Personal Communications Services (PCSs) will enable all users to economically transfer any form of information between any desired location. The new network will be built on and interface with the separate first- and second-generation cordless and cellular services and will also encompass other means of wireline and wireless access to Local Area Networks (LANs) and Specialized Mobile Radio (SMR) systems. Virtually any electronic communications device imaginable can become a part of the new wireless frontier.

The major problems faced with mobile telephone and data services are a limited number of users within the allowable frequency band, limited data rates within the allowable bandwidth, and signal fading. User numbers and data rates are limited by the very limited radio spectrum available to mobile telephone services. Efforts are underway to reduce these problems.

11.1   MOBILE SERVICES HISTORY

The first major mobile radio system designed to work with the public telephone system was Mobile Telephone Service (MTS), inaugurated in 1946. The central transmitter/receiver had a coverage radius of about 20 miles, which is considerably more than today's cellular transmitters. The transmitter had a 250 watt output, which resulted in a relatively large coverage area. This is a disadvantage to subscribers, however, because it limits the total number of customers and makes expansion of the service very difficult. MTS was not a cellular system.

Improved Mobile Phone Service (IMPS) replaced MTS. A user could dial a number, eliminating the operator that was required by the MTS system. The basic problems of limited channels and difficulty of expansion, however, remained. Many more users wanted mobile service than could be provided by the system.

Advanced Mobile Phone Service (AMPS) was the first cellular telephone system in North America and was conceptually different from MTS and IMPS. Integrated Circuits (ICs) with powerful features were a major factor in making cellular telephone systems affordable. These ICs are needed in both the cellular base stations and the mobile cell phone to keep down the price of cellular, make the size of the cell phone manageable, and minimize battery drain.

11.2   WIRELESS TRANSMISSION

Wireless transmissions involve electromagnetic waves, which are oscillating electromagnetic radiation caused by inducing a current in a transmitting antenna. The waves then travel through the air, or free space, where they are sensed by a receiving antenna. Free radio and television transmit signals in this manner. Figure 11–1 shows the electromagnetic spectrum.

This spectrum is measured in terms of the frequency of the wave forms used for communication. It is measured in Cycles Per Second (cps), usually expressed in **hertz (Hz),** in honor of Robert Hertz, one of the inventors of the radio. The spectrum starts with low-frequency waves, such as those for electrical power (60 Hz) and telephone systems (0 to 3 kHz), and goes all the way through the spectra associated with visible light. These frequency bands are assigned by the **Federal Communications Commission (FCC).**

In wireless communications, the frequency affects the amount of data and the speed at which data may be transmitted. The strength of power of the transmission determines the distance that broadcast data can travel and still remain intelligible. In general, the principles that govern wireless transmissions dictate that lower frequency transmission can carry less data more slowly over longer distances, whereas higher frequency transmissions can carry more data faster over short distances [Bates & Gregory, 1997].

The middle part of the electromagnetic spectrum is commonly divided into several named frequency ranges, or bands. The most commonly used frequencies for wireless data communications are as follows:

- Radio   —10 kHz to 1 GHz
- Microwave   —1 GHz to 500 GHz
- Infrared   —500 GHz to 1 THz

hertz (Hz)

federal communications commission FCC

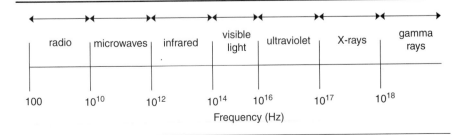

Figure 11–1
Electromagnetic Waves

Table 11–1
Unguided Communications

Frequency Band	Name	Analog Data Bandwidth	Digital Data Rate	Principal Applications
300–3,000 kHz	LF (Low Frequency)	To 4 kHz	10 to 1,000 bps	Commercial AM Radio
3–30 MHz	MF (Medium Frequency)	To 4 kHz	10 to 3,000 bps	Shortwave Radio
30–300 MHz	HF (High Frequency)	5 kHz to 5 MHz	To 100 kbps	VHF Television, FM Radio
300–3,000 MHz	VHF (Very High Frequency)	To 20 MHz	To 10 Mbps	UHF Television, Terrestrial microwave
3–30 GHz	UHF (Ultra High Frequency)	To 500 MHz	To 100 Mbps	Terrestrial microwave, Satellite microwave
30–300 GHz	SHF (Extremely High Frequency)	To 1 GHz	To 750 Mbps	Experimental short Point-to-point

The important principles to remember about a broadcast medium focus on the inverse relationship between the frequency and distance, and the direct relationship between frequency and data transfer rate and bandwidth. It is also important to understand that higher frequency technologies will often use tight-beam broadcasts and require a line-of sight between the sender and the receiver to ensure correct data delivery.

guided
unguided
antenna

There are two types of media, **guided** and **unguided.** For unguided media, transmission and reception are achieved by means of an **antenna.** The antenna radiates electromagnetic energy into the medium (air), and for reception, the antenna picks up electromagnetic waves from the surrounding medium. The two basic types of configurations for wireless transmission are directional and omnidirectional. For the directional configuration, the transmitting antenna emits a focused electromagnetic beam, which requires that the transmitting and receiving antennas be carefully aligned. In the omnidirectional case, the transmitted signal spreads out in all directions and can be received by many antennas. In general, the higher the frequency of a signal, the more it is possible to focus it into a directional beam.

terrestrial microwave
satellite microwave
broadcast radio

The three general ranges of frequencies that will be discussed in this chapter are associated with **terrestrial microwave, satellite microwave,** and **broadcast radio.** Table 11–1 provides a list of the characteristics of unguided communications bands.

11.3   WIRELESS NETWORK CONFIGURATIONS

local area networks
(LANs),
extended LANs
mobile computing

Depending on the role that wireless components play in a network, wireless networks can be subdivided into three primary categories: **Local Area Networks (LANs), extended LANs,** and **mobile computing.**

Service	Usage	Network
Cellular	Circuit switched roaming modem and telephone connections	WAN
Microwave	Short-range data transmission	LAN
Mobile satellite	Highly mobile messaging and location services	WAN
Packet radio	Packet-switched roaming data transmission	WAN, messaging services
Infrared	Line-of-sight data transmission	LAN, peripheral sharing device
Spread spectrum	Short-range data transmission	LAN

Table 11–2
Wireless Technologies

An easy way to differentiate among these uses is to distinguish in-house from carrier-based facilities. Both the LAN and extended LAN uses of wireless networking involve equipment that a user owns and controls. Mobile computing typically involves a third party that supplies the necessary transmission and reception facilities to link the mobile part of a network with the wired part. Table 11–2 lists a number of wireless technologies.

11.4 LOCAL AREA NETWORKS

In local area networks, wireless components act as part of an ordinary LAN, usually to provide connectivity for roving users or changing environments. They may provide connectivity across areas that may not otherwise be networkable, such as in older buildings where wiring would be impractical or across right-of-ways where wire runs might not be permitted.

The wireless components of most LANs behave like their wired counterparts, except for the media and related hardware involved in the physical connectivity. The operational principles are almost the same. It is still necessary to attach a network interface of some type to a computer device, but the interface attaches to an antenna and an emitter, rather than a wire or cable. Users still access the network just as if they were wired into it.

An additional item of equipment is required to link wireless users with wired users or resources. It is still necessary to install a transceiver or an **access point** that translates between the wired and wireless networks. This access point broadcasts messages in wireless format, which must be directed to wireless users, and relay messages sent by wireless users directed to resources or users on the wired side of its connection. An access point device includes an antenna and transmitter to send and receive wireless traffic, but is also connected to the wired side of the network. This

access point

Figure 11–2
Wireless Portable Computer
Connected to a Wired
Network Access Point

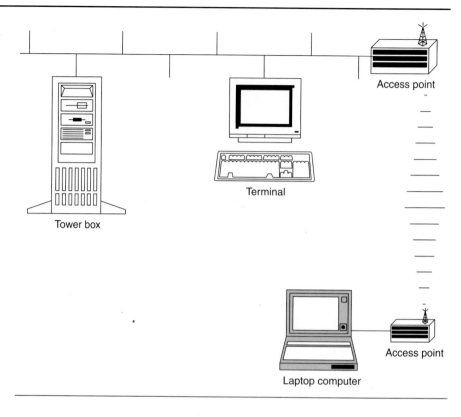

Access point

Tower box

Terminal

Laptop computer

Access point

permits the device to shuttle back and forth between the wired and wireless sides of the network. Figure 11–2 shows an access point device that is connected to a wireless portable personal computer.

Some wireless LANs use small individual transceivers, wall mounted or free standing, to attach individual computers or devices to a wired network. This permits some limited mobility, with an unobstructed view of the transceiver for such devices.

11.5   EXTENDED LANs

In extended local area networks, an organization might use wireless components to extend the span of a LAN beyond normal distance limitations for wire- or fiber-based systems. Certain kinds of wireless networking equipment are available that extend LANs beyond their normal wire-based distance limitations or provide connectivity across areas where wireless bridges   wire might not be allowed or able to traverse. **Wireless bridges** are available that can connect networks up to 3 miles apart.

Such LAN bridges permit linking of locations, such as buildings or facilities, using line-of-sight broadcast transmissions. Such devices may make it unnecessary to route dedicated digital communications lines from one site to another through a communications carrier. Normally, up-front charges for this technology will be considerably higher, but the user will avoid the recurring monthly service charges from a carrier that can quickly make up this difference. Spread spectrum radio, infrared, and laser-based equipment of this kind is readily available from equipment suppliers.

Longer range wireless bridges are also available, including spread spectrum solutions that work with either Ethernet or Token Ring over distances up to 25 miles. As with shorter range wireless bridges, the cost of a long-range wireless bridge may be justified because of the savings in communications costs it can realize over time. Where appropriately connected, such equipment can transport both voice and data traffic.

A low-power service, called *Personal Communication Network (PCN)*, is available for wireless LAN access. The following applications are available on a PCN:

- Person-to-person calling
- Messaging
- Fax
- Wireless PBX
- Cordless telephone
- Mobile communications
- Laptop interfaces
- Closed user group communications
- Backbone access to existing services

PCN has the same characteristics as a PCS system, including portable handsets, handoff processes, rural area coverage in cells, and availability for networking purposes in inside or outside structures.

11.6 WIRELESS LAN TECHNOLOGIES

Wireless LANs make use of these four primary technologies for transmitting and receiving data:

- Infrared
- Laser
- Narrowband, single-frequency radio
- Spread spectrum radio

Infrared light can be used as a medium for network communications. It transmits in the light frequency ranges of 100 gigahertz (GHz) to 1,000

infrared

Figure 11–3
Line of Sight Infrared
System

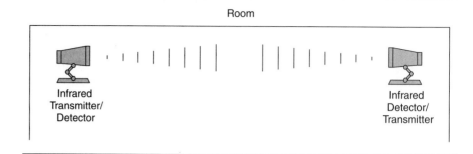

terahertz (THz). This technology is used in the remote control devices for television and stereo devices. Infrared can be broadcast in a single direction or in all directions, using a Light Emitting Diode (LED) to transmit and a photodiode to receive.

line of sight
diffuse

Infrared connections are limited to two types, **line of sight** and **diffuse.** Line-of-sight infrared systems (Figure 11–3) require that the transmitter and receiver of the infrared light be straight and in line with each other. Diffuse systems allow an infrared signal to bounce off a wall or other object.

Like radio waves, infrared can be an inexpensive solution in areas where transmission cable is difficult or impossible to install. It can also be used where there are mobile users, with the advantage that the signal is difficult to intercept without someone knowing. Transmission rates only reach up to 16 Mbps for directional communications, and are less than 1 Mbps for omnidirectional communications. Infrared does not penetrate walls and can experience interference from strong light sources.

Computer networks can use infrared technology for data communications. For example, it is possible to equip a large room with a single infrared connection that provides network access to all computers in the room. Computers can remain in contact with the network while they are moved within the room. Infrared networks are especially convenient for small, portable computers, because infrared offers the advantages of wireless communication without the use of antennas.

direct sequence

Two different methods of implementing spread spectrum are in common use in wireless LANs. In the **direct sequence** method, the radio signal is broadcast over the entire bandwidth of the allocated spectrum. The transmitter and receiver both include a synchronized pseudo-noise generator, which the receiver uses to detect the desired signal out of the resulting jumble. Most of the early wireless LAN products used direct sequence.

frequency hopping

The second method is **frequency hopping.** The transmitter and receiver are synchronized to hop between frequencies, stopping on each frequency for a few milliseconds. The amount of time per hop is called

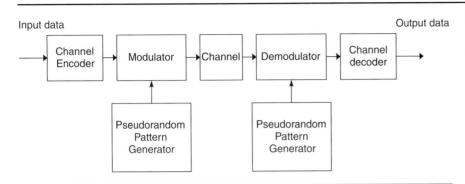

Figure 11–4
General Model of Spread Spectrum

dwell time. Later-generation LANs used frequency hopping, which is an excellent method of handling interference. If the equipment finds an interfering signal, it marks that portion of the spectrum as busy and skips it. Both frequency hopping and direct sequence provide excellent security. Figure 11–4 highlights the key characteristics of a spread spectrum system. Input data enters the channel encoder, is modulated, and passed over the channel where it is demodulated and passed to the channel decoder.

11.7 RADIO TECHNOLOGIES

In addition to its uses for the public broadcast of radio and television programs and for private communications with devices such as portable phones, electromagnetic radiation can be used to transmit computer data. A network that uses electromagnetic radio waves is said to operate at radio frequency, and the transmissions are referred to as **Radio Frequency (RF)** transmissions. Unlike networks that use wires or optical fibers, networks using RF transmissions do not require a direct physical connection between computers. Instead, each participating computer attaches to an antenna, which can both transmit and receive RF.

radio frequency (RF)

Network signals are transmitted over radio waves similar to the way your radio station broadcasts, but network applications use much higher frequencies. AM radio stations transmit in a frequency range of 535 kHz to 1605 kHz, and FM radio stations transmit in a frequency range of 88 MHz to 108 MHz. In the United States, network signals are transmitted at much higher frequencies of 902 MHz to 5.85 GHz.

A radio network transmits a signal in one or multiple directions, depending on the type of antenna that is utilized. The wavelength is very short with a low transmission strength, which means that it is best suited for line-of-sight transmissions. A line-of-sight transmission is one in which the signal goes from point to point, rather than bouncing off the atmosphere to skip across the country or across continents. A limitation of line-of-sight transmissions is that they are interrupted by tall land masses such as mountains.

Most wireless network equipment employs spread spectrum technology for packet transmissions. This technology uses one or more adjoining frequencies to transmit the signal across greater bandwidth. Spread spectrum frequency ranges are in the 902 MHz to 928 MHz range and higher. Spread spectrum transmissions typically send data at a rate of 2 MHz to 6 Mbps.

Bluetooth Wireless Technology

bluetooth

A new wireless technology allows users to make wireless connections between various communications devices, such as mobile phones and desktop computers. This technology is called **Bluetooth,** and it uses radio transmission to transfer both voice and data in real time. The Bluetooth radio is built into a small microchip and operates in a globally available frequency band.

The Bluetooth specification is a joint venture involving Ericsson, IBM, Intel, and Toshiba. It is designed to be an open standard for short-range systems. The sophisticated mode of transmission adopted ensures protection from interference and security of data. The Bluetooth specification has two power levels defined: a lower power level that covers a shorter personal area within a room and a higher power level that can cover a medium range, such as within a home. Information concerning the specification is available on Bluetooth's Website: www.bluetooth.com/developer/specification/overview.asp [Bluetooth Special Interest Group [SIG], retrieved 11-27-2000].

The Bluetooth technology supports both point-to-point and point-to-multipoint connections. Up to seven "slave" devices can communicate with a "master" radio. These so-called *piconets* can be established and linked together. All devices in the same piconet have priority synchronization; however, other devices can be set to enter at any time. The topology can be described as a flexible, multiple piconet structure.

11.8   MOBILE TELEPHONY

Mobile telephony provides the luxury and convenience of placing calls through the telephone company directly from one's own vehicle. The conventional mobile configuration depicted in Figure 11–5 includes a single base station connected to a local CO and the mobile vehicle in which the mobile telephone set is installed. The base station is capable of transmitting and receiving on several UHF channels in succession. A high-power transmitter delivers 200 to 250 watts (W) to the base station's antenna,

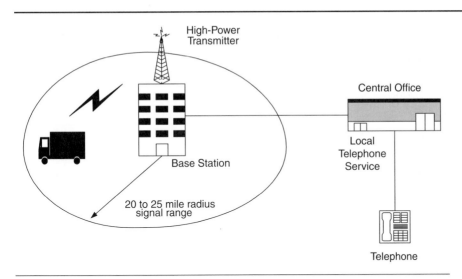

Figure 11–5
Conventional Mobile Telephone Service

which is typically located on a tower or tall building. The mobile unit can travel within a 30 mile radius of the base station and reliably communicate with a transmission power output of up to 25 watts. Such large amounts of signal power have caused interference between adjacent channels when mobile units are in proximity to each other or the base. Only one conversation can be held at a time on the limited number of frequency channels available within a given service area. Under these circumstances, the use of the mobile telephone has been limited to public safety services such as fire and police, forestry services, construction companies, and other private organizations.

11.9 MICROWAVE TECHNOLOGIES

Electromagnetic radiation beyond the frequency range used for radio and television can also be used to transport information. Microwave is a form of radio transmission using ultra high frequencies in the gigahertz (GHz) range to complete a link from one point to another. These signals are propagated from the transmitter to the antenna using **wave guides,** which are hollow rectangular tubes that form a resonant cavity for passing a selected frequency signal. These cavities are tuned by inserting or extracting a "tuning slug" into the cavity, causing its resonance to change. Frequencies other than the resonant frequency are high attenuated. A figure of merit for wave guides is the standing wave ratio, which is a measure of the ratio of the standing wave power to the main signal power. The lower the ratio, the less attenuated the transmitted main signal. Transmitted signals are

wave guides

Figure 11–6
Line of Sight Microwave
Transmission

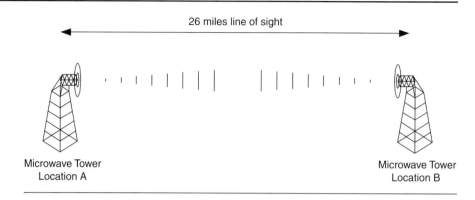

often degenerated by weather and other factors. Temperature layers can cause the signal to be refracted and portions of the signal can be reflected from the earth's surface. These effects are called **multipath fading.**

multipath fading

Although microwaves are merely a higher frequency version of radio waves, they behave differently. Microwave is considered to be a line-of-sight transmission method, because the transmission and reception facilities must be within a line of sight. Distances are usually limited to less than 50 miles because of the curvature of the earth, and are actually placed within 26 miles of each other, as depicted in Figure 11–6. Each tower receives the signal, amplifies and otherwise regenerates it, and sends it to the next tower.

Microwave systems work in one of two ways. Terrestrial microwave transmits the signal between two directional antennas, shaped like dishes. These transmissions are performed in the frequency ranges of 4 GHz to 6 GHz and 21 GHz to 23 GHz, and require the operator to obtain a FCC license. As with other wireless media, microwave solutions are applied where transmission cable costs are too high or where cabling is impossible. Terrestrial microwave may be a good solution for communications between two tall buildings in a metro area. Satellite microwave, discussed next, is another solution for joining networks across a country or across continents.

Both types of microwave media transmit at speeds from 1 Mbps to 10 Mbps, which is a limitation where higher network speeds are desired.

11.10   SATELLITE TRANSMISSION

Much of today's communications between distant localities involves the use of communications satellites. Worldwide area networks such as the Internet are being established and maintained through an interface of

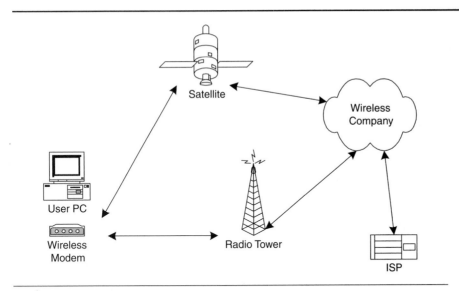

Figure 11–7
Wireless Internet Access

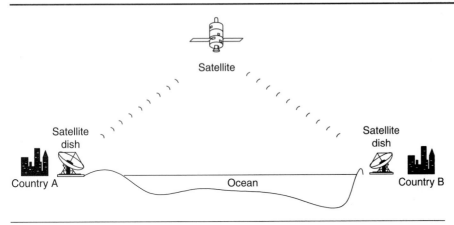

Figure 11–8
Satellite with Two Ground Stations

satellites and ground communications systems. Figure 11–7 depicts such an example of a wireless Internet access. The user has access to the Internet Service Provider (ISP) through a wireless company that utilizes both satellite and terrestrial communication links.

Although radio transmissions do not bend around the surface of the earth, RF technology can be combined with satellites to provide communication across longer distances. Figure 11–8 shows how a communications satellite in orbit around the earth can provide a network connection across an ocean.

transponder

Communications satellites are essentially electronic repeaters located many miles above the earth's surface. A satellite link is a channel connection using radio frequency waves between an earth station transmitter and a device on a satellite called a **transponder.** The transponder is a receiver and amplifier coupled to a transmitter which receives the incoming signal, amplifies it, and then retransmits it to another receiving earth station or satellite. The satellite receives the transmission at one carrier frequency, amplifies it, and retransmits the information at a different carrier frequency. Satellites receive their operating power from solar cells that are attached to the satellite body.

Satellites act as relay stations for very-high-speed communications from one point to another. Typically, a satellite provides for multiple channels, each having a capacity of 1.544 Mbps. Satellite microwave transmits the signal between three directional antennas, one of which is on a satellite in space. These transmissions are in the 11 GHz to 14 GHz frequency range.

footprint

Data communications connections using a satellite link require ground stations for transmission and reception of the high-speed signals. Like microwave, satellite transmission is considered a line-of-sight transmission method. Because of the distance involved and satellite location and placement, however, its signal can cover a wide area. This wide area is referred to as its **footprint** and is determined by the altitude and position of the satellite.

geosynchronous

Communications satellites can be grouped into categories according to the height at which they orbit. The easiest types to understand are the **geosynchronous,** or geostationary, satellites. The name arises because a geosynchronous satellite is placed in an orbit that is exactly synchronized with the rotation of the earth. Such an orbit is classified as a Geostationary Earth Orbit (GEO) because, when viewed from the ground, the satellite appears to remain at exactly the same point in the sky at all times. The distance required for geosynchronous orbit is approximately 36,000 km (20,000 miles) and is defined as a high earth orbit.

low earth orbit (LEO)

The second category of communications satellites operate in **Low Earth Orbit (LEO),** which means that they orbit several hundred miles above the earth. The chief disadvantage of a LEO is the speed at which a satellite must travel. Because their period of rotation is faster than the rotation of the earth, satellites in lower earth orbits do not stay above a single point on the earth's surface. The satellite can only be used during the time that its orbit passes between two ground stations, and the ground stations must track the satellite for signal alignment.

Satellite and terrestrial microwave systems can be implemented for sending and receiving multiplexed channel signals on their high-frequency carriers, which provides a way of handling large volumes of data transfers without using wires. They are also used as an extension of the cellular mobile communications telephone system.

11.11 CELLULAR TELEPHONY

The cellular mobile telephone system was developed to allow dial-up telephone service using mobile telephone handsets and radio transmissions. The cellular system is characterized by intermittent connections between users and flexible communication times. Cellular technology's most common use is to provide cellular telephone services, although the cellular network is also being used more and more by data communications devices such as pagers and Personal Digital Assistants (PDAs). Cellular technology is a form of high-frequency radio in which antennas are spaced strategically throughout a metropolitan area. A service area or city is divided into many coverage areas, each with its own antenna. This arrangement generally provides subscribers with reliable mobile service of a quality that almost equals the hardwired telephone system.

Cellular radio systems, more familiar to many as cellular telephone systems, utilize a large number of FM radio links, Frequency Division Multiplexed (FDM), to accommodate a large number of simultaneous calls on the same frequency. This is accomplished by splitting up the region, typically a city, into a number of roughly hexagonal cells (Figure 11–9) that resemble a honeycomb. In each cell there is one base station with a transmitter capable of spanning the entire area of the cell, but not much more. Thus, each FM radio link, operating in one of the four groups of possible frequencies, is limited to a single cell and its adjacent partners. Since the power output of the base station is low, those frequencies can be used by another cell's base station to communicate with another user just several miles away in another nonadjacent cell.

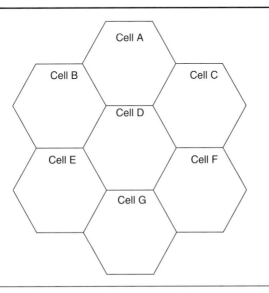

Figure 11–9
Cellular Honeycomb Structure

Table 11–3
Cellular Frequencies in the
United States

Mobile unit transmit—824 to 851 MHz/25 MHz total divided between two providers
Mobile unit receive—869 to 896 MHz/25 MHz total divided between two providers
Mobile receive frequency = mobile transmit frequency + 45 MHz
Total Channels = 25 MHz/0.030 Mhz/channel = 832 total channels (416 channels per provider)
21 signaling channels
1 signaling channel per cell site
395 voice channels per provider

Therefore, the same four groups of frequencies are used many times over in any cellular system. As long as no two sides of the honeycomb cells touch, any frequency that has been used previously can be used again. The central concept of cellular lies in the frequency channels' reuse in the same geographic area. This is necessary because the allocation of bandwidth is expensive for an industry to purchase, which follows that the carriers are looking for ways to accommodate more users in the same frequency bands.

Cellular Frequencies

The cellular frequencies in the United States are shown in Table 11–3. Note that the frequencies are from 824 MHz to 896 MHz. The transmit and receive frequency channels each occupy 25 MHz. The mobile unit's frequency (down-link or forward channel) is always 45 MHz above the transmit (up-link or reverse channel) frequency.

The up-link always occupies the lower band, because radio propagation is slightly better for lower frequencies, and system designers want to give the maximum advantage to the mobile transmitter. This saves batteries and allows for smaller, less efficient antennas. The base station usually operates from the electric power grid and has higher gain antennas than the mobile units.

AMPS

mobile telephone switch-
ing office (MTSO)

Advanced Mobile Phone Service (AMPS) was developed by AT&T and has become known as cellular telephone. The basic concept behind this wireless technology is to divide heavily populated areas into many small regions called cells. Each cell is linked to a central location called the **Mobile Telephone Switching Office (MTSO).** Each MTSO is connected to a central office and other MTSOs. This connection is by coax, twisted pair, optical fiber, or microwave. The MTSOs are connected to the Public Switched Telephone Network (PSTN) via a central office. The MTSO coordinates all mobile calls between an area comprised of several cell sites and the local central office. In this way, calls from fixed telephone locations can be routed to a mobile subscriber, or vice versa. Time and

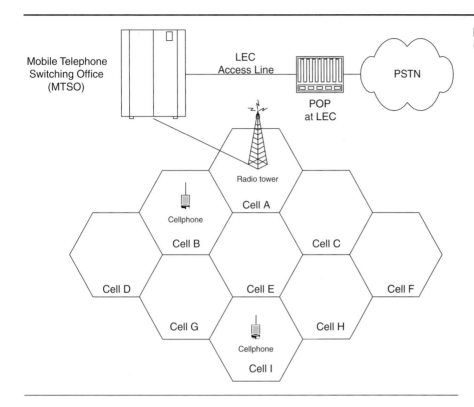

Figure 11–10
Cellular Telephone System

billing information for each mobile unit is accounted for by the MTSO. Figure 11–10 provides an example of the cell environment.

At the cell site, a base station is equipped to transmit, receive, and switch calls to and from any mobile unit operating within the cell to the MTSO. The cell itself encompasses only a few square miles, thus reducing the power requirements necessary to communicate with cellular telephones. This structure permits the same frequencies to be used by other cells since the power levels emitted diminish to a level that does not interfere with other cells. Heavily populated areas can be serviced by several transmission towers, rather than one, as used by conventional mobile techniques.

Cellular systems are full duplex. Different frequencies are used for base-to-mobile (down-link) and mobile-to-base (up-link) transmissions. Full duplex allows the MTSO to send control information to the mobile unit at any time. This control information is piggybacked onto the voice channel, and the user is often not aware of this transfer. The MTSO controls the transmitter power and frequency channel of the mobile unit. The MTSO finds an unused frequency channel to use and controls the mobile transmitter power to keep all received signals at the base station at the same level.

When a cellular phone is turned on, its microprocessor samples dedicated setup channels and tunes to the channel with the strongest signal. A closed loop is effectively established between the mobile unit and the cell site at all times. If the cellular phone's signal strength significantly diminishes as a result of traveling outside one cell and entering another, the MTSO polls through all of its cell sites to determine the cellular phone's new cell location. This will be the cell in which the maximum signal strength is received.

When a mobile unit leaves one cell and enters another, the process called handoff occurs and is designed to be transparent to the user. This process includes the following activities:

- The MTSO receives signal strength reports from both cells. When the mobile's signal strength is consistently stronger in the new cell, the MTSO begins the handoff procedure.
- The mobile unit is told to use a new frequency via a data burst on the voice channel (down-link). The new frequency is the one assigned to the new cell.
- The mobile unit responds with an acknowledgment (ACK) on the reverse (up-link) setup channel.
- The conversation on the mobile unit continues uninterrupted.

The handoff procedure when a mobile unit travels from one MTSO's region to another MTSO's region is as follows:

- MTSO A notes that the signal is getting weaker and asks the new MTSO (which will be designated as MTSO B) for a signal strength report from the mobile unit.
- When the mobile unit's signal strength is significantly greater in MTSO B's region, MTSO A will transfer control to MTSO B.
- MTSO B will assign a new frequency to the mobile unit and adjust its power via a data burst on the voice (down-link) channel.
- The mobile unit responds with an acknowledgment (ACK) on the reverse (up-link) channel.
- The conversation on the mobile unit continues uninterrupted.

When everything is working properly, the handoffs between cells and between MTSOs should be transparent to the mobile user, although it is possible to hear *clicks* from the up-link and down-link control commands.

Normally, the cellular phone is used only within the metropolitan area where it is registered, which may include several counties or cities. Frequently, a user needs to operate a cellular phone outside of the home area. This **roaming** capability is possible anywhere throughout the country provided that cellular services are available and a prearranged agreement has been made between telephone companies and their users.

roaming

11.12   SECURITY AND PRIVACY IN WIRELESS SYSTEMS

Communications on shared media can be intercepted by any user of the media, and anyone with access to the media can receive or transmit on the media. When the media are shared, privacy and authentication are lost unless some method is established to regain it. **Cryptography** is a method that provides the means to regain control over privacy and authentication. Some of the cryptographic requirements are in the air interface between the Personal Station (PS) and the Radio System (RS). Other requirements are on databases stored in the network and on information shared between systems in the process of handoffs or giving service for roaming units.

     There are four levels of voice privacy which are used to identify security and privacy requirements. These four privacy levels are as follows:

cryptography

- Level 0—No privacy
- Level 1—Equivalent to wireline
- Level 2—Commercially secure
- Level 3—Military and government secure

The security and privacy requirements for these four levels include the following categories:

- Radio system performance
- Theft resistance
- Physical requirements
- System lifetime
- Privacy requirements
- Law enforcement needs

Radio System Performance
When a cryptographic system is designed for wireless systems, it must function in a hostile radio environment characterized by the following impediments:

- Interference
- Jamming
- Handoff activities
- Multipath fading
- Thermal noise

Theft Resistance
The system operator may or may not care if a call is placed from a stolen personal station as long as the call is billed to the correct account. The owner of a personal station, however, will care if the unit is stolen. Requirements needed to accomplish the reduction of theft are:

- Unique user ID
- Unique personal station ID

- Clone-resistant design
- Elimination of repair and installation fraud

Physical Requirements

Any cryptographic system used in a personal station must work in a mass-produced consumer product. The cryptographic system must meet the following physical requirements:

- Basic handset requirements
- Low-cost wireline compliant
- Mass produced
- Import/export specifications

System Lifetime

It has been estimated that computing power doubles every two years, which means that a cryptographic algorithm that is secure today may be breakable in 5–10 years. A reasonable security requirement is that procedures must be viable for many years. It is necessary for the security designers to consider the best cracking algorithms today and have provisions for field upgradeability in the future.

Privacy Requirements

A user of a PCS personal station needs privacy in the following areas:

- Call setup information
- Speech
- Data
- User location
- User identification
- Calling patterns

Every use of the PCS needs to be private so that the user can send information on any channel—voice, data, or control—and be assured that the transmission is secure.

Law Enforcement Needs

When a valid court order is obtained, current analog telephones are relatively easy to tap by the law enforcement community. There are several methods that a PCS system operator can use to meet the requirements of the order. It is essential that the method used does not compromise the security of the system.

11.13   PERSONAL COMMUNICATION SERVICES

personal communication service (PCS)

Personal Communication Service (PCS) is a multibillion-dollar industry representing the latest advancements in digital wireless telecommu-

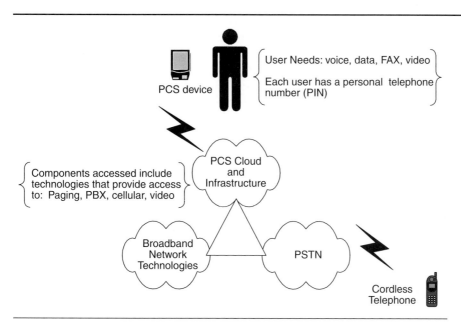

Figure 11–11
Personal Communication Environment

PCS device

User Needs: voice, data, FAX, video

Each user has a personal telephone number (PIN)

PCS Cloud and Infrastructure

Components accessed include technologies that provide access to: Paging, PBX, cellular, video

Broadband Network Technologies

PSTN

Cordless Telephone

nications. A broad goal of PCS is for calls to follow a user as the user travels. A single cordless handset that could be used at home, on a mobile basis, and in the office is being widely discussed. At home, the handset would operate as a cordless phone. Outside, or in a vehicle, it would automatically switch to a cellular service, and at work, it would talk to the company's PBX. Users could also direct their calls to traditional wired phones when they reach their destination. Although most of the technology for providing this type of service exists, there are still topics of debate for choosing the methods to make the service work. The premise of PCS is for a user to have a personal phone number that would be the user's interface to PCS and the vast array of transparently available telecommunications services.

The cellular system uses many different types of media to complete calls, including microwave and satellite transmissions. This is all transparent to the user, who has the ability to call anywhere, anytime. Figure 11–11 depicts this PCS environment.

Many products and services marketed under the name of PCS are now available through long-distance and local exchange carriers. The broad range of services include digital cellular telephony with voice mail, e-mail, caller ID, and alphanumeric two-way paging systems, all available on a handheld wireless PCS telephone or two-way alphanumeric pager. In many cases, these products are just a combination of existing services that use the conventional digital cellular technology, like TDMA and CDMA systems [Garg & Wilkes, 1996; Wesel, 1998].

global system for mobile (GSM)

The **Global System for Mobile (GSM)** standards offer the most promise as a global infrastructure for carrying the PCS. Wireless data communications schemes will probably be greatly affected by the evolution of PCS standards in the future. Although widespread deployment of PCS is envisioned, it is not yet clear if demand will be sufficient to support this deployment in less-developed areas of the world.

GSM is a digital mobile telephone system that is widely used in Europe and other areas of the world. It uses a variation of Time Division Multiple Access (TDMA). GSM digitizes and compresses data before it sends it down a channel with two other streams of user data, each in its own time slot. It operates in either the 900 MHz or 1,800 MHz frequency band.

cellular digital packet data (CDPD)

microcells

cell splitting

Although GSM and **Cellular Digital Packet Data (CDPD)** will offer many improvements, they still use large cells and have a limited bandwidth. The most important gains in digital systems is their use of much smaller cells called **microcells.** Instead of being a mile or more in diameter, a PCS microcell may only be a quarter of a mile in diameter or even smaller. Figure 11–12 provides an example of **cell splitting.** The number of

Figure 11–12
Cellular Cell Splitting

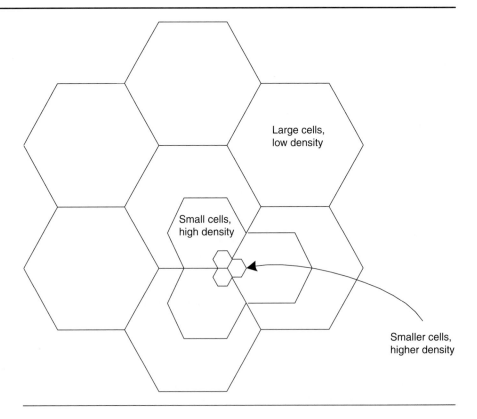

Large cells, low density

Small cells, high density

Smaller cells, higher density

cells increase as the inverse square of the cell size. Although the gains in capacity are not completely proportional, microcells can support about ten times as many subscribers as large cells. Having many microcells and the capacity for massive amounts of reuse is the real key to capacity increases in PCS. Power requirements will reduce by the cube of the cell size. This means that reduced power requirements will make PCS cell phones very small, light, and inexpensive. It should also reduce concerns about radiation.

The GSM Association is the world's leading wireless industry representative body consisting of more than 500 second- and third-generation GSM network operators and key manufacturers and suppliers to the GSM industry. The GSM system accounts for a major percent of the total digital wireless market (www.gsmworld.com).

The next generation of wireless service in the communications revolution is called **Personal Communications Satellite Services (PCSS)**. Advances in the satellite systems and wireless technology will put the cellular phone system in the sky. Some satellite-based wireless telephone systems have been launched that allow practically universal coverage with a single network.

personal communications satellite services (PCSS)

The FCC has allocated frequency bands for nationwide PCS coverage in the 902 MHz to 928 MHz and 1.8 GHz to 2.2 GHz frequency bands. Narrowband FM is used in the 902 to 928 MHz band. Broadband PCS operating in the 1.8 GHz to 2.2 GHz band employ 100 percent digital technology, which is much more efficient in terms of bandwidth than conventional narrowband FM. Modulation techniques such as TDMA, CDMA, and FDMA, as discussed in the following sections, are utilized.

FDMA

Frequency Division Multiple Access (FDMA) is a technology used to separate multiple transmissions over a finite frequency allocation. FDMA refers to the method of allocating a discrete amount of frequency bandwidth to each user to permit many simultaneous conversations. In cellular telephony, each caller occupies approximately 25 kHz of the frequency spectrum. The cellular telephone frequency band consists of 416 total channels, or frequency slots, available for conversations. Within each cell, approximately 48 channels are available for mobile users. Different channels are allocated for neighboring cell sites, allowing for re-use of frequencies with a minimum of interference. FDMA assigns individual frequency slots and reuses them throughout the system.

frequency division multiple access (FDMA)

TDMA

In North America, PCS 1900 **Time Division Multiple Access (TDMA)** has been chosen by a number of service providers as the second-generation technology for mobile wireless networks. TDMA is similar to GSM

time division multiple access (TDMA)

900/DCS 1800, which operates in the 1,900 MHz spectrum, and uses the same protocol. TDMA, as with FDMA, is a technology that is used to separate multiple conversation transmissions over a finite frequency allocation of through-the-air bandwidth. As with FDMA, TDMA is used to allocate a discrete amount of frequency bandwidth to each user in order to permit many simultaneous conversations. Each caller is assigned a specific timeslot for transmission. A digital cellular telephone system using TDMA assigns ten timeslots for each frequency channel, and cellular telephones send bursts, or packets, of information during each timeslot. The packets of information are reassembled by the receiving equipment into the original voice components. TDMA promises to significantly increase the efficiency of cellular telephone systems, allowing a greater number of simultaneous conversations.

CDMA

code division multiple access (CDMA)

Code Division Multiple Access (CDMA) is the newest and most advanced technique for maximizing the number of calls transmitted within a limited bandwidth by using a spread spectrum transmission technique [CDMA Development Group, www.cdg.org/frame.tech.html, 2000]. Rather than allocate specific frequency channels within the allocated bandwidth to specific conversations as in the case of TDMA, CDMA transmits digitized packets from numerous calls at different frequencies spread over the entire allocated bandwidth spectrum.

The code part of CDMA lies in the fact that to keep track of these various digitized voice packets from various conversations amidst the spectrum of allocated bandwidth, a code is appended to each packet indicating which voice conversation it owns. This technique is not unlike the datagram connectionless service used by packet-switched networks to send data over numerous switched virtual circuits within the packet-switched network. Identifying the source and sequence of each packet maintains the original message integrity and maximizes the overall performance of the network. Additional CDMA features include the following:

- Transmission of data is up to 14.4 kbps.
- Handoff is soft in that two cells share the call during hand-off.
- A power control system adjusts transmit power to enhance the quality of the signal.
- Multipath receptions of signals are coherently combined at the receiver, which enhances the signal quality.

In addition to the previously discussed modulation techniques, there exist new standards and systems for transporting wireless communications. These include GPRS, Third Generation, EDGE, UMTS, CDMA2000, and WCDMA.

GPRS

General Packet Radio Service (GPRS) is a standard for wireless communications. It operates at speeds of up to 150 kbps, compared with the current GSM Communications, which operates at 9.6 kbps. GPRS is an essential stepping stone to Third-Generation (3G) personal multimedia services. The higher data rates will allow users to participate in video conferences and interact with multimedia Websites and similar applications using mobile devices. GPRS will also complement Bluetooth and will support IP and X.25. GPRS is a step toward Enhanced Data GSM Evolution (EDGE) and Universal Mobile Telephone Service (UMTS). It is a way of giving handheld devices, mobile phones, or laptop computers an "always on" connection to the network, whether it is a corporate network or the Internet. Unlike the GSM network, which has to create a dedicated connection whenever exchanging data, GPRS sends bits of data only when needed. This results in a cheaper, more efficient technique of exchanging data; and, because GPRS uses existing mobile phone base stations, it is relatively inexpensive to install.

general packet radio service (GPRS)

Third Generation

A new radio communications technology called **Third Generation (3G)** will provide high-speed access to Internet-based services. These include Internet videoconferencing and sharing of database information. Applications targeted include shopping, banking, reservations, computer games, and health care. Examples include videoconferencing from a limo, a subway, or a train or providing on-the-spot vacation coverage to your friends. 3G is generally considered applicable to mobile wireless; however, it is also relevant to fixed wireless and portable devices. Proponents of 3G say that it will now be possible to provide connectivity to all users at all times in all places.

third generation (3G)

EDGE

Enhanced Data GSM Evolution (EDGE) is a fast version of GSM and can deliver data at rates up to 384 kbps, which provides for multimedia and other broadband applications on mobile devices. EDGE is basically a new modulation scheme that is more bandwidth efficient than the scheme utilized in the GSM standard. The technology defines a new physical layer that has the potential of increasing the data rate of existing GSM systems by a factor of three.

enhanced data GSM evolution (EDGE)

UMTS

Universal Mobile Telecommunications System (UMTS) is a 3G broadband, packet- based transmission of text, digitized voice, video, and multimedia at data rates up to 2 Mbps. It is based on the GSM communication standard. It will provide users with mobile access through a combination of terrestrial wireless and satellite facilities. The higher bandwidth of UMTS will provide for such services as videoconferencing.

universal mobile telecommunications system (UMTS)

CDMA2000

CDMA2000

CDMA2000, also called IMT-CDMA MultiCarrier, is a Code Division Multiple Access (CDMA) version of the IMT2000 standard that was developed by the ITU. It is a 3G mobile wireless technology. CDMA2000 can support mobile data communications at speeds ranging from 144 kbps to 2 Mbps. Deployment has been initiated by a number of mobile vendors.

WCDMA

wideband code division multiple access (WCDMA)

Wideband Code Division Multiple Access (WCDMA) is an ITU standard derived from Code Division Multiple Access (CDMA). It is officially known as IMT2000 direct spread. WCDMA is a 3G mobile wireless technology that offers much higher data speeds to mobile and portable wireless devices than is presently available in today's market. WCDMA can support mobile voice, images, data, and video communications from 384 kbps (wide area) to 2 Mbps (local area). The input signals are digitized and transmitted in coded, spread-spectrum mode over a broad range of frequencies. A 5 MHz carrier is utilized compared with a 200 kHz carrier for narrowband CDMA.

11.14   TWO-WAY MESSAGING

personal digital assistants (PDAs)
alphanumeric pagers

Two-way messaging, sometimes referred to as enhanced paging, allows short text messages to be transmitted between devices such as **Personal Digital Assistants (PDAs)** and **alphanumeric pagers.** Two distinct architectures and associated protocols have the potential to deliver these services. These protocols are Cellular Digital Packet Data (CDPD) and telocator data protocol (TDP). CDPD uses idle capacity in the circuit switched cellular network to transmit IP-based data packets. TDP is a suite of protocols that define an end-to-end system for two-way messaging to and from paging devices. Figure 11–13 depicts two-way messaging components using telocator data protocol. The activities are as follows:

- A message is entered according to the Telocator Message Entry (TME) format.
- The paging service's message processor breaks the message into smaller packets.
- The processor transmits it to the destination via Telocator Radio Transport (TRT).
- The remote user's alphanumeric pager receives packets via TRT.
- The pager reassembles packets into the full message.
- The pager outputs the full message via a Telocator Mobile Computer (TMC).

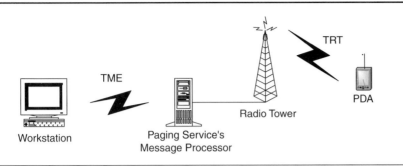

Figure 11–13
Two-way Messaging
Utilizing TDP

TME

TRT

PDA

Radio Tower

Workstation

Paging Service's
Message Processor

Pagers

Wireless messaging is best represented by the ever-present **pager.** A beeper device once worn only by the medical community, the pager has grown into one of the most visible and widely used message retrieval systems today. Today's pagers are available with numerous display and alert features as well as PCS options [Hioki, 1998].

pager

Pagers are available in a variety of shapes, colors, features, and sizes. Pagers can be classified into several categories:

- Tone-only pagers
- Numeric pagers
- Alphanumeric pagers
- Two-way pagers

The most basic function of the pager is to alert its user of an incoming message. To do this, the pager must receive the transmitted message from the calling source. Figure 11–14 illustrates a paging system. Transmitters throughout the service area are linked to a device called a paging terminal, which is also linked to the PSTN. Its function is to encode the caller's message into a unique paging code that in turn is sent to the various transmitters throughout the coverage area.

Figure 11–14 also illustrates an alphanumeric paging system. The same procedure, as outlined previously, is used to page a person with an alphanumeric pager. However, alphanumeric messages, in addition to numeric messages, can be received on the pager.

Alphanumeric paging is made easier with a paging server that can be operated from a LAN. Any user on the network can access the server, send the message to it, and the server out-dials to the paging system. Features available on available products include sending messages to e-mail, screening calls, answer-back paging, and out-dialing messages from only selected stations.

Service providers that operate paging systems for the general public are called Radio Common Carriers (RCCs). A RCC must be licensed by

Figure 11–14
Radio Paging a "Roaming Individual"

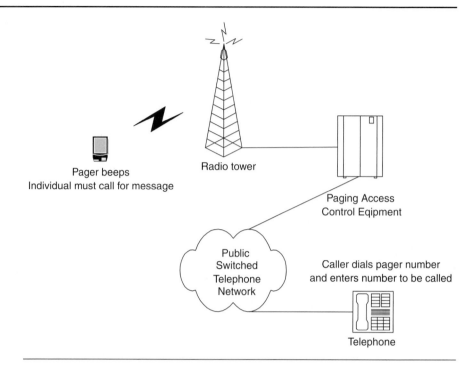

Pager beeps
Individual must call for message

Radio tower

Paging Access Control Eqipment

Public Switched Telephone Network

Caller dials pager number and enters number to be called

Telephone

the FCC to provide radio paging services throughout its service area. Pagers use numerous frequencies in both the VHF and UHF bands. Some of the variables that determine the operating frequency band are the use, service area, country of use, and paging protocol. Those devices designed for PCS will continue to be developed as pager technology advances.

11.15   WIRELESS DATA TECHNOLOGIES

A wireless data link can be used to provide end-to-end communication between a pair of end-user systems, or it can be used in conjunction with conventional telephone circuits. Wireless WAN data transmission is most often handled using some form of radio transmission.

Wireless WAN services using radio transmission technology are most commonly used to provide a data communication channel to computer users who move frequently from one location to another. As long as the users are within range of the radio network, they are able to communicate with the enterprise network from any location.

Wireless data links are particularly useful for implementing mobile computing applications in which it would be difficult or impossible to connect systems using conventional land-based telecommunication

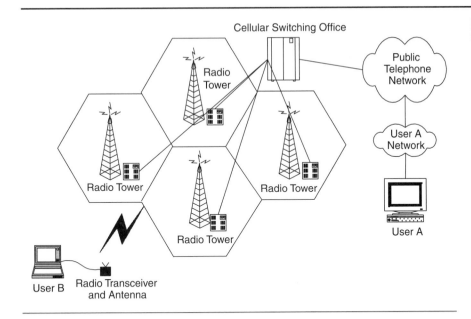

Figure 11–15
Cellular Network

lines. The three principal technologies used to implement wireless WAN data transmission are cellular telephone networks, cellular digital packet data, and packet radio networks.

Cellular Data Telephone Networks

Cellular networks were originally developed, and have come into widespread use, for voice communication using portable telephones. These networks are also being used for data transmission. Many cellular networks use analog transmission between the mobile device and a local tower that serves a particular area or cell. Data are transmitted between the towers and switching facility connected to the common carrier telephone network. Connections between users are established using a combination of wireless transmission and standard telephone circuits. For data transmission over a cellular network using analog technology, the computing device uses a cellular modem to access the service, and a standard voice telephone circuit is allocated for the duration of the transmission. The typical range of data rates supported by cellular networks is 2,400 kbps through 28.8 kbps. Figure 11–15 depicts a cellular data telephone network with data communication between user A on the user A private network and user B with a portable laptop computer.

Circuit Switched Analog

Still the dominant method of sending data, circuit switched analog is the optimal way to handle the movement of lengthy amounts of information, such

Figure 11–16
Circuit Switched Analog
Cellular Network

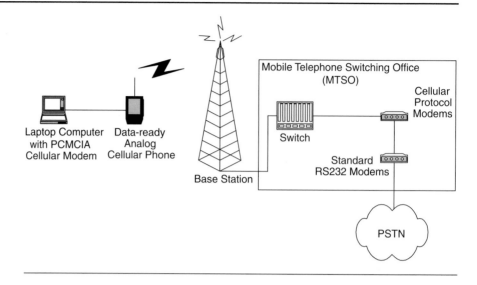

as file transfers or lengthy messages. It can be complex to set up because the customer needs a data-compatible phone and the correct modem, and may need to write setup strings to let the phone and computer communicate. The system has only moderate reliability, although coverage improvements and the installation of wireless modems at the cellular switch are helping. Figure 11–16 shows the components that are part of the circuit switched analog cellular data network. The MTSO consists of a switch, various modems, and a base station that provides access to the PSTN.

Cellular Digital Packet Data

Cellular Digital Packet Data (CDPD) can operate over the same channels as the existing analog cellular network, much as X.25 networks can operate over analog phone lines using modems; however, it must be modified to accommodate digital traffic [CDPD, www.cdpd.org/cdpd, 2000]. Instead of using a modem and connecting the computer to a cellular phone, the user connects the computer to a special device that combines the modem, radio transceiver, and other functions. This device operates independently and does not need to be connected to a cellular phone, although it does communicate with the network operated by the cellular carrier or other provider. This device formats the data according to the CDPD protocol and encrypts it for security.

The CDPD network and mobile device can check the data for errors and request retransmission if necessary. If the destination is not a mobile user, most CDPD networks convert the data back to the original format and then transmit the data over standard modems on the land-based telephone network. Most CDPD networks take advantage of spare band-

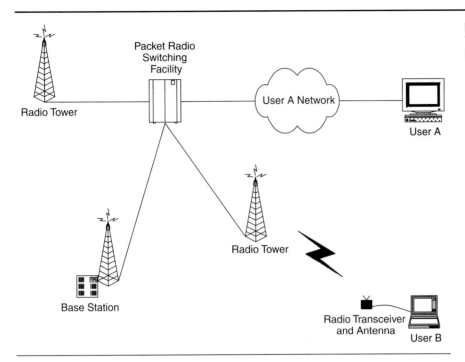

Figure 11–17
Typical Packet Radio
Network

width on voice channels in the existing cellular networks, minimizing the cost of transmission. CDPD essentially provides the benefits of a packet switching network for the wireless portion of the connection. In simple terms, CDPD works by breaking a message into digital packets of data and sending the packets over idle voice channels. The cellular network does not allocate a dedicated circuit to the transmission.

CDPD promises to be a very cost-effective and reliable way to send short messages, but has been hampered by a slow roll out nationwide (like all packet data services, it is not cost-effective for large data file transfers). A new hybrid technology of circuit switched and CDPD is said to allow customers to use circuit switched analog where there is no CDPD coverage, then switch to CDPD when it becomes available. A major benefit is that the end user can use the same cell phone for voice and data.

Radio Packet Data

Packet radio networks provide digital data transmission services in a way similar to X.25 networks. Each mobile computing device used with a packet radio network must include, or be attached to, an appropriate radio transceiver. Figure 11–17 depicts a typical packet radio network.

When data are to be sent from stationary user A to mobile user B, the data travel from user A's network to a packet radio switching facility, which sends the data to the appropriate broadcasting facility. The facility (base station) then uses radio transmission to forward the data to user B. When data are transmitted from user B to user A, the process is reversed. Because of the high cost of the base stations and the limited range of small portable transceivers, packet radio networks typically serve only selected geographic areas such as large cities.

ARDIS

A packet data network in the 800 MHz range was originally developed for use by IBM service technicians and is now owned by Motorola. Although it is similar to CDPD in the use of packet architecture, it has several important distinctions. It is data only, but nationwide, so there are no issues of intercarrier roaming; in addition, it existed in over 90 percent of the U.S. business areas in the fall of 1995, as opposed to the early-stage deployment of CDPD. ARDIS service is primarily used for near real-time messaging, which can be done to anyone with an Internet address worldwide, and for sending faxes wirelessly. It is believed to have the best in-building coverage due to a unique architecture with significant cell overlapping. It works with a range of products from Motorola and other manufacturers, including the pen-based Envoy for executives and salespeople, the InfoTAC for dispatch and messaging applications, and the new PM 100D PCMCIA card that directly links to a portable computer.

RAM Mobile Data

Similar to ARDIS in that it is also a nationwide network that handles packet data exclusively, RAM is partly owned by BellSouth Corporation, and has taken a more vertical solutions approach to the marketplace. The architecture is open, and many independent software vendors develop customized applications for companies in the area of field sales and service, transportation, dispatch, and the like.

Digital Specialized Mobile Radio

digital specialized mobile radio (DSMR)

A digital service that has evolved out of the two-way radio business is **Digital Specialized Mobile Radio (DSMR).** Digital SMR came about when Nextel, formerly FleetCall, a service provider for a variety of dispatch-oriented fleets such as taxicabs, was granted approval to operate a nationwide network over two-way frequencies. Nextel bought out many small operators and began the process of building a network that is similar to digital cellular in having multiple cell sites and handoffs, but is more limited in terms of spectrum allocation. A recent competitor for Nextel is Geotek, which has begun to roll out a wide range of local markets starting in the Northeast using a digital technology called **Frequency Hopping Multiple Access (FHMA).** End users can perform many data functions such as two-way messaging, vehicle location, paging, and even integration with an in-house computer system—all from a single handset.

frequency hopping multiple access (FHMA)

Nextel had once aspired to be the third cellular carrier, but its focus seems to have changed once Craig McCaw came aboard, and Nextel and Geotek may both focus on work groups and vertical applications that require intensive voice and data communications.

11.16 SATELLITE SERVICES

The International Telecommunications Satellite Organization (INTELSAT) series began with the launch of the Early Bird satellite in 1965. INTELSAT arose from a United Nations resolution to develop worldwide satellite communications. Several versions of these satellites have been launched, the most current being Intelsat V. It consists of 12,500 channels with a 2,300 MHz bandwidth system. This version has twenty-seven transponders and weighs less than 1,000 kilograms.

The U.S. Communications Satellite Corporation (COMSAT), although owned through stockholder shares, regulates the use and operation and sets tariffs for U.S. satellites. It operates a monopoly and sells user time on satellites to many diverse users.

Data are being sent today over global satellite services such as Inmarsat. The customer connects the computer to a satellite phone and can transmit at relatively low data speeds. Data and fax services will be available before the end of 2001 on American Mobile Satellite Corp's SKYCELL service, which covers the United States and Canada, Mexico north of the Panama Canal, and the Caribbean. Customers will be able to fax from a facsimile machine or a computer, and send data from a computer at 2,400 kbps.

11.17 WIRELESS STANDARDS

Wireless technologies continue to evolve and expand, while simultaneously becoming less expensive. Some new wireless networking technologies have recently emerged [IEEE Spectrum, www.spectrum.ieee.org, 2000]. Most significant is the IEEE's **802.11** Wireless Networking Standard, completed in 1997, which is the specification for wireless LANs. These standards legitimize wireless networking for corporate use.

802.11

The basis for local area wireless networks is the Ethernet specifications of 802.3, using **Carrier Sense Multiple Access/Collision Avoidance (CSMA/CA)** instead of CSMA/CD. The reason behind collision avoidance rather than collision detection is that it is difficult to detect collisions on a medium such as air. Access uses a form of arbitration to settle who gets immediate access. The controlling station monitors the airways for a signal from any of the network nodes. Upon detecting a request for access, the central node makes sure that no one else is also trying to gain

carrier sense multiple access/collision avoidance (CSMA/CA)

access. This being the case, the central node grants access to the requesting station.

Four major cellular standards in the world are as follows.

- Advanced Mobile Phone Service (AMPS)—Analog
- Interim Standard 54 (IS54)—Digital, TDM and FDM, Replacing AMPS
- Global System for Mobile (GSM) communications—Digital, TDM and FDM, most used system in the world except in North America
- Interim Standard 95 (IS95)—CMDA

The present analog system in North America is AMPS, which replaced MPS. Both MTS and AMPS use FDM. The digital system IS54 is replacing AMPS on an incremental basis. AMPS and IS54 must share the same frequency spectrum even while the changeover occurs. IS54 uses both FDM and TDM.

TDMA and CDMA are the two methodologies currently being researched in Personal Computer Services (PCS) field trials. The CDMA standard defined by the Telecommunications Industry Association (TIA) is IS99: data services option for wideband spread spectrum digital cellular systems. TDMA digital standards to handle call setup, maintenance, and termination have been defined by the TIA as follows:

- IS130: TDMA radio interface and radio link protocol 1
- IS135: TDMA services, asynchronous data, and fax

The Universal Wireless Communications Consortium (UWCC) has included TDMA interoperability with GSM as a core component of the GSM Global Roaming Forum. The Roaming Forum is the central body representing the interests of mobile operators and suppliers that are working toward interoperability between GSM and other technologies. GSM and TDMA interoperability, via the GSM/ANSI-136 Interoperability Team (GAIT), has been under development since 1999, via the combined efforts of the North American GSM operators and the UWCC. The mission of the Roaming Forum is to develop technical requirements for terminals and networking standards for a number of services. The UWCC (www.tdma-edge.org) is based in Bellevue, Washington. This international consortium of more than 100 wireless carriers and vendors supports TDMA, EDGE, and other technology standards.

The International Telecommunications Union—Radiocommunications Sector (ITU-R) approved an update to the UWCC's option for IMT-2000 wireless services. The update, which includes EDGE technology as a wireless data delivery standard, was approved in February 2001. It provides specifications for TDMA single-carrier technology and includes a radio interface called Universal Wireless Communications-136 (UWC-136).

11.18 WIRELESS ISSUES AND CONSIDERATIONS

Following are several limitations to wireless WAN communications given today's technology:

- Security—Radio transmission is easily intercepted and more susceptible to eavesdropping.
- Coverage—Radio networks cover most major metropolitan areas, but do not yet have the ubiquitous reach of the global telephone network.
- Interoperability—There have been few standards developed for radio networks. Most devices used with the radio network, such as transceivers and radio modems, will work only with a specific vendor's network.

Most wireless LAN products use a standard Ethernet frame format. This makes it relatively easy to design equipment such as NICs, bridges, and routers, to interconnect wireless LAN equipment with wired networks. Network interconnection equipment must be designed specifically for connecting a particular vendor's wireless LAN equipment with a particular standard form of wired LAN technology.

Interference of electrical signals is one of the prime factors limiting the performance of telecommunications systems. The subject has become one of major concern, and extensive technical standards have been developed to define the measurement of stray emissions from devices and the sensitivity of devices from interference. This interference comes from ambient electromagnetic radiation from power supplies, lightning, and radio signals.

The issue of health effects from radiation is an ongoing source of interest. The jury is apparently still out concerning any health effects from the wireless devices.

New products and technologies continue to be introduced into the marketplace. Bluetooth is the latest candidate for entry into the wireless environment. Current information concerning this technology is available at the IEEE's Website (www.spectrum.ieee.org).

11.19 ADVANTAGES AND DISADVANTAGES

Wireless LAN Advantages

LAN transmission that does not depend on a physical wire is becoming more prevalent in the LAN environment. Wireless LAN can take advantage of the situation in a variety of ways. One way is to interconnect individual LAN wire segments where it is difficult to physically interconnect the two sites. A more flexible way in which wireless transmission is sometimes used is through radio transmission or infrared links to replace the physical wires that are used to connect individual computer systems to the LAN. Wireless forms of LANs make it very easy to move computer systems and other types of network devices from one location to another without having to change physical wiring.

Wireless extended LAN has the following additional advantages:

- Installation is not prohibitively expensive, but can range from easy to difficult.
- It is highly resistant to interference.
- It is not very susceptible to eavesdropping.

Radio Advantages

Radio communications can save money when it is difficult or expensive to run wire or cable. Radio wave installations are used in situations when portable computers are used and need to be moved around frequently. Compared with other wireless options, it is relatively inexpensive and easy to install.

Microwave Advantages

Microwave solutions are an advantage and an alternative when communications cannot be easily installed, such as over long distances.

Satellite Advantages

Satellites have the following advantages over terrestrial communications:

- Costs of satellite circuits are independent of distance within the coverage range of a single satellite.
- Impairments that accumulate on a per-hop basis on terrestrial microwave circuits are avoided with satellites.
- Sparsely populated or inaccessible areas can be covered by a satellite signal.
- Satellites can broadcast signals that can be received in wide areas simultaneously.
- Large amounts of bandwidth are available over satellite circuits.
- Local telephone facilities can be bypassed by the satellite signals.
- Multipath reflections that impair terrestrial microwave communications have little effect on satellite radio paths.

Infrared Advantages

Infrared products have special appeal since they do not require any form of FCC licensing. Line-of-sight infrared has a wider bandwidth than spread spectrum. The spectrum for infrared is virtually unlimited, which presents the possibility of achieving extremely high data rates.

Infrared light is diffusely reflected by light-colored objects, which means that it is possible to use ceiling reflection to achieve coverage of an entire room.

The equipment for infrared communications is relatively inexpensive and simple [Infrared Data Association, www.irda.org/products/index.asp, 2000]. Infrared data transmission uses intensity modulation, so that IR receivers need to detect only the amplitude of optical signals, whereas most microwave receivers must detect frequency and phases.

Wireless Disadvantages

An important limitation to note when using wireless is that the signals can experience problems from other signals using the same media, and from sun spots, ionospheric changes, and other atmospheric disturbances.

Radio Disadvantages

Radio communications involve some significant disadvantages. Many network installations implement high-speed communications of 100 Mbps to handle heavy traffic, including transmission of large files. Radio-based networks do not yet have the speeds to accommodate these requirements. Another disadvantage is that the frequencies in use are un-regulated. Amateur radio operators, the military, and cell phone companies also use these frequencies and may be a source of interference. Natural obstacles, such as hills and tall buildings, can also diminish or interfere with the signal transmission.

Microwave Disadvantages

Microwave may not be feasible when high-speed communications are needed. It is expensive to install and maintain and is subject to interference from bad weather, EMI, and atmospheric conditions. Following are more specific disadvantages for terrestrial and satellite microwaves.

Terrestrial Microwave Disadvantages   Terrestrial microwave is subject to the following disadvantages:

- Interference varies with respect to power and distance.
- The longer distances covered cause it to be more susceptible to weather disturbances.
- It is difficult to install and maintain.
- It is expensive.
- It is susceptible to security breaches, but the signals are usually encrypted.

Satellite Microwave Disadvantages   Satellite microwave is also subject to the following disadvantages:

- Installation and maintenance is prohibitively difficult.
- It is prone to Elecromagnetic Interference (EMI) and jamming.
- Atmosphere disturbances and rain absorption affect path loss.
- The cost is prohibitive.
- The delay from earth station to satellite and back is about one-quarter second.
- Path loss is high from earth to satellite.
- Frequency crowding is high with the potential of interference between satellites and terrestrial microwave operating on the same frequency.

Infrared Disadvantages

Infrared disadvantages include the following:

- May not be feasible when high-speed communications are needed.
- Subject to interference from other light sources.
- Concerns for eye safety.
- Excessive power consumption for higher-power transmitters.
- Not able to penetrate through walls.

11.20   WIRELESS TECHNOLOGY ALTERNATIVES

Wireless can be utilized to improve the functionality of a number of Enterprise networking technologies and provide additional access to applications that are currently accessible via wire or cable. Historically, access to data and data-processing resources required the user to be connected via some local network connection. Fixed-connection access is now possible wherever there is an access point. This means that Internet/Web access, voice communication, message services, database access, fax services, and any other process that was historically available only through a local hard-wired connection is now possible over a wireless network.

Wireless is the alternative to the older, wire-based technologies. It is essential that the network developer look at every alternative that leads to wireless and make decisions based on sound information. Planning for a migration to these new wireless technologies is a must, as considerable resources may need to be deployed for such a conversion from wire to wireless.

11.21   WIRELESS APPLICATIONS

Wireless networking has considerable appeal in many circumstances. Commercial applications for wireless networking technologies include the following:

- Ready access to data for mobile professionals, such as medical personnel in hospitals or delivery personnel in their vehicles.
- Delivery of network access into isolated facilities or even into disaster areas.
- Access in environments in which layout and settings change constantly.
- Network connectivity in facilities where in-wall wiring would be prohibitive or impossible.

Wireless technology can provide capabilities for related technologies, as follows:

- Creating temporary connections into existing wired networks.
- Establishing a backup or contingency connectivity for existing wired networks.
- Extending a network's span beyond the reach of wire- or fiber-optic-based cabling.
- Permitting certain users to roam with their devices, within certain limitations.

Each of these capabilities supports uses that allow the benefits of networking to expand or extend beyond conventional limits. Although wireless networking is more expensive than wire-based alternatives, sometimes these benefits can more than repay the extra costs involved.

A new service offering, called cellular packet radio, provides wireless networking where users can carry laptop PCs anywhere within the coverage area and establish a 2 Mb connection at will. Another new service offering, CDPD, provides a 19.2 kbps connectivity in most major U.S. metropolitan areas.

Wireless Application Protocol

The technology that links wireless devices to the Internet is called **Wireless Application Protocol (WAP).** This is made possible by a process of translating Internet information so that it can be viewed on the display screen of a mobile telephone or other portable device. Devices that can utilize WAP include mobile phones, pagers, two-way radios, and smartphones. A WAP gateway is required as an intermediary between the Internet and the mobile network. It converts the WAP request into a Web request when information is sent from a mobile telephone to the Internet and vice versa. WAP-enabled mobile phones vary little from conventional mobile phones.

wireless application protocol (WAP)

The most notable difference is a larger screen that allows for easier use of the Internet. Web mobiles need access to Web pages that are written in Wireless Markup Language (WML) instead of the usual HTML utilized on laptops and workstations. This means that websites that require a WAP format must convert the original HTML format to WML. WAP is designed to work with most wireless networks, including CDPD, CDMA, TDMA, and GSM. WAP provides a universal open standard for bringing Internet content and advanced services to mobile devices. Applications can be developed for e-mail, news, sports, entertainment, electronic commerce, and travel.

Another development that will impact the wireless network environment is the narrowband socket. This specification will permit pagers, cell phones, and wireless computers to communicate more readily with the Internet. This specification accommodates the unique requirements of cellular and other wireless communications much more readily than IP and provides a way to bridge the wireless world into Internet information and resources.

As wireless technologies get less expensive, the number of service uses will increase dramatically. Figure 11–18 depicts the wireless environment that can provide access to many services and utilize a personal telephone number worldwide for this access. As depicted in the figure, user requirements include wireless access for voice, data, fax, and video applications. Middleware provides transparent telecommunication services anywhere, anytime. Wireless telecommunications software interfaces with the various wireless networks through the various wireless topologies to interact with the message service databases and applications.

11.22 PRODUCT AND SERVICE PROVIDERS

Numerous companies supply wireless service and the devices that are necessary to install it [CTIA, www.wow-com.com, 2000]. The Web provides a wealth of information concerning wireless products and services. These URLs are subject to change due to the numerous mergers that frequently occur. Service providers include:

Airnote Technologies	www.airnote.net/
Ameritech	www.ameritech.com/
Bell Atlantic	www.bellatlantic.com/largebiz/smds.htm/
BellSouth Mobility	www.bellsouth.com/
GTE	www.gte.com/
MCIWorldcom	www.wcom.com/main.phtml/
NortelNetworks	www.nortelnetworks.com/
NYNEX	www.nynex.com/
Pacific Bell	www.pacbell.com/

Figure 11–18
The Wireless Environment

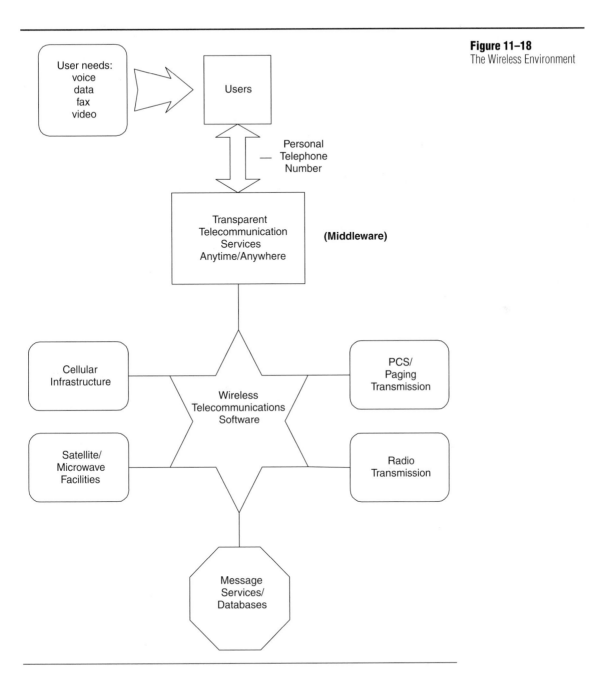

SBC Communications   www.sbc.com/
Skytel Communications www.skytel.com/
Sprint PCS   www.sprintpcs.com/
US West   www.uswest.com/

Some companies that supply mobile, cellular, and PCS infrastructure and end-user products and services are:

3Com	www.3com.com/
Alcatel, Inc.	www.alcatel.com/
Corsair Communication	www.corsair.com/
Ericsson, Inc.	www.ericsson.com/
Lucent Technologies	www.lucent.com/
Motorola Inc.	www.motorola.com/General/index.html/
NEC America, Inc.	www.nec_global.com/
Nokia	www.nokia.com/main.html
Paradyne Corp.	www.paradyne.com/
Qualcomm, Inc.	www.qualcomm.com/
Racal-Datacom, Inc.	www.racal.com/
Toshiba America, Inc.	www.toshiba.com/

■ SUMMARY

Wireless networking is taking over an increasing portion of the networking load. Wireless technologies work well to provide wire-free LAN access, to extend the span of LANs, to provide WAN links, and to support mobile computing requirements.

A typical wireless network acts like its wired counterpart. A network adapter facilitates communications transfer across the networking medium similar to a wired network, except the wires are not needed to carry the signals involved. Otherwise, users communicate as they would on any other network.

Wireless networks use a variety of electromagnetic frequency ranges, including narrowband and spread spectrum radio, microwave, and laser transmission techniques. LANs also can be extended using a pair of devices called a wireless bridge. Short-range wireless bridges can span distances up to 3 miles; long-range wireless bridges can span up to 25 miles.

Mobile computing involves using broadcast frequencies and communications carriers to transmit and receive signals using packet radio, cellular, or satellite communications techniques. It requires specialized software including mobile-aware operating systems, mobile-aware applications, and mobile middleware to interface between multiple applications and multiple possible wireless WAN services.

Wireless WAN services vary widely in terms of availability, bandwidth, reliability, and cost. No single wireless WAN service is appropriate for all mobile-computing applications. Limitations often associated with wireless WAN data links include lack of security, insufficient range of coverage, and lack of standards for interoperability. The three principal technologies used to implement wireless WAN data transmission are packet radio networks, cellular telephone networks, and Cellular Digital Packet Data. Available wireless WAN services include the following:

- Personal Communication Service (PCS)
- Private Packet Radio
- Circuit-switched analog cellular
- Cellular Digital Packet Data (CDPD)
- Enhanced paging and two-way messaging
- Enhanced Specialized Mobile Radio (ESMR)
- Microcellular spread spectrum

Wireless networking appears poised to grab an increasing share of network installations as newer and more powerful technologies and standards come online. To gain the advantage of mobility, it will be necessary to update the routing protocols and to enhance the security of the Internet, so that the network identifies the mobiles in a trusted manner.

Key Terms

802.11

Access Point

Advanced Mobile Phone Service (AMPS)

Alphanumeric Pager

Antenna

Bluetooth

Broadcast Radio

Carrier Sense Multiple Access/Collision Avoidance (CSMA/CA)

CDMA2000

Cell Splitting

Cellular Digital Packet Data (CDPD)

Code Division Multiple Access (CDMA)

Cryptography

Diffuse

Digital Specialized Mobile Radio (DSMR)

Direct Sequence

Enhanced Data GSM Evolution (EDGE)

Extended LANs

Federal Communications Commission (FCC)

Footprint

Frequency Division Multiple Access (FDMA)

Frequency Hopping

Frequency Hopping Multiple Access (FHMA)

General Packet Radio Service (GPRS)

Geosynchronous

Global System for Mobile (GSM)

Guided

Hertz (Hz)

Infrared

Line of Sight

Local Area Networks (LANs)

Low Earth Orbit (LEO)

Microcells

Mobile Computing

Mobile Telephone Switching Office (MTSO)

Multipath Fading

Pager

Personal Communication Service (PCS)

Personal Communications Satellite Services (PCSS)

Personal Digital Assistant (PDA)

Radio Frequency (RF)

Roaming

Satellite Microwave

Terrestrial Microwave

Third Generation (3G)

Time Division Multiple Access (TDMA)

Transponder

Unguided

Universal Mobile Telecommunications System (UMTS)

Wave Guides

Wideband Code Division Multiple Access (WCDMA)

Wireless Application Protocol (WAP)

Wireless Bridge

REVIEW QUESTIONS

1. What was the first major mobile radio system? What were the limitations of this system?

2. What was IMPS? What service did it replace?

3. What is a hybrid network?

4. What was the first cellular telephone system in the United States? How was it different from the previously developed mobile radio systems?

5. Define wireless transmission.

6. What are the two types of media?

7. What are the two types of configurations for wireless transmission?

8. What are the three general ranges of frequencies discussed in this chapter?

9. What are the three subdivisions or categories of wireless networks? How do you distinguish among the uses of these networks?

10. List the different wireless technologies. How are each of these technologies utilized?

11. What is an access point or transceiver that is utilized in a wireless LAN? Why is it required?

12. What are the most commonly used frequencies for wireless data communications?

13. What is an extended LAN? Where would an extended LAN be deployed?

14. What is line of sight? Define line-of-site networks.

15. What is a PCN and what are some applications for the service?

16. Wireless LANs make use of which technologies?

17. What is the difference between reflective wireless networks and scatter infrared networks?

18. Describe infrared network communications.

19. What are the two different methods for implementing spread spectrum?

20. Describe the general model for spread spectrum operation.

21. Describe RF transmissions.

22. Give an overview of mobile telephony.

23. What is microwave technology? How are the signals propagated?

24. Describe terrestrial microwave.

25. Describe satellite microwave.

26. What is a geosynchronous satellite?

27. What is a LEO satellite? How is it different from a geosynchronous satellite?

28. What is cellular telephony? Describe the cellular radio systems.

29. What is a cell? How do they interact with each other? Why is this arrangement necessary?

30. What is AMPS?

31. What is a MTSO?

32. What is a handoff process?

33. How is the handoff process different when the user is in the same MTSO than when the user travels into another MTSO region?

34. What is PCS? What is a personal phone number?

35. What is GSM?

36. What is FDMA?

37. What is TDMA? How is it different from FDMA?

38. Describe CDMA. What are the CDMA features?

39. What is two-way messaging?

40. Describe the difference between a standard pager and an alphanumeric pager.

41. Describe the wireless data technologies environment.

42. What are the three principal technologies for wireless WAN transmission?

43. Describe the CDPD technology.

44. Describe some packet data systems.

45. Comment on satellite wireless services. What phenomenon causes satellite microwave signals as long as 5 seconds to travel from sender to receiver? Why is this so?

46. What protocol forms the basis for LAN wireless networks? What is the IEEE specification?

47. Identify some issues and considerations when deploying a wireless network.

48. Identify advantages for the various wireless networks.

49. Identify disadvantages for the various wireless networks.

50. What are some applications for commercial wireless networks? What capabilities can these networks provide?

51. Describe the Bluetooth wireless technology.

ACTIVITIES

The first two activities can be enhanced through role play, where one group offers the questions, and the other group provides the answers. Brainstorming and boarding is an effective, in-class group exercise.

1. Identify applications that would benefit from the wireless technology. Also identify applications that are allied or interface with the wireless technology. Examples are as follows:

   a.   To support a population of mobile computing users, which wireless technology is most appropriate?

   b.   Which wireless technologies would be most appropriate to link two buildings together?

2. Produce a list of the advantages and disadvantages for deploying the various wireless technologies.

3. Use the list of RFCs in Appendix C to develop a matrix of standards for Wireless service and PCS. Use the IETF web page for additional references.

4. Identify the various product and service providers in your specific area. Use the Internet or the DataPro for Telecommunications service.

5. Identify the devices that would be utilized in the various wireless technologies. Create a matrix of devices, prices, features, and other specifications.

6. Create a matrix of the different wireless technologies. Provide information concerning frequencies, bandwidth, speed, usage, and any other pertinent information for each technology.

7. Arrange for a facility tour of a mobile telephony provider. Arrange for a mobility provider to visit the classroom. Look to this person for pricing options and features.

8. There is considerable activity surrounding the Bluetooth technology. Research this subject and develop a report. Give a five minute oral presentation.

9. There is considerable interest concerning the health hazards that surround the use of wireless products and services. Research this issue and report on the pros and cons of health concerns.

CASE STUDY/PROJECT

Develop a wireless network design with all of the physical components. This design should apply to one of the applications identified in Activity 1. Show the general flow of communications between the endpoint users. Services include cellular, microwave, mobile satellite, packet radio, infrared, and spread spectrum systems. A GSM-integrated network design is an example of a wireless network design project.

Since wireless networks can interface with a number of other networking technologies, there are a considerable number of networking alternatives available. Several alternatives are presented that require a networking design.

Local Area Network Wireless

DocuPrep develops training packages for several corporations. One department is responsible for reviewing and compiling the training packages. A pool of document reviewers require access to common documents that are located in the database of a centrally located file server. Ten reviewers are located in a room divided by cloth-covered 4 ft partitions. The file server is located in an adjoining room. There are no provisions for LAN wiring in the room.

Requirement

Suggest a solution to this situation. Create a graphic showing all necessary components. Price out the solution. Provide an alternative solution along with a cost/benefit analysis. Show the advantages and disadvantages of the solutions.

Wide Area Network Wireless

Speedy is a local delivery service that utilizes fifty trucks to deliver packages for a long-haul carrier. There have been a number of complaints of shipments being lost or late. Damaged shipments are common. Speedy may lose this contract if service does not improve quickly.

Requirement

Your supervisor has given you the responsibility of designing a solution for Speedy. All that you know about the project is what you see in the above description. Your mission is to determine the requirements needed to develop a viable wireless system for this customer. Your first step is to develop a list of questions that may be asked in order to design such a wireless system. The second step is for you to answer these questions with any assumption that you desire. The third step is to design the wireless system, document it, price it out, and present it.

12

Fibre Channel

■ INTRODUCTION

The information explosion during the 1990s and the need for high-performance communications has spawned a need for data-intensive and high-speed networking applications. To interconnect these client-server architectures demands a new level of performance in reliability, speed, and distance. What is needed is a highly reliable, gigabit interconnect technology that allows for concurrent communications among workstations, mainframes, servers, data storage systems, and other peripherals using the Small Computer System Interface (SCSI) and Internet Protocol (IP). It also needs to provide interconnect systems for multiple topologies that could scale to a total system bandwidth on the order of a terabit per second.

This alternative network architecture is known as **Fibre Channel** (ANSI Standard X.3T9.3). It has been defined to run at speeds of 133 Mbps to 1.062 Gbps over optical fiber and copper cable. Support for speeds up to 4.268 Gbps is expected in the future. Fibre Channel is often used to connect high-performance storage devices and a **Redundant Array of Inexpensive Disk (RAID)** subsystems to computers. Fibre Channel switches, storage systems, storage devices, hubs, and network interface cards are available today for network implementation.

Networks typically manage transfers between end systems over local, metropolitan, or wide area distances. Fibre Channel is designed to combine the best features of two technologies—I/O (input/output) channel communications technology and protocol-based network communications. This fusion of approaches allows system designers to combine traditional peripheral connection, host-to-host internetworking, loosely coupled processor clustering, and multimedia applications in a single multiprotocol interface. The Fibre Channel Industry Association (FCIA) [http://www.fibrechannel.com, 2000] is the industry consortium for promoting Fibre Channel in today's broadband networking environment.

OBJECTIVES
Material included in this chapter should enable you to:

● understand Fibre Channel technology and how it is utilized to provide performance improvements in data transport within the broadband environment. Identify the various classes of service.

● become familiar with the terms and definitions of the Fibre Channel technology which are relevant in the Enterprise Networking environment. Understand the concept of compute clusters and storage pools. Become familiar with frames, sequences, and exchanges.

● understand the function and interaction of Fibre Channel in the Enterprise Network environment. Become familiar with the various standards associated with Fibre Channel.

fibre channel
redundant array of inexpensive disk (RAID)

fabric

small computer system interface (SCSI)

12.1   FIBRE CHANNEL OVERVIEW

The emerging Fibre Channel standard, as developed by the ANSI X.3T9.3 Committee, is a high-performance standard designed to transfer data at speeds up to 1 Gbps over distances of 10 km using fiber-optic transmission media.

Hewlett Packard, IBM, and Sun Microsystems Computer Corp. have announced the Fibre Channel Systems Initiative, which is a joint effort to do the following:

- Advance the Fibre Channel as an affordable, high-speed interconnection standard for workstations and systems
- Promote open systems for distributed computing
- Propose selected sets of Fibre Channel options for manufacturers to build conforming products

The Fibre Channel defines a matrix of switches (nodes), called a **fabric,** that performs network switching functions similar to that of a telephone system. Each computer system attaches to the fabric with dedicated send and receive lines designed for point-to-point, bidirectional serial communications. The switch can route any incoming signal to any output port. These elements are interconnected by point-to-point links between ports on the individual nodes and switches. Communication consists of the transmission of frames across the point-to-point links. Frames may be buffered within the fabric, making it possible for different nodes to connect to the fabric at different data rates.

Facilities are provided for both I/O channel connectivity and network connectivity. The types of channel-oriented facilities incorporated into the Fibre Channel protocol architecture include the following:

- Data-type qualifiers for routing frame payload into particular interface buffers
- Link-level constructs associated with individual I/O operations
- Protocol interface specifications to allow support for existing I/O channel architectures, such as **Small Computer System Interface (SCSI).**

The types of network-oriented facilities incorporated into the Fibre Channel protocol architecture include the following:

- Full multiplexing of traffic between multiple destinations
- Peer-to-peer connectivity between any pair of ports on a Fibre Channel network
- Capabilities for internetworking to other connection technologies

The Fibre Channel is one of the key technologies of the twenty-first century to provide high-speed performance at an affordable cost. It is expected that a significant share of the workstations and servers will be using the Fibre Channel interface in the future.

12.2   FIBRE CHANNEL TECHNOLOGY

Fibre Channel is a relatively new networking technology, although it has been used inside computers and Direct Access Storage Devices (DASDs) for several years (ANSI X.3T11). Fibre channel is designed to provide the best elements of a high-speed hardware channel with a multiprotocol network.

The Fibre Channel network is quite different from the IEEE 802 LANs. Fibre Channel is more like a traditional circuit-switched or packet-switched network, in contrast with the typical shared-medium LAN. Fibre Channel need not be concerned with MAC issues.

Fibre Channel consists of a set of full-duplex point-to-point switched circuits between devices. Networks can be designed using a traditional switch or a ring topology called an arbitrated loop topology. Figure 12–1 illustrates the two Fibre Channel topologies.

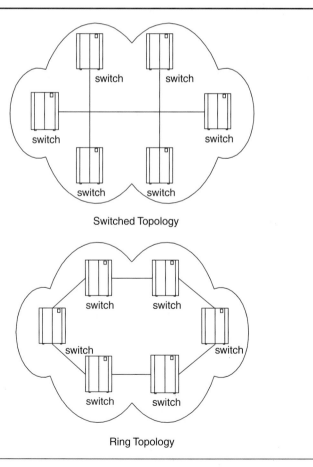

Figure 12–1
Fibre Channel Topologies

Media	100 Mbps	200 Mbps	400 Mbps	800 Mbps
STP	80 m	57 m	46 m	28 m
Miniature coaxial cable	42 m	28 m	19 m	14 m
Video coaxial cable	100 m	100 m	71 m	50 m
Multimode fiber—62.5	—	1 km	1 km	175 m
Multimode fiber—50	—	2 km	1 km	.5 km
Single Mode Fiber	—	10 km	10 km	10 km

12.3   FIBRE CHANNEL PHYSICAL MEDIUM

A major strength of Fibre Channel is that it provides a range of options for the physical medium, the data rate on that medium, and the topology of the network. Table 12–1 summarizes the options that are available under Fibre Channel for the physical transmission medium and data rate. Each entry specifies the maximum point-to-point link distance between ports that is defined for a given transmission medium at a given data rate. These media may be mixed in an overall configuration. Note that FC-0 referred to in this section is similar to the physical layer of the OSI model.

Fibre Channel was originally designed to provide high-speed transmission over fiber-optic cable. The maximum data rate today is 1.062 Gbps over fiber-optic cable up to 10 km, with data rates of 2.13 Gbps and 4.26 Gbps currently under development. Slower-speed versions are also available and include 133 Mbps, 266 Mbps, and 533 Mbps. Several versions have also been defined for use on category 5 (CAT 5) twisted-pair cable, allowing 1.062 Gbps up to 24 m or 266 Mbps up to 47 m.

Optical Fiber Transmission Media

FC-0, which is the functional Layer 1 of the Fibre Channel model, specifies both single- and multimode optical fiber alternatives. For single mode, specifications are provided for optical fiber at up to 800 Mbps and distances up to 10 km. The differences among the various single-mode options primarily deal with the spectral characteristics of the signal produced by the transmitter. There are a number of options for multimode fiber. These include the use of 62.5 micron and 50 micron diameter fiber and the use of a wavelength of 780 nm or 1,300 nm. For a given data rate, greater distances can be achieved with 50 micron fiber. For a given diameter fiber, greater data rates can be achieved with the 1,300 nm wavelength.

Coaxial Cable Transmission Media

Coaxial cable provides a lower cost alternative to optical fiber when long distances are not required. The two types of coaxial cable that are specified, video and miniature, are both 75 ohm cables. The video coaxial is what is referred to as RG-6/U or RG-59/U type. These are flexible cables, generally with an outside diameter of 0.332 inch and 0.242 inch, respectively. The miniature coaxial cable has an outside diameter of 0.1 inch. The thinner the cable, the more attenuation is experienced, and consequently the lower the data rate and/or shorter the distance that can be supported.

Shield Twisted-Pair Transmission Media

The final media specified in FC-0 are two types of 150 Ohm shielded twisted-pair (STP), designated as EIA 568 Type 1 and Type 2. These cables can only be used over very short distances and only at the 100 Mbps and 200 Mbps data rates. Type 1 consists of two twisted pairs enclosed in a metallic shield and covered in an appropriate sheath. Type 2 contains four twisted pairs with two pairs used for voice and two pairs used for data.

Fibre Channel also has much in common with Ethernet. Fibre Channel uses variable-length data link layer packets with a maximum of 2,048 bytes of user data in each, and performs routine error control using Cyclic Redundancy Check (CRC-32). Like Ethernet, all Fibre Channel network interface cards are assigned a permanent 8 byte (64 bit) data link MAC address, compared with Ethernet's 6 bytes (48 bits). Data link layer address resolution is accomplished using the same Address Resolution Protocol (ARP) broadcast approach as Transmission Control Protocol/Internet Protocol (TCP/IP).

12.4   FIBRE CHANNEL FEATURES

Fibre Channel features include storage interfaces, storage devices and systems, storage networks, and networks. The major features include:

- Ability to carry multiple existing interface command sets, including Internet Protocol (IP), SCSI, IPI, HIPPI-FP, and audio/video
- Broad availability (i.e., standard components)
- Greater connectivity than existing multidrop channels
- High-bandwidth utilization with distance insensitivity
- Performance from 266 Mbps to over 4 Gbps
- Support for distances up to 10 km
- Small connectors
- Support for multiple cost/performance levels, from small systems to supercomputers

Storage

Fibre Channel is the next storage interface. It has been adopted by the major computer systems and storage manufacturers as the next technology for enterprise storage. It eliminates distance, bandwidth, scalability, and reliability issues for SCSI.

Storage Devices and Systems

Fibre Channel is being provided as a standard disk interface. Industry leading RAID manufacturers are shipping Fibre Channel systems. Soon, RAID providers will not be regarded as viable vendors unless they offer Fibre Channel.

Storage Area Network

The network behind Fibre Channel consists of linking one or more servers to one or more storage systems. Each storage system could be RAID, tape backup, tape library, CD-ROM library, or *Just a Bunch of Disks (JBOD)*. Fibre Channel networks are robust and resilient with these features:

- Fast data access and backup
- High performance
- Robust data integrity and reliability
- Shared storage among systems
- Scalable network

In a Fibre Channel network, legacy (embedded hardware) storage systems are interfaced using a Fibre Channel to a SCSI bridge. IP is used for server to server and client-server communications. Storage networks operate with both SCSI and networking (IP) protocols. Servers and workstations use the Fibre Channel network for shared access to the same storage device or system. Legacy SCSI systems are also interfaced using a Fibre Channel to a SCSI bridge.

Fibre Channel products have defined a new standard of performance, delivering a sustained bandwidth of over 97 Mbps for large file transfers and tens of thousands I/Os per second for business-critical database applications on a Gigabit link. This new capability for open systems storage is the reason why Fibre Channel is the connectivity standard for storage access. Note that a disk drive is both an input and an output device, since it can provide information to the computer and receive information from the computer.

Networks

Fibre Channel networks provide to enterprises new levels of performance and reliability. The many network applications for Fibre Channel include:

- High-performance Computer Assisted Design/Engineering (CAD/CAE) network
- Movie animation and postproduction projects, to reduce the time to market
- Nonstop corporate backbone
- Quick-response network for imaging applications

Fibre Channel was developed by the computer industry for Information technology (IT) applications. Its development focused on removing the performance barriers of legacy LANs. Performance-enhancing features of Fibre Channel for networking are as follows:

- Automatic self-discovery of Fibre Channel topology
- Complete support for traditional network self-discovery; full support of ARP, RARP, and other self-discovery protocols
- Confirmed delivery, enhancing the reliability of the protocol stack or the option of bypassing the protocol stack for increased performance
- Efficient, high-bandwidth, low latency transfers using variable length (0 to 2 KB) frames; highly effective for protocol frames of less than 100 bytes as well as bulk data transfer using the maximum frame size
- Instant circuit setup time measured in microseconds using hardware enhanced Fibre Channel protocol
- Extremely low latency connection and connectionless service
- Full support for time synchronous applications such as video, using fractional bandwidth virtual circuits
- Support for dedicated bandwidth point-to-point circuits, shared bandwidth loop circuits, or scalable bandwidth switched circuits
- True connection service or fractional bandwidth, connection-oriented virtual circuits to guarantee quality of service for critical backups or other operations
- The option of real circuits or virtual circuits

In the early days, a single computer vendor provided a proprietary solution to a single buyer—the data-processing manager. With the minicomputer, the process changed, and departments bought their own computing solution. The market transitioned to multiple solutions sold to multiple buyers, resulting in incompatible, proprietary data processing systems. Over time, users realized they needed to combine all data processing into an integrated environment. This requirement opened the door for open standards-based solutions. Now, companies are connecting their mainframes with enterprise and department servers for distributed client-server architectures. Distributed computing and parallel processing has resulted in a significant increase in process-to-process communications. At the same time, the data storage requirements have exploded. This new paradigm only works if data can be moved and shared quickly. The need for very high bandwidth and extremely low latency I/O is paramount. Fibre Channel is the solution that delivers.

12.5   FIBRE CHANNEL PORTS

Each node includes one or more ports, called **N_ports,** for interconnection. Similarly, each fabric switching element includes one or more ports, called **F_ports.** Interconnection is by means of bidirectional links between

N_ports

F_ports

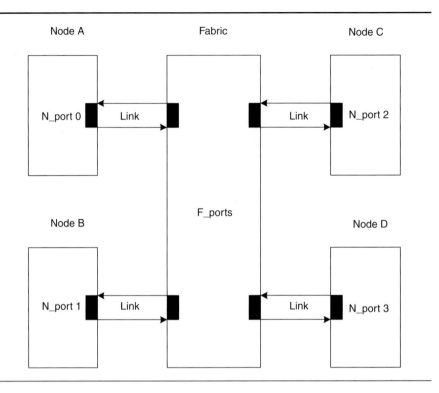

Figure 12–2
Fibre Channel N_ and F_
Port Types

ports. Any node can communicate with any other node connected to the same fabric using the services of the fabric. All routing of frames between N_ports is done by the fabric. Frames may be buffered within the fabric, making it possible for different nodes to connect to the fabric at different data rates. A fabric can be implemented as a single fabric element or as a more general network of fabric elements. In either case, the fabric is responsible for buffering and routing frames between source and destination nodes. Figure 12–2 depicts a Fibre Channel fabric with both N_ports and F_ports. The F_ports in the Fabric are accessible via links to the N_ports in the various nodes.

12.6   FABRICS

Switched Fibre Channel networks are called fabrics and are used to interconnect devices such as workstations, personal computers, servers, routers, mainframes, and storage devices. An originating port calls the fabric switch by entering the destination port address into a Fibre Channel frame header. The fabric switch sets up the desired connection based on the source and destination address. There is no permanent virtual circuits hogging any of the bandwidth, allowing for fabric switches to handle 16 million addresses within the network. Because the switch sets up connections between stations, there is no possibility of contention for, or blocking of access to, the system.

12.7   FIBRE CHANNEL TOPOLOGY

The most general topology supported by Fibre Channel is referred to as a fabric or switched topology. This arbitrary topology includes at least one switch to interconnect a number of N_ports.

Arbitrated loops are Fibre Channel ring connections that provide shared access to bandwidth via a form of arbitration between stations on the loop. A possible 127 **L_ports** may be attached to a single network. However, only two ports may communicate at any given time, forcing the remaining ports to be idle until the loop becomes available. When a L_port wants to gain access to the loop, it sends a request onto the loop when it is free of traffic. The establishing port establishes a bidirectional connection with the destination port and communication can then commence. If more than one L_port attempts to gain access simultaneously, the access is resolved using arbitration based on the sending stations' addresses. The L_port with the lowest IEEE 48-bit address has the highest priority for the purposes of loop access, which is similar to the 802.5 ring-monitor selection process. These are reserved for host stations, followed by switched fabric connections, and finally Node (N) ports with the highest address and lowest priority.

Routing in the fabric topology is transparent to the nodes. Each port in the configuration has a unique address. When data from a node are transmitted into the fabric, the edge switch to which the node is attached uses the destination port address in the incoming data frame to determine the destination port location. The switch then either delivers the frame to another node attached to the same switch or transfers the frame to an adjacent switch to begin the routing of the frame to a remote destination.

The fabric topology provides scalability of capacity, which means as additional ports are added, the aggregate capacity of the network increases, thus minimizing congestion and contention and increasing throughput. The fabric is protocol independent and largely distance insensitive.

In addition to the fabric topology, the Fibre Channel standard defines two other topologies: point-to-point and arbitrated loop. With the point-to-point topology there are only two N_ports, and these are directly connected, with no intervening fabric switches, which means there is no routing. The arbitrated loop topology is a simple, low-cost topology for connecting up to 126 nodes in a loop. The ports on an arbitrated loop must contain the functions of both N_ports and F_ports, and are called

arbitrated loops

L_ports

Figure 12–3
Basic Fibre Channel
Topologies

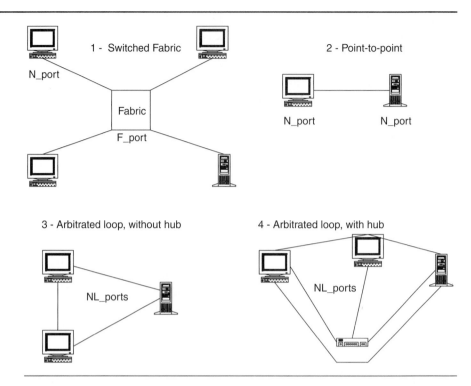

NL_ports

NL_ports. The arbitrated loop operates in a manner similar to the Token Ring/802.5 protocols. Each port sees all frames and passes and ignores those not addressed to itself. There is a token acquisition protocol to control access to the loop. Figure 12–3 depicts the following four basic Fibre Channel topologies.

- Arbitrated loop—without hub
- Arbitrated loop—with hub
- Point-to-point
- Switched Fabric

FL_port

The fabric and arbitrated loop topologies may be combined in one configuration to optimize the cost of the configuration. In this case, one of the nodes on the arbitrated loop must be a Fabric-Loop **(FL_port)** node so that it participates in routing with the other switches in the fabric configuration. Figure 12–4 illustrates the following principal applications of Fibre Channel.

- Connecting mainframes to each other
- Clustering disk farms
- Giving server farms high-speed pipes
- Linking LANs and WANs to the backbone
- Linking high-performance workstation clusters

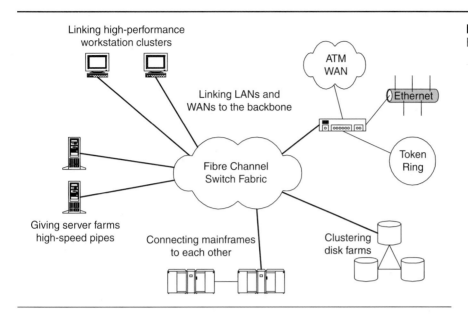

Figure 12–4
Fibre Channel Applications

Linking high-performance workstation clusters

Linking LANs and WANs to the backbone

ATM WAN

Ethernet

Token Ring

Fibre Channel Switch Fabric

Giving server farms high-speed pipes

Connecting mainframes to each other

Clustering disk farms

12.8 FIBRE CHANNEL PROTOCOL

The Fibre Channel protocol is subdivided into several functional layers, similar to the layers of the OSI model. Layer *FC-0* addresses the physical layer of the network. Three types of cabling are specified under FC-0, including 2 km of multimode fiber carrying data at 200 Mbps. Other cables specified include 10 km of single-mode fiber and 100 m maximum distance using shield twisted-pair copper wire.

The transmission protocol level *(FC-1)* defines the signal encoding technique used for transmission and for synchronization across the point-to-point link. The encoding scheme used is **8B/10B**, in which 8 bits of data from level FC-2 are converted into 10 bits for transmission. This scheme has a similar philosophy to the 4B/5B scheme used by FDDI. The 8B/10B scheme was developed and patented by IBM for use in its 200-megabaud interconnect system.

8B/10B

The transport layer of the OSI model is incorporated into the *FC-2* layer, which deals with framing protocols and flow control. Level FC-2, referred to as the framing protocol level, deals with the transmission of data between N_ports in the form of frames. Concepts and functions defined at this level include:

- Classes of service provided by the fabric
- Grouping of frames into logical entities called sequences and exchanges
- Node and N_port and their identifiers

- Segmentation of data into frames and reassembly
- Topologies

FC-2 sets up addresses and signaling functions such as those that define the connection between two ports—one an originator and one a responder. Sequencing of packets is an additional responsibility of this layer. Traffic management, including flow control, link management, buffer memory management, CRC-32 error detection resolution, and managing full-duplex operation of the network completes the FC-2 layer.

The *FC-3* layer specifies areas associated with the network layer of the OSI model. This layer, called the common services layer, uses multiple links to transmit data boosting the use of available bandwidth. Functions defined at this level include:

- Hunt groups—Consists of a set of associated N_ports at a single node.
- Multicast—Delivers a transmission to multiple locations.
- Striping—Makes use of multiple N_ports in parallel to transmit a single information unit across multiple links simultaneously.

The last layer, *FC-4,* is known as the multiple-services interconnect layer and addresses the remaining upper levels of the OSI model. This layer can handle any incoming protocol payload and convert it into fibre channel use and format. This layer deals with both individual channels and network traffic. Following are examples of upper level protocols that can be mapped into the Fibre Channel.

- ATM Adaptation Layer 5 (AAL-5)
- High-Performance Parallel Interface (HIPPI)
- Internet Protocol (IP)
- IEEE 802.3/Ethernet
- Small Computer Systems Interface (SCSI)

The Fibre Channel standard does not dictate a correspondence between levels and actual implementations, with a specific interface between adjacent levels. The levels are utilized to group-related functions. Table 12–2 summarizes the definitions of the five levels.

Table 12–2
Fibre Channel Protocol Architecture

Fibre Channel Layer	Definition
FC-4 Mapping	Defines the mapping of various channel and network protocols to Fibre Channel, including IEEE 802, ATM, IP, and SCSI.
FC-3 Common Services	Includes multicasting
FC-2 Framing Protocol	Deals with defining topologies, frame format, flow and error control, and grouping of frames into logical entities called sequences and exchanges
FC-1 Transmission Protocol	Defines 8B/10B signal encoding scheme
FC-0 Physical Media	Includes optical fiber for long-distance applications, coaxial cable for high-speeds over short distances, and STP for lower speeds over short distances.

12.9   FIBRE CHANNEL CLASSES OF SERVICE

Fibre Channel is a switched technology like ATM, and like ATM, all circuits are full duplex and provide several classes of service. These classes are determined by the way communication is established between two ports and the flow and error control features of the communications channel. These classes of service are defined in the FC-2 level (similar to 802.2 LLC) of Fibre Channel.

- Class 1 (also called circuit-switched service) is similar to ATM's Constant Bit Rate (CBR) service. With Class 1 service, a virtual circuit is established that is guaranteed to provide the requested data transmission rate. All packets arrive in the order in which they were transmitted. Class 1 provides acknowledged connection services, which guarantee the delivery of packets. Class 1 service is useful when the connection setup time is short relative to the data transmission time. It is especially useful if large blocks of data are to be transmitted or if various throughput rates are required.
- Class 2 (also called connectionless service) is similar to ATM's Variable Bit Rate (VBR-NRT) service. All packets are guaranteed to be transmitted error-free to the destination, although they may suffer traffic delays and may arrive out of order. Class 2 handles switched frame service, which is connectionless, but still guarantees the delivery of packets. This class guarantees the delivery by using confirmations of the receipt of data as acknowledgment of packets sent. This class is appealing for a Storage Area Network (SAN).
- Class 3 (also called unacknowledged connectionless service) is similar to ATM's Available Bit Rate (ABR) service. Packets are transmitted when possible, but error-free delivery is not guaranteed and no acknowledgments are sent. This is commonly used for broadcast messages. Class 3 is a one-to-many connectionless datagram service that does not guarantee delivery of data. Class 3 is useful for internetworking and for multicast and broadcast transmission.
- Class 4, as with Class 1, provides an acknowledged connection-oriented service. In addition to the reliability of such a service, it enables the establishment of virtual connections with bandwidth reservation for a predictable quality of service. Class 4 service—which is connection-based and guarantees bandwidth allocation and specific latency factors—is used to support isochronous services. Class 4 is appropriate for time-critical and real-time applications, including audio and video.
- Class 6 provides the reliable unicast delivery of Class 1. In addition, Class 6 supports reliable multicast and preemption Unidirectional Connection Service. This class is appropriate for video broadcast applications and real-time systems that move large amounts of data.

12.10   FRAMES, SEQUENCES, AND EXCHANGES

The FC-2 standard defines a hierarchy of building blocks that support upper level functions in a natural fashion. At any given node, FC-2 may offer a number of different classes of service. Each class of service may be provided by one or more N_ports. FC-2 also defines basic protocols, or procedures, used to implement a service at a port. These include procedures for setting up a connection, for transferring data, and for terminating a connection. Each of these procedures is defined as part of an exchange of information between N_ports. An exchange consists of one or more unidirectional sequences, and each *sequence* consists of one or more frames.

Frames

frames

All traffic between N_ports over Fibre Channel is in the form of a stream of **frames.** There are two categories of frames:

- Data frames transfer higher level information between source and destination N_ports.
- Link control frames are used to manage frame transfer and to provide some control for FC-2 Class 1 and 2 services. Link control frames are used to indicate receipt or loss of a frame, to provide flow control mechanisms, and to indicate when a destination N_port or the fabric is busy.

Sequences

In Fibre Channel, a maximum frame size for each direction is negotiated between two communicating ports and between communicating ports and the fabric. FC-2 performs two functions related to sequences: segmentation and reassembly, and error control.

On transmission, FC-2 accepts a sequence of data and segments this into one or more data frames. Each data frame includes a sequence identifier in its header that uniquely identifies the frame as part of a particular sequence. Each frame also includes a sequence count that numbers the frames within a sequence so that they may be reassembled by FC-2 at the receiving point.

Exchanges

exchange

The **exchange** is a mechanism for organizing multiple sequences into a higher level construct for the convenience of applications. An exchange may involve either the unidirectional or bidirectional transfer of sequences. It consists of one or more nonconcurrent sequences.

12.11   FIBRE CHANNEL SYSTEMS

Fibre Channel systems are assembled from adapters, hubs, storage, and switches. Host bus adapters are installed into hosts like any other SCSI host bus adapter. Hubs link individual elements to form a shared bandwidth loop. Disk systems integrate a loop into the backplane. A port bypass circuit provides the ability to hot-swap Fibre Channel disks and Fibre Channel links to a hub. Fibre Channel switches provide scalable systems of almost any size.

Information Technology (IT) systems today require an order of magnitude improvement in performance. High-performance, gigabit Fibre Channel meets this requirement. Fibre Channel is the most reliable, scalable, gigabit communications technology today. It was designed by the computer industry for high-performance communications, and no other technology matches its total system solution.

Gigabit Fibre Channel networks put to work previously untapped power in storage, servers, and workstations. Fibre Channel networks make clusters possible with:

- High-bandwidth transfers
- Highly available systems
- Low latency messaging
- Reliable communications
- Scalable networks

Fibre Channel clusters provide IT managers high availability and improved performance. Instead of replacing systems to meet expanded requirements, Fibre Channel networks enhance capabilities of the installed base and preserve a company's investment.

IT managers use Fibre Channel's flexibility and reliability to add incremental processing and storage for just-in-time computing and storage. When compared with alternate technologies, Fibre Channel clusters are easier on the budget and easier to install, integrate, and manage.

Fibre Channel is an open ANSI standard, connecting a heterogeneous mix of servers, workstations, and storage. Any topology (point-to-point, loop, or switched) is implemented as required, meeting the various needs of each application.

12.12 HIGH-AVAILABILITY ARCHITECTURES

Gigabit Fibre Channel networks enable assured communications, with servers and storage devices backing up each other. If a device fails or is taken down for maintenance, Fibre Channel connectivity enables its activity to be performed by another device without losing functional capability.

Clusters Defined

Compute clusters link multiple servers and workstations into an integrated network for processing. **Storage clusters** are pools of storage that

compute clusters
storage clusters

serve multiple servers and workstations. The net result of clusters is the ability to utilize multiple resources for high availability and storage pools.

IT organizations' success starts with service to corporate customers. Corporate customers are satisfied when processing and storage resources are always available, delivering information when and where it is required. Clustering is a cost-effective way of linking corporate computing resources into a highly available, scalable processing system.

Compute Clusters   Servers and workstations linked on a Fibre Channel network cooperate to process an application. Fibre Channel networks provide low latency messaging for coordination and high-bandwidth storage interfaces. Load sharing dispatches tasks to idle units, putting unused compute cycles to work. Cluster power grows by adding more elements to the cluster.

Storage Pools   Fibre Channel is protocol independent. SCSI and IP are both effectively used to provide shared storage. Direct access uses the SCSI protocol. Network storage using File Transfer Protocol (FTP) or Network File System (NFS) is faster using gigabit Fibre Channel Networks. Fibre Channel-powered clusters effectively share data, share processing, put unused capacity to work, and provide high availability. The net result is more effective use of IT resources.

Fibre Channel Clusters

Servers and workstations communicate using network protocols, primarily IP. Storage communicates using the SCSI protocol. In the past, IT systems used channels to communicate with storage and networks to link servers and workstations. Today, a single Fibre Channel network combines SCSI and networking protocols on a single, reliable, gigabit network. It has the speed and reliability of a channel with the flexibility and scalability of a network. Fibre Channel was designed by the computer industry as an ANSI standard to remove the barriers of performance posed by legacy channels and networks.

Fibre Channel networks have the option of dedicated bandwidth using point-to-point links, shared bandwidth with loops, or scaled bandwidth with switched networks. Fibre Channel delivers gigabit communications up to 10 Km over a campus, up to 30 Km using extenders, or across a WAN in the future. The benefit is the ability to cluster processors and storage regardless of location.

High-Availability Systems

Today, corporate operations span the globe and rely on IT services that are always available. Without access to data, it would be impossible to service customers, know the inventory, or make shipments. Fibre Channel brings the power of high-availability clustering to all organizations. Fibre

Channel provides high-availability architectures without the cost of duplicating a complete system. Simply by adding redundant connections, the user can choose from different high-availability architectures, which deliver the performance and scalability required. Point-to-point links provide a simple way of linking servers and storage systems together. Redundant gigabit Fibre Channel arbitrated loops deliver 200 Mbps, shared among multiple servers and storage systems. Switched systems provide scalability to virtually any size.

In high-availability systems, special software monitors the status of each storage or processor element. If an element should fail or be taken off-line, then the system reallocates resources to retain constant capability. The capacity of the total system will vary depending upon the status of all the processing and storage elements.

Fibre Channel systems are built without restrictions. Virtually any topology that an IT organization requires is possible. The basic building blocks are point-to-point dedicated bandwidth, loop-shared bandwidth, and switched-scaled bandwidth. Switches and hubs are stackable.

12.13   FIBRE CHANNEL COMPONENTS

Fibre Channel networks and storage are built from products that are familiar to IT professionals. These products, and a short description, are identified by separate categories of hardware, cable, adapters, and software.

Fibre Channel Hardware

Fibre Channel hardware includes hubs, routers, switches, bridges, gateways, and disks. Pertinent information follows for each of these devices.

Hubs   Fibre Channel hubs are used to connect nodes in a loop. Logically, the hub is similar to a Token Ring hub with ring in and ring out. Each port on a hub contains a Port Bypass Circuit (PBC) to automatically open and close the loop. Hubs support hot insertion and removal from the loop. If an attached node is not operational, a hub will detect this and bypass the node. Typically, a hub has seven to ten ports and can be stacked to the maximum loop size of 127 ports.

Routers/LAN Switches   Routers/LAN switches interface Fibre Channel with legacy LANs. These are Layer 2 and/or Layer 3 devices that use Fibre Channel for a reliable, gigabit backbone.

SCSI Bridge   Fibre Channel provides the ability to link existing SCSI-based storage and peripherals using a SCSI bridge. SCSI-based peripherals appear to the server or workstation as if they were connected directly on Fibre Channel.

Systems Network Architecture Gateway   SNA gateways interface Fibre Channel to SNA. Fibre Channel host bus adapters are integrated into standard products such as the Novell SAA and Microsoft SNA gateways.

Static Switches   Static switches, or link switches, provide point-to-point connections and are externally controlled. They offer a low-cost option for applications not requiring the fast, dynamic switching capability inherent in the Fibre Channel protocol.

Switches   Fibre Channel switches are among the highest performing switches available for high-bandwidth and low latency communications. The secret is in the Fibre Channel protocol, designed specifically by the computer industry to remove the barriers of performance with legacy channels and networks. Today, a Fibre Channel switch provides connection and connectionless service (Classes 1, 2, and 3) or only connectionless service (Classes 2 and 3). Typical connection setup or frame switching time is less than 1 microsecond. Switches are stackable to meet the most demanding applications. The number of addresses available is 224 or over 16 million. Switch options provide high-availability features.

Disk Enclosures   Fibre Channel disk enclosures utilize a backplane with a built-in Fibre Channel loop. At each disk location in the backplane loop is a port bypass circuit which permits hot swapping of disks. If a disk is not present, the circuit automatically closes the loop. When a disk is inserted, the loop is opened to accommodate the disk.

Fibre Channel Disks   Fibre Channel disks have the highest capacity and transfer capability available. Typically, these disks have a capacity of 9 GB and support redundant Fibre Channel loop interfaces.

Fibre Channel Cable

A variety of cable is utilized in support of the Fibre Channel network. Cable types include multimode, single mode, copper cable, and a number of associated connectors [Fiber Optics Association, www.std.com/fotec/foa.htm, 2000)].

Multimode Cable   Multimode cable is dominant for short distances of 2 km or less. Multimode has an inner diameter of either 62.5 microns or 50 microns, allowing light to enter the cable in multiple modes, either straight or at different angles. The many light beams tend to lose shape as they move down the cable. This loss of shape is called dispersion, and limits the distance for multimode cable. Cable quality is measured by the product of bandwidth and distance. Existing 62.5 micron FDDI cable is usually rated at 100 or 200 MHz/km, providing gigabit communications up to 200 m.

Single mode fiber—SC connector   Multimode fiber—ST connector

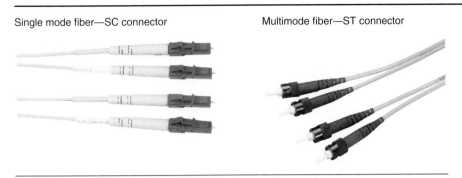

Figure 12–5
Fiber-Optic Cable
(Reproduced with
permission from Black Box
Corp.)

Single Mode Cable   Single mode cable is used for long distance cable runs. Its distance is limited by the power of the laser at the transmitter and by the sensitivity of the receiver. Single mode cable has an inner diameter of 7 or 9 microns and only allows a single ray of light to enter the cable. Therefore, with single mode cables there is no dispersion.

Copper Cables   Four kinds of copper cables are defined in the Fibre Channel standard. The most popular implementations are twin-ax using DB-9 or HSSD connectors.

Galaxy Connector   Fibre Channel has recently adopted a new connector called the Galaxy. It reduces the size of the connector by 50 percent, doubling the connector density for hubs and switches. The Galaxy should be generally available.

Fiber-Optic Cable Connector   The SC connector is the standard connector for Fibre Channel fiber-optic cables. It is a push-pull connector and is favored over the ST connector. If the cable is pulled, the tip of the cable in the connector does not move out, resulting in the loss of signal quality. As depicted in Figure 12–5, the ST connector is round, whereas the SC connector is square.

Fibre Channel Adapters
Supporting equipment for the cable plant includes extenders, converters, and adapters.

Extenders   Extenders are used to provide longer cable distances. Most optical interfaces are multimode cable. Extenders convert the multimode interface to single mode and boost the power on the laser. Typically, an extender will provide a single mode cable distance of 30 km or 18 mi.

Gigabit Interface Converters   Distances in a data center are supported with twin-ax copper circuits; therefore, hubs, disks, and many host bus adapters come standard with a copper interface. Gigabit Interface Converters (GBIC) and media interface converters plug into the copper interface and convert it to an optical interface. GBICs use an HSSD connector for the copper interface, and media interface converters use the DB-9 copper interface. The benefit is a low-cost copper link and optics for longer distance when required.

Gigabit Link Modules   Gigabit Link Modules (GLMs) are pluggable modules providing either a copper or fiber-optic interface. GLMs include the Serializer/Deserializer (SERDES) and have a media-independent parallel interface to the host bus adapter. Users can easily change the media interface from copper to fiber optics.

Host Bus Adapters   Host Bus Adapters (HBAs) are similar to SCSI host bus adapters and Network Interface Cards (NICs). Fibre Channel HBAs are available for copper and optical media. A typical Fibre Channel PCI HBA is half length, and utilizes a highly integrated Fibre Channel ASIC for processing the Fibre Channel protocol and managing the I/O with the host. Adapters are also available for SBus, PCI, MCA, EISA, GIO, HIO, PMC, and Compact PCI.

Switch WAN Extender   Fibre Channel switches can be connected over WANs using an Interworking Unit (IWU). Expansion ports on switches are linked using either ATM or STM services. Since Fibre Channel may be faster than a single ATM or STM interface, multiple WAN channels can be used for full Fibre Channel bandwidth.

Fibre Channel Software

Fibre Channel software includes several drivers and a link analyzer to support the topology.

Drivers   If software drivers for the host bus adapter vendor are not resident in the server or workstation, they are installed into the Operating system (OS) using its standard procedures. Fibre Channel drivers support multiple protocols, typically SCSI and IP. Most popular operating systems are supported including Windows NT, AIX, Solaris, IRIX, and HPUX.

Link Analyzer   Fibre Channel Link Analyzers capture the cause and effect of data errors. Specific frame headers can be monitored and captured for analysis.

12.14 FIBRE CHANNEL MANAGEMENT SYSTEMS

Many companies are returning to the concepts of centralized management of data storage, even within distributed IT architectures. Storage management requires a global perspective. An IT manager must look at the whole enterprise as well as the parts that make up the whole. Fibre Channel supports both needs. Fibre Channel networks provide an integrated storage system with the performance and efficiency of a distributed architecture and the reliability and manageability of centralized data storage. Fibre Channel devices provide information to manage them individually.

Storage must be viewed as a system, delivering services and protecting data assets. Proper management of this system provides highly available data access, improved performance, complete data security, and storage growth at a reasonable cost. A storage system consists of online storage, near-line storage, archival storage, and backup storage. Storage management software moves data among these elements as required to meet the enterprise's storage management strategy. Fibre Channel removes barriers associated with implementation of this strategy.

Fibre Channel devices use *Simple Network Management Protocol (SNMP)* and *SCSI Enclosure Standard (SES)* for management. The Fibre Channel standard supports SNMP over IP or directly using the Fibre Channel native protocol. Fibre Channel manufacturers normally provide a point solution for SNMP that can be integrated into an enterprise management system. Fibre Channel is designed to be self-managing, and most of this management activity is status monitoring.

SES is similar to SNMP but is designed to obtain information from storage devices that only use SCSI. Typical information from a storage device includes the following:

- Manufacturer's name
- Status on what slots are occupied
- Cooling information
- Power supply status

Centralized storage simplifies management. Fibre Channel provides the ability to view storage as if it were centralized, even if it is physically distributed. Fibre Channel supports critical business applications with performance, reliability, fast data access and transfer, and managed storage and storage networks.

Storage Management

Storage is a service that users want available at all times as an unlimited resource. It is more than a collection of disks and tapes. It is an interconnected system of storage devices, software, and servers receiving and delivering data. The role of storage management is to maximize performance, availability, and security at a reasonable cost.

Fibre Channel brings new levels of performance for reliability, connectivity, security, and availability that enhance an IT organization's ability to manage storage. This involves the following techniques:

- Automated backup and archival is made faster and safer with gigabit Fibre Channel networks that span long distances.
- Diagnostics can be run on-line to determine the health of a storage system using Fibre Channel.
- Enterprise data management provides backup and restoration across a heterogeneous environment using a Fibre Channel network.
- Fibre Channel provides improved reliability over SCSI-connected storage.
- Fault-tolerant Fibre Channel architectures enable highly available storage systems.
- Hierarchical storage management can be implemented across the enterprise with Fibre Channel networks.
- Media and library management software can report results of media monitoring to a central location over the Fibre Channel network.
- System management is enabled using SNMP or SES.
- Storage networks link existing SCSI-based storage and new Fibre Channel storage with servers and workstations for data storage and retrieval.

Fibre Channel storage and networks bring new capabilities to network storage. Multiple servers and workstations directly share storage devices without an intermediate device such as an NFS server. Thus, direct storage access benefits expand beyond the mainframe to the entire enterprise.

The benefits of central management and distributed computing are made possible with Fibre Channel. Storage management and network management work together using gigabit Fibre Channel solutions for increased levels of performance, security, and availability. Following are examples of FC benefits.

- Management of Fibre Channel systems melds storage and network management, bringing new capabilities to IT systems.
- Fibre Channel combines the speed and reliability of a channel with the flexibility and distance of a network.
- Fibre Channel storage devices connect with servers and workstations similar to network devices.
- The protocol used to link with these storage devices is SCSI. Fibre Channel is protocol independent and runs SCSI as easily as IP or digital audio/video. This brings new levels of management capability to IT managers. They can effectively use SCSI or IP to implement functions like automated backup on physically distributed devices as if they were centrally located.

12.15   FIBRE CHANNEL STANDARDS

Fibre Channel, a family of ANSI standards, is a common efficient transport system supporting multiple protocols or raw data using native Fibre Channel guaranteed delivery services. Profiles define interoperable standards for using Fibre Channel for different protocols or applications. Specifications are outlined in RFC2625, IP, and ARP over Fibre Channel.

Fibre Channel, a channel/network standard, contains network features that provide the required connectivity, distance, and protocol multiplexing. It also supports traditional channel features for simplicity, repeatable performance, and guaranteed delivery. Fibre Channel also works as a generic transport mechanism.

Fibre Channel architecture represents a true channel/network integration with an active, intelligent interconnection among devices. Fibre Channel is only required to manage a simple point-to-point connection. The transmission is isolated from the control protocol, so point-to-point links, arbitrated loops, and switched topologies are used to meet the specific needs of an application. The fabric is self-managing, and nodes do not need station management, which greatly simplifies implementation.

Fibre Channel provides for interoperability between a number of systems. Certain organizations test for interoperability compliance. The Fibre Channel Industry Association has two such independent laboratories:

- Interoperability Laboratory (ILO) at the University of New Hampshire
- Computational Science and Engineering Laboratory at the University of Minnesota

To accelerate speeds from 100 Mbps to 1 Gbps required changes to be made at the physical interface. This has been resolved by merging standards for IEEE 802.3 and ANSI X.3T11 for Fibre Channel. This allows for compatibility between Gigabit Ethernet and Fibre Channel. Figure 12–6 shows these two protocol stacks and how they merge into the Gigabit protocol stack.

Fibre Channel Frame Format

Figure 12–7 illustrates the basic Fibre Channel frame format. The Start-Of-Frame (SOF) delimiter consists of 4 bytes and is followed by the 24-byte frame header. The data field consists of 0 to 2,112 bytes and can contain optional headers. The frame ends with a 4 byte CRC and a 4 byte End-Of-Frame (EOF) delimiter.

Figure 12–6
Fibre Channel and Gigabit
Ethernet Protocol Stacks

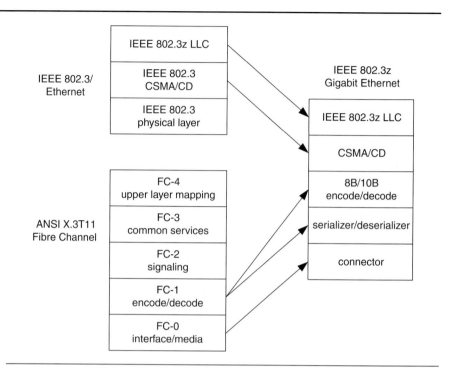

Figure 12–7
Fibre Channel Frame
Format

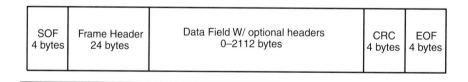

SOF 4 bytes	Frame Header 24 bytes	Data Field W/ optional headers 0–2112 bytes	CRC 4 bytes	EOF 4 bytes

12.16   FIBRE CHANNEL APPLICATIONS

Fibre Channel networks have found a use as backbones for larger networks, for handling imaging and multimedia data, and for video and
teleconferencing. Fibre Channel is ideal for the following applications:

- Campus backbones
- Digital audio/video networks
- High-performance work groups
- High-performance storage
- Large databases and data warehouses
- Network-based storage
- Storage backup systems and recovery
- Server clusters

An interesting application has arisen for Fibre Channel technology. It is being used to allow fast access to large storage devices connected in a **Storage Area Network (SAN).** These storage devices include disks, disk arrays, tape drives, and so forth. The SAN's purpose is to link these devices to their own network, thereby freeing much-needed bandwidth on the communications network. SAN provides faster, unencumbered access to these storage devices compared with accessing them as a node on a general-purpose network. Fibre Channel technology can be used for a SAN to interconnect devices through hubs or switches and transfer data at a rate of 100 Mbps over copper cable up to 30 m or at 1.06 Gbps using fiber-optic cable up to 10 km in length.

storage area network (SAN)

Compute Clusters Application

Compute clusters divide a computing task among several processors and use interprocessor messaging or shared memory. Results are joined when processing is complete. Most IT systems consist of heterogeneous products that are effectively clustered using Fibre Channel. In fact, Fibre Channel can be used to cluster Symmetric Multiprocessor (SMP) servers, delivering a high-availability cluster of multiprocessors. An application example is data mining. Groups of rows or columns are assigned to independent processors. Servers and workstations that are already in place can work together on the problem and deliver the results.

Fibre Channel is ideal for this application. Storage is linked to each of the servers or workstations using gigabit Fibre Channel networks. Fibre Channel delivers low latency for short coordination messages between the processors and high bandwidth to move data to and from storage. Even with SMP systems, Fibre Channel is used for high-bandwidth storage access.

One of the earliest cluster computing models for Fibre Channel was load sharing. Load sharing or load leveling with Fibre Channel has been providing full-time service since 1995. Under load sharing, a master server dispatches tasks to idle workstations or servers. In most cases this processing is accomplished in the background and does not interfere with the primary tasks assigned to these workstations or servers.

Load sharing is based upon the premise that unused compute cycles are a corporate resource and need to be utilized. A master server monitors the activity of other servers and workstations, dispatching tasks to idle units. Fibre Channel makes this model practical and effective with low latency communications for short status messages and high-bandwidth delivery of the data for the task and return of the results.

Storage Pools Application

Fibre Channel enables storage pools or clusters, a new paradigm for disk storage. Because Fibre Channel provides the ability to use SCSI and IP on the same network, new dimensions have emerged. Multiple servers and workstations can access disk storage directly using SCSI. Traditional shared storage using FTP or NFS is enhanced with gigabit Fibre Channel networks. Pooled storage is a better use of resources. Storage only needs to be added when space is exhausted in the pool. Considerable savings have been reported.

12.17   TECHNOLOGY COMPARISONS

Fibre Channel is a product of the computer industry, specifically designed to remove the barriers of performance existing in legacy LANs and channels. In addition to providing scalable gigabit technology, the architects provided flow control, self-management, and ultrareliability.

Gigabit Ethernet is designed to enable a common frame from the desktop to the backbone [Gigabit Ethernet Alliance, www.gigabit-ethernet.org/technology, 2000]. However, Fibre Channel is designed to be a transport service independent of protocol. Fibre Channel's ability to use a single technology for storage, networks, audio/video, or to move raw data is superior to the common frame feature.

ATM was designed as a WAN with the ability to provide QoS for fractional bandwidth service. The feature of fractional bandwidth with assured QoS is attractive for some applications. For the more demanding applications, Class 4 Fibre Channel provides guaranteed delivery, gigabit bandwidth, and fractional bandwidth QoS.

Fibre Channel's use in both networks and storage provides a price savings due to economies of scale associated with larger volumes. Users can expect their most cost-effective, highest performance solutions to be built using Fibre Channel. As shown in Table 12–3, Fibre Channel is the

Table 12–3
Technology Comparison

Category	Fibre Channel	Gigabit Ethernet	ATM
Technology application	Storage, network, video, clusters	Network	Network, video
Topologies	Point-to-point loop hub, switched	Point-to-point hub, switched	Switched
Baud rate	1.06 Gbps	1.25 Gbps	622 Mbps
Scalability to higher data rates	2.12 Gbps, 4.24 Gbps	Not defined	1.24 Gbps
Guaranteed delivery	Yes	No	No
Congestion data loss	None	Yes	Yes
Frame size	Variable, 0-2 KB	Variable, 0-1.5 KB	Fixed, 53B
Flow control	Credit based	Rate based	Rate based
Physical media	Copper and fiber	Copper and fiber	Copper and fiber
Protocols supported	Network, SCSI, video	Network	Network, video

best technology for applications that require high-bandwidth, reliable solutions that scale from small to very large.

12.18   FIBRE CHANNEL ADVANTAGES

Fibre Channel is the solution for IT professionals who need reliable, cost-effective information storage and delivery at blazing speeds. With development started in 1988 and ANSI standard approval in 1994, Fibre Channel is the mature, safe solution for gigabit communications. Today's data explosion presents unprecedented challenges incorporating data warehousing, imaging, integrated audio/video, networked storage, real-time computing, collaborative projects and CAD/CAE. Fibre Channel is simply the easiest, most reliable solution for information storage and retrieval.

Fibre Channel, a powerful ANSI standard, economically and practically meets the challenge with the following advantages.

- Congestion free—Fibre Channel's credit-based flow control delivers data as fast as the destination buffer is able to receive it.
- Gigabit bandwidth now—Gigabit solutions are in place today! On the horizon is 2 Gbps data delivery.
- High efficiency—Real price performance is directly correlated to the efficiency of the technology. Fibre Channel has very little transmission overhead. Most important, the Fibre Channel protocol is specifically designed for highly efficient operation using hardware.
- Multiple topologies—Dedicated point-to-point, shared loops, and scaled-switched topologies meet application requirements.
- Multiple protocols—Fibre Channel delivers data. SCSI, TCP/IP, video, or raw data can all take advantage of high-performance, reliable Fibre Channel technology.
- Price performance leadership—Fibre Channel delivers cost-effective solutions for storage and networks.
- Reliability—Fibre Channel sustains an enterprise with assured information delivery.
- Solutions leadership—Fibre Channel provides versatile connectivity with scalable performance.
- Scalability—From single point-to-point gigabit links to integrated enterprises with hundreds of servers, Fibre Channel delivers unmatched performance.

Corporate information is a key competitive factor, and Fibre Channel enhances IT departments' ability to access and protect it more efficiently.

In fact, multiple terabytes of Fibre Channel interfaced storage are installed every day. Fibre Channel works equally well for storage, networks, video, data acquisition, and many other applications. Fibre Channel is ideal for reliable, high-speed transport of digital audio or video. Aerospace developers are using Fibre Channel for ultrareliable, real-time networking.

Fibre Channel is a fast, reliable data transport system that scales to meet the requirements of any enterprise. Today, installations range from small post-production systems on Fibre Channel loop to huge CAD systems linking thousands of users into a switched Fibre Channel network.

Fibre Channel is attractive because it offers a standards-based solution. With the emphasis on open systems, end users are shying away from proprietary solutions and vertically integrated, single provider solutions. Today, users are integrating the best industry offers into integrated, seamless systems.

These new systems are being driven by the technology and marketing forces associated with client-server implementations. Fibre Channel is the only technology available with the reliability, responsiveness, scalability, high throughput, and low latency needed to meet the broad range of market and technology requirements. Users enjoy these advantages:

- Continued support of legacy systems
- More cost-effective systems
- Graceful upward migration
- Scalable systems
- Straightforward migration to Fibre Channel

Fibre Channel's scalability provides a continued return on investment long into the future.

12.19   PRODUCT AND SERVICE PROVIDERS

The different categories of Fibre Channel products are available from manufacturers and equipment suppliers. These categories include the following:

- Bridges/Routers/networking devices
- Cables and connectors
- Converters
- Host adapters
- Hubs
- RAID devices
- Storage devices
- Storage systems
- Switches

Bridge, router, and networking suppliers include the following:

ATTO Technology

Chaparral Network Storage, Inc.

Computer Network Technology

Crossroads Systems

Finisar Corp.

McData Corp.

Storage Tek

TD Systems

Vicom Systems, Inc.

■ SUMMARY

Fibre Channel is designed to provide a common, efficient transport system so that a variety of devices and applications can be supported through a single port type. The Fibre Channel Industry Association (FCIA), the industry consortium promoting Fibre Channel, sets forth the following requirements that Fibre Channel is intended to satisfy.

- Ability to carry multiple existing interface command sets for existing channel and network protocols
- Broad availability for standard components
- Full-duplex links with two fibers per link
- Greater connectivity than existing multidrop channels
- High-capacity utilization with distance insensitivity
- Performance from 100 Mbps to 3.2 Gbps on a single link
- Support for multiple cost/performance levels, from small systems to super computers
- Support for distances up to 10 km

The Fibre Channel solution was to develop a simple transport mechanism based on point-to-point links with a switching network. This underlying infrastructure supports a simple encoding and framing scheme that in turn supports a variety of channel and network protocols. Considerable information concerning the industry is available at www.fibrechannel.com.

Key Terms

8B/10B	Fabric
Arbitrated Loops	F_ports
Compute Cluster	Fibre Channel
Exchange	FL_port

Frames

L_ports

N_ports

NL_ports

Redundant Array of Inexpensive Disk (RAID)

Small Computer System Interface (SCSI)

Storage Area Network (SAN)

Storage Cluster

REVIEW QUESTIONS

1. Provide a general description of Fibre Channel.

2. What are the types of channel-oriented facilities that are part of the Fibre Channel architecture?

3. What are the types of network-oriented facilities that are part of the Fibre Channel architecture?

4. How is Fibre Channel different from IEEE LANs?

5. What are the different types of media supported by Fibre Channel?

6. What are the basic rates on the Fibre Channel?

7. What are the different types of fiber optics supported by Fibre Channel?

8. What are the common attributes that Fibre Channel has in common with Ethernet?

9. What are the implications for storage and the RAID architecture?

10. What are the major features of Fibre Channel?

11. How does Fibre Channel interoperate with SCSI?

12. What are some Fibre Channel applications?

13. What are some performance-enhancing features of Fibre Channel?

14. Discuss the ports that are used for interconnection. What is the difference between N_Ports and L_Ports?

15. What is an arbitrated loop? How does it work?

16. What is the difference between the point-to-point and arbitrated loop topologies?

17. What is a fabric? What is the function of a fabric?

18. What are the functional layers of the Fibre Channel protocol?

19. What concepts and functions are defined at the FC-2 layer?

20. What functions are defined at the FC-3 layer?

21. What are examples of upper level protocols that can be mapped into the FC-4 layer?

22. How is Fibre Channel similar to ATM?

23. What are the classes of service for Fibre Channel? How does each relate to ATM?

24. Layer FC-2 defines a number of procedures for the exchange of information. Describe each of these procedures.

25. What are the categories of frames in the FC-2 layer?

26. What functions are performed by the sequence procedure?

27. What happens in the exchange procedure?

28. Describe a Fibre Channel System. What are some of the components?

29. What is a cluster? How does Fibre Channel make clusters possible?

30. What are the components of a cluster? What is the difference between a computer cluster and a storage pool?

31. What is meant by a high-availability system?

32. What are the major component categories of Fibre Channel? Provide a brief description of each category. Identify some specific devices in each category.

33. What management protocols are utilized by Fibre Channel? Describe each.

34. What is the function of storage management?

35. How does Fibre Channel improve performance of storage management? Identify the various techniques that are involved.

36. There are a number of benefits from storage and network management. How are these accomplished?

37. Discuss the area of standards that affect Fibre Channel.

38. Provide a general description of the Fibre Channel frame format.

39. What are the applications for Fibre Channel?

40. Discuss the computer clusters application.

41. How does the storage pools application relate to the compute clusters application?

42. Provide a general comparison of Fibre Channel with Gigabit Ethernet and ATM.

43. Identify a number of advantages for implementing Fibre Channel.

44. Describe the encoding scheme for Fibre Channel. Is it similar to the scheme for Gigabit Ethernet?

45. What is a Fibre Channel datagram?

ACTIVITIES

1. Develop a comprehensive comparison of advantages and disadvantages of implementing and utilizing the Fibre Channel.

2. Develop a comparison matrix of the technologies that require a high bandwidth. Show the various applications that could be supported.

3. Look at the information available on the Internet at the Fibre Channel Industry Association Website. Provide a general overview on the state of the technology.

4. Research industry publications for Fibre Channel information. Provide an overview of the products that are available to support the technology.

5. Identify the standards associated with Fibre Channel.

6. Research the products and services that are available to support the Fibre Channel technology.

7. Prepare a five-minute overview of the Fibre Channel technology, with graphics.

8. Develop a comprehensive comparison between Fibre Channel and Gigabit Ethernet.

CASE STUDY/PROJECT

Fibre Channel is utilized to connect mainframes to each other, provide server farms with high-speed pipes, link high-performance workstations, provide for clustering of disk farms, and link LANs and WANs to the backbone infrastructure. This is accomplished by utilizing a number of different media.

Details

Graphics Ltd. produces high-quality graphics for the magazine industry. This operation requires a considerable amount of storage due to the quality requirements. The graphics workstations have a need for very-high-speed access to the various servers that contain the various components of the graphics. All network components will be located in the same department.

Requirements

Develop a network design that implements the Fibre Channel technology. Show all of the components that go to make up this design. Include the various types of media that might be utilized to connect the various components.

Alternative Requirement

Develop Gigabit Ethernet as an alternative to Fibre Channel. Include all of the components necessary to implement this network. Develop a graphic design. Provide a comparison between Fibre Channel and Gigabit Ethernet. Show advantages and disadvantages and a cost/benefit analysis.

13

Internet/Intranet/Extranet

■ INTRODUCTION

The **Internet,** sometimes called simply "the Net," is a worldwide system of computer networks—a network of networks in which users at any one computer can, if they have permission, get information from any other computer. Its development was funded by the Advanced Research Projects Agency (ARPA) of the U.S. government in 1969 and was first known as the **ARPANET.** The original aim was to create a network that would allow users of a research computer at one university to be able to "talk to" research computers at other universities. A side benefit of ARPANET's design was that, because messages could be routed or rerouted in more than one direction, the network could continue to function even if parts of it were destroyed in the event of a military attack or other disaster.

Today, the Internet is a public, cooperative, and self-sustaining facility accessible to hundreds of millions of people worldwide. Physically, the Internet uses a portion of the total resources of the currently existing public telecommunications networks. Technically, what distinguishes the Internet is its use of a set of protocols called TCP/IP (Transmission Control Protocol/Internet Protocol). Two recent adaptations of Internet technology, the intranet and the extranet, also make use of the TCP/IP suite.

For many Internet users, Electronic Mail (e-mail) has practically replaced the Postal Service for short written transactions. Electronic mail is the most widely used application on the Net. You can also carry on live "conversations" with other computer users, using Internet Relay Chat (IRC). More recently, Internet hardware and software allows real-time voice conversations.

The most widely used part of the Internet is the **World Wide Web** (also **WWW** or the **Web**). Its outstanding feature is hypertext, a method of instant cross referencing. Most Websites contain certain words or phrases that appear in text as a different color than the rest; often this text is underlined. When users select one of these words or phrases, they will

OBJECTIVES
Material included in this chapter should enable you to:

- become familiar with the history of the Internet and its implications in today's business environment. Look at how the business environment is affected by this technology.

- understand the technologies and infrastructure of the Internet. Identify components of the Internet. Become familiar with the numerous terms and definitions that are part of this environment.

- understand how Internet clients and servers are able to connect to the Internet. Look at the various search tools that are available to the user. Become familiar with browsers, cookies, and search engines.

- understand where the intranet and extranet fit in the Internet environment.

internet
ARPANET
world wide web (www)
web

- become familiar with the functions and services of ISPs and NSPs. Look at Internet connectivity.
- identify applications that fit the Internet environment. Become familiar with electronic data interexchange (EDI), Web publishing, and computer telephone integration (CTI).
- look at the advantages and disadvantages of the Internet environment. Look at the pros and cons for Intranets and Extranets.
- become familiar with the various standards associated with the Internet environment.

transmission control protocol/internet protocol (TCP/IP)

simple mail transport protocol (SMTP)
trivial file transfer protocol (TFTP)
file transfer protocol (FTP)
Internet Service Providers (ISP)

browsers

Hypertext Markup Language (HTML)

be transferred to the site or page that is relevant to this word or phrase. Sometimes there are buttons, images, or portions of images that are "clickable." If users move the pointer over a spot on a Website and the pointer changes into a hand, this indicates that they can click and be transferred to another site.

13.1 INTERNET HISTORY

The Internet emerged from a U.S. government and military initiative to enable the interconnection of different, mainly UNIX-based computer systems for intercommunication. As UNIX was proclaimed to be the first portable operating system for computers, enabling software developed on a particular manufacturer's computer hardware to be easily ported to another manufacturer's hardware, it was also natural to develop a means for easy transfer of data between systems. This led to **Transmission Control Protocol/Internet Protocol (TCP/IP).**

TCP/IP was quickly adopted by the academic community in the United States, and soon thereafter by academics worldwide, because it allowed for rapid sharing of scientific information and the electronic mail communication necessary for its rapid discussion and analysis, both within and between university campuses.

In its original form, the Internet and the Internet Protocol (IP) provided a means for interconnecting computer servers together. The Internet addressing scheme allowed individual workstations, personal computers, or software applications running on either the server or any of the workstations to direct and send information to other applications on other servers or distant LANs. Being a unique address, the Internet address allowed an end user to be identified, no matter how many transit servers, routers, or networks would have to be traversed along the way.

A suite of new protocols arrived with the Internet. These included **Simple Mail Transport Protocol (SMTP), Trivial File Transfer Protocol (TFTP),** and **File Transfer Protocol (FTP),** which together form the basis of the Internet electronic mail service.

As the popularity of the Internet has grown, so have the numbers of servers and routers making up the network. **Internet Service Providers (ISPs)** have provided for dial-up access to the servers from private individuals using their PCs at home, and new information providers have furnished more Internet pages of information. **Browsers,** computer software that allows users to "surf" the Internet and the WWW, allow users to seek information from any of the connected servers by means of menu-driven screen software. These browsers, using the **Hypertext Markup Language (HTML)** tool, provide a hypertext search and browse capability.

It was not until 1986 that the dramatic growth of the Internet began. At that time, the National Science Foundation (NSF) linked five of its regional supercomputer centers together to provide a national high-speed backbone network across the United States. The world's fastest and most powerful computers were made available to the academic and scientific community. The collection of these networks linked together to form what we call the Internet. Each network's host computer shared the common TCP/IP suite.

The World Wide Web or Web was originally developed by Tim Berners-Lee in 1989 at the European Laboratory for Particle Physics (CERN) in Geneva, Switzerland. The original intent was to develop a global information system that combined text, graphics, and sound on a series of computer-displayed documents called web pages.

The WWW is a distributed hypermedia repository of information that is accessed with an interactive browser. A browser displays a page of information and allows the user to move to another page by making a mouse selection. Embedded into the Web page are hypertext or hypermedia documents, which are underlined text or highlighted graphical icons. By simply pointing and clicking on the hypertext or hypermedia, a link is made to that particular source document, which can also lead to the pathways of other hypertext links.

In 1994, a major international initiative to develop a universal computer framework for the global information society was announced by European and American governments and educational leaders. The combined efforts of MIT and CERN were to further the development and standardization of the Web to make the WWW easier to use for research, commercial, and future applications.

IBM, MCI, and Merit formed a nonprofit corporation that operates the Internet backbone in the United States. IBM develops hardware and software, MCI operates the fiber-optic backbone, and Merit provides the administrative functions. In 1995, the heart of the Internet was created with four Network Access Points (NAPs) located in San Francisco; Chicago; Washington, D.C.; and Pennsauken, New Jersey. With these four access points anyone, in theory, can interconnect with the rest of the Internet.

13.2 USING THE WEB

Using the Web, users have access to millions of pages of information. Web "surfing" is done with a Web browser—the most popular being Netscape Navigator and Microsoft Internet Explorer. The appearance of a particular website may vary slightly depending on the browser used. Also, later versions of a particular browser are able to render more "bells and whistles" such as animation, virtual reality, sound, and music files than earlier versions. **Java** and **JavaScript** are two common languages utilized in web application design [Shay, 1999].

Java
JavaScript

Figure 13–1
Internet Access

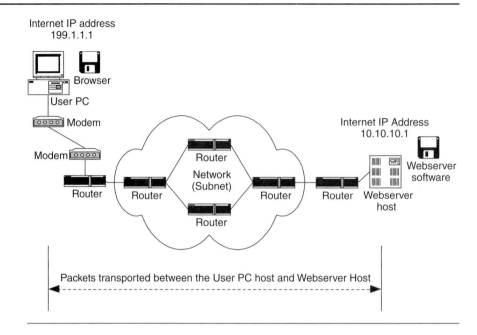

Packets transported between the User PC host and Webserver Host

packets

A good place to start with understanding the Internet is to identify all of the components that make it up. The Internet is not a single network, but thousands of networks scattered around the globe. Universities, corporations, and other organizations own individual networks, which are all linked together so that everyone can communicate with each other. Messages of the Internet are called **packets.** A device called a router makes it possible to send packets from one computer to another. Packets are transported from one network to another through a router at that network. The packet hops from one router to another across multiple networks until it eventually reaches the destination computer. The path it travels is called its route.

Figure 13–1 shows the components and data flow involved in accessing the Internet from a user's PC. The personal computer user at 199.1.1.1 uses a dial-up modem to access the network that is usually represented by an ISP. The ISP routes the user's request to the appropriate Web server host (10.10.10.1) and then provides the response that is received from the Web server back to the user. The function of an ISP will be explained later in more detail.

13.3   INTERNET SEARCH TOOLS

A variety of tools have been developed to help one navigate the Internet [Panko, 1999]. Although many of these tools have rendered themselves

obsolete due to the advent of the WWW browsers, they are still exten-
sively used as reference tools by the WWW. These tools include the FTP,
Anonymous FTP, Telnet, Gopher, Veronica, Archie, and Wide Area Information Servers (WAISs).

Web Browsers

Navigating the Web requires the use of a program called a client. The
client software provides the communications protocol to interface with
the various host computers on the network. The more common name for
client software is a web browser. A web browser is a program that handles most of the details of document access and display and permits the
user to navigate the Web by accessing Web documents that have been
coded in a language called HTML.

In addition to text, an HTML document contains tags that specify
document layout and formatting. Some tags cause an immediate change;
others are used in pairs to apply an action to multiple items. To make document retrieval efficient, a browser uses a **cache.** The browser places a   cache
copy of each document or image the user views on the local disk. When
a document is needed, the browser checks the cache before requesting the
document from a server on the network. The latest Web Browser products
display all reachable resources, from the local PC to the worldwide Internet, on a single hierarchical file tree display.

HTML was also invented by Tim Berners-Lee at CERN. Documents
written in HTML are plain-text ASCII files that can be generated with a
text editor. They reside on the Web server and can be identified with an
".html" or ".htm" extension. Several Web browsers are used to navigate
the Web (e.g., Netscape Navigator and Internet Explorer). Each browser
must contain an HTML interpreter to display documents. One of the most
important functions in an HTML interpreter involves selectable items.
The interpreter must store information about the relationship between
positions on the display and anchored items in the HTML document.
When the user selects an item with the mouse, the browser uses the current cursor position and the stored position information to determine
which item the user has selected.

One thing to remember about Web browsers and their ability to
search the Web for a particular Web page or topic is that Web browsers
still require a master Web index service or home page. The Lycos home
page at Carnegie-Mellon University is an example of such a master Web
indexing service.

Push Technology

One of the most radical changes in the Web has been the creation of
push technology. Traditionally, if several web pages were of importance, there was no way of knowing when they changed. Users were required to download each page periodically to check for changes. Often,

there would be no changes, so the downloading would be a waste of time.

In push technology, the effect is as if the web servers periodically downloaded updated versions of subscribed web pages to the user without any conscious effort. When reading a web page, the user would know it was a very recent version. The web servers would push the web pages to the user, instead of having to pull them.

States and Cookies

HTTP was created with the concept that each request-response cycle would be separate. A web server host would not remember the user from request to request: It would simply get the file that was requested and send it.

More complex actions, however, such as financial transactions, consist of a sequence of HTTP request-response cycles that build upon one another. Each action in the sequence creates a state (condition). The next action builds upon that state to create a new state.

To help web servers track a user's history, browsers allow web servers to store brief text files on the user's PC hard disk drive. These short text messages are called **cookies.** When the user sends a message to the web server, the web server can also get the cookie to see what has been done in the past.

cookies

Cookies are somewhat controversial because they allow a web server to store files on the hard disk drive. They are text files, however, not programs that could contain viruses.

Uniform Resource Locator

universal resource locator (URL)

All Web pages and Internet resources have a unique address called a **Universal Resource Locator (URL).** A URL is analogous to a card catalog number that references a library book. It permits its user, through the use of a client program, to retrieve documents from the Internet server. A browser extracts from the URL the protocol used to access the item, the name of the computer on which the item resides, and the name of the item.

hypertext transfer protocol (HTTP)

The first part of the URL specifies the protocol **Hypertext Transfer Protocol (HTTP).** It is the Web's main protocol for transporting HTML documents between clients and servers. The colon and two slashes follow the protocol. The "www" and remainder of the URL indicate you are accessing a Web server in search of a particular file and also its path, which is specified after any single slashes. An example of a URL is http://www.weather.com.

Search Engines

search engines

The most common tool used for navigating the Web is called a search engine. **Search engines** are programs used to search for virtually any form of intelligence imaginable.

Domain	Description
com	Commercial organization
edu	Educational institute
gov	Government organization
int	International organization
mil	Military organization
net	Network or service provider
org	General organization

Table 13–1
U.S. Internet Domains

Access to search engines is provided by ISPs and commercial online services. When a connection to the Web is first established, the home page typically has the ISP's proprietary search engine and may list links to other popular search engines such as Yahoo, Lycos, Alta Vista, Dogpile, and WebCrawler.

Domain Name System

The **Domain Name System (DNS)** is a method of mapping domain names and IP addresses for computers linked to the Internet. This means that DNS resolves symbolic names to their corresponding IP addresses. The system was developed in 1986 to replace a central registry of IP addresses used to identify host computers connected to the ARPANET. There are currently nine "root" DNS servers on the Internet. The DNS structure is a hierarchical tree consisting of domains. A **domain** is a node and its descendent nodes on a network. A domain name is the unique name of a particular node within the domain, and domain names are made up of subdomains, which are nodes leading to a particular node. Subdomains are separated by a period. On the Internet, URLs refer to the domain names for a computer or a group of computers. These domains are separated into categories, which are denoted by the last element in the domain name. For example, in the domain name "john.faculty.atl.devry.edu," the user *john* is a faculty member on the Atlanta DeVry campus. The root node edu is at the top of the hierarchy. Table 13–1 lists various types of Internet domains used in the United States.

domain name system (DNS)

domain

A complete listing of country top-level domain names can be viewed at http://www.uninett.no/navn/domreg.html.

13.4 INTERNET COMPONENTS

To connect to an Internet site, the user normally first connects to an ISP. When dialing into an ISP, access is usually via a router owned by that ISP. The ISP also has a router connected to the Internet. This second router is the interface into the entire Internet. The backbone of the Internet in the United States consists of companies called **Network Service Providers (NSPs).** The NSPs are interconnected via a number of networks.

The levels of Internet access can be separated into five categories:

- Level 1: Interconnection—Network Access Points (NAPs).
- Level 2: National Backbone
- Level 3: Regional Networks
- Level 4: Internet Service Providers (ISPs)
- Level 5: Consumer and Business Market

On the top level are the NAPs where major backbone operators or NSPs interconnect to establish the core concept of an Internet.

Level 2 consists of the national backbone operators. The network of networks spreads out from this point.

The third level of the Internet consists of regional networks and the companies that operate regional backbones. Typically, these companies operate backbones within a state or among several adjoining states much like the NSPs. They typically connect to a national backbone operator, or increasingly to several NSPs, to be on the Internet. Some have a presence at a NAP, which is used to extend this network to smaller cities and towns in their areas. Businesses are connected to those points with direct access connections, where they usually maintain dial-up terminal banks to offer 28.8 to 56 kbps dial-up SLIP/PPP connections to consumers. In many cases, regional networks are much more extensive than national backbones, but on a smaller geographic scale.

At the fourth level of the Internet are the individual ISPs. These vary from small operations to large organizations, some with up to 100,000 dial-up customers. They generally do not operate a backbone or even a regional network. Connections are leased from a national backbone provider or a regional network operator. They may offer service nationally, but use the POPs and backbone structure of their larger backbone operator associate. Several large providers, such as EarthLink and MindSpring, operate at this level. Generally, these companies operate an equipment room in a single area code, lease connections to a national backbone provider, and offer dial-up connections and leased connections to consumers and businesses in the local area. The focus is on customer service, configuration, training, and often lower rates.

The fifth level of the Internet is the consumer and business market. Many companies set up dial-up ports at their offices for employees to connect to from home or on the road.

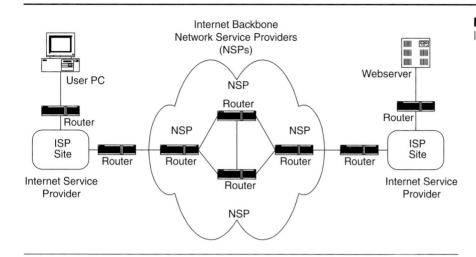

Figure 13–2
ISPs and NSPs

13.5 ACCESS AND TRANSPORT

A carrier is an organization that transmits telecommunications traffic for a price. Figure 13–2 shows the two types of carriers involved in the United States—ISPs and NSPs [WWW Consortium, www.w3.org/, 2000].

Accessing the Internet

To access the resources available on the Internet, the user must connect the computer to the network. For most users today, this means subscribing to an ISP. ISPs provide access using modems, ISDN service, ADSL, or other dial-up connection services. Other types of connections are possible, but may be beyond the means of most individuals. These include the large-bandwidth connections of DS-3 (44.736 Mbps) or OC-3 (155 Mbps) as provided to large corporations and governmental organizations.

Figure 13–3 shows various access technologies which can be utilized to connect to the Internet. The ISP can provide one-stop shopping for all network access, including client-server hardware and software, and security requirements. Access from the full-service client to the ISP is 56 kbps Frame Relay service. The Internet Presence Provider (IPP) provides design, management, and maintenance. The IPP connects to the ISP via 56 kbps Frame Relay service and provides dial-up access to client PC-5. Client PC-2 utilizes ISDN service. Client PC-3 has router access via T-1 service to the ISP. Client PC-4 has router access to the ADSL service.

Figure 13–3
Client-Server Internet
Connectivity

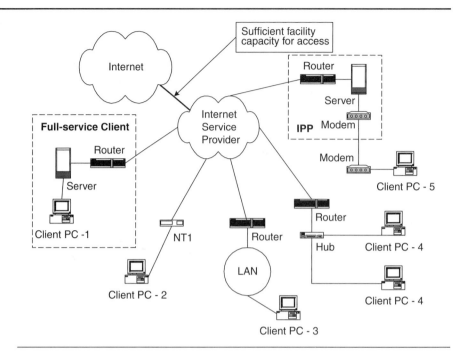

Tunneling

tunneling

Tunneling provides virtual private networking capabilities, using the Internet as an enterprise network backbone. Basically, private, secure channels are created between systems. A tunnel is simply a relay point between two TCP connections. The HTTP messages pass unchanged as if there were a single HTTP connection between the user agent and the origin server. Tunnels are used when there must be an intermediary system between client and server, but it is not necessary for that system to understand the contents of the messages. An example is a firewall in which a client or server external to a protected network can establish an authenticated connection and then maintain that connection for purposes of HTTP transactions. Tunneling is explained in detail in Chapter 6 (VPN).

Internet Service Providers

When users connect to the Internet, they first connect to an ISP. Figure 13–4 shows that when dialing into an ISP, the user dials into a router owned by that ISP. The ISP also has a router connected to the Internet backbone. This second router is the gateway to the entire Internet. Several

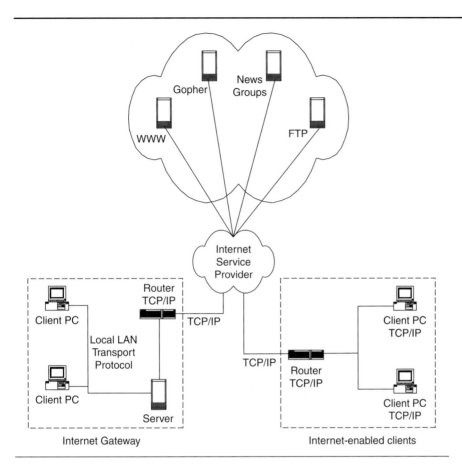

Figure 13–4
Internet Gateways, Services, and Clients

sites on the Web provide listings of ISPs. At least one site indicates more than 8,000 ISPs worldwide.

Network Service Providers

As described earlier, the backbone of the Internet consists of several competing NSPs, which are interconnected. Although NSPs compete for ISP business, they must work together to provide total interconnection. To connect to a NSP, the ISP must pay the NSP a monthly fee.

Figure 13–4 provides a view of the overall client-server architecture for Internet connectivity. The Internet gateway configuration has no special software loaded on the client PCs. The server provides local LAN transport protocol to TCP/IP translation and client software for supported Internet services. In the internet-enabled client configuration, every client has TCP/IP and local LAN transport protocol loaded. Internet client software is also loaded on each client.

OSI Layer	Function	Used for
Application	Browser	HTTP
Transport	Transport Layer Process	TCP
Internet	Internet Layer Process	IP
Data Link	Data Link Layer Process	PPP
Physical	Physical Layer Process	Serial port, modem

13.6   INTERNET STANDARDS

Table 13–2 shows that Web access requires five levels or layers of standards that align with some of the OSI model layers. These include the Physical Layer Process, the Data Link Layer Process, the Internet Layer Process, the Transport Layer Process, and the Browser. The layers work together to provide the functionality needed to connect an applications program on the user's PC to an applications program on another PC (Internet Engineering Task Force [IETF], www.ietf.cnri.reston.va.us/, 2000; IETF, www.rfc-editor.org/, 2000).

This standard allows for two applications programs on different machines to be able to communicate effectively. This process of interoperability represents the TCP/IP OSI model. Note that this is a subset of the standard seven-layer OSI model shown in Appendix A. Table 13–2 lists the layers that apply to the Internet.

Internet Protocol

The IP is a network layer protocol that contains addressing information and some control information to enable packets to be routed. IP is documented in RFC791 and is the primary network layer protocol in the IP suite. Along with the Transmission Control Protocol (TCP), IP represents the heart of the Internet protocols. IP has two primary responsibilities:

- Providing connectionless, best-effort delivery of datagrams through an internetwork
- Providing fragmentation and reassembly of datagrams to support data links with different Maximum Transmission Unit (MTU) sizes

E-mail Standards

The standard for text messages on the Internet is RFC822. This standard specifies that a message must have two parts: a header and a body. The header must contain specific fields, such as date sent, to, from, and subject. Other fields are optional. Each field has a keyword, a colon (:), and the content of the field. The rigid header structure allows a mail program to search for messages with a specific e-mail address, sort mes-

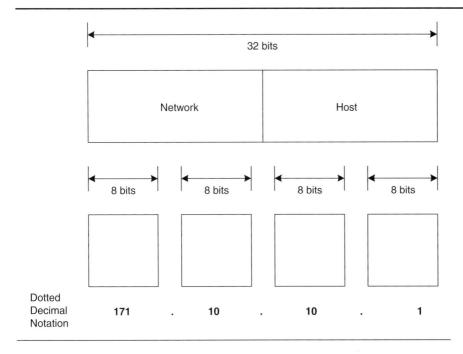

Figure 13–5
IPv4 Address Format

sages by author or date, and do other things to make message retrieval easy. The body contains free-form text.

13.7 IP ADDRESSING

As with any other network layer protocol, the IP addressing scheme is integral to the process of routing IP datagrams through an internetwork. Each IP address has specific components and follows a basic format. These IP addresses can be subdivided and used to create addresses for subnetworks.

The 32 bit IPv4 address is grouped 8 bits at a time, separated by dots, and represented in decimal format, otherwise referred to as dotted decimal notation. The minimum value for an octet is zero and the maximum value is 255. Figure 13–5 provides the basic format and an example of an IP address (171.10.10.1). Appendix D provides information concerning the manipulation of IP addresses for routing tables.

Internet Addressing

Any computer attached to the Internet is called a host. When users are connected to the Internet with a home PC, they become a temporary WWW host. To contact someone via telephone requires a telephone number. On the Internet the equivalent is the Internet address. Currently, the Internet address (IPv4) is a string of 32 bits (0,1s). A new version (IPv6) will allow for 128 bits to accommodate the enormous growth expected. This address is the IP address. Each host has its own unique IP address just like each telephone has its own unique number.

The IP address is converted into four 8 bit strings (segments). An example of an address is 188.23.44.5. The four decimal segments are separated by decimal points, called dots, and thus the name dotted-decimal notation. Unique IP addresses are assigned by the **Internet Assigned Numbers Authority (IANA).** A host name is defined by the Internet, which consists of several labels of text separated by dots. This substitutes for the Internet address. The user can use either the Internet address or the host name when accessing the Internet site.

13.8   INTERNET APPLICATIONS

The Internet has four primary purposes:

- To provide electronic mail service to the users
- To support file transfer between hosts
- To permit users to log on to remote computers
- To provide users with access to information databases

The Internet provides the following services:

- Business services
- File Transfer Protocol (FTP) for file transfer
- Electronic journals
- HyperText Transfer Protocol (HTTP) for the WWW
- Internet Relay Chat (IRC) for chat
- Network News Transfer Protocol (NNTP) for newsgroups and the distribution of news
- Simple Mail Transfer Protocol (SMTP) for e-mail
- Telnet for remote logon
- Miscellaneous services

Some of the most popular Internet services include Electronic Mail (e-mail), File Transfer Protocol (FTP), Gopher, newsgroups, Telnet, and the WWW.

Electronic Mail
Although e-mail primarily remains character oriented, its ability to permit individuals to easily exchange information and files makes it the most popular networked application of any kind.

Millions of users are connected worldwide to the Internet via the global e-mail subsystem. From a business perspective, Internet e-mail offers one method of sending intercompany e-mail. Most companies have private networks that support e-mail transport to fellow employees, but not necessarily to employees of other companies. By adding Internet e-mail gateways to its private network, a company can send e-mail to users almost anywhere.

Internet electronic mail hosts exchange messages through the SMTP, which is in the TCP/IP architecture. E-mail hosts hold the mail until the subscriber is ready to read it. This allows the receiver to work on a client PC, which will often be turned off. E-mail hosts also transmit outgoing messages to other mail hosts. Together, mail hosts function as a network of electronic post offices.

File Transfer Protocol

To download or transfer information back to their client PCs, users would access another TCP/IP protocol called File Transfer Protocol (FTP) or anonymous FTP servers. Users can access FTP servers directly or through Telnet sessions (explained shortly). The difficulty with searching for information in this manner is that users must know the Internet address of the specific information server that they wish to access.

FTP makes it possible to move files across the Internet and handles some of the details involved in moving text and other forms of data between different types of computers. Although FTP is not a highly graphical application, it remains an important tool for individuals and organizations that must exchange files containing data or documents.

Gopher

Gopher is a menu-based client/server that features search engines which comb through all of the information in all of the information servers looking for a user's specific request. Gopher provides a way to index and organize all kinds of different collections of textual data, as well as other kinds of documents. Before the introduction of the WWW, Gopher was the premier tool for browsing the Internet to look for information. Gopher was named after the mascot of the University of Minnesota where the system was developed.

Gopher client software is most often installed on a client PC and interacts with software running on a particular Gopher server, which transparently searches multiple FTP sites for requested information and delivers that information to the Gopher client. Gopher users do not need to know the exact Internet address of the information servers that they wish to access.

Newsgroups

Based on a TCP/IP service known as **USENET, newsgroups** provide a way for individuals to exchange information on specific, identifiable topics or areas of interest. This technology lets users read information on a variety of subtopics that are pertinent to a newsgroups' focus. Over

USENET
newsgroup

10,000 newsgroups covering selected topics are available. USENET servers update each other on a regular basis with news items that are pertinent to the newsgroups housed on a particular server. For technical matters, this is an especially useful way to exchange opinions and information on a broad range of topics.

Telnet

telnet

Text-based information stored in Internet-connected servers can be accessed by remote users logging into these servers via a TCP/IP known as **telnet.** Telnet permits a user on one computer to establish a session on another computer elsewhere on the Internet, as if the local computer was directly attached to the remote computer. Given the proper access to remote machines, this program lets users achieve many tasks remotely that they might ordinarily only be able to accomplish locally. Telnet is an application of choice for configuring all kinds of networking equipment, especially routers and hubs.

13.9   WORLD WIDE WEB

Today, the WWW is the premier application for most Internet users [Haynal, 2000]. The WWW is a collection of servers accessed via the Internet that offer graphical or multimedia presentations about a company's products, personnel, or services. Web browsers integrate e-mail and newsreaders and support an increasingly interactive, visual, and even animated interface to the Internet. Companies wishing to use the WWW as a marketing tool establish a Website on the Internet to publicize the address of that location. The Web site and Web server presentation design, implementation, and management can be done in-house or be contracted out to a professional Website development and management service.

Almost any of the current data processing/data communications applications are candidates to ride the Internet. Some major application areas are as follows:

- Customer Service
- Distance Learning
- Electronic Commerce
- Internet Access
- Telemarketing
- Web design
- Web hosting

Reasons why people use the Web include the following:

- Browsing
- Business

- Chatting socially with other Web users
- Entertainment

Virtually all classes of businesses can use the Internet in the following ways:

- Advertising via Web pages and e-mail
- Communicating via e-mail
- Selling goods and services
- Selling stocks or other financial instruments
- Transferring documents via electronic data interchange (EDI)
- Providing travel services
- Providing reference materials

13.10   INTRANET OVERVIEW

An emerging network paradigm arising out of the enormous growth of the Internet is the intranet. Many of the components and concepts that have been presented in the Internet section will apply for both the intranet and the extranet. The main difference that the user will encounter is the utilization of the various applications.

An **intranet** is a network of networks that is contained within an enterprise. It may consist of many interlinked LANs and utilize leased lines in the WAN. Typically, an intranet includes connections through one or more gateway (firewall) computers to the outside Internet. The main purpose of an intranet is to share company information and computing resources among employees. An intranet can also be used to facilitate group or team working and for interactive teleconferences. Figure 13–6 shows the intranet environment. Employees can access information such as directories, schedules, messages, and other corporate data that are useful, but need not be available to nonemployees.

intranet

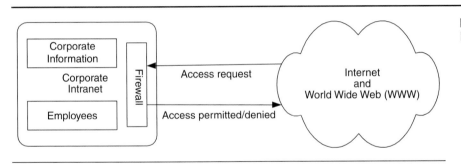

Figure 13–6
Intranet Environment

13.11   INTRANET TECHNOLOGIES

An intranet uses open standards such as TCP/IP, HTTP, HTML, and other Internet protocols, and in general looks like a private version of the Internet. Today, an open design is much better than a closed proprietary design, because of the flexibility it gives customers.

Typically, larger enterprises allow users within their intranet to access the public Internet through firewall servers. These have the ability to screen messages in both directions to maintain company security. When part of an intranet is made accessible to customers, partners, suppliers, or others outside the company, that part is called an extranet. The extranet will be discussed in detail later in this chapter.

Intranets use the TCP/IP standards architecture for internal corporate transmissions. They use routers to deliver messages within the company. They use Internet standards for e-mail, the Web, and other applications.

The popularity of intranets comes from a combination of the demands on businesses today and the ability of intranets to help companies meet those demands. The first step is to look at the demands on businesses today. Then it will be necessary to take a detailed look at the advantages of intranets and the leverage that vendors bring to the table.

There are several reasons why organizations are using intranets. One is the lure of the TCP/IP transmission technology, which is standardized, proven, highly scalable, and inexpensive. Having an Internet access to the intranet will allow companies to move from proprietary applications, which lock them into a single vendor, to open Internet applications such as e-mail. With open application standards, companies can buy products from multiple vendors that will work together.

13.12   FIREWALLS

firewall

Although the premise of the intranet is to support the sharing of corporate information with a limited or well-defined group, some information (e.g., products and services) is often shared with the general public over the Internet. Therefore, intranet sites have rapidly become commonplace over the Internet and the Web. A **firewall** is an intranet security system designed to limit Internet access to a company's intranet. The firewall permits the flow of e-mail traffic, product and service information, job postings, and so forth, while restricting unauthorized access to private information residing on the intranet. Two firewalls are depicted in Figure 13–7.

Desk-top firewall

Rack-mount firewall

Figure 13–7
Firewalls

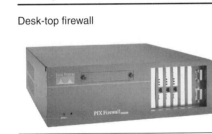

13.13 INTRANET APPLICATIONS

There are numerous justifications for an intranet. Most companies need to distribute vast amounts of internal documents to their employees, such as personnel directories, newsletters, memorandums, work schedules, and office policies. By making this information available to employees over an intranet, most hard-copy procedures and dissemination of information in the traditional manner becomes obsolete. The "paperless workforce" is justified through cost savings and efficiencies. Some common uses of intranets are:

- Departmental scheduling
- Education and training
- Human resource policies
- Job postings

Two other applications that have a potential for Intranets are Workflow and GroupWare. **Workflow** is the movement of workload through seamless processes. Several software packages are available to enable work products to flow among the various people who have responsibilities for different parts of the process. Proprietary workflow tools include Lotus Notes, IBM FlowMark, and Action Technologies Action Workflow. GroupWare utilizes multiple tools such as group composition, group memory, electronic document management, scheduling, workflow, and project management systems. Proprietary GroupWare tools include Lotus Notes, Microsoft Exchange, and Novell Groupwise.

workflow

   The growth of the Internet, including its use on the Web, has created new possibilities for GroupWare that are just beginning to be explored. Using standard browsers, these systems bring many of the benefits of traditional systems while adding the features of the Web, including Common Gateway Interface (CGI) processing on the web server and downloading applications to the user's PC via Java.

Another advantage of groupwise is its ability to facilitate videoconferencing. It offers options for room-to-room and desktop systems. The same benefits that other Internet technologies have brought to the Internet and to corporate intranets now are designed to accelerate Business to Businesses (B2B).

13.14   EXTRANET ENVIRONMENT

extranet

The term **extranet** was coined by Bob MetCalfe in the April 8, 1996 issue of *InfoWorld*. It is an Internet-like network that a company runs to conduct business with its employees, customers, vendors, and suppliers. Extranets include Websites that provide corporate information to internal employees and also to external partners, large customers, and particular suppliers through a secure access arrangement. An extranet is not available to the general public. It is called an extranet because it uses the technology of the public Internet (TCP/IP) and customers and suppliers often access it through the Internet via an ISP.

13.15   EXTRANET TECHNOLOGIES

An extranet is a private network that uses the Internet protocols and the public telecommunications system to securely share part of an organization's information or operations with suppliers, vendors, partners, customers, or other businesses.

An extranet can be viewed as part of a company's intranet that is extended to users outside the company. It has also been described as a "state of mind" in which the Internet is perceived as a way to do business with other companies as well as to sell products to customers. The same benefits that HTML, HTTP, and SMTP/POP3 provide for the Internet users can be applied to the extranet users.

13.16   VIRTUAL PRIVATE NETWORKS

In the past, WANs were largely implemented by leasing transmission facilities from a common carrier. The transmission facilities were owned by the common carrier but dedicated to the lessor. The transmission media leasing expense was large for companies with big WANs. If a company has its systems connected to the Internet, the nodes are able to communicate over the Internet lines. The problems inherent in this situation are security and availability. Because an extranet requires security and privacy, it needs firewall server management, the issuance and use of digital certificates or similar means of user authentication, encryption of messages, and the use of Virtual

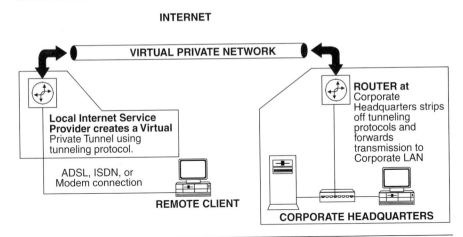

Figure 13–8
VPN Access

Private Networks (VPNs) that tunnel through the public network. Figure 13–8 shows a VPN access between a remote client and a corporate headquarters LAN. The transport between these two locations is transparent to the users. In this example the remote client accesses an ISP which provides the VPN connectivity to the corporate headquarters.

A **Virtual Private Network (VPN)** is customer connectivity deployed on a shared infrastructure with the same policies as a private network. The shared infrastructure can leverage a service provider IP, Frame Relay, ATM, or the Internet. The three types of VPNs, which align with how businesses and organizations use VPNs, are access VPN, intranet VPN, and extranet VPN.

virtual private network
(VPN)

Access VPN
Access VPN provides remote access to a corporate intranet or extranet over a shared infrastructure with the same policies as a private network. They enable users to access corporate resources when, where, and however they require. Access VPNs encompass analog, dial, DSL, mobile IP, and cable technologies to securely connect mobile users, telecommuters, or branch offices.

Intranet VPN
The **intranet VPN** links corporate headquarters, remote offices, and branch offices over a shared infrastructure using dedicated connections. Businesses enjoy the same policies as a private network, including security, QoS, manageability, and reliability.

intranet VPN

Extranet VPN
The **extranet VPN** links customers, suppliers, partners, or communities of interest to a corporate intranet over a shared infrastructure using dedicated

extranet VPN

connections. Businesses enjoy the same policies as a private network, including security, QoS, manageability, and reliability.

A key component of the VPN access is tunneling, a vehicle for encapsulating packets inside a protocol as understood at the entry and exit point of a given network. The entry and exit points are defined as a tunnel interface. VPNs and tunneling are described in detail in Chapter 6.

13.17   INTRANET AND EXTRANET APPLICATIONS

As with the intranet, a wealth of applications also can ride the extranet. These include:

Computer Telephone Integration

Customer service

Distance learning

Electronic Commerce

Electronic Data Interchange

Inventory

Manufacturing

Order Fulfillment

Professional Services

Sales and Marketing

Telemarketing

Web page publishing

Electronic Data Interchange

electronic data interchange (EDI)

Historically, buyers and sellers have communicated through **Electronic Data Interchange (EDI).** EDI standards specify standard formats for a number of business documents such as invoices and purchase orders. Vendors have allowed their customers to log into a vendor host, check stock for availability and price, and create a purchase order. For example, pharmacists can place orders with local drug suppliers and the order will be delivered to the pharmacists the next morning. This process is typical of the extranet environment.

EDI is a series of standards that provide a computer-to-computer exchange of business documents between different companies' computers over phone lines and the Internet [Goldman, 1998]. These standards allow for the transmission of items such as purchase orders, shipping docu-

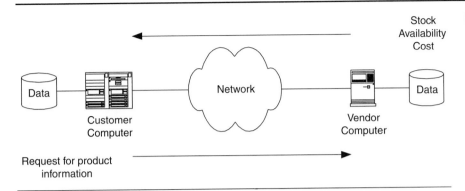

Figure 13–9
EDI Environment

ments, and invoices between an enterprise and its suppliers. Figure 13–9 depicts an EDI enterprise environment. In this example, EDI provides the ability of the customer to request product information from a vendor and allows the vendor to respond to the request with stock availability and cost.

Some benefits of EDI are cost savings, reduction of errors, security, integration with office automation applications, and just-in-time deliveries.

Applications for the extranet allow the user to do the following:

- Exchange large volumes of data using EDI
- Share product catalogs exclusively with wholesalers or those "in the trade"
- Collaborate with other companies on joint development efforts
- Jointly develop and use training programs with other companies
- Provide or access services provided by one company to a group of other companies, such as an online banking application managed by one company on behalf of affiliated banks
- Share news of common interest exclusively with partner companies
- Facilitate direct transfer of orders from one company to another

Computer Telephone Integration

Computer Telephone Integration (CTI) is a term for connecting a computer to a telephone switch. It has the computer issue the telephone switch commands for further processing of the call [Enterprise Computer Telephony Forum, www.ectf.org/, 2000]. CTI is currently implemented in three architectures—PBX-to-host interfaces, desktop CTI, and client-server CTI [Goldman, 1998]. Figure 13–10 is an example of a desktop implementation. There are different requirements to deliver CTI services for each of the three methods.

computer telephone integration (CTI)

Figure 13–10
Desktop CTI Environment

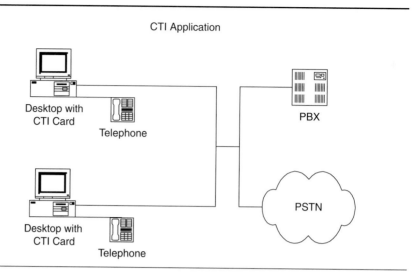

CTI Application

Desktop with
CTI Card

Telephone

PBX

Desktop with
CTI Card

Telephone

PSTN

Web Page Publishing

One application that can be leveraged across the enterprise is the development of Web pages.

As outlined in Figure 13–11, the intranet is the vehicle for development of Web pages that can be utilized in intranet, extranet, and Internet applications. The common element between these networks is, of course, the Internet. The process is as follows:

- A client PC requests a Web page publishing activity
- A Web page is published using Web publishing software running on a client or a server.
- A Web page is formatted using HTML.
- A stand-alone PC runs HTML-based publishing software which creates a HTML-formatted Web page. This page is then forwarded and stored on the Web.
- A LAN-attached client runs the server-based Web page publishing software to create a HTML-formatted Web page which is stored on the Web server. The Web server is the common point between the Internet and the Web publishing components.
- The Web page is delivered from the server through the Internet to the client for display.
- Note: The client PC has Web browser software and the Web server supports HTTP. TCP/IP protocol is running on the network.

Web pages can be developed and prototyped on the Intranet, using the local database resources and then exported to the extranet users who will be accessing the corporate resources.

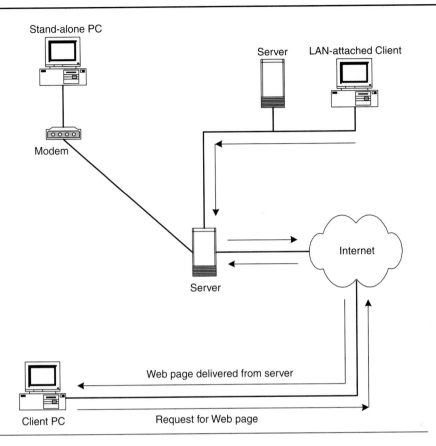

Figure 13–11
Web Clients, Web Services, and Web Publishers

With the proper firewalls installed for the extranet Web pages, this system becomes a useful tool for corporate users as well as buyers, contractors, and suppliers who will be using the extranet. Similar to the utilization of the Internet, both the extranet and the Intranet share in the access to the various networks and user communities.

13.18   ISSUES AND CONSIDERATIONS

As the organization places all of its data on an intranet, certain issues must be addressed. These include but are not limited to the following:

- When can users access the data; and will they have the ability to edit and delete the information, or will they be restricted to just reading it?

- Will there be sufficient bandwidth for the increased traffic load that will include GUI, GIF, JPEG, and TIFF files?
- Will the response time to users be acceptable with this additional traffic load?
- What would be the result if a disgruntled employee decided to destroy or modify the information?
- Have the issues of security, manageability, and recoverability in the event of loss or destruction of information been addressed?
- Are there sufficient personnel available for managing the infrastructure that includes the integrity of the system?

Using an intranet can be an opportunity to provide high-speed access and easy-to-use functionality to the desktop user. Substantial growth is a surety in this area.

13.19   TECHNOLOGY ALTERNATIVES AND COMPETITION

Internet Products and Services

Several broadband technologies can be utilized to provide access to the user's business. Some of these technologies are in competition with the Internet; however, many of them can be used in conjunction with the Internet to enhance the network application. These include the following:

- CTI technology
- EDI technology
- Private/leased Line
- Frame Relay
- X.25
- ATM
- ADSL
- SDSL
- ISDN
- VPN
- Wireless

Product and Service Providers

The Web is an excellent place to identify product and service providers. A sample of some providers is as follows:

Altavista	www.altavista.com/
America Online	www.aol.com/
Excite	www.excite.com/
Lycos	www.lycos.com/
Microsoft Internet Explorer	www.microsoft.com/
Netscape Navigator	www.netscape.com/
Yahoo	www.yahoo.com/

■ SUMMARY

The long-predicted merger of telecommunications and the computer is becoming a reality. Nowhere is it better illustrated than on the Internet, which has leapt into the public consciousness over the past few years. Internet users are interested in the ability to access information wherever it resides and to communicate with others effortlessly, with the network making distance and protocol irrelevant.

The Internet has emerged as a major force in shaping telecommunications. However, many key issues remain to be solved. Among them are funding, security, control of obscene material, and the impact of high-bandwidth applications such as voice and video traffic. The original concept of the Internet as a means of open information interchange among government, universities, and scientists has been swamped by the deluge of the World Wide Web.

Access to the Internet is viewed as a right in many businesses. It is also becoming essential for businesses when keeping in touch with their public and customers. E-mail is the most widely used function/application on the WWW. It utilizes Internet Relay Chat for "live" conversations with other users. Some typical uses of the Internet are as follows:

- Communications (E-mail)
- Downloading software
- Interactive discussions (chat room)
- E-commerce
- Information lookup
- Online research
- Real-time audio and video
- Using another computer

An intranet is a network that is contained within an organization. Organizations use intranets to share information across departments and widespread offices. Because intranets provide company-wide access to proprietary information, such an intranet must be protected from outside intruders. It is typical for companies to install firewalls around their intranets to protect their data resources. A firewall is a term that refers to special hardware and software that protect the intranet from unauthorized outside access.

An extranet is a network outside of an organization, yet it has access to the organization's intranet. An extranet facilitates intercompany relationships. Extranets typically link companies and businesses with their customers, suppliers, and partners over the Internet.

Key Terms

ARPANET

Browser

Cache

Computer Telephone Integration (CTI)

Cookies

Domain

Domain Name System (DNS)

Electronic Data Interchange (EDI)

Extranet

Extranet VPN

File Transfer Protocol (FTP)

Firewall

HyperText Markup Language (HTML)

HyperText Transfer Protocol (HTTP)

Internet

Internet Assigned Numbers Authority (IANA)

Internet Service Provider (ISP)

Intranet

Intranet VPN

Java

JavaScript

Network Service Provider (NSPs)

Newsgroup

Packet

Search Engine

Simple Mail Transport Protocol (SMTP)

Telnet

Transmission Control Protocol/Internet Protocol (TCP/IP)

Trivial File Transfer Protocol (TFTP)

Tunneling

Universal Resource Locator (URL)

USENET

Virtual Private Network (VPN)

Web

Workflow

World Wide Web (WWW)

REVIEW QUESTIONS

1. Where did the Internet originate? What were the different organizations involved?

2. What are the new protocols that arrived with the Internet?

3. What functions and services are provided by ISPs and NSPs?

4. What is the Web? What can you do on the Web?

5. What is the function of a Web Browser? Name several browsers.

6. What is HTML and how is it used?

7. What is push technology and how is it used?

8. What is the function of a cookie?

9. Why do you need a URL? Give an example.

10. What is the function of a search engine? Name several.

11. What is the purpose of the domain name system? Provide a list of the various Internet domains in the United States.

12. How is tunneling utilized with the Internet?

13. What are the relationships between the OSI model and the internet layers?

14. How is the Internet Protocol used in the Internet? What are the two responsibilities?

15. Give an example of an IP address. How is the IP address constructed? What are the components?

16. What are some broadband technologies that can be utilized with the Internet?

17. Provide a list of Internet applications.

18. What are the primary purposes for the Internet?

19. Name six Internet services that are available.

20. Provide a list of the Internet search tools.

21. How does Gopher differ from FTP?

22. What is SMTP and how does it relate to e-mail?

23. Explain the use of electronic mail.

24. Describe the file transfer capability of the Internet.

25. Give an overview of Gopher.

26. Who would be interested in the Newsgroups service? Give examples of the information available.

27. Why is Telnet important to network technicians?

28. Provide a list of applications for the World Wide Web.

29. Why do people use the WWW?

30. How is the WWW utilized?

31. Describe the Intranet environment.

32. What is the function of a firewall?

33. What applications would be accessible over an intranet?

34. What is Workflow and GroupWare? Why are they useful to business applications?

35. Describe the extranet environment.

36. What are the differences between intranets and extranets?

37. What is EDI? How is it used?

38. What is CTI? Why is it useful to businesses?

39. What is Web page publishing?

40. Where is the use of the Web and Internet headed?

ACTIVITIES

1. Prepare a list of proposal elements to present a solution for a customer or an organization that currently has Internet access but does not have intranet or extranet access. Develop a solution that would fit one of these applications. Be sure to list tangible and intangible benefits.

2. This exercise involves identifying several elements that exist in the Internet environment.

- Identify customers/users who might benefit from Internet access.
- Identify those who already have Internet access, but might be underutilized.
- Identify opportunities that might match the applications presented in this chapter.
- What other applications can be identified that might ride the Internet Service?

- What additional questions might be used to generate opportunities to sell Internet services?

Note: This activity can best be conducted during the lecture session of a class. It is most effective to divide into workgroups of four or five members. Utilize brainstorming and boarding activities to generate a lot of interaction.

3. To study major network components, draw an enterprise network diagram showing the following components with some type of connectivity: WAN switch, LAN switch, router, hub, bridge, gateway, firewall, modem, DSU, and NT-1.

4. Use the CTI graphic and ask students to extend it to utilize EDI by adding intranet/extranet access.

5. Utilize DataPro, the Web, or other sources to identify Internet Service Providers.

6. Utilize the same sources in question 5 to identify Network Service providers.

7. Identify any industry standards that would apply to the Internet/intranet/extranet environment.

8. Research and develop a five-minute overview of a topic that relates to the Internet environment. Present this topic with a graphic.

CASE STUDY/PROJECT

Case Study 1

Situation

LoTofStuff is a telemarketing organization that is part of a toy manufacturing company. The toy manufacturing company headquarters and production facility is located in Nashville, Tennessee. The telemarketing organization has locations in ten major cities across the continental United States. Each of these telemarketing locations is served by a private LAN. The company headquarters has access to the Internet via an ISP. You have an interview with the CEO of the headquarters organization.

Details

As far as you know the company does not have a Web presence. LoTofStuff has been losing market share to a competitor. The CEO does not have a PC on the desk. There is an IBM mainframe located at the company headquarters. The CIO reports to the CFO.

Requirements

Determine a method of approach for your presentation. This will entail the development of a series of questions that are designed to solicit information that will be used to design the network solution. Present benefits of a Web presence on the Internet. Develop a network design that would support a Web presence. Utilize a graphic package to depict this design. If you are HTML literate, develop a sample home page.

Case Study 2

Situation

National Pharmacies, Inc. (NPI) is a drug warehouse supplier for the neighborhood pharmacy. NPI currently supplies products to 1,000 different pharmacies located primarily east of the Mississippi River. The marketing VP has asked for your assistance in identifying how the Internet can be utilized to increase the sales volume for NPI.

Details

- None of the neighborhood pharmacies currently have Internet access.
- Many of the pharmacies complete a manual order for drugs.
- Many of the pharmacies call in an order to the NPI operation.

- Most of the pharmacies complete at least one order per day (6 days/week).
- Emergency orders are frequent.
- Drug recalls are possible.
- Drug interaction/problem notifications are frequent.

Requirements

- Develop an integrated extranet/intranet solution that would provide for a real-time E-Commerce application.
- Create a diagram of the proposed solution. Produce a graphic of this design.
- Elaborate on the advantages of vendor service and support.

PART FOUR

CHAPTER 14
Network Management

Chapter 14 presents an overview of network management and provides a look at the various network management systems. Described are the network management activities that include monitoring, reporting, and controlling the network. The chapter discussion includes network management architecture and the network management system requirements. The chapter presents considerable information on CMIP and SNMP protocols, as well as the management information base (MIB) and the abstract syntax notation system. Other topics include network management standards, the protocol data units, and network management products and services. The chapter concludes with a discussion about Remote Network Monitoring (RMON).

CHAPTER 15
Problem Solving and Troubleshooting

Chapter 15 is oriented toward the issues of prevention or troubleshooting when managing an enterprise network. The chapter details the pros and cons for each approach to network management and the implications for the grade of service that will be provided to the network users. Information includes techniques and procedures for maintaining network integrity, network troubleshooting tools for both hardware and software solutions, and in-depth details for troubleshooting and problem-solving a number of common and not-so-common network issues. This section also includes information specific to the broadband technologies as presented in this book.

Network Management

■ INTRODUCTION

Any network, whether it is a LAN, MAN, or WAN, is really a collection of individual components working together. Network management helps maintain this harmony, ensuring consistent reliability and availability of the network, as well as timely transmission and routing of data. A **Network Management System (NMS)** is defined as the systems or actions that help maintain, characterize, or troubleshoot a network. The three primary objectives to network management are to support systems users, to keep the network operating efficiently, and to provide cost-effective solutions to an organization's telecommunications requirements.

A large network cannot be engineered and managed by human effort alone. The complexity of such a system dictates the use of automated network management tools. The urgency of the need for such tools and the difficulty of supplying such tools increase if the network includes equipment from multiple vendors and manufacturers.

Network management can be accomplished by utilizing dedicated devices, by host computers on the network, by people, or by some combination of these. No matter how network management is performed, it usually includes the following key functions:

- Network Control
- Network Monitoring
- Network Troubleshooting
- Network Statistical Reporting

These functions assume the role of network watchdog, boss, diagnostician, and statistician. These functions are closely interrelated and are often performed on the same device.

This chapter begins with an overview of network management, with a focus on hardware and software tools, and organized systems of such

OBJECTIVES
Material included in this chapter should enable you to:

- understand the network management issues related to today's enterprise networking environment. Understand the function and interaction of network management in the enterprise network.

- become familiar with the resources that are available to accomplish network management in the enterprise networking environment. Look at the activities of monitoring, reporting, and controlling.

- understand the use of network management techniques for problem solving in the LAN, MAN, and WAN environments.

- identify network management software for troubleshooting LAN, MAN, and WAN networking environments.

network management system (NMS)

- become familiar with the various standards associated with the Network Management System environment. Look at the key areas of network management as proposed by the ISO.
- understand the relationship between a MIB, an agent, and Simple Network Management Protocol (SNMP). See how SNMP plays a major part in network management. Look at the structure of a MIB.
- become familiar with the aspects of remote network monitoring (RMON).

tools, that aid the human network manager in this difficult task. Considerable information concerns the requirements for network management, the general architecture of a network management system, and the Simple Network Management Protocol (SNMP), which is the standardized software package for supporting network management. The chapter concludes with a presentation on network management support systems and network management products and services, including NetView, OpenView, SunNet Manager, and RMON.

14.1   NETWORK MANAGEMENT

The term network management has traditionally been used to specify real-time network surveillance and control; network traffic management; functions necessary to plan, implement, operate, and maintain the network; and systems management [Stamper, 1999].

Network management supports the users' needs for activities that enable managers to plan, organize, supervise, control, and account for the use of interconnection services. It also provides for the ability to respond to changing requirements, such as ensuring that facilities are available for predictable communications behavior and providing for information protection, which includes the authentication of sources and destinations of transmitted data.

The task of network management involves setting up and running a network, monitoring network activities, controlling the network to provide acceptable performance, and assuring high availability and fast response time to the network users.

Network equipment manufacturers have developed an impressive array of network management products over the past several years. As with most telecommunications technologies, network management began as a proprietary system, but the real growth started when the Internet community introduced Simple Network Management Protocol (SNMP) in the late 1990s.

14.2   NETWORK MANAGEMENT SYSTEMS

An effective network management system requires trained personnel to interpret the results. An effective network management system will include most of the following elements:

- An inventory of circuits and equipment
- A trouble report receiving and logging process
- A trouble history file
- A trouble diagnostic, testing, and isolation facility and procedures
- A hierarchy of trouble clearance and escalation procedures

- An activity log for retaining records of all major changes
- An alarm reporting and processing facility

Not every network management system will have all of these elements, but most systems will contain the above functions to some degree. The more complex the network, the more likely the functions will be automated on a mechanized system.

Network Management System Functions

Network management is a collection of activities that are required to plan, design, organize, maintain, and expand the network. A network management and control system consists of a collection of techniques, policies, procedures, and systems that are integrated to ensure that the network delivers its intended functions. At the heart of the system is a database of information, either on paper or mechanized. The database consists of several related files that allow the network managers to have the information they need to exercise control over its functions. A network control system has five major functions:

- Managing network information
- Managing network performance
- Monitoring circuits and equipment on the network
- Isolating trouble when it occurs
- Restoring service to end users

14.3 NETWORK MANAGEMENT ACTIVITIES

Network management is generally concerned with monitoring the operation of components in the network, reporting on the events that occur during the network operation, and controlling the operational characteristics of the network and its components.

Monitoring

Monitoring involves determining the status and processing characteristics currently associated with the different physical and logical components of the network. Depending on the type of component in question, monitoring can be done either by continuously checking the operation of the component or by detecting the occurrence of extraordinary events that occur during the network operations.

Reporting

The results of monitoring activities must be reported, or made available, to either a network administrator or to network management software operating on some machine in the network.

Controlling

Based on the results of the monitoring and reporting functions, the network administrator or network management software should be able to modify the operational characteristics of the network and its components. These modifications should make it possible to resolve problems, improve network performance, and continue normal operation of the network.

14.4   NETWORK MANAGEMENT ARCHITECTURE

Most network management architectures use the same basic structure and set of relationships. Managed devices such as routers and other network devices run software that enables them to send alerts when they recognize problems or predefined situations that might warrant attention. Upon receiving these alerts, management entities are programmed to react by executing some action including operator notification, event logging, system shutdown, or automatic attempts at system repair.

Management entities can also poll end-network devices to check the values of certain variables. Polling can be automatic or user initiated, but agents in the managed devices respond to all polls. Agents are software modules that first compile information about the managed devices where they reside, then store this information in a management database, and then provide it to management entities within the NMS. These subjects will be discussed in detail later in this chapter.

Network Management Requirements

An important requirement for enterprise networking is the ability to manage large networks that consist of hardware and software from a number of providers and manufacturers. This section describes the protocols and standards that are available to the enterprise network manager.

simple network management protocol (SNMP)
common management information protocol (CMIP)

Well-known network management protocols include the **Simple Network Management Protocol (SNMP)** and **Common Management Information Protocol (CMIP).** Management proxies are entities that provide management information on behalf of other entities. Figure 14–1 depicts a typical network management architecture. Following are brief descriptions of each of the SNMP architectural components:

- Network Management Station (NMS)—an end system in the network that executes a network management application. A system administrator typically uses a NMS to monitor and control the network.
- Network Management Application (NMA)—the program running in a network management station that monitors and controls one or more network elements.

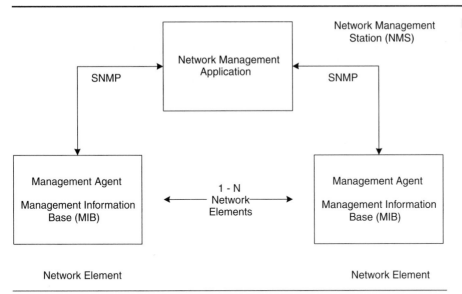

Network Management
Station (NMS)

Figure 14–1
Management Architecture

* Network Element—a component in the network that maintains a
  management agent and a portion of the **Management Information
  Base (MIB)** that contains manageable objects. Examples of network elements are end systems and routers.
* Management Agent—a program running in a network element
  that is responsible for performing the network management functions requested by a network management application. A management agent operates on the objects stored in the portion of the
  MIB that is maintained in that network element.

management information
base (MIB)

14.5   COMMON MANAGEMENT INFORMATION PROTOCOL

As mentioned, the International Standards Organization's approach to network management is the Common Management Information Protocol
(CMIP). This protocol defines the notion of objects, which are elements to be
managed. The key areas of network management as proposed by the ISO
are divided into five **Specific Management Functional Areas (SMFAs);**
which include Fault Management, Configuration Management, Performance Management, Security Management, and Accounting Management.

specific management
functional areas (SMFAs)

Fault Management
Fault management concerns the ability to detect, isolate, and correct abnormal conditions that occur in the network environment. Central to the

definition of fault management is the fundamental concept of a fault. Faults are to be distinguished from errors. A fault is an abnormal condition that requires management's attention to fix. A fault is usually indicated by a failure to operate or by excessive errors. It is usually possible to compensate for errors using the error-control mechanisms of the various protocols.

Because faults cause downtime or unacceptable network degradation, fault management's is the most widely implemented of the ISO network management elements. Fault management involves (1) determining symptoms and isolating the problem, (2) fixing the problem and testing the solution on all-important subsystems, and (3) recording the detection and resolution of the problem.

Configuration Management

Modern data communications networks consist of individual components and logical subsystems that can be configured to perform many different applications. Configuration management is concerned with the ability to identify the various components that comprise the network configuration; to process additions, changes, or deletions to the configuration; to report on the status of components; and to start up or shut down any or all parts of the network.

Each network device has a variety of version information associated with it. A workstation could have version information concerning the operating system, communications hardware and software, and network protocols. Configuration management subsystems store this information in a database for easy access. When a problem surfaces, this database can be searched for clues that may help solve the problem.

Performance Management

Data communications networks are composed of many and varied components, which must intercommunicate and share data and resources. Performance management concerns the ability to evaluate activities of the network and to make adjustments to improve the network's performance. Performance variables that might be provided include network throughput, user response times, and line utilization.

Performance management involves three main steps:

- Performance data are gathered on variables of interest to network administrators.
- The data are analyzed to determine normal (baseline) levels.
- Appropriate performance thresholds are determined for each important variable so that exceeding these thresholds indicates a situation requiring attention.

Performance management of a computer network comprises two broad functional categories: monitoring and controlling. Monitoring is the function that tracks activities on the network. The controlling function enables performance management to make adjustments to improve network performance.

Management entities continually monitor performance variables. When a performance threshold is exceeded, an alert is generated and sent to the network management system. Performance management also permits proactive methods using network simulation software. Such simulation can alert administrators to impending problems so that counteractive measures can be taken.

Two measurements that are necessary in performance management are system effectiveness and system availability. A system is effective if it provides good performance, is available when needed, and is reliable when it is used.

Security Management

Security management concerns the ability to monitor and control access to network resources, including generating, distributing, and storing encryption keys. Passwords and other authorization or access-control information must be maintained and distributed. Security management is involved with the collection, storage, and examination of audit records and security logs.

Security management subsystems work by partitioning network resources into authorized and unauthorized areas. For some users, access to any network resources is inappropriate, because such users are usually company outsiders. For internal users, access to information originating from a particular department, such as payroll, is inappropriate.

Security management subsystems perform several functions. They identify sensitive network resources and determine mappings between sensitive network resources and user sets. They also monitor access points to sensitive network resources and log inappropriate access to sensitive network resources.

Accounting Management

Accounting management concerns the ability to identify costs and establish charges related to the use of network resources. In many enterprise networks, individual divisions or cost centers are charged for the use of network services. These are internal accounting issues rather than actual cash transfers, but are important to the participating users. The network manager needs to be able to track the use of network resources to ensure that the network is functioning efficiently and effectively.

Steps toward appropriate accounting management include:

- Measuring utilization of all-important network resources
- Analysis of the results for insights into current usage patterns
- Setting of usage quotas

- Corrections made to reach optimal access practices
- Ongoing measurement of resource use to yield billing information

Thirteen CMIP supporting functions are mapped onto the five SMFAs. These supporting functions are:

Access Control	Security-alarm Reporting
Accounting Meter	Security-audit Trail
Alarm Reporting	State Management
Event-report Management	Summarization
Log Control	Test Management
Object Management	Workload Monitoring
Relationship Management	

14.6   SIMPLE NETWORK MANAGEMENT PROTOCOL

simple network management protocol (SNMP)

Simple Network Management Protocol (SNMP) is the means by which the management station and the managed nodes exchange information [Rose, 1996]. In 1988, SNMP was defined by the Internet Engineering Task Force (IETF). Although SNMP has several deficiencies, it has become the de facto standard for the management of data networks. *SNMPv2* was produced in 1993 to tackle the deficiencies. Work is underway on the next generation, called *SNMPv3*.

SNMP was originally designed to provide an easy-to-implement, but comprehensive, approach to network management in the TCP/IP environment. SNMP, however, is now being applied in networks that conform to many other network architectures in addition to TCP/IP. SNMP is an industry-standard protocol that is supported by most networking equipment manufacturers. In a network environment like the one depicted in

agents

Figure 14–2, software **agents** are loaded on each managed network device that will be using SNMP. Each agent monitors network traffic and device status and stores information in a MIB.

To use the information gathered by the software agents, a computer with a SNMP network management program must be present on the network. This management station communicates with software agents and collects data stored in the MIB component on the managed network devices. This information is then combined with the information obtained from all networking devices. Statistics and charts are then generated detailing current network conditions. With most SNMP managers, thresholds can be set, and alert messages generated for network administrators when thresholds are reached or exceeded.

SNMP defines the formats of a set of network management messages and rules by which messages are exchanged. Many network components can be managed using SNMP. Through their software agents, it is possi-

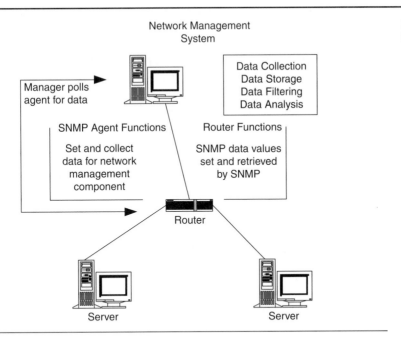

Network Management
System

Manager polls
agent for data

Data Collection
Data Storage
Data Filtering
Data Analysis

SNMP Agent Functions

Set and collect
data for network
management
component

Router Functions

SNMP data values
set and retrieved
by SNMP

Router

Server   Server

Figure 14–2
SNMP Environment

ble to configure networking devices, and in some cases, reset them from the network management station. SNMP can manage network devices such as bridges, routers, and servers. A SNMP network management program can interrogate these devices, and make configuration changes remotely to help managers control their networks from a single application from a central location.

The SNMP approach to network management was developed at a time when considerable work had already been done concerning the Common Management Information Protocol (CMIP) approach to network management. The developers of SNMP used the same basic concepts regarding how management information should be described and defined as those concepts that were developed for CMIP.

SNMP has a number of advantages over CMIP. An important one is that it operates independently on the network, which means that it does not depend on a two-way connection at the protocol level with other network entities. This feature enables SNMP to analyze network activity without depending on faulty information from a failing node. SNMP also has an advantage in that management functions are carried out at a centralized network management station.

SNMP is based on the premise that all devices on a network are able to provide information about themselves. SNMP includes a **Structure of Management Information (SMI)** document that defines the allowable

structure of management
information (SMI)

object identifier (ID)
abstract syntax notation
one (ASN.1)

data types for the MIB. The MIB is a hierarchical structure of information relevant to the specific device and is defined in object-oriented terminology as a collection of objects, relations, and operations among objects. Each object has an **Object Identifier (ID)** to uniquely identify it among its siblings. Like ISO CMIP, SNMP uses **Abstract Syntax Notation One (ASN.1)** to define and identify objects in a MIB.

SNMP Components

SNMP provides a means to monitor and control network devices, and manage configurations, statistics collection, performance, and security. SNMP is an application layer protocol that facilitates the exchange of management information between network devices. Three versions of SNMP exist: SNMPv1, SNMPv2, and SNMPv3.

The original version of SNMP had some shortcomings that were addressed in a second version, called SNMPv2. One of the most important of these shortcomings was SNMP's lack of strict security measures. When SNMP is used, the community name is sent by the NMS without encryption. If captured, this password can be used to gain access to sensitive network management commands, providing the ability to remotely configure a managed network device, such as a router or gateway, comprising the security on a network. SNMPv2 provides an encrypted community name (password), improved error handling, and multiprotocol support. It also adds support for IPX and Appletalk, and provides the ability to retrieve more **MIB-II** information at one time.

MIB-II

Many of the deficiencies in SNMP were corrected with SNMPv2; however, additional enhancements were made in the latest version, SNMPv3. SNMPv3 was issued as a set of proposed standards in January 1998. This set of documents does not provide a complete SNMP capability, but rather defines an overall SNMP architecture and a set of security capabilities. These capabilities are intended to be used with the existing SNMPv2.

SNMP and SNMPv2 can be used to monitor LANs, MANs, and WANs. An important SNMP-based tool used to monitor LANs connected through WANs is **Remote Network Monitoring (RMON),** an IETF standard developed in the early 1990s.

remote network monitoring (RMON)

RMON not only employs SNMP, but also incorporates a special database for remote monitoring, called RMON MIB-II. This database enables remote network nodes to gather network analysis data at virtually any point on a LAN or WAN. The remote nodes are agents or probes. Information gathered by the probes can be sent to a management station that compiles it into a database. RMON/MIB-II standards are currently in place for FDDI, Ethernet, and Token Ring networks.

As presented in the network management architecture section, a SNMP-managed network consists of four key components. The relationship between these components is shown in Figure 14–3.

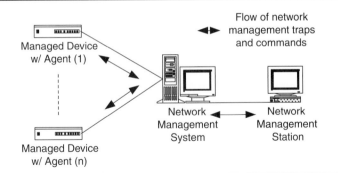

Figure 14–3
SNMP Key Components

These components interact with each other to provide the data and information necessary to manage the enterprise network. To understand how this works, a discussion follows that shows the interaction between the managed device, the agent, the network management system, and the network management station.

A *managed* device is a network node that contains an SNMP agent and resides on a managed network. Managed devices collect and store management information and make this information available to the network management systems using SNMP. Managed devices, called network elements, can be routers, access servers, switches, bridges, wiring hubs, and computers.

A managed object is a variable of a managed node. This variable contains one piece of information about the node. Each node can have several objects. The management station monitors and controls the node by reading and changing the values of these objects.

An agent is a network management software module that resides in a managed device. An agent has local knowledge of management information and translates that information into a form compatible with SNMP. Each agent keeps a database of information, including the number of packets sent, the number of packets received, packet errors, the number of connections, and others. An agent's database is MIB. The agent responds to requests for information from a management station for actions from the management station. The agent may also provide the management station with important, but unsolicited, information. The agent can either be resident on the node or be a proxy residing elsewhere that acts on behalf of the node.

The network management system consists of hardware and software implemented in existing network components. The software used in accomplishing the network management task resides in the host computers or servers and communications processors. A network management system is designed to view the entire network as a unified structure. It contains addresses and labels assigned to each network interface and the specific attributes of each element and link known to the system. HP Openview is an example of such a software system.

The network management station is typically a stand-alone device that serves as the interface for the human network manager into the network management system. The management station contains the following components:

- An interface by which the network manager can monitor and control the network
- The capability of translating the network manager's requirements into the actual monitoring and control of remote network elements
- A set of management applications for data analysis and fault recovery
- A database of management information extracted from the databases of all the managed entities in the network

In a traditional centralized network management scheme, the network management station is located at the central site. Remote locations are accessible to the network management system via a number of routers. MIB agents are resident in the routers and other managed devices that are components on the various topologies. There may be, however, one or two other management stations in a backup role. As networks grow in size and traffic load, the centralized approach to network management becomes unworkable. Figure 14–4 depicts an example of a complex distributed network management configuration. This decentralized approach provides for two network management stations. This type of architecture spreads the processing burden and reduces total network traffic.

SNMP Management

SNMP is a distributed-management protocol. A system can operate exclusively as with a network management station or an agent, or it can perform the functions of both.

SNMP Security

SNMP lacks any authentication capabilities, which results in vulnerability to a variety of security threats. Because SNMP does not implement authentication, many vendors do not implement "Set" operations, thereby reducing SNMP to a monitoring facility.

SNMP Operations

SNMP is a simple request/response protocol. Responses to messages from agents do not have a set send/receive structure. The sender can send multiple requests without receiving a response.

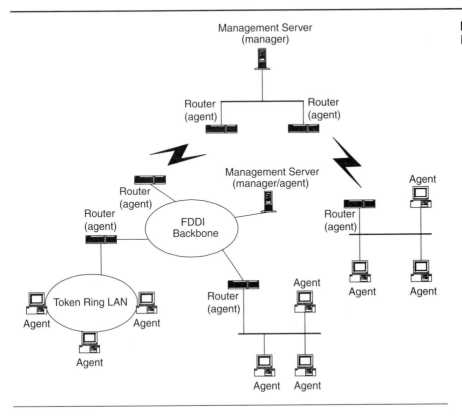

Figure 14–4
Distributed NMS

Six SNMP operations are defined:

- Get—Allows the management system to retrieve a piece of management information from the agent.
- GetNext—Allows the management system to retrieve the next piece of management information from a table or list within an agent. In SNMPv1, when a management system wants to retrieve all elements of a table from an agent, it initiates a Get operation, followed by a series of GetNext operations.
- GetBulk (SNMPv2)—New for SNMPv2, this operation makes it easier to acquire large amounts of related information without initiating repeated GetNext operations. GetBulk was designed to eliminate the need for GetNext operations.
- **Set**—Allows the management system to set values for object instances within an agent.
- **Trap**—Used by the agent to asynchronously inform the enterprise management software of some event. The SNMPv2 trap message is designed to replace the SNMPv1 trap message.

set

trap

Figure 14–5
Manager-Agent
Communication Through
SNMP

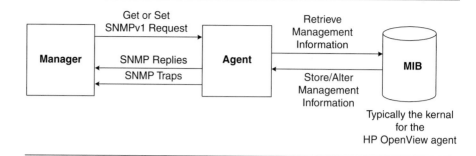

- Inform (SNMPv2)—New for SNMPv2, this operation was added to allow one management system to send trap information to another.

These commands are used by enterprise management software and are the basis of SNMP management. Enterprise management software must be able to contact each device that is managed. SNMP agents may be a software processor or be built into hardware microcode. Regardless, it will be listening and it will have the ability to respond to the SNMP commands. Figure 14–5 shows a simplified diagram of the manager-agent interactions. The process is as follows:

- The manager issues a Get or Set SNMP request.
- The agent retrieves information from the management information database.
- The agent replies with SNMP replies and traps.

Management Information Base

A MIB is a collection of information that is organized hierarchically. MIBS are accessed using a network management protocol such as SNMP. They are comprised of managed objects and are identified by object identifiers. Table 14–1 provides information on the MIB variables.

Devices that support the MIB are manageable by SNMP. The names for all objects in the MIB are defined by either the Internet-standard MIB or by other documents that conform to SMI conventions. Each "instance" of an object type is defined by a unique variable name.

A managed object is one of any number of specific characteristics of a managed device. Managed objects consist of one or more object instances, which are essentially variables. The two types of managed objects are scalar and tabular. Scalar objects define a single object instance. Tabular objects define multiple related object instances that are grouped in MIB tables. Figure 14–6 provides an example of the scalar and tabular format using RFC1157 Trap-PDU.

Table 14–1
MIB Variables

MIB Variables	Purpose
Address Translation group—AT	Converts network addresses to subnet or physical addresses
Electronic Gateway Protocol group—EGP	Provides information about nodes on the same segment as the network agent
Interfaces group—interface	Tracks the number of network NICs and the number of subnets
Internet Control Message Protocol group—ICMP	Gathers data on the number of messages sent and received through the agent
Internet Protocol group—IP	Tracks the number of input datagrams received and the number rejected
SNMP group	Gathers data about communications with the MIB
System group—system	Contains information about the network agent
Transmission group	
Transmission Control Protocol group—TCP	Provides information about TCP connections on the network, including address and time-out information
User Datagram Protocol group—UDP	Contains information about the listening agent that the NMS is currently contacting.

Figure 14–6
SNMP Trap-PDU

```
Trap-PDU ::=
        IMPLICIT SEQUENCE {
                enterprise

                        OBJECT IDENTIFIER

                agent-addr
                        NetworkAddress

                generic-trap
                        INTEGER {
                                coldstart (0),
                                warmstart (1),
                                linkdown (2),
                                linkup (3),
                                authenticationFailure (4),
                                egpNeighborLoss (5),
                                enterpriseSpecific (6)
                        }.
                specific-trap
                        INTEGER,

                time-stamp
                        TimeTicks,

                variable-bindings
                        VarBindList
        }
```

Figure 14–7
MIB Tree

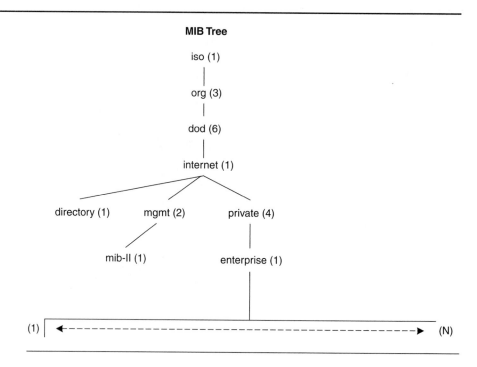

MIB Tree

MIB tree

node

subtree

An object identifier uniquely identifies a managed object in the MIB hierarchy. The MIB hierarchy can be depicted as a tree with a nameless root, the levels of which are assigned by different organizations. Figure 14–7 depicts a portion of the **MIB tree.** Each branch of the tree consists of logical groupings used to generate unique object IDs. The branch is referred to as a **node.** A node can have both "parents" and "children." A node that does not have children is referred to as a leaf node. The leaf node is the actual object. Only leaf nodes return MIB values from agents or have their MIB values altered. A **subtree** is used to refer to all nodes and children under a branch of the tree.

The top-level object MIBs belong to different standards organizations, whereas lower-level object IDs are allocated by associated organizations. An object ID uniquely identifies a managed object in the MIB hierarchy. The MIB hierarchy can be depicted as a tree with a nameless root, the levels of which are assigned by different organizations.

A MIB object is named by concatenating the numerical names of each node when traversing the MIB tree from iso (1) to the particular node. A full object ID name contains all the nodes, including the lead nodes. The nodes are concatenated and separated by periods. This MIB object notation follows the standard notation defined in ASN.1.

Object identifier = 1.3.6.1.4.1.9.3

This decodes as: iso.org.dod.internet.private.cisco.temporary variables

Figure 14–8
MIB Object Identifier
Example

ASN.1 Term	Description
Abstract Syntax	Describes the generic structure of data independent of any encoding technique used to represent the data. The syntax allows data types to be defined and values of those types to be specified.
Data Type	A named set of values. A type may be simple, which is defined by specifying the set of its values, or structured, which is defined in terms of other types.
Encoding	The complete sequence of octets used to represent a data value.
Encoding Rules	A specification of the mapping from one syntax to another. Encoding rules determine algorithmically, for any set of data values defined in an abstract syntax, the representation of those values in a transfer syntax.
Transfer Syntax	The way in which data are actually represented in terms of bit patterns while in transit between presentation entities.

Table 14–2
ASN.1 Notation

Vendors can define private branches that include managed objects for their own products. Figure 14–8 shows an object associated with a Cisco device. The enterprise number associated with Cisco (9) is found in the SMI Network Management Private Enterprise Codes. This IEEE Organizationally Unique Identifier (OUI) assignment list can be found on the Internet or can be obtained from the IEEE Registration Authority.

Abstract Syntax Notation One

Managed objects are named and described using the notation ASN.1. ASN.1 defines a powerful data description notation that allows the format and meaning of data structures to be defined without specifying how those data structures are represented in a computer or how they are encoded for transmission through a network. The advantage to using a notation like ASN.1 to define management information is that it can be described independently of any particular form of information processing technology.

The objective of ASN.1 is the transfer of information from the internal representation of one machine to another machine's internal representation by way of a machine independent abstract syntax. Data types in ASN.1 are defined in a programming language similar to the Pascal programming language.

Table 14–2 provides a list the terms that are relevant to ASN.1. ASN.1 is used to define the format of PDUs, the representation of distributed information, and operations performed on transmitted data.

Figure 14–9
SNMP Message and PDU
Format

SNMP Message Format

Message Header	PDU

SNMP PDU Format

PDU Type	Request ID	Error Status	Error Index	Object 1 Value 1	Object 2 Value 2	Object n Value n

Variable Bindings

14.7 TRAP PROTOCOL DATA UNIT

As mentioned in previous chapters, a Protocol Data Unit (PDU) is a set of data specified in a protocol of a given layer and consists of protocol control information, and possibly user data of that layer. Basically, a PDU is the OSI model terminology for packet. The PDU is also known as a header or a trailer which contains specific information that is sent from one layer on the source computer to the same layer on the destination computer. This type of communication is called peer-to-peer communication. The SNMP message contains the message header and the PDU. The SNMP PDU format and the SNMP message format are depicted in Figure 14–9.

The field descriptions for the SNMPv1 message are:

- Version number—Specifies the version of SNMP used
- Community Name—Defines an access environment for a group of network management stations.

The field descriptions for the SNMPv1 PDU are:

- PDU Type—Specifies the type of PDU transmitted
- Request ID—Associates SNMP requests with responses
- Error Status—Indicates one of a number of errors and error types
- Error Index—Associates an error with a particular object instance
- Variable Bindings—Serves as the data field of the SNMPv1 PDU.

Enterprise	Agent Address	Generic Trap Type	Specific Trap Code	Time Stamp	Object 1 Value 1	Object 2 Value 2	Object n Value n

Figure 14–10
Trap PDU Format

Variable Bindings

Traps

A trap is a message sent without an explicit request by an SNMP agent to a network management station, console, or terminal to indicate the occurrence of a significant event, such as a specifically defined condition or a threshold that has been reached.

Traps are defined in RFC1157. Each SNMP trap contains a generic trap number and a specific trap number. The generic trap numbers range from 0 to 6, where numbers 0 to 5 are defined by SNMP (Figure 14–6). The user can define an enterprise trap by combining the generic enterprise-specific trap 6 with a specific user-generated trap number.

Figure 14–10 depicts the SNMP Trap PDU format. The managed device responds to commands from the network management station with a Get response. When the managed device needs to report an event in response to some "trigger," it initiates a trap message to the network management station. A trigger is some attribute that has been set in the managed device that is being monitored. An example would include such attributes as interface status (up/down) and operational status (errors/traffic).

The fields associated with the Trap PDU are as follows:

- Enterprise—Identifies the type of managed object generating the trap.
- Agent Address—Provides the address of the managed object generating the trap.
- Generic Trap Type—Indicates one of a number of generic trap types.
- Specific Trap Code—Indicates one of a number of specific trap codes.
- Time Stamp—Provides the amount of time that has elapsed between the last network reinitialization and generation of the trap.
- Variable Bindings—Indicates the data field of the SNMPv1 Trap PDU.

14.8   NETWORK MANAGEMENT STANDARDS

Many organizations today deal with network management standardization. The roles played by these organizations range from setting the network management standards to promoting acceptance of the standards. The organizations that play a role in network management include:

- American National Standards Institute (ANSI)
- Corporate for Open Systems (COS)
- International Organization for Standards (ISO)
- Institute of Electrical and Electronic Engineers 802 Committee (IEEE 802)
- Internet Activities Board (IAB)
- International Telecommunications Union Telecommunications Sector (ITU-T)
- National Institute of Standards and Technology (NIST)
- **Open Systems Foundation (OSF)**

open systems foundation
(OSF)

The OSF has developed several network management standards, including DMI, DME, and DCE, as discussed in the following sections.

Desktop Management Interface

The Desktop Management Interface (DMI) is a proposed network management standard being developed by the Desktop Management Task Force, a group of vendors in the desktop computing marketplace. DMI is designed to provide a standardized way to manage the various hardware and software components that are part of a network consisting of primarily personal computers and desktop workstations. DMI is designed to be complementary to standard network management protocols such as CMIP and SNMP.

Distributed Management Environment

The OSF has defined a set of standards for management services called the Distributed Management Environment (DME). DME Services are complementary to those defined by the OSF DCE. DME Services provide common services that are useful to management applications of all kinds. In a distributed computing environment, varying types of hardware and software, located in widely disparate locations, must all interoperate. To achieve efficient and effective operation, management services must be available for monitoring and controlling all the sources connected to the network. DME Services are designed to provide the tools necessary to manage network resources, and include software distribution services, event services, license management services, subsystem management services, and personal computer services.

Distributed Computing Environment

The OSF has developed an important architecture for client-server, network-based computing called the Distributed Computing Environment (DCE). The OSF DCE defines services that fall into many of the OSI application layer service categories such as print, file, electronic mail, directory, and network management services.

DCE is intended to provide a set of standardized services that can be made available across a variety of different system environments, so a distributed application developed using DCE services can support different operating systems, different network software subsystems, and different transport protocols.

The six services included in the DCE are:

- Directory Service
- Distributed File Service
- Distributed Time Service
- Remote Procedure Call Service
- Security Service
- Threads Service

14.9   NETWORK MANAGEMENT SUPPORT SYSTEMS

A Network Management System executes applications that monitor and control managed devices. It provides the processing and memory resources required for network management.

Enterprise network management systems must be able to gather information from a variety of sources throughout the enterprise network and display that information in a clear and meaningful format. One of the difficulties with implementing enterprise network management systems is a lack of interoperability between different software systems. In addition to the network management systems are third-party or vendor-specific (proprietary) applications that provide NMS support.

Performance management information can be communicated to enterprise management systems such as HP OpenView, SunSoft Solstice, or IBM SystemView in the proper SNMP format. Figure 14–11 depicts the interaction between these three systems and third-party network management software. SNMP interfaces with, and provides a flow of network management system information between, these software systems and the agents in the managed devices. Note the common management protocol between the various NMS and an **Application Program Interface (API)** between the NMS and third-party software. API is software that an application program uses to request and execute basic services performed by a computer.

application program interface (API)

Information from the standard MIB is of limited use. The number and diversity of devices necessitate extensions to this MIB. Private MIBs are often provided by hardware manufacturers and can be seen in the SNMP MIB tree structure under Private Enterprises. Each manufacturer is issued a unique number to define MIB information.

Figure 14–11
Network Management
Interaction

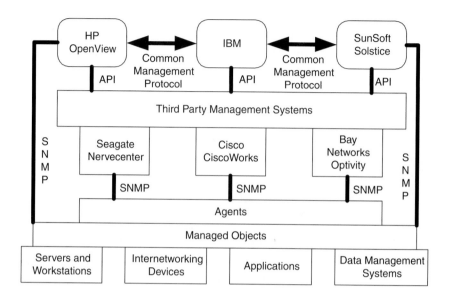

Management software analyzes the data returned by the agent. First a device must be discovered or created, and then it must be determined if it contains an SNMP agent.

MIB definitions are "compiled" into the enterprise management software allowing the management software to query the agent for information specific to that device.

Historically, the SNMP management protocol has been utilized as a monitoring tool. SNMP management stations feature an icon that represents a computing component which turns red to indicate it has gone offline. From there, control is turned over to the application associated with administering the device and making the necessary fix. The monitoring role of SNMP reflects the simplicity of the SNMP message-based protocol.

Unique data requests are used by the administrator to confirm a device's state. Data logging allows operators to gather values for selected attributes over time. Once captured, the data may be displayed or logged for future analysis.

Event requests allow operators to monitor changes in attribute values based on predefined thresholds. If a threshold is exceeded, the management software alerts the operators with audio, visual, e-mail, or programmatic responses.

Traps are events that might represent the crossing of a threshold boundary (e.g., "interface s01 receive errors > 1000") or an event that impacts some MIB variable.

Enterprise management software offers a graphical representation of the network topology. This is a device-based view with little or no connectivity information or monitoring, because the SNMP protocol is device-based, not connection-oriented. The Remote Network Monitoring (RMON) standard allows connection monitoring.

Web-Based Management

The Web-Based Enterprise Management (WBEM) consortium is currently developing a series of standards to enable active management and to monitor network-based elements. Their proposal has been endorsed by over sixty vendors of hardware and systems management software including CA, IBM/Tivoli, and HP. The key purpose of the WBEM initiative is to consolidate and unify data provided by existing management technologies. At the heart of WBEM are these two initiatives:

- The definition of an implementation-independent, extensible, common data description/schema called HyperMedia Management Schema (HMMS), allowing data from a variety of sources to be described and accessed in real time regardless of the data source. HMMS development is under the auspices of the Desktop Management Task Force (DMTF).
- The definition of a standard protocol called HyperMedia Management Protocol (HMMP), over which these data may be published and accessed, allowing enterprise management to be platform independent and physically distributed across the enterprise. HMMP development is under the auspices of the Internet Engineering Task Force (IETF).

14.10 NETWORK MANAGEMENT PRODUCTS AND SERVICES

Three network management products that are commonly used in today's networking environment are NetView, OpenView, and SunNet Manager [Telecommunications Industry Association [TIA], 2000].

NetView

NetView is the name of IBM's network management products. The NetView architecture represents a proprietary view of network management. Many NetView specifications are closely aligned with the ISO CMIP standards for network management. The major functions of NetView include problem management, configuration management, change management, and performance and accounting management. Some of the NetView products include NetView/390, NetView/6000, and LAN NetView.

OpenView

Hewlett Packard's OpenView family of network management products conforms to the SNMP approach to network management and also includes support for OSI CMIP standards [HP Guide, 1995; Huntington, Terplan, & Gibson, 1997]. OpenView products can be used to manage networks conforming to TCP/IP, and also networks that employ Novell NetWare IPX/SPX protocols. OpenView includes a facility that can be used to monitor IBM SNA networks. The OpenView approach to network management provides fault, configuration, and performance management functions.

SunNet Manager

SunNet Manager network management products from Sun Microsystems provide management capabilities for TCP/IP internets. SunNet products support SNMP manageable devices. Sun also provides add-on products that support CMIP standards and handle two-way interface with NetView products for managing IBM SNA networks. SunNet also supports Novell NetWare networks.

Additional network management products and enhancements to the current products are becoming available as the need for network management services increases.

14.11   REMOTE MONITORING

Remote monitoring (RMON) is a standard monitoring specification that enables various network monitors and network management systems to exchange network monitoring data. RMON provides network administrators with more freedom in selecting network-monitoring probes and network management stations with features that meet their particular networking needs.

The RMON specification defines a set of statistics and functions that can be exchanged between RMON-compliant console managers and network probes. RMON provides network administrators with comprehensive network-fault diagnosis, planning, and performance-tuning information. The specifications are set out in RFC1757. This MIB complements the existing MIB-II. Special devices are attached to various subnets in or-

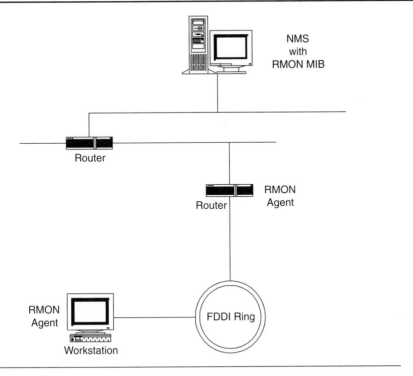

Figure 14–12
RMON Agents

der to collect information specific to the LAN. Examples of the types of data collected include:

- Delivered packets per second
- Number of collisions
- Runt packet numbers (A runt is an Ethernet frame that is shorter than the valid minimum length, usually caused by a collision.)
- Giants (Opposite of a runt.)

Each subnet is required to have an attached monitor collecting data. The devices can be servers, workstations, and routers. These remote monitors need to communicate with one or more network management stations. Figure 14–12 illustrates the components and connectivity to a RMON agent.

Table 14–3 provides information on the functions that are part of the RMON specification. The RMON structure is comprised of nine groupings.

Table 14–3
RMON Specifications

RMON Group	Function	Elements
Statistics	Contains statistics measured by the probe for each monitored interface on the device	Packets dropped, packets sent, bytes sent, broadcast packets, multicast packets, CRC errors, runts, giants, jabbers, and collisions
History	Records periodic statistical samples and stores them for later retrieval	Sample period, number of samples, item(s) sampled
Alarm	Periodically takes statistical samples from variables in the probe and compares them with previously configured thresholds	Includes the alarm table Alarm type, interval, starting threshold, stop threshold
Host	Contains statistics associated with each host discovered on the network	Host address, packets, and bytes transmitted and received; broadcast, multicast, and error packets
HostTopN	Prepares tables that describe the hosts that top a list ordered by one of their statistics	Statistics, host(s), sample start and stop periods, rate base, duration
Matrix	Stores statistics for conversations between sets of two addresses.	Source and destination address pairs and packets, bytes, and errors for each pair
Filters	Enables packets to be matched by a filter equation	Bit-filter type, filter expression, conditional expression
Packet Capture	Enables packets to be captured after they flow through a channel	Size of buffer for captured packets, full status, number of captured packets
Events	Controls the generation and notification of events from the device	Event type, description, last time event sent

■ SUMMARY

Network management includes the subject of how to use various tools to automate much of the effort involved in collecting statistics, querying status, remotely configuring devices, and maintaining other activities involved in the day-to-day operation of a network.

Network management is probably the most critical tool used in the everyday operation of networks. As networks have grown, the availability of tools to effectively manage, query, and control network devices is critical. The three core elements to Simple Network Management Protocol (SNMP), in addition to the protocol definition itself, are:

1. Agents, present in the managed devices themselves
2. Network Management Station (NMS), typically a dedicated computer for this task

3. Management Information Base (MIB), contains the database of network information

The agents are small software components that are placed in the network devices. Essentially, the agent responds to requests from the NMS or responds independently to the NMS. The NMS runs software that builds the MIB through querying network devices. It typically provides a variety of graphical displays of the data that it is collecting. It also allows a method for configuration of any managed network device.

The MIB is described as a collection of objects with unique identifying tags, but really are a collection of data items that are present in any managed object. Each class of network device has its own set of data items or objects.

SNMP has three basic operations, and for the most part it operates as a simple request-response protocol using the MIB and the NMS operation. The NMS requests information exchange from a network device, and the network device responds with one or more of its data values. The NMS can also enable an agent to notify it upon the occurrence of some specific event. These NMS exchanges take the form of three commands: Get, Set, and Notify (Trap).

There are two versions of SNMP on the market, SNMP and SNMPv2. SNMPv2 is an improved protocol with increased security functions. It is backward compatible to SNMP. SNMPv2 provides a way for the agent to verify that the authorized NMS is making a request and allows encryption of the data exchanged. There is a new proposal (SNMPv3) in development that will provide enhancements to SNMPv2.

Key Terms

Abstract Syntax Notation One (ASN.1)

Agent

Application Program Interface (API)

Common Management Information Protocol (CMIP)

Management Information Base (MIB)

MIB-II

MIB Tree

Network Management Station (NMS)

Node

Object Identifier (ID)

Open System Foundation (OSF)

Remote Network Monitoring (RMON)

Set

Simple Network Management Protocol (SNMP)

Specific Management Functional Area (SMFA)

Structure of Management Information (SMI)

Subtree

Trap

REVIEW QUESTIONS

1. What are the primary objectives for network management?

2. What is the function of an enterprise NMS? What is the difference between network availability and reliability?

3. What are the key functions of network management?

4. Define network management.

5. An effective network management system will include what elements?

6. A network control system has five major functions. Describe each.

7. What is the function of CMIP?

8. What is the function of a management agent? Where does the agent reside in a network?

9. What are the ISO's five specific management functional areas?

10. What are the three main steps for performance management?

11. What activities occur at accounting management?

12. What are the thirteen CMIP supporting functions of SMFA?

13. What is SNMP? Why was it designed?

14. What is the basis of SNMP as it relates to managed devices?

15. What are the components of SNMP? Describe each and identify how they relate to each other.

16. How many versions exist of SNMP? Provide an overview of each version.

17. What activities occur at the network management station? Describe the flow of information between a managed network device and the network management station.

18. What are the operations allowed under SNMP?

19. What is a MIB and why is it necessary?

20. What is a trap and how does it relate to RFC1157?

21. What are the generic traps depicted in the trap PDU? What are their values?

22. What is the structure of the MIB tree? How does this relate to ASN.1?

23. What is the relationship between the following terms: agent, MIB, trap, object, and SNMP?

24. What is ASN.1 and what is its function?

25. What is a protocol data unit (PDU)? What are the formats of the PDU?

26. What are the components of a trap PDU format? Why would a network manager want to poll a managed device after receiving a trap?

27. Identify the various standards organizations that are involved in network management.

28. What is the difference between DMI, DME, and DCE standards?

29. What is the significance of the Open Software Foundation (OSF)

30. What is the function of a network management software system?

31. Identify the major network management software systems. How are they similar?

32. What are some of the most important functional characteristics of enterprise network management systems?

33. Differentiate between point products, frameworks, and integrated suites as alternate enterprise network management technology architectures. Does integrated network management require all of the devices in the network to be supplied by a single vendor? Why or why not?

34. What is RMON? What information can be collected by a RMON?

35. What is a distributed network probe and how does it differ from a SNMP agent or a RMON probe? What is the

difference between distributed device management and centralized enterprise network management?

36. What is an object ID (OID)? What is the dotted decimal notation for the first five layers of the MIB tree?

37. What is the difference between MIB and MIB-II?

38. What is the function of SMI?

39. What is a Request for Comment (RFC)?

40. What is the difference between SNMPv1 and SNMPv2? Why is SNMPv3 required?

ACTIVITIES

1. Utilize the Internet to look up the various Requests For Comments (RFCs) that are associated with the specific MIBs. These include RFC1157 and RFC1213. Develop a one page overview of each MIB.

2. Research specific products and product groups, such as routers, and identify whether they are managed devices. DataPro for Telecommunications is one source that can be utilized for the research. Identify in which layer of the OSI model these devices operate.

3. Arrange with a local carrier to visit its Network Management Operations Center. Such centers contain the network management stations that are utilized to monitor and manage the network. Identify the software that is utilized to manage this carrier's network.

4. Look up the IANA database on the Internet and identify the number of organizations that are currently represented. These are located under SMI Network Management Private Enterprise Codes.

5. Schedule a field trip to a network management control center or surveillance site. The major ILECs and IXCs have these facilities in most capital cities or major metropolitan areas. Identify the software that is used to provide the monitoring and surveillance functions.

6. Research network management software topics and develop a five-minute oral presentation, with accompanying graphics.

CASE STUDY/PROJECT

Case studies and projects for network management require a significant amount of hardware and software. Most network management centers will not let outsiders, and only a few insiders, access the system. On-the-job training is the common way to learn how to work in the network management environment. There are, however, activities that can help the student prepare for an assignment in the network environment.

1. Identify a company that utilizes network management and document the hardware and software components of that network. Show the flow of the network management information.

2. Research the current market for network management software. Produce a comparison analysis of cost, features, and functionality.

3. Utilize a graphics package such as Visio to illustrate a complex network management environment. Figure 14–13 is a skeleton of an enterprise network that consists of the

Figure 14–13
Overview of the Enterprise
Network

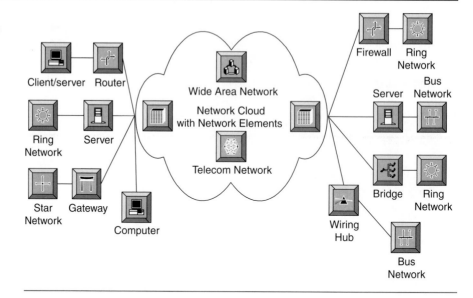

WAN and telephone infrastructure that can be utilized as a basis for this project. The devices that are available for network management access include bridges, PBXs, routers, firewalls, client-servers, and gateways. Note that the devices must contain an agent and have SNMP capabilities.

15

Problem Solving and Troubleshooting

OBJECTIVES
Material included in this chapter should enable you to:

- understand the service issues related to today's enterprise networking environment.

- become familiar with the resources for troubleshooting that are available in the enterprise networking environment.

- look at the various procedures and processes that assist the network manager in preventing and identifying network problems.

- understand techniques for problem solving in the LAN, MAN, and WAN environments.

- identify devices for troubleshooting LAN, MAN, and WAN networking environments. Become familiar with the features of network monitors, network analyzers, and cable testers.

■ INTRODUCTION

Internetworks may involve large-scale, multiprotocol WANs that span multiple time zones or may involve simple single-protocol, point-to-point connections in a local environment. The trend is toward an increasingly complex environment, which involves multiple media types, multiple protocols, and often some connectivity to a private network service or an ISP. It is, therefore, possible that control of the components, throughput, and management of these networks is in someone else's hands. More complex network environments mean that the potential for performance and availability issues in internetworks is high, and the source of problems is often elusive. The keys to maintaining a problem-free network environment, as well as maintaining the ability to isolate and fix a network fault quickly, are documentation, planning, and communication. This requires a framework of procedures and personnel in place before the requirement for problem solving and troubleshooting occurs.

Network problems can typically be resolved in one of two ways, either through prevention or troubleshooting. It is possible to prevent problems from occurring by a program of network planning and management. The alternative is repair and control of damage that can be accomplished by troubleshooting. Network management and troubleshooting should combine to form an overall network plan [Tittel & Johnson, 1998]. This network plan combination should include:

- Cable diagrams
- Cable layouts
- Documentation on computer and network device configurations
- Important files and their layouts
- Listings of protocols and network standards in use
- Network capacity information
- Software

- look at typical network problems and more specific issues relating to broadband technologies. Learn how to identify a wide range of incidents and situations that could impact network performance. Look at issues that are relevant to specific broadband technologies.
- understand how to develop a baseline and become familiar with its importance in the network environment.

Troubleshooting is often characterized as an art that can only be gained from experience in the trenches. This experience, along with some documented methodology, allows the technician to successfully solve difficult network issues. The methodology keeps the troubleshooter from skipping or overlooking the obvious, and the experience helps to draw appropriate conclusions and to check for obscure problems.

Before someone starts this troubleshooting process, it is essential for the network professional to establish a baseline or reference point for system comparisons [Oppenheimer, 1999]. Information should be gathered and documented on the network devices, facility components, network traffic, performance levels, and security systems. This information will prove invaluable later when a problem develops in the network.

Establishment of policies and procedures that apply to the network must be developed during its planning stages and continue throughout the network's life. Such policies should include security, hardware and software standards, upgrade guidelines, backup methods, and documentation requirements. Through careful planning, it is possible to minimize the damage that results from most predictable events and control and manage their impact on the organization.

15.1   MAINTAINING NETWORK INTEGRITY

When troubleshooting a network environment, a systematic approach is the most effective. The administrator should be able to identify problems from symptoms and initiate corrective action based on this systematic approach. If change is not managed, a great deal of time will be spent fire fighting instead of fire preventing. A major part of this structured approach is network management and planning, which can be accomplished by addressing the following issues:

- Data backup
- Documentation procedures and methodology
- Hardware and software standards
- Network baseline
- Preemptive troubleshooting
- Security policies
- Upgrade guidelines

Network Baseline

baseline

A baseline for network performance must be established if network monitoring is going to be used as a preemptive troubleshooting tool. A **baseline** defines a point of reference against which to measure network performance and behavior when problems occur. This baseline is established during a period when no problems are evident on the network. A baseline is useful when identifying daily network utilization patterns, possible network bottlenecks, protocol traffic patterns, and heavy-user usage patterns and time frames. A baseline can also indicate whether a network needs to be partitioned or segmented, or whether the network access speed should be increased. The three components that must be created to establish a baseline are:

- Current topology diagrams
- Response time measurements of regular events
- Statistical characterization of the critical segments

These three components will require some effort in developing; however, the payoff will come when a problem occurs in the network. A small amount of time spent each week by a number of personnel who are assigned portions of the network can accomplish this task in a relatively short time.

Security Policies

All security policies set forth in a network plan should be detailed and followed closely. The security policies will depend on the network size, the organization's security standards, and the value and sensitivity of the data. The security plan must include both physical security and computer security.

Network security can be enhanced by a number of user-name and password requirements and resource access requirements. Standards for user name and passwords include the following suggestions:

- Establish minimum and maximum password lengths for user accounts
- Provide the users with the detailed reference to character restrictions
- Determine the frequency for changing passwords
- Decide if and when passwords can be reused
- Decide if there will be exceptions to the policy

Resource access is generally granted only to those who specifically require it. It is always easier to grant new access to users than to take it away. For dial-in users, special security arrangements will probably be necessary. Many organizations require a security card which provides a code that must be entered for dial-up access. It is essential that the number of users who perform network administration tasks be limited to the absolute minimum. The more users with access to administrative functions, the more likely security problems will occur.

Hardware and Software Standards

To make hardware and software easier to manage, all network components should follow established standards. Several different levels of configurations can be established, depending on the requirements of the desktop users. These standards should cover both hardware and software configurations.

Standards should also be established for networking devices, including manufacturers, and operating systems, including versions. This also includes standards for server configurations and server types.

When establishing hardware and software standards, keep in mind the pace of industry change and obsolescence. Regular evaluations of network standards will be required to ensure that the network remains current.

Upgrade Guidelines

Upgrades for hardware and software and new networking products are a fact of life. It is necessary to establish guidelines for handling these upgrades. They can be handled easier if the user community has advance notice of such an upgrade, and they should not be performed during normal working hours.

It is a good idea to test upgrades through stand-alone platforms or through a group of technically astute network users. When performing upgrades, always have a plan for backing out the upgrade if it fails to perform as expected. Through careful planning and testing, the upgrade process can be relatively painless.

Preemptive Troubleshooting

Preemptive troubleshooting may be costly in the short term; however, it saves time and resources when problems arise, it prevents equipment problems, and it ensures data security. A preemptive approach can prevent additional expense and frustration when trying to identify the causes of failures. The ISO has identified five preemptive troubleshooting network management categories:

- Accounting management—records and reports usage of network resources
- Configuration management—defines and controls network component configurations and parameters
- Fault management—detects and isolates network problems
- Performance management—monitors, analyzes, and controls network data production
- Security management—monitors and controls access to network resources

See Chapter 14 for a detailed explanation of each of these categories.

Data Backup

A comprehensive backup program can prevent significant data losses. A backup plan is an important part of the network plan and should be revised as the needs of the network grow. Six major action parts of this plan include the following:

- Determine what data should be backed up and how often
- Develop a schedule for backing up the data
- Develop a plan for storing data in a secure location
- Identify the personnel responsible for performing backups
- Maintain a backup log listing of what data are backed up
- Test the backup system regularly

Types of backups include the full backup, copies of files, incremental backups, daily copies, and differential (changed) backups. Each of these methods has advantages and disadvantages, so the administrator may elect to use different methods for different data files.

Documentation Procedures

A well-documented network includes everything necessary to review history, understand the current status, plan for growth, and provide comparisons when problems occur. The following list outlines a set of documents that should be included in a network plan:

- Address list
- Cable map
- Contact list
- Equipment list
- Network history
- Network map
- Server configuration
- Software configuration
- Software licenses
- User administration

It is essential to take the time during installation to ensure that all hardware and software components of the network are correctly installed and accurately recorded in the master network record. This documentation also applies to the hardware and software configurations for each of the network components. Being able to quickly find a faulty network component and have documentation on the configuration will save a considerable amount of personnel effort and time.

Documentation should be kept in both hard copy and electronic form so that it is readily available to anyone who needs it. Complete, accurate, and up-to-date documentation will aid in troubleshooting the network, planning for growth, and training new employees.

Methodology

The first step in solving a problem is to collect and document as much information as possible in order to have an accurate description of the

electromagnetic interference (EMI)
radio frequency interference (RFI)

situation. It is necessary to collect peripheral information that might not appear to have any impact on the situation. This includes environmental conditions that could contribute to the problem, such as heat, humidity, and possible sources of **Electromagnetic Interference (EMI)** and **Radio Frequency Interference (RFI).**

If the problem involves user interaction, it will be necessary to check the sequence of steps the user took. Before troubleshooting the physical devices and system software, attempt to recreate or confirm the problem to ensure that the problem is not user (cockpit) error.

From the information gathered for the trouble report, a structured approach can begin to list, prioritize, and examine possible causes of the reported incident. It is essential that the problem be logically evaluated, starting with the most basic causes. The first step would be to check if the network cable and power cable is connected.

After checking the obvious, it would be very useful to have a troubleshooting device, such as a Flukemeter, that can be utilized to check the network cable, the Network Interface Card (NIC), the port (interface) of the device, and the wiring in the installation. Another option, if such a device is not available, is to troubleshoot the situation by using a replacement. Replacement can take longer to troubleshoot a situation because of the time necessary to remove and replace the components to see if this would resolve the problem. Since it is necessary to make only one change at a time, this process could become time consuming and, most likely, a replacement component will not be readily available. Both replacement and network management tools can solve problems, but the network management tool will probably allow for a quicker resolution.

Troubleshooting should progress logically and experience will be the best teacher [Thompson, 2000]. If a single device is malfunctioning, start troubleshooting with that device. If multiple devices are malfunctioning, look for a common cause, such as a centralized wiring hub. If the entire network is malfunctioning, check the network statistics against the baseline. In every case, start with the most basic and work toward the most complex. Once the problem is solved, ensure that documentation is completed and includes a list of the conditions observed, the steps taken to solve the problem, and a summary of the entire service call.

Figure 15–1 illustrates the process flow for troubleshooting and problem solving some network issue. This model assumes that the organization has formally established and documented all elements of the plan as presented in this chapter. This process flow is recursive in that the flow repeats until the problem is solved. The following steps detail the activities that occur at each element of the process flow.

- Definition of the problem: The problem should be defined in terms of a set of symptoms and potential causes. By using the baseline, it

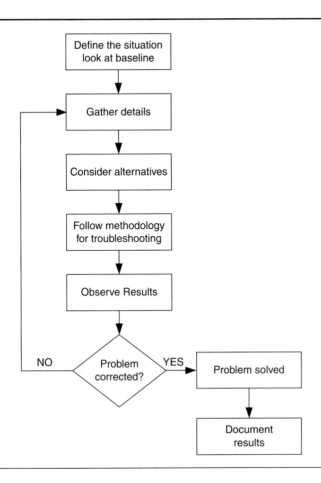

Figure 15–1
Standard Process Flow for Troubleshooting

is possible to identify where the problem resides or at least know where to start the troubleshooting effort.

- Details of the problem: Information can be collected from sources such as network management systems, protocol analyzer traces, network monitors, and network surveillance personnel.

- Alternatives assessment: Now is the time to consider the possible problems based on the facts that have been gathered. The obvious non-issues can be eliminated at this time.

- Problem-solving methodology: An action plan can be developed and the most likely cause of the problem can be identified. The plan will allow for one variable to be changed at a time until all options are exhausted. Changing one variable at a time allows for the reproduction of a given solution for a specific problem. Also, if more than one change is made before a test is made, it will be impossible to determine which change made the problem disappear.

- Results observation: Tests must be made at this module to ensure that the modifications are correcting the problem and not creating other problems. This process reiterates the process for each test and may require additional details for each cycle.
- Problem resolution: When the problem has been solved, the next and last step of the process flow is to document the situation. If the problem has not been solved, the process is repeated.
- Documentation: The final step is important because the problem may well occur again and it is essential that the troubleshooting efforts are not duplicated.

15.2   NETWORK MANAGEMENT ELEMENTS

Some basic infrastructure elements are part of the network troubleshooting and problem-solving environment. First is the requirement to understand the various interface standards. Second, tools, both hardware and software, can be utilized to accomplish the task at hand. Troubleshooting documentation is required to support the use of the various tools. Finally, service level agreements are used to provide a base for the level of service that is required.

Troubleshooting Interface Standards

The importance of the descriptions of the interface standards is that to fix a problem, it first must be located. The easiest way is to isolate on which side of the DCE/DTE interface the problem is occurring. To do this, it is necessary to understand what signals are on what pins of the interface and from which direction they originate. Remember that Data Communication Equipment (DCE) components are devices such as modems and DSUs and Data Terminal Equipment (DTE) is Customer Premise Equipment (CPE).

Two important rules concern interface standards and the cables that implement them:

- The male connector is always associated with the DTE, and the female connector is always associated with the DCE.
- The DTE provider is normally required to supply the cable for interconnection use. These cables are often proprietary components and not available as commodities. The DCE devices are rarely shipped with cables included.

Interface standards that are implemented on many of the devices today include the following:

- EIA-232-D: 25 pin/9 pin
- EIA-449: 37 pin

- EIA-530: 25 pin
- V.35: 34 pin

Network Management and Monitoring

Network management is a collection of activities that are required to plan, design, organize, control, maintain, and expand the communications network. A network control system consists of a collection of techniques, policies and procedures, and integrated systems that are utilized to ensure that the network delivers its intended functions. A network control system provides for five major functions:

- Managing network performance
- Monitoring circuits and devices on the network
- Isolating trouble situations
- Restoring service to end users
- Managing network information

There are many software solutions available to assist with network management. This software can help identify conditions that may lead to problems, prevent network failures, and troubleshoot problems when they occur.

After a baseline for the network has been developed, it will be possible to monitor the network for changes that could indicate potential problems. It is essential to establish what is "normal" in the network so that "abnormal" situations can be readily identified. Network monitoring software can gather information on events, system usage statistics, and system performance statistics. Information gathered from these monitors can assist the network administrator in the following ways:

- Monitoring trends in network traffic and utilization
- Developing plans to improve network performance
- Providing forecasting information for growth
- Identifying those network devices that create bottlenecks
- Monitoring events that result from upgrades

Network Management Tools

Network management tools allow for the performance of management functions such as monitoring network traffic levels, monitoring software usage, finding efficiencies, and finding bottlenecks. Chapter 14 provides an in-depth review of the various components of network management, and should be reviewed before attacking this chapter. Table 15–1 shows the top nine network management platforms currently used to support network management.

Platform	% Utilized
HP OpenView	21.3
IBM NetView (mainframe)	15.7
Sun Soltice/SunNet Manager	11.2
Tivoli Enterprise	7.9
IBM NetView (AIX)	6.7
CA Unicenter TNG	5.6
Compuware Ecotools	3.4
BMC Patrol	3.4
Boole & Baggage Command/Post	1.1
Datapro 1999 Network Management Survey	

This information is the result of data provided to DataPro Information Services late in 1999.

Additional tools utilized to conduct network management operations are:

- CiscoWorks
- Network General Sniffer
- RMON Probes
- Bay Optivity
- Cabletron Spectrum
- Locally developed software

Network managers need a comprehensive set of tools to help them perform the various network tasks. Most tools can be classified as primarily hardware or software, and most hardware test instruments are supported by software. Network management software is categorized into three different types: device management, system management, and application management.

The most common categories of hardware tools for network management are cable testers, network monitors, and network analyzers. Other sophisticated network management tools can be used for daily network management and controls. These management tools, discussed in Chapter 14, typically have three components:

- Agent—the client software part of the management tool. The agent resides on each managed network device.
- Manager—a centralized software component that manages the network. The management software stores the information collected from the managed devices in a standardized database.
- Administration system—the centralized management component that collects and analyzes the information from the managers. Most administration systems provide information, alerts, traps, and the ability to make programmable modifications to the network components.

The two main management protocols used with network management systems are **Simple Network Management Protocol (SNMP)** and **Common Management Information Protocol (CMIP).** SNMP allows management agents that reside on the managed devices to provide information to SNMP management software. SNMP utilizes a management information base (MIB) for maintaining the statistics and information that SNMP reports and uses. Typically stored measurement items include network errors, system-utilization statistics, packets transmitted/received, and numerous other items of information that may be useful for a particular network component.

Network administrators use SNMP to manage devices such as wiring hubs and routers. Management tasks might include:

- Network traffic monitoring
- Remote management capabilities
- Port isolation for testing purposes
- Automatic disconnection of nodes
- Automatic reconfiguration based on time-of-day

Network managers can provide network topology maps, historical management information, traps and alerts, and traffic monitoring throughout the network.

Troubleshooting Documentation
Some type of electronic tracking or journal can be maintained to accumulate troubleshooting and problem-solving information. This will ensure that time will not be wasted repeating work that has already been complete and an audit trail will be developed for each problem. The information developed can also be utilized in requests for additional equipment, personnel, and training. It will also be a useful tool for training future network support personnel.

A typical method for administering such a database is to assign some unique identifier to each problem or trouble report. A trouble report would be generated for each incident and cross-referenced for recurring incidents to the same network element. The trouble report documents issues and requests for service from network users, and can be used to ensure that a consistent troubleshooting methodology is being followed. A typical trouble report might include the following elements:

- A trouble report identifier
- A preliminary description of the situation
- Investigation and analysis of the situation
- The service actions taken to resolve the issue
- A summarization of the incident

Another source of information that can be useful in troubleshooting the network is the data, usually in the form of traps or alerts, which are collected

simple network management protocol (SNMP)
common management information protocol (CMIP)

from the managed network devices. It is essential that data collected from the managed devices be stored so that problems can be tracked and trends can be analyzed.

Service Level Agreements

To determine whether the performance of a network is adequate requires preestablished performance objectives. Performance objectives may be established for the entire network or for a specified set of equipment or users. When the objectives are established for a group of users, such as a department, the objectives are commonly called a **Service Level Agreement (SLA).** Service level agreements often involve stating the number of hours of a day or week that a user's devices or applications will be available, and the reliability of the same components. SLAs also contain a statement about the response time that the user can expect to receive at the device. A SLA must be negotiated between the user and the network and computers operations personnel, and often involves penalties for service outages and poor performance.

service level agreement (SLA)

The performance of a network is typically judged by comparing it with acceptable service levels in the key areas of availability, reliability, response time, and throughput.

mean time between failures (MTBF)

- Availability is the ratio of uptime to total uptime and downtime. **Mean Time Between Failures (MTBF)** is the uptime, or average time the network works properly.

bit error rate test (BERT)
block error rate test (BLERT)
response time

- A data communications network's reliability is usually measured by its ability to pass data without errors. Acceptable error rates are usually specified for each communications link and device in the network. Two measures for reliability include the **Bit Error Rate Test (BERT)** and **Block Error Rate Test (BLERT).**
- **Response time** can be defined as the time between the instant when the terminal user presses the Enter key and when the reply message is received from the host computer. Although response time may be important to a user, a two-second response time is considered adequate.

throughput

- **Throughput** is even more important than response time as a measure of network transmission speed. Throughput is the net bandwidth of a network, which is the number of information bits per second that can be accepted and transmitted by a network.

15.3   NETWORK TROUBLESHOOTING

Many network problems can be solved by first verifying the status of the affected computers or networking components [Badgett, Palmer, &

Jonker, 1999]. Taking several initial steps can help the administrator in resolving network problems. These are:

- Identify possible cockpit problems and user errors
- Ensure that all physical connections are in place
- Verify that the NIC is working
- Warm-start the device (reload the software)
- Cold-start the device (power cycle off and on)

It is essential that a structured approach be taken when troubleshooting. These simple steps include the following:

- Prioritize the problem in relation to all other problems in the network
- Develop information about the problem
- Identify possible causes
- Eliminate the possibilities, one at a time
- Ensure that the fix does not cause other problems
- Document the solution

PSTN Impairments

The PSTN contains several types of noise, including thermal noise, electrical noise, and transients. Noise power is measured by the C-message notch-filter testing technique. This is accomplished by sending a 1004 Hz tone down the circuit. A signal is needed to ensure that all switches and connections are properly configured. At the receiver, both the noise and the tone are present. By using the notch filter to remove the tone, only the noise remains. The signal amplitude can be measured and the Signal-to-Noise Ratio (SNR) calculated. A SNR of 28 dB is considered acceptable.

Point of Presence

The **Point Of Presence (POP)** is the legal boundary for the responsibility of maintaining communications equipment and transmission lines. From a troubleshooting point of view, if a problem has been isolated and it is not on the user's side of the POP, then the carrier or LEC is obligated to provide troubleshooting and repair according to the terms of the maintenance contract. It is, therefore, very important that the user determine that the problem is not part of CPE before calling in a problem to the communications company. It is the policy of many carriers to charge for repair service when it has been determined that the problem was on the user's side of the POP.

point of presence (POP)

The network administrator can use special troubleshooting tools in addition to the all-important experience that must be gained in the trenches. Since many networking problems occur at the lower layers of the OSI model, it is essential that physical layer tools be employed for problem isolation and identification. These tools include digital volt-meters, Time Domain Reflectometers (TDRs), cable testers, oscilloscopes, network monitors, and protocol analyzers.

15.4   TEST EQUIPMENT AND RESOURCES

Test equipment is very important in the development of telecommunications equipment and, in particular, the implementation of communications protocols. First, it is necessary to ensure that protocols work correctly and that the medium is capable of carrying the bits without errors. Second, it is important to test the interoperability of different vendor's equipment, because not all implementations of a "standard" will interoperate. Additionally, during the design of one piece of equipment, it is often necessary to simulate the other network devices with which this device will communicate.

There is no substitute for experience and the proper test equipment when troubleshooting or tuning networks. Techniques that work well in one situation may not provide the same results in others. Changes of one parameter may well have an adverse effect on some other parameter. In many cases it is necessary for empirical data to be collected over a long time period.

Analysis and monitoring of the physical media can take many forms. Break-out boxes can be used to ensure that the interface and physical line are operating properly. A BERT or BLERT can be used to provide information on the number of bit errors and the patterns in which they occur. Physical media testers determine whether an appropriate signal quality is being maintained over a given transmission facility. Media simulators, on the other hand, allow real or prototype devices to be tested over simulated transmission lines.

volt-ohm meter (VOM)

A digital voltmeter or a **Volt-Ohm Meter (VOM)** can be utilized to determine a cable break by measuring resistance in the cable. The VOM can also be used to determine if a ground exists between the central core and the shielding in thinnet or thicknet cabling.

time domain reflectome-ter (TDR)

A **Time Domain Reflectometer (TDR)** can also be used to determine the existence of a short or break in a cable. A TDR is more sophisticated and can be used to pinpoint the distance from the device where the break is located, and to determine the length of installed cables. TDRs are available for both electrical and fiber-optic cables.

oscilloscopes

Oscilloscopes contain a video screen that shows information on signal voltages over time. When used with a TDR, an oscilloscope can help to identify cable breaks, shorts, crimps or sharp bends in a cable, and attenuation problems.

protocol analyzers

Protocol analyzers monitor the traffic being carried on in a facility or test implementations of a protocol. Protocol analysis and monitoring is essential to track the performance of a network and to ensure that different vendor's equipment will interoperate. Protocol analyzers can also be used for conformance testing, which is typically done during the product development cycle, prior to connecting a new device onto a live network, to assess the implementation of a protocol.

Because routers and gateways operate at both LAN and WAN layers of the OSI model, network analyzers must have the capability of capturing all of the protocols used, and present an analysis that determines the health of the network. Problems that occur with routers and gateways, which can be addressed with network analyzers, include the following:

- Filtering problems
- Protocol mismatches
- Routing path problems
- Routing protocol configuration
- Security problems

These analyzers can look at the contents of a data signal and allow an experienced operator to diagnose and troubleshoot problems. There are basically two levels of analyzers: Low-end devices consist of a circuit board that plugs into a personal computer, and expensive stand-alone devices that have their own chassis and monitor. The major functions of these devices are:

- Data trapping
- Device emulation
- Interface lead monitoring
- Performance measurements

Network test equipment has the capability to perform analog tests on both voice and data circuits. These devices can contain multiple function sets and can measure a wide range of network circuit variables and parameters. The following is a list of the principal analog circuit tests that can be performed by these devices:

- Circuit continuity
- Circuit transmission loss
- Capability of the circuit to handle high-speed data signals
- Capability of the circuit to handle data
- Delay of different frequencies propagated over an analog circuit
- Steady-state noise of the circuit
- Frequency response
- Noise burst measurements
- Variation in the phase of a signal on the circuit

Monitoring with SNMP

The Simple Network Management Protocol (SNMP), as presented in Chapter 14, is part of the TCP/IP suite that is used for network management.

Figure 15–2
SNMP Environment

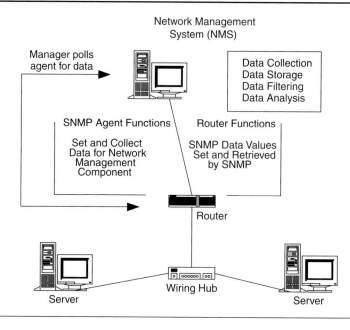

Figure 15–2
SNMP Environment

SNMP is an industry standard that is supported by most networking equipment manufacturers. A common network environment, presented in Figure 15–2, illustrates managed network devices that contain software agents that are managed by SNMP. Each agent monitors network traffic and device status and stores the information in a MIB.

The information collected by the agents is provided to a network management computer system that is present on the network. A management station on the network communicates with the software agents and collects data that has been stored in the MIBs on the managed network devices. The NMS can combine information from all networking devices and generate statistics or graphics detailing current network conditions. SNMP provides for the setting of thresholds and the generation of messages that provide information on the status of the network and the devices.

Network Monitors, Network Analyzers, and Cable Testers

Network monitors, network analyzers, and cable testers each provide the network manager and technician with a different suite of tools. The capabilities of these devices are listed here.

Network monitors provide the following features:

- Collects network statistics
- Provides diagnostic tools
- Stores and opens network packets

Network analyzers provide the following features:

- Collect network statistics
- Use SNMP to query managed devices
- Generate traffic for an analysis
- Analyze network statistics
- Analyze types of devices on the network
- Analyze types of collisions
- Analyze types of errors
- Identify top broadcasting hosts
- Provide diagnostic tools
- Troubleshoot managed network devices

Cable testers provide for the following measurements:

- Wire map—provides the basis for checking improperly connected wires, including crossed wires, reversed pairs, and crossed wire pairs.
- **Attenuation**—the loss of signal power over the distance of the cable. To measure attenuation on a cable, the cable tester must be assisted by a second test unit called a signal injector, or remote.

  attenuation

- **Noise**—unwanted electrical signals that alter the shape of the transmitted signal on a network. Noise can be produced by fluorescent lights, radios, electronic devices, heaters, and air-conditioning units.

  noise

- **Near End Crosstalk (NEXT)**—a measure of interference from other wire pairs. The signal bleeds from one set of wires to the other, causing crosstalk.

  near end crosstalk (NEXT)

- Distance measure—EIA/TIA-568A specifies maximum cable lengths for network media. Cables that are too long can cause delays in transmission and network errors.

Figure 15–3 shows three examples of fiber optic testing equipment.

EIA/TIA and IEEE specify standards for these measurements, which gives a reference point for acceptable measurements. There is a requirement to perform tests on various types of media including twisted pair, coaxial cable, and optical fiber. The tests previously mentioned can be performed with an instrument such as a Fluke DSP-100 Cable Meter. This handheld tester can perform crosstalk measures and use a TDR function to identify faults on the cable. An Optical TDR (OTDR), such as a Tektronix Ranger TFS 3031 OTDR, can be utilized for troubleshooting optical fiber cable. Figure 15–4 shows various pieces of testing equipment, including a LAN meter, LAN tester, and protocol analyzer.

Network Support Resources

Several resources are available for troubleshooting a network problem. These include online services, subscription services, printed documentation,

Figure 15–3
Fiber Optic Testing
Equipment

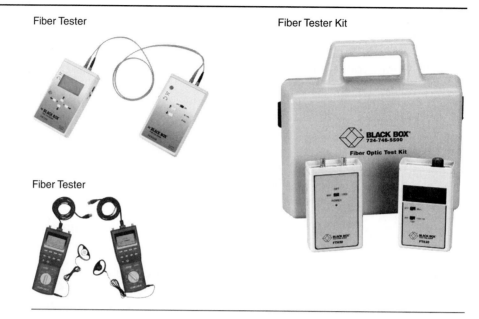

Fiber Tester

Fiber Tester Kit

Fiber Tester

Figure 15–4
LAN Testing Equipment

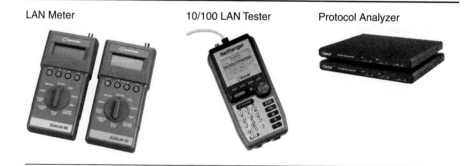

LAN Meter

10/100 LAN Tester

Protocol Analyzer

and software products. Networking journals and magazines available include:

- *Communications Week*
- *Info World*
- *LAN Magazine*
- *LAN Times*
- *Network Computing*
- *PC Magazine*
- *PC Week*

15.5   COMMON PROBLEM AREAS

Four areas are candidates to start the troubleshooting process. These include cabling and related components, power fluctuations, poor network performance, and upgrades [Cisco Systems, 2000]. A common approach when isolating a problem is to use a checklist of elements. Take a structured approach and check the component parts in this order:

- Cables and connectors
- Transceivers or hub ports
- Interface card
- Configuration files
- Driver software
- Network software
- Application software

Cabling and Electrical Components

There are common problems that will occur with the facilities cabling and electrical power components [Thurwachter, Jr., 2000]. A first step might be to consider the following:

- Disconnected network cable
- Disconnected power cable
- Power cable is cut
- Circuit breaker has flipped
- Bad network cable or connector
- Incorrect cable
- Cable is too long
- Improper wiring at the patch panel or wall outlet
- Two many devices on the same circuit
- Noisy devices on the same circuit
- UPS failure
- No UPS
- No surge protector

A good Uninterruptable Power Supply (UPS) can save not only time but also the cost of replacing equipment damaged by voltage surges or brownouts.

Interface Cards and Adjunct Components

If a cabling problem has been ruled out, then look at the physical layer connection devices, which include the following:

- Bad network card
- NIC configuration problem
- IP address conflict

- Faulty NIC slot
- Options set incorrectly on the NIC
- Bad port on the hubbing device
- Defective transceiver

System Upgrade Issues

Since network technology is constantly changing, it is necessary to upgrade network hardware and software frequently. The operating systems on the various servers in the network must be upgraded periodically with new versions of code. During these upgrade processes, it is not uncommon for multiple versions of software to be running in the network. Possible problems with upgrading may include:

- Upgrade was not tested/prototyped before installation.
- Users were not provided with upgrade information.
- Users were not aware that there was an upgrade.
- Wrong version of the software has been loaded on the system.

Network Performance Issues

Network performance is likely to be a concern of the network users. They will be the first to notice a deterioration of service and the first to complain when this happens. It will be necessary to identify the culprit by looking at anything that has recently changed in the network environment. The administrator must determine whether the following occurred:

- Sudden network performance degradation
- Sporadic network performance
- Gradual performance deterioration
- No network throughput

The following questions may assist in the problem-solving effort:

- Has new equipment been added to the network?
- Are there new users on the network?
- Have any new peripherals been added to the network?
- Have any new applications been added to the network?
- Are there any security breaches in the network?
- Are any users playing electronic games on the network?

If new equipment, new applications, or additional users have been added to the network, and it appears that they are having an adverse impact on performance, then it might be time to consider expanding or reconfiguring the network. The addition of network equipment and segmentation of the network might limit and constrain traffic. Higher-speed backbone facilities, additional servers, bridges, gateways, firewalls, and routers are alternatives to consider when the increased capacity has become a performance issue.

15.6 TYPICAL NETWORK PROBLEMS

Typical problems that will reoccur on the enterprise network [Network General Corporation, 1996] include:

- Response time is slow.
- A single station cannot access the network.
- A specific application does not function.
- No stations on a single segment can access another segment.
- No stations on a single segment can access the network.
- Stations on a single segment intermittently cannot access the network.

Each of these situations may be caused by a number of conditions. Several corrective measures are listed for each situation.

Response Time is Slow There are numerous reasons why the network could experience a slow response time, and they become more complex when introducing the components of the WAN. Excessive traffic and/or an infrastructure that is under-engineered is the most common contributing factor to poor network performance.

Slow response to a device on a LAN or a MAN to devices on segments that are remote can be attributed to an undersize access line, excessive traffic on the line, or congestion in the network switching systems. In many instances, this is a time-of-day problem, and will resolve itself when the traffic returns to normal. Time of day refers to the busy hour(s) for traffic transmitted by the user.

A number of response time and performance issues can be corrected by simple proactive efforts of the administrator. Several of these include:

- A workstation running Windows NT provides the capability to specify the order in which the workstation handles protocols on a multiprotocol network. One way to improve response time on the workstation is to set the protocol order so that the most frequently used protocol is handled first.
- A problem with NICs is that they can start saturating the network with repeated packet broadcasts, which can be corrected by updating the NIC drivers. Network administrators can regularly monitor the network and individual nodes to ensure none are creating excessive traffic.

Response time difficulties can be identified by utilizing either a Network Management System (NMS), a monitor, an analyzer, or a combination of the three. As is usually the case, without a trained technician or network administrator, the testing equipment is useless.

A Single Station Cannot Access the Network A first step is to check the station's configuration, its own data link control and network address, and the address of the router it uses to reach the Internet.

The next step is to check if the station is being used for the first time, and to look for jumper and dip-switch settings. There may be settings for thick, thin, twisted pair, or fast Ethernet media or 4/16/100 Mbps speeds for Token Ring.

If these steps fail to identify the cause, check the interface card for interrupt or address conflicts with other recently installed cards. Verify that software settings match with the NICs.

If none of these steps correct the situation, then the following options are available:

- Run the manufacturer's diagnostic software that is supplied with the station's NIC.
- Reboot the station and use a protocol analyzer to watch the initialization frames, which include duplicate address test and service request broadcast.
- Power cycle the device and see if this clears up the problem.
- Reinstall the software.
- Use a protocol analyzer or monitor to see if many errors are coming from this station. If so, replace the NIC.

A Specific Application Does Not Function The first step here is to find out if all users are affected. If not, determine what is different about this user's configuration. Observe the user to determine if there is a cockpit problem.

The next step is to check if a new version of the software has been installed. If this is a global problem, the old version of the software might need to be reinstalled and the new version checked for bugs.

If the application software is provided by a vendor, check with the provider about version incompatibilities with the NIC or with operating system software. Gather as much information as possible before contacting the vendor, because it may be necessary to prove that the vendor is at fault. An analyzer trace printout may also be required.

No Stations on a Single Segment Can Access Another Segment These situations suggest problems with interconnect devices such as repeaters, bridges, routers, or gateways. Several troubleshooting options involve the following activities.

- Investigate configuration files and use a protocol analyzer to see if there is a network layer addressing problem.
- Study the bridge, router, and gateway configurations to see if they are improperly filtering frames. Use the network management system logs in this endeavor.

- Check if a repeater stopped forwarding frames due to jabbering on one side.
- Look at other users that are attached to that segment to see if their devices are communicating.
- Use a distributed analysis system to verify that frames are reaching the other segment.

No Stations on a Single Segment Can Access the Network Use a protocol analyzer to see if there are damaged frames due to excessive **Radio Frequency Interference (RFI)** or **Electromagnetic Interference (EMI)** caused by fluorescent lights, motorized equipment, or two-way radios.

For Ethernet environments use a protocol analyzer to see if there are excessive collisions, which could be caused by a jabbering station, a cable that is too long, or too many stations on a segment. A malfunctioning transceiver will cause jabbering.

A cable tester, such as an ohmmeter, can be used to determine if the Ethernet network is no longer terminated, if there are shorts or opens in the cable, or if there is excessive signal interference.

A protocol analyzer can be used on a Token Ring network to determine if there is beaconing caused by bad cable or a faulty network card.

Stations on a Single Segment Intermittently Cannot Access the Network A regularly scheduled (or maybe not scheduled) download or backup operation could congest the network periodically.

With an Ethernet network, a TDR can be used to determine if the cable is too long, which causes late collisions.

Use a protocol analyzer to determine if there are:

- Excessive frames to multicast and broadcast addresses
- Intermittent retransmissions caused by EMI and/or RFI
- Duplicate locally administrated data link or network addressing errors

This type of problem points to station-specific physical and data link problems. It is frequently caused by the user pushing the upper limit of the specifications.

15.7 TECHNOLOGY-SPECIFIC TROUBLESHOOTING AND PROBLEM SOLVING

The remainder of this chapter is devoted to problem solving and troubleshooting for the specific wide area technologies. Specific details for each of these technologies can be obtained from the referenced chapters. Sections are devoted to the following technologies:

- Asynchronous Transfer Mode (Chapter 4)
- Digital Subscriber Line (Chapter 8)

- Fiber Distributed Data Interface (Chapter 7)
- Fibre Channel (Chapter 12)
- Frame Relay (Chapter 3)
- Integrated Services Digital Network (Chapter 9)
- Internet/Intranet/Extranet (Chapter 13)
- Private Line/POTS (Chapter 2)
- Synchronous Optical Network (Chapter 5)
- Switched Multimegabit Data Service/MAN (Chapter 10)
- Virtual Private Network (Chapter 6)
- Wireless/Personal Communications Service (Chapter 11)

There may be some overlap of information in these sections because troubleshooting and problem-solving techniques will apply to multiple technologies. In addition, test equipment, analyzers, and monitors are usually designed to work for numerous technologies, topologies, and applications.

Several problem situations can be identified with the high-speed, powerful test tools that are available today. Some of these common features include the following activities:

- Decode and analysis of different protocols
- Isolation of causes of network inefficiencies
- Latency and loss analysis
- Jitter analysis
- Precise capture of network events
- Tracking of network events
- Network problems discovery
- Reports generation

Testing instruments have the capability of decoding and analyzing the following protocols.

- AppleTalk
- ATM
- Banyan
- CDMA
- DECnet
- Frame Relay
- GPRS
- GSM
- Ipsilon
- ISDN
- LAN
- LAN Emulation
- MPOA
- Novell
- PPP
- SMDS
- SNA
- SNMP
- SS7
- SUN
- TCP/IP
- Token Ring
- V5
- X.25
- XNS
- WAN data link

Certain testing instruments can provide support for many of the broad-band technologies [Sveum, 2000]. One such device that provides protocol support for WANs, LANs, ISDN, and ATM is the RADCOM family of products. This analyzer is available from several sources.

15.8   ASYNCHRONOUS TRANSFER MODE

The most popular use of ATM in LAN environments is as a backbone network. Using ATM as the backbone technology, when properly designed, simplifies network management. If the network is not well designed and implemented, however, ATM networks can be extremely complex and difficult to manage.

Problems, such as link failure or congestion, are intensified in these high-speed networks. Consequently, the network must have the ability to perform preventive measures to avoid congestion and implement mechanisms for rapid dissemination of control information about the problems and potential problems. Even though there are built-in routines oriented toward integrity of the network, there must be an online network management system for monitoring and observing the status of the ATM backbones.

As networks evolve, network designers tend to overlook the need to increase backbone speeds, for example, from 100 Mbps Ethernet to 155 Mbps or faster. Capacity management is especially essential in large multi-LAN segments, where interconnection over long distances is required, and redundancy is necessary to ensure uninterrupted communications.

The five elements of ATM testing include:

- Standards conformance testing as specified by ISO-9646
- Interoperability testing of equipment from different manufacturers
- Regression testing for version compatibility
- Performance testing of the operational network for data transfer and signaling conformance
- Diagnostic testing to find and diagnose network operational problems

These tests can be performed in an in-service (promiscuous) or out-of-service (intrusive) mode. Sample functions include the ability to:

- Generate and monitor ATM cell traffic
- Capture received ATM cells for analysis
- Filter virtual path and channel traffic
- Insert and detect alarms and errors
- Trigger output for external devices

The University of New Hampshire's Interoperability Laboratory (IOL) formed an ATM Consortium in 1993 for the purpose of testing ATM equipment for interoperabilty, protocol conformance, and functionality. Conformance testing is of utmost importance when implementing ATM equipment from different manufacturers. If each piece of equipment conforms to the standards, installation and running of the systems will likely be error free.

ATM Testing Instruments

ATM test equipment that is available includes protocol simulators, protocol analyzers, traffic generators, and protocol conformance testers. Manufacturers of such equipment are:

- Adtech, Inc. (AX/4000 Broadband Test System)   www.adtech-inc.com/
- Digitech-LeCroy (ATM900)   www.tekelec.com/
- Fluke, Corp. Analyzer Series (DSP4000)   www.fluke.com/
- GN Nettest (interWATCH, WinPharaoh, FastNet)   www.gnnettest.com/
- Hewlett Packard (Broadband Series Test System Internet Advisor)   www.hp.com
- Hewlett Packard (HP 75000 ATM Analyzer)   www.hp.com/
- Network General (ATM Sniffer)   www.datacomsystems.com/
- Radcom (WireSpeed 622 Mbps ATM Analyzer)   www.radcom-inc.com/
- Tekelec (Open Chameleon)   www.tekelec.com/

Today's backbones and high-speed server links run at increasingly higher speeds, carrying a multitude of data, voice, and video services. Downtime of these high-bandwidth links is not an option. As management and maintenance of these connections becomes a bigger challenge, more is required of test equipment.

An example of the test instruments available today, the Adtech AX/4000, was first developed as a tool for testing ATM networks and devices; however, it has evolved into a system for testing the performance and QoS of broadband transmission technologies. This instrument can generate real-time broadband test traffic and allows for full-rate traffic analysis.

Another device, the WireSpeed 622 Mbps ATM Analyzer, can perform fault and performance tests when installing and troubleshooting high-speed ATM backbone networks. This device can manage high-speed links, perform preventative maintenance, and solve network problems.

There are various plug-in hardware and software modules available for these devices, which can be utilized to test for signaling, interoperability, and conformance issues.

15.9   DIGITAL SUBSCRIBER LINE

Local loops
Because DSL technologies are designed to run over the local loop, certain issues must be addressed when installing digital subscriber lines. When the telephone system was digitized in the early 1970s, the local loop was left with analog circuits. T-1 service was initiated that used repeaters to clean up a local loop and render it capable of passing digital signals up to 1.544 Mbps and extend the length of a T-1 line. The following components may be encountered in the local loop and each will have an impact upon the successful implementation of DSL:

- Loading coils
- Bridge taps
- Amplifiers
- Wire gauges

An issue that must be addressed before any DSL service is installed concerns loading coils and Digital Loop Carrier (DLC) local loops. Approximately 20 percent of local loops have loading coils and a greater percent are on DLC systems. To use DSL on the local loops, the loading coils must be removed for DSL to work; however, this job is labor intensive and time consuming. This problem also impacted the installation of ISDN.

Digital Loop Carriers
The problem with DLCs is that they only digitize and pass along the analog voice passband of 300 Hz to 3.3 kHz, so no xDSL can take advantage of the 1.1 MHz usually used on the xDSL twisted pair without doing something about the DLC system. The simplest answer would be to run something else to these loading coil and DLC customers.

Testing Issues
How are DSL links to be tested, repaired, and managed? It has taken a long time for personnel to be trained and become comfortable with ISDN DSL links. It was quickly apparent that this was no longer an analog environment and that radically different testing, repair, and management techniques were needed, but test equipment was lacking for the DSL environment. The following questions need to be answered:

- What is the standard procedure for splicing a break in an ADSL link?

- What monitoring and analysis equipment will be needed in the DSL environment?
- Where are the uniform network management packages for multi-vendor DSL environments?
- How is the network evaluated for data throughput and network efficiency?

ADSL

Carriers may need to recondition or replace some lines within the 18,000 ft. distance limitations to provide ADSL services. The wiring required is 24 AWG; however, the distance is reduced to 14,000 ft. when using 26 AWG. As with all xDSL, ADSL equipment cannot work through bridge taps and loading coils that carriers have installed over the years to boost signals to residences.

Performance Issues

A fundamental performance issue with store-and-forward packet switching occurs when a long frame gets ahead of a short one, particularly on a lower speed WAN circuit. If an application that is sensitive to variations in delay, such as voice, generated the short packet, then it may not be able to wait until the long packet completes transmission. As an example, consider a DS-1 with a high-speed downstream connection on an ADSL to a residential subscriber. Typically, a maximum-length Ethernet package is 1,500 bytes, which takes approximately 8 ms to transmit on a DS-1. A long Ethernet packet getting ahead of a short VoIP packet could delay it by up to 8 ms. This effect means that real-time gaming, circuit emulation, or interactive videoconferencing cannot be effectively performed over store-and-forward packet switching systems operating even at modest speeds.

Several approaches can be utilized to compensate for this situation. One solution requires a preempt-resume protocol, similar to that used in most multitasking operating systems. In essence, the packet switch must stop transmission of the long packet by reliably communicating this interrupt to the next node, and then transmit the high-priority packet. Not all vendors support this capability, but development is underway to provide this functionality.

A simple solution to this problem in packet networks is to increase the link speed and/or decrease the maximum packet size. The threshold of human perception of delay variations is on the order of 10 ms for auditory and visual stimuli.

Splitter Modem

Performance can be affected by the installation of the splitter on the subscriber's premise. The ADSL Terminal Unit-Remote (ATU-R) should be installed close to the Network Interface Device (NID), which is the telephone company demarcation point. The quality of the wiring job from the NID to the set-top box or PC can ensure a minimum of appliance or POTS coupled noise. Note that an Alternating Current (AC) power supply is required where the splitter is to be located.

DSL Test Instruments

The quality of the local loop has the largest impact on the operation, performance, and stability of xDSL service. Products are available to help service providers quickly identify and locate loop faults, which reduces troubleshooting time.

Fluke xDSL OneTouch Installation Assistant provides for insertion loss testing which is used to qualify the local loop for ADSL and HDSL. The wideband loop qualification option can identify the following:

- Incorrect circuit design
- Incorrect cable lengths
- Incorrect cable gauges
- Presence of load coils
- Bridge taps left on the cable pair
- DC resistance faults

A Time Domain Reflectometer (TDR) function is available which helps locate the following:

- Cable faults
- Cable characteristics
- Network elements
- Bridge taps/laterals
- Partial opens caused by poor splice joints
- Low-resistance faults
- Splits and resplits
- Water damage
- Load coils and build-out capacitors

The HP 7900 ADSL Test Station is designed for volume testing of ADSL modems for adherence to ANSI T1.433 and ITU T1.133 standards. It can also be used to test the central office ADSL transceiver unit or a remote-end transceiver unit (ATU-R). The conformance tests performed on these ADSL modems include both component testing and signal testing. This device can also test for compatibility with ISDN, POTS, E-1, T-1, and HDSL.

Consultronics provides test equipment for xDSL, Cable, and Dataline analysis. Consultronics test sets are capable of qualifying circuits for HDSL, SDSL, HDSL2, G.Lite, ADSL, and several other xDSL technologies.

15.10   FIBER DISTRIBUTED DATA INTERFACE

As with most networks, the FDDI cabling system can be a major source of trouble, particularly if not installed by experienced technicians. Terminating fiber and installing connectors requires training and practice to develop the necessary skills. Each end of a fiber must be cut at exactly 90 degrees and highly polished before a connector can be epoxied onto the fiber end.

A special magnifying glass is necessary to inspect the polished end to ensure its smoothness. Fiber cable testers are available that ensure the confidence of each finished cable.

Fiber-optic cables are subject to kinking and shiners. To accurately ensure the quality of the cable at all points along its length, an optical time domain reflectometer is required. The bend radius of fiber should not be less than 12 inches. An optical loss detector (light meter) can measure the loss of light in decibels (dB) from a calibrated source to the opposite end of the fiber. The primary test is light signal loss which is also measured in decibels. The less loss, the better. A level of negative 25 dB would be considered within the proper range.

Another issue with fiber-optic cable involves the attachment of connectors and cable splices. An insertion loss can occur from the splicing and connector assembly process, which is caused by the user or the equipment that is being used for the assembly. The problems can be grouped into intrinsic losses, extrinsic losses, and mechanical offsets. Table 15–2 summarizes these problems.

An aspect that can affect the results of data transmission using fiber optics is the propagation time through the core. Light that enters the core at the centerline travels in an unimpeded straight line through the core. Light entering at any other angle will eventually hit the cladding and be bounced down the cable. These light rays travel at greater distances than one traveling down the center of the core. The rays, therefore, emerge at different times, producing a phenomenon

Table 15–2
Fiber-Optic Cable Insertion Losses

Intrinsic Losses	Extrinsic Losses	Mechanical Offsets
Core diameter	Mechanical offsets between fiber ends	Angular misalignment
Numerical aperture	Contaminants between fiber ends	Lateral misalignment
Index profile	Improper fusion, bonding, and crimping methods	End separation
Core/cladding eccentricity	End finishes	End face roughness

known as pulse spreading, which causes the replicated electrical information to be distorted by varying arrival times of the light rays. This distortion is usually small, but it presents a limiting factor in the length of the fiber cable and the data rates that can propagate through it. If the cable is too long, the spreading can cause loss of data and reduce the data throughput rates.

It may be necessary to view the FDDI frames. If so, network sniffers (analyzers) have attachments that allow direct connection to the fiber. This device will allow the technician to look for an excess of voided or beacon frames. FDDI has no active monitor, unlike Token Ring.

Troubleshooting Equipment

The following testing devices can be utilized to test and troubleshoot the FDDI network.

Fotec LCT-mm/sm FDDI Link	www.fontronic.com
Tekelec WAN9000 Protocol Analyzer	www.tekelec.com/products

15.11   FIBRE CHANNEL

Particular hardware devices, cabling, and Fibre Channel components can cause service issues. Where data stored on servers are critical, it might be necessary to justify the cost of drive-mirroring or RAID systems that contain hot-swappable drives and power supplies. Replicating data to other servers or server mirroring can prevent downtime and data loss.

Storage and access have been essential parts of large-volume data handling in computer applications. The amount of information to be stored and retrieved has been growing larger and larger over the years. When the storage capacity is increased, and the amount of information to be stored is increased at the same time, both storage and retrieval of data become more difficult. It is necessary to have the system designer answer the following questions when developing data storage requirements:

- What is the optimal unit of storage?
- How large can storage be?
- How can Net transfer rates be improved?
- What factors influence Net transfer rates?

Test Instruments

Specialized testing instruments are required in the Fibre Channel environment. Mainframe and I/O channel testing devices that operate at very high transfer rates are also required.

15.12   FRAME RELAY

Installation

Turning up a Frame Relay link can be difficult if there is no advance preparation. Usually, a Frame Relay PVC will be spanning a number of UNI and NNI links, across multiple LATAs, and even multiple states. The technicians working the installation will need to be communicating with someone at the user locations.

Both DCE and DTE devices must be onsite and optioned at the appropriate speeds with the proper IP addresses and DLCIs. This means that the subscriber and the carrier have communicated before cutover day to agree on the options to be installed on the routers or FRADs. The carrier technicians will be using LMI as a tool to verify that there is communication between the DCE and DTE devices. In most cases, carriers do not troubleshoot behind the DCE devices. The carrier, through the network management system, can determine the number of packets that are being transmitted and received on each of the PVCs, the number of errors encountered, and the current actual CIR. They do not provide any content on the packets, however, which means that it is up to the user to validate the data being transmitted and received.

Congestion/Throughput

In situations of congestion or when the allocated bandwidth is exceeded, Frame Relay allows frames to be discarded without notifying the end stations. There are, however, two mechanisms employed by Frame Relay to alert the user nodes and Frame Relay switches about congestion and initiate corrective action. Both capabilities are achieved by the Backward Explicit Congestion Notification (BECN) bit and the Forward Explicit Congestion Notification (FECN) bit. Because congestion can be a problem in a demand-driven network, Frame Relay may discard or reduce its throughput level to avoid congestion problems. The discard eligibility (DE) bit is utilized to discern which of the user's traffic should be discarded. This can be accomplished by using a technique called the Committed Information Rate (CIR). The end user will estimate the amount of traffic that will be sent during a normal period of time, and select the appropriate CIR level that will accomplish these requirements. There is usually a cost for CIR, so the user will need to determine the actual or estimated traffic and busy hour(s) on the PVCs. Although several procedures can produce this information, the hardware and software may not be readily available to the end user. The alternative is not to purchase CIR and test the throughput on the network. If the response time is acceptable, then no CIR may be required, and it is an easy proposition to add CIR if response time is not acceptable.

Committed Information Rate

An example of using CIR to improve price performance is as follows:

A host site utilizes a T-1 access line which is part of a Frame Relay network that also consists of twenty-four remote locations, where each remote location has a 64 kbps access line. Assume that this is a one-to-many configuration where each remote location has a 64 kbps PVC to the host site. This means that the T-1 (24 DS-0s) at the host site has sufficient bandwidth to carry all twenty-four PVCs. Additional 64 kbps PVCs can be added on the T-1 as long as CIR has been added to the 64 kbps PVCs to keep the aggregate speed less than the host T-1. As long as all of the remote locations are not transmitting at the same time, oversubscribing will be possible. If response time degrades, another look at the CIR will be required.

Network Management Software

If a network management system is available or if the carrier has a system such as OpenView, NetView, or SunNet Manager, then it will be possible to use the ITU-T performance parameters for Frame Relay to isolate QoS issues. The statistics and performance criteria that may be available are:

- Delivered erred frames
- Delivered duplicate frames
- Delivered out-of-sequence frames
- Lost frames
- Misdelivered frames
- Premature disconnect

These data can be processed with QoS applications that are part of the Service Level Agreements (SLAs).

X.25 Issues

Packets that can be utilized for troubleshooting the X.25 topology include flow control, reject, diagnostic, interrupt, and registration packets. The status or contents of these packets can be viewed with a protocol analyzer.

Frame Relay Test Instruments

Most of the major protocol analyzer vendors provide equipment for Frame Relay protocol analysis, emulation, and testing. A partial list is as follows:

- Fluke 660 Frame Relay Assistant www.flukenetworks.com/
- GN Navtel   www.gnnettest.com
- Hewlett Packard   www.hp.com/
- Odin TeleSystems   www.odints.com/
- Radcom   www.radcom-inc/products/

- Tekelec   www.tekelec.com/
- Telenetworks   www.telenetworks.com/
- Telecommunications   www.ttc.com/ttc/home.nsf/
  Techniques
- Trend Communications   www.trendcomms.com/index.htm/
- Visual Networks   www.visualnetworks.com/
- Wandel & Goltermann   www.wwgsolutions.com/

Whereas a number of single-purpose test instruments operate in the WAN, LAN, and ISDN environments, devices also operate on a number of technologies and topologies. One such example is the RADCOM line of products that offers a wide range of protocol analyzers for Frame Relay, X.25, ISDN, Internet, and LAN/WAN networks.

The RADCOM RC-100W/100WL provides for the following Frame Relay analysis and emulation:

- Frame Relay Expert Troubleshooter for identifying protocol violations
- DLCI map and load applications for online activity and utilization per DLCI
- Network monitoring and simulation
- X.25 to FR connectivity
- Automatic recognition of encapsulated protocols
- On- and offline filtering with different criteria including erroneous frames

15.13   INTERNET/INTRANET/EXTRANET

Simple network design guidelines and defined network performance measures create a formula for success. The Internet, however, is a complex network, made of complex topologies with sophisticated automatic control and routing mechanisms. It will be a challenge to provide a better guarantee of successful message delivery and acceptable propagation times over the Internet.

TCP/IP network management and troubleshooting is involved with controlling many different packet switches which are interconnected by IP gateways. The IP network management must run at the application layer because of the many possible protocols and types of systems within an IP network. The IP administrator cannot fix a problem that occurs beneath the management and troubleshooting software. Examples of these problems include a corrupted routing table, a system that must be rebooted, or a failed operating system. These types of problems must be fixed by the network personnel.

The most popular tool for TCP/IP management is Simple Network Management Protocol (SNMP). Each gateway maintains statistics on its operation in a database called a Management Information Base (MIB). Some MIB statistics available for analysis are:

- Number of datagrams fragmented
- Number of datagrams reassembled
- The IP routing table
- Number of routing failures
- Number of messages received and type of message
- Number of datagrams received
- Number of datagrams forwarded

A **trap** is the result of a particular event, such as too many routing failures. When a preset number of routing failures occur, the trap is activated and a message is sent to the network management system. The number and variety of traps sent to the NMS is programmable by the network administrator.

trap

Routers
Another issue affecting performance in large Internets occurs in responses to frequent state changes to routers, their ports, and the links that interconnect them. Whenever a state change occurs, the routing protocol distributes the updated topology information, and the packets may begin following another path. A change in routed path usually results in jittered, lost, missequenced, or duplicated packets. These impairments will likely cause interruptions in the video playback.

Most Internet routers operate in a "first-come, first-served" fashion, which means that packets arriving from the input lines are immediately examined by the routing process, queued on an outgoing interface, and then transmitted in the order of arrival. This behavior has the advantage of being easy to understand and implement; however, delays will then vary as a function of the Internet network load. As the traffic increases, the delay becomes longer and more variable. If the traffic load exceeds the server's capacity, then the queue can increase indefinitely. The server will not have enough buffers and will be forced to drop packets, which results in poor response time. The only way to restore lower delays is to increase the server's capacity or to reduce the load.

Increasing the capacity requires capital investments such as buying more powerful equipment or acquiring more expensive bandwidth. Network managers will need to implement traffic measurement processes to determine the cost versus benefits of these additional costs, but this is a long-term endeavor. In the short term, the network capacity may be fixed, which means that the user must adapt to the traffic load.

Shared Media
Problems occur on some high-speed networks, particularly on shared-media LANs. The collisions that still occur on 100BaseT Ethernets can

create significant delay variations during periods of relatively high load. The 100VG-AnyLAN technology overcomes this problem by prohibiting shared media and requiring that intelligent hubs use a demand priority scheme to serve high-priority traffic before low-priority traffic.

Segments

As LANs grow, the installation of bridges or switches to separate segments and routers to connect to other networks introduces a new set of problems for network designers and administrators. Both bridges and routers introduce delay into the network. Delay can be caused by the wide area link operating at a slower speed than the local network or by the delay caused by insufficient buffers in the bridge or router itself. This situation can be corrected by:

- Adding more memory to the bridge or router
- Adding additional lines between bridges or routers so that each device is capable of passing frames from its input queue to the first available line
- Filtering the source and destination addresses so that the addresses on a local segment cannot be passed to a remote segment

Most of the cures to improve throughput and response time over wide area links must be weighted against significantly increased costs.

15.14   INTEGRATED SERVICES DIGITAL NETWORK

Local Loop

One concern with ISDN-BRI implementations is whether current local loop facilities are capable of transporting the 160 kbps 2B1Q digital signals over a sufficient distance without incurring too much signal loss or interference. Digital signal quality is also a concern over in-building wiring. Physical media testers can determine whether an appropriate signal quality is being maintained over a given transmission facility.

ISDN Adapters

There are requirements for DCE devices such as terminal adapters and network terminators that connect to the local loop. Chapter 9 provides an in-depth analysis into the ISDN technology. Some serial ports have difficulty communicating through PC software, which may cause a throughput problem. The speed for ISDN is 64 kbps, however RS232-C, as implemented in the 16550A UART, found a serial port maximum speed of 38.4 kbps in PCs built after 1992. In many cases, there is insufficient cable shielding and grounding to allow speeds greater than 19.2 kbps. The easiest solution to this problem is to install the terminal adapter in an expansion slot and avoid the need for RS232-C. Users of external terminal adapters are advised to interface to the PC with a V.35 cable.

ISDN Test Instruments

A wealth of test equipment is available for troubleshooting in the ISDN environment. A number of these devices perform simulator and emulator functions in addition to the normal ISDN test suite. Some of these devices are as follows:

- Consultronics CoBRA ISDN Basic   www.consultronics.com/
  & Primary Rate Analyzer
- Frederick Engineering   www.fetest.co/
  FELINE/PARASCOPE
  WAN/ISDN Analyzer
- Harris/Dracon biTS-1 Test Set   www.harris.com/
- Processing Telecom Technologies   www.cxr.com/products/
  Model 5200
  ISDN/DDS Cable Simulator and
  Model 5260
  26 Gauge Cable Simulator
- RADCOM WAN/LAN/ISDN/   www.radcom-inc.com/products/
  ATM analyzer
- Telebyte Technology Model 451
  Wire Line Simulator
- Telecom Analysis Systems TAS   www.taskit.com/products.htm/
  2200 A/I Loop
- Emulator and TAS 2270
  Subscriber Loop Emulator
- Telecommunications Techniques www.ttc.com/ttc/home.nsf/
  Firebird 6000
- Wandel & Goltermann IBT-1V   www.wwgsolutions.com/
  ISDN Data Tester and ILS-1/
  ILS-2 ISDN Line Simulator

It is important to realize that a protocol tester does not contain an ISDN protocol implementation, but only a script of test frames and messages and the expected responses. Attempts are made to test all legal possibilities and determine if illegal frames, messages, and other procedural errors are being handled correctly by the device being tested.

A good instrument for testing BRI ISDN is the HP Internet Advisor WAN. It can use the test module HP J2905B for BRI S/T and U Interfaces. This combination provides LAPD analysis (CCITT Q.921 specification), S/T and U testing at the customer's premises, SAPI, full X.25, Frame Relay, data-link control decodes on the D channel, and monitors encapsulated LAN traffic. It can log and filter network statistics for troubleshooting and monitor error rates. An associated device is the simulation and BERT tester, which is the HPJ2904B Simulation/BERT. It is used to test the S and T interfaces for BRI.

15.15   PRIVATE LINE/POTS

The major problems in the private line and POTs environment usually occur in the DCE devices. The modems and Data Service Units (DSUs) can fail due to component failure from lightning strikes and other voltage surges. It is therefore important that Universal Power Supplies (UPS) and/or surge protectors be in place between the devices and the power grid.

Central Office

There can be problems with central office equipment that will cause an interruption in service. These will usually occur in Office Channel Units (OCU) or multiplexors. A trouble call to the LEC or carrier will be required to resolve this type of problem. Before making this call, however, the network administrator should investigate for subscriber CPE-oriented problems. Note that a number of carriers keep a record of "chronic" problems for facilities that have a high problem rate.

Line Impairments

There are a number of line impairments that can cause problems in the enterprise network. These include:

- Impulse gain hit
- Impulse phase hit
- Cross-talk
- Impedance mismatch
- Echoes
- Jitter

These impairments are usually random and last for short periods of time, which makes them difficult to troubleshoot. Many of these problems are caused by induced currents into the lines from lightning strikes, power surges from the power station, an atomic explosion, or magnetic disturbances from a solar flare. An accurate log that shows time-of-day for network problems can assist the technician in isolating these elusive incidents.

Customer Premises Equipment Problems

A structured approach to troubleshooting can save time and avoid a service call. The first step is to look at the common network problems listed in Section 15.5. Utilize the following checklist:

- Are the Data Communications Equipment (DCE) devices plugged in and powered up?
- Are the indicator lights illuminated on the DCE devices?

- Is the serial port operational? A voltage surge can destroy the serial interface card but it will not be evident to the naked eye. Swap the DCE device with one that is working.
- Check the RS232 interface with a pin-out indicator. If there is no Data Terminal Ready (DTR), replace the serial card with one that is working. A pin-out checker costs less than $100.
- Is the transmission cable plugged into the wall jack (demarcation point)?
- Swap the transmission cable (RJ11/RJ45 connectors) with one that is working.
- Place a trouble call.

Modem Impairments

Because modems use the PSTN, imperfections of the PSTN affect modem performance. The impairments that can be encountered include:

- Attenuation
- Noise
- Delay distortion
- Harmonic distortion
- Intermodulation distortion
- Echoes
- Amplitude jitter
- Phase jitter
- Signal differentiation

The Signal-to-Noise Ratio (SNR) is about 30 dB for a local call in the United States and about 25 dB for a long-distance call. PSTN noise tends to occur in bursts and corrupt several symbols, resulting in a large number of bit errors. The Bit Error Rate Test (BERT) is the number of bit errors received divided by the total number of bits sent. Modern error control requires retransmission of a block, so block checking is performed.

The TIA and the EIA organizations have formulated RS496-A, which provides for six tests on modems. These tests include:

- Noise
- Attenuation and attenuation distortion
- Delay distortion
- Phase jitter
- Frequency offset
- Intermodulation distortion

Test results may differ from one time to another, depending upon the routing of the call, since the PSTN environment is dynamic.

Figure 15–5
Modem (DCE) Loopback
Paths

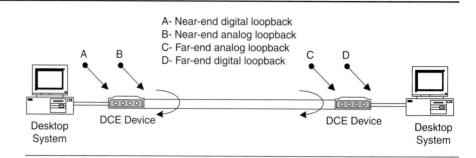

A- Near-end digital loopback
B- Near-end analog loopback
C- Far-end analog loopback
D- Far-end digital loopback

Loopback Testing

If there is no communication across the carrier's network, a simple loopback test can be conducted. Figure 15–5 shows both a far-end and near-end loopback test. A loopback test on a full-duplex circuit is an effective way of locating faults and impairments in a data circuit. Many DCE devices include integral loopback capabilities in addition to other simple tests, such as bit error rates. Tests can be conducted by sending a short message or a standard test pattern. By looping the circuit at progressively further points, the element causing the complaint can be identified. These tests are usually described in the device's user manual.

If the loopback test runs successfully, then the problem is usually not in the DCE device or circuit, but is likely in the terminal, the computer, or the cables that connect them to the DCE devices.

Although loopback tests are useful in locating hard faults, some impairments such as data errors, phase jitter, and envelope delay are cumulative over the length of the circuit. These tests are more effective when conducted end to end. Digital tests are of little value in finding totally failed circuit conditions and of no value in voice frequency circuit tests. Analog tests are useful only on dedicated circuits or in verifying the condition of a circuit up to the point of interface with a common carrier.

Private Line Testing Instruments

The HP T-1 Test Advisor is a versatile testing instrument. There are alarm indicators for signal loss, frame loss, pattern loss, coding type, and density violations. The test summary displays the framing type, status, number of frame errors, and the number of bit errors. The technician can insert errors as a cross-check on the system test. BERTs can be done using pseudo random number sequences. The test set has the capability to find bridges and taps. Signal levels can be monitored and signals inserted at precise levels and frequencies.

Wilcom, Inc. provides telecommunications test instruments for both two-wire and four-wire facilities. These devices can be utilized for conducting the following network tests:

- Evaluation of circuit performance
- Measuring ground return currents in power systems
- Spectrum Analysis
- Noise Measurement
- Conduct loop parameter measurements
- Simulates demarcation point network functions

15.16 SYNCHRONOUS OPTICAL NETWORK

SONET service is not normally available for subscriber usage. In most cases the network user will be unaware of problems in the SONET backbone network. Because SONET consists of dual counterrotating fiber rings, the possibility of a subscriber-effecting outage is remote. There is sufficient redundancy in the backbone network facilities to maintain integrity of the WAN, although safeguards have been built into the SONET infrastructure. The built-in fault management system detects and reports the following faults:

- Network element hardware failure
- Cable failure
- Performance degradation
- Routing errors

Frequently, cable failure is attributed to utility company contractors who cut the network fiber-optic cables when trenching with a backhoe.

SONET Test Instruments
The HP SONET Maintenance Test Station (MTS) Lite, which handles speeds up to OC-12, is designed for cases where the fault management system is inadequate. ANDO Corp. provides optical test instruments and optical multimeters for analyzing the optical spectrum.

15.17 SWITCHED MULTIMEGABIT DATA SERVICE

Issues relating to troubleshooting and problem solving in the SMDS environment are similar to those in the Frame Relay technology. There are, however, a number of differences because SMDS is a connectionless technology and thus has a different addressing scheme.

A common problem that occurs with SMDS is corrupted or incorrect group or screening tables which reside in the SMDS switch. If these are not configured correctly, users will not be able to communicate with each other.

SMDS Test Instruments

SMDS protocol test equipment can be used to test protocol implementations, analyze network traffic, generate traffic for product testing, and provide a general troubleshooting tool. The following devices can test both the SNI and DXI protocols:

- GN Navtel 9470 Protocol Analyzer   www.gnnettest.com/
- HP PT502 Protocol Analyzer and   www.hp.com/
  Broadband Series Test Systems
- INTERVIEW 8000 TURBO   www.generalsignal.com/
  Protocol Analyzer/Emulator
- Tekelec Chameleon 32 and   www.tekelec.com/products/
  Chameleon Open
- TTC FIREBERD 6000   www.ttc.com/ttc/home.nsf/
- W&G SMDS DXI Decoder and   www.wwgsolutions.com/
  DA-30 Internetwork Analyzer

15.18   VIRTUAL PRIVATE NETWORK

Troubleshooting in the virtual network can be difficult because of the number of technologies involved. The first step is to look at the common problems presented earlier and then look at the problem-solving and troubleshooting sections for the various protocols, technologies, and topologies involved.

A common point of failure might be the firewall, gateway, or router that is attached to each end of the tunnel. If traffic is flowing up to these devices from each direction, then the problem is probably in the network that is providing the tunnel.

15.19   WIRELESS/PERSONAL COMMUNICATIONS SERVICE

Service-related issues with wireless and personal communication systems and devices are impacted by the various protocols and algorithms that have been implemented in the system. Certain problems occur due to noise and interference in the system, and the electrical limitations of the technology. Other issues can emerge when the system requires upgrading or it becomes necessary to enlarge the geographical coverage areas.

Multi-Hopping, Forwarding, and Routing

Network issues that must be addressed include multi-hopping, forwarding, and routing. Radio propagation effects such as path loss and multipath limit the coverage area of a radio transceiver. Nodes that are too far apart cannot hear each other's transmissions. The greater the distance between the transmitter and receiver, the greater the number of nodes needed to forward the message. In such cases, several nodes may need to forward the packet sequentially, taking multiple hops to reach its destination, in a scheme called multi-hop.

Multi-hop requires that a forwarding node maintain a routing table of all the nodes it can hear, and that it update the table often enough to take into account fading effects and the mobility of its end nodes. To serve neighboring nodes, the forwarding node needs to broadcast its routing table frequently. This overhead information decreases the efficiency of the overall system by using some of the channel resources for control information.

When there are multiple nodes between a transmitter and a receiver, the packet may have several different routes to take to reach its destination. The network may decide the routing at the source node, or on a hop-by-hop basis at each intermediate forwarder. Poor routing decisions are a major source of congestion. It will be necessary to look at the routing algorithms to identify a possible strategy to counter these congestion issues.

PCS Voice to Wireline Network

When a user on a PCS system makes a voice call to another user on the PCS system, both Personal Stations (PS) will use the same speech coding algorithm, and no speech encoding is needed. Some systems may convert to PCM at the base station for uniformity.

When a PCS system user makes a voice call to the wireline network, the voice coding system must be converted to 64 kbps PCM. When a PCS system user makes a voice call to another PCS system, the speech will be converted to 64 kbps PCM for transmission over standard transmission facilities.

When two PCS systems use the same speech coding system and have a significant amount of traffic, then the efficiency of the transmission facilities can be improved. For example, if the two PCS systems that use 32 kbps ADPCM communicate, then each 64 kbps transmission facility between the switches can carry two voice calls rather than one.

Radio Performance

The objective of RF engineering for cellular systems is to provide the highest possible performance in the radio spectrum while minimizing costs. Radio performance includes the quality of both the control transmission path and the voice transmission path. The measure of transmission performance is the RF signal-to-impairment ratio, $S/(I + N)$, where impairment refers to the power sum of noise (N) and co-channel interference (I). In some cases, it may be necessary to trade off different types of system impairments. Reducing power at one cell site to solve an interference problem at a nearby cell site may result in degraded noise or interference at the cell site where the power was reduced. There may be other situations where it may be necessary to reduce transmission problems in relatively small geographical areas at the expense of slightly degraded performance over a relatively large area and vice versa.

Several customer satisfaction surveys indicate a signal-to-noise (SNR) plus interference ratio equal to 17 dB or better in 90 percent of the geographical area of a system at the recommended performance level. If this QoS level is maintained, a good coverage of the area will be achieved.

Packet Radio Capacity

In both the ALOHA and the slotted ALOHA channels, the transmitter sends a packet without checking for a busy or idle channel status. In many ALOHA systems, the transmitter cannot determine if the channel is being used. In Carrier Sense Multiple Access (CSMA) systems, the transmitter senses the state of the channel before transmitting. If the channel is busy, the transmitter waits until the next slot. The collisions during a transmission are avoided, but not at the start of a slot, and the capacity of the channel improves. When a CSMA system does not work because some of the transmitters are hidden, then some of the channel capacity must be used to send the status of the reverse channel. This is seen in cellular and PCS systems that send busy-idle bits on the forward control channel to indicate reverse channel status.

Any transmitters receiving a packet for transmission during a slot will transmit the packet in the next slot. If all transmitters delay by a random delay before transmitting, the traffic spreads out and the channel capacity improves.

System Upgrades

After a system has been in operation, traffic in the system will grow and require additional infrastructure and communication channels. Segmentation and dualization issues must be addressed when adding additional cells to the system. It is possible to add an additional cell at less than the reuse distance without using a complete cell-splitting process. This method might be used to fill a coverage gap in the system, but it can result in co-channel interference. The most straightforward method to avoid an increase in co-channel interference is not to use them.

Segmentation divides a channel group into segments of mutually exclusive voice channel frequencies. By assigning different segments to particular cell sites, co-channel interference between these cell sites is avoided. The disadvantage of segmentation is that the capacity of the segmented cells is lower.

When a cellular system is growing, there may be cells of different radii in the same region of the coverage area, which can also result in co-channel interference. By dividing the radios at the cell site into two separate server groups—one for the larger cell and one for the smaller cell—the interference can be minimized. Dualization provides for control of co-channel interference and also gives a real increase in engineered traffic capacity.

Wireless and PCS Testing Instruments

Manufacturers and suppliers are responding to the demand for wireless testing equipment and systems. The following devices are currently available on the market.

Sage CDR2000 Cellular Diagnostic System	www.sageinst.com
Summit SI-800A/900A AMPS/GSM	www.summitekinstruments.com
Pixelmetrix DVStation Digital Video Monitoring	www.pixelmetrix.com

■ SUMMARY

The network administrator has a broad range of responsibilities that include network planning, monitoring, and maintenance. Typical activities revolve around network configurations, user connectivity, data protection, problem solving, and troubleshooting.

Network problems and issues can be resolved in one of two ways preventing the situation before it happens through network planning and management, or fixing the problem after it happens through troubleshooting techniques.

Troubleshooting techniques that are effective in solving problems in the enterprise network include the following:

- Implement a program for system upgrades and change control.
- Develop a baseline for the enterprise network.
- Use a systematic approach to isolate and correct network problems.
- Look at various alternatives and develop a hypothesis rather than getting tunnel vision.
- Change only one attribute at a time when troubleshooting and test each change thoroughly.
- Document the entire troubleshooting incident with the discoveries and conclusions.

Hardware and software solutions that can be utilized in the troubleshooting and problem-solving environment include protocol analyzers, monitors, cable testers, specialty tools, and Network Management Systems (NMSs).

Key Terms

Attenuation	Point Of Presence (POP)
Baseline	Protocol Analyzer
Bit Error Rate Test (Bert)	Response Time
Block Error Rate Test (Blert)	Radio Frequency Intereference (RFI)
Common Management Information Protocol (CMPI)	Service Level Agreement (SLA)
Electromagnetic Interface (EMI)	Simple Network Management Protocol (SNMP)
Mean Time Between Failures (MTBF)	Throughput
Near End Crosstalk (NEXT)	Time Domain Reflectometer (TDR)
Noise	Trap
Oscilloscope	Volt-Ohm Meter (VOM)

REVIEW QUESTIONS

1. What are two methods for resolving network problems?

2. What are the components of the network plan?

3. What issues must be addressed with network planning and management?

4. Describe a network baseline.

5. What are the baseline components?

6. What issues need to be considered when establishing standards for user name and passwords?

7. Why is an upgrade guideline and procedure required?

8. What are the five preemptive troubleshooting management categories?

9. What are the major action parts of the data backup plan?

10. What documentation should be included in the network plan?

11. Discuss the methodology used for problem solving. Draw a flow chart of the standard process for troubleshooting.

12. Why is it necessary to understand interface standards? What are the major interface standards?

13. A network control system provides for what functions?

14. Discuss the subject of network management tools. What are the categories of tools? Differentiate between hardware and software tools.

15. What tasks can occur in network management? How does this relate to SNMP?

16. What is a trouble report? What information would be found in a trouble report? How is it used?

17. What is a service level agreement?

18. Discuss MTBF.

19. What are the initial steps for performing network troubleshooting? Discuss the structured approach that can be taken. What is a PSTN impairment?

20. What is the significance of a Point of Presence?

21. Identify several test equipment devices and describe each.

22. What problems can be identified with a network analyzer?

23. What are the major functions performed by network test equipment?

24. What analog circuit tests can be performed with network test equipment?

25. Describe the interaction between SNMP, MIBs, and agents in a Network Management System (NMS).

26. Categories of network management devices include monitors, analyzers, and cable testers. Identify the types of features that might be a part of each of these devices.

27. What are the four areas that are candidates to start the troubleshooting process? What information should the checkoff list contain?

28. What common problems might occur with facility cabling?

29. What common problems might occur with the electrical power components?

30. What should the technician look for when testing network interface cards and transceivers?

31. How can a system upgrade impact the network? What can be done to alleviate this situation?

32. What clues can the network administration use to identify network performance issues?

33. What questions can assist in this endeavor?

34. What typical network problems can happen in the enterprise network? Give a brief description of each situation.

35. What issues might emerge in the ATM network?

36. What network components must be considered when implementing the DSL technology?

37. Discuss the issues related to fiber-optic cable that must be addressed when deploying a FDDI network.

38. What are some of the issues for a Fibre Channel environment?

39. What are the components of the Frame Relay congestion mechanism? How do they relate and how can the network administrator utilize them?

40. How do routers impact the Internet? How can these problems be corrected?

41. What are the major issues when installing and maintaining an ISDN network?

42. What are the major impairments in a private line environment?

43. Describe the loopback process. Why is it useful?

44. List common customer premises equipment (CPE) problems.

45. Is it necessary to be concerned with the quality of the SONET network? Why/why not?

46. SMDS networks utilize similar CO-based switches to provide service. What issues are related to SMDS service? What techniques can be used to troubleshoot SMDS?

47. What are the issues that can be encountered when troubleshooting a virtual private network?

48. What are the security issues that must be addressed in the wireless/PCS environment?

49. What is the process for troubleshooting a wireless system?

ACTIVITIES

1. A field trip to a Network Operations Center (NOC) is an excellent place to see how the troubleshooting and problem environment really works. This trip could also be tied to the Network Management activity discussed in Chapter 14.

2. A worthwhile research project is to use the Internet, trade magazines, trade shows, and DataPro to develop a test instrument spreadsheet of competing troubleshooting tools. This comparison would include the technology and the various features and functionality provided by the hardware and software.

3. Develop a matrix of network management systems. Show the technologies supported and the functionality of each system. Develop a one page overview for each system.

4. If a lab setting and equipment are available, use breakout boxes and VOMs to demonstrate shorts and cable breaks.

5. RS232 and V.35 breakout boxes can also be used to demonstrate the interaction of the various leads on the interfaces.

6. If available, use a cable meter to determine cable lengths and also if wires have been crossed.

7. Use a LAN meter to quantify the LAN utilization. Check the LAN during different time intervals and create a spreadsheet with the results.

CASE STUDY/PROJECT

A case study for this area of troubleshooting and problem solving will be time consuming and will require a number of hardware and software devices in support of the effort. At a minimum, a cable meter and a LAN meter will be required. For more complex activities, a protocol analyzer will be required.

The first step is to identify a problem or situation and develop a problem statement. It would then be necessary to collect information on the situation, using this chapter as a guide (See Figure 15–1). Documentation of all efforts and steps taken in addressing the situation would be required.

A simple test that can be conducted in most educational labs is to detect cable breaks. Replace a good cable such as a RJ-45 patch cable with one that has a short or a break. Use a cable meter or TDR in this test to identify the defective cable.

If a LAN meter is available, connect it to different places in the LAN and obtain readings. Record the results and explain the differences noted.

APPENDICES

APPENDIX A
OSI Model

The seven layers of the OSI model are described in this appendix. Relevant information concerning the model is presented in each technology chapter.

APPENDIX B
Voice over Internet Protocol

An introduction and overview is provided for Voice over IP (VoIP) in this appendix. Various other textbooks cover IP telephony in detail.

APPENDIX C
Standards References

RFCs and other standards references are listed by technology in this appendix. Some references may be superceded by newer versions.

APPENDIX D
Number Systems

Examples of decimal, binary, and hexadecimal conversions are provided in this appendix.

APPENDIX E
Broadband Case Study/Project

Details are presented for a comprehensive case study and project. This effort will attempt to incorporate most of the broadband technologies into one massive broadband network.

APPENDIX F
Acronyms

Acronyms that are in common use in the telecommunications networking environment are listed in this appendix. Additional acronyms and their definitions may be found in RFC1983—Internet User's Glossary.

OSI Model

OSI OVERVIEW

Computer software and hardware has been developed and introduced by many separate and independent computer and network vendors worldwide over the past decades. Many user application problems were introduced, most based on the complexity of emerging computer and network technologies which were lacking in compatibility. To overcome these incompatibilities, the International Standards Organization (ISO) adopted a layered functional approach for data communications and computer networking.

In the context of data communications, a protocol, or set of rules, is necessary for two devices to communicate. In this context, these rules do expedite communication, but they must be followed exactly in order for any communication to take place. Because there are so many functions that a communications architecture must provide, it was necessary to develop an architecture called a protocol stack.

A widely accepted structuring technique, and the one chosen by the ISO, is layering. The communications functions are partitioned into a hierarchical set of layers. Each layer performs a related subset of the functions required to communicate with another system. It relies on the next lower layer to perform more primitive functions and to conceal the details of the functions. It provides services to the next higher layer. Communication equipment manufacturers could then develop software implementations of protocols for one layer that are independent of implementations for any other layer. Standardized interfaces are then used to integrate these protocols into a working architecture. Standards are needed to promote interoperability among vendor equipment and to encourage economies of scale.

The task of ISO was to define a set of layers and the services performed by each layer. The partitioning should group functions logically and should have enough layers to make each manageably small, but should not have so many layers that the processing overhead imposed by the collection of layers is burdensome. Figure 1 provides a general view of the OSI architecture.

The seven-layer stack of protocols is called the Open Systems Interconnection (OSI) Reference Model. This reference model was established by the ISO in 1983 to provide a target for the future development of interoperable communications standards [Fitzgerald & Dennis, 1999]. The ISO charter is to encourage, facilitate, and document the cooperative development of standards that meet with international approval.

Protocols have been developed that do not conform exactly with the OSI model, and the long-term goal of the standards organizations has been to have all developers convert to this model. As an example, the TCP/IP stack consists of four layers, rather than the seven layers of the OSI model.

An overview of each of the seven layers will be discussed in the following sections.

PHYSICAL LAYER—LAYER 1

The physical layer covers the physical interface between devices and the rules by which bits are passed from one to another. The physical layer has four important characteristics:

- Mechanical
- Electrical
- Functional
- Procedural

The Physical Layer provides the procedural characteristics to activate, maintain, and deactivate the physical link through the transmission media. This layer is concerned

Figure 1
OSI Model

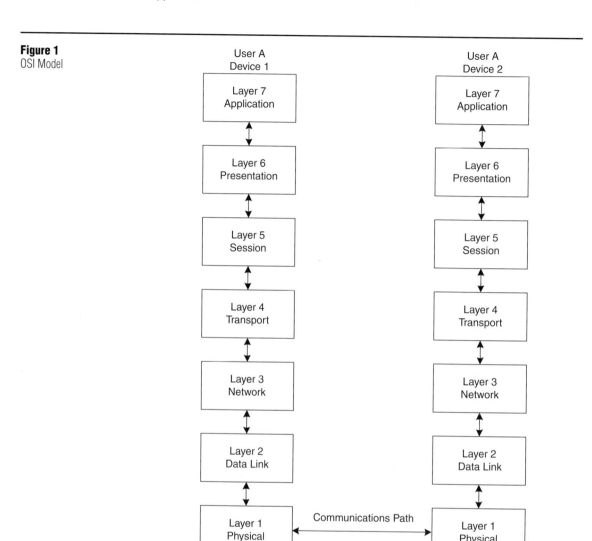

with both the physical and electrical interface between the user equipment and the network terminating equipment (NTE). The Physical Layer also defines the electrical and mechanical characteristics such as voltage levels, pin connection design, cable lengths, and other functions associated with the physical connection of data communications equipment. Devices that operate at this layer are repeaters, hubs, and multistation access units (MAUs).

The Electronics Industries Association/Telecommunications Industry Association (EIA/TIA) standard 233-E is a good example of a Physical layer specification.

It defines precisely the connectors and cable that connect a computer (DTE) and a modem (DCE).

The physical layer provides the Data Link layer with a means of transmitting a serial bit stream between two corresponding systems.

DATA LINK LAYER—LAYER 2

The purpose of the Data Link layer is to ensure the orderly and reliable exchange of information across the physical link. The principal service provided by the data link layer to higher layers is that of error detection and control, and retransmission of messages. To accomplish this, the data units carry synchronization, sequence number, error-detection fields, in addition to other control fields, and data. This layer is also responsible for the rate of data flow on the link.

The ISO/IEC standard 8802-3 specifies an implementation of the OSI Data Link layer protocols. In this implementation, the Data Link layer is divided into two sublayers: the Logical Link Control (LLC) and the Medium Access Control (MAC) sublayers.

Examples of devices that operate at this layer are bridges and LAN switches. These devices are basically used for segmentation of a LAN.

NETWORK LAYER—LAYER 3

The Network layer provides for the transfer of information between end systems across a communications network. It is responsible for establishing, maintaining, and terminating the connection between the communicating end systems. The principal example of a device that operates at this layer is a router. Network functions include:

- Network connection—The establishment of the path between the communication end systems (call setup)
- Data transfer—The exchange of data between the connected end systems via the established path (Data multiplexing and error control are employed, with a means to monitor and ensure the sequential delivery (flow control) of data units.)
- Connection release—The termination of the connection upon completion of the communication (call disconnect)

The Network layer relieves the higher layers of the need to know about the underlying data transmission and switching techniques used to connect systems. At this layer, the computer system engages in a dialogue with the network to specify the destination address and to request certain network facilities such as a priority.

TRANSPORT LAYER—LAYER 4

The transport layer provides a mechanism for the exchange of data between end systems. The connection-oriented transport service ensures that the data are delivered error free, in sequence, with no losses or duplications. The Transport layer and the three lower layers are collectively referred to as the Network or Transport service.

This layer provides both reliable and unreliable transport protocols. The reliable protocol is the Transport Control Protocol (TCP) in the TCP/IP stack. Also in the TCP/IP stack is the unreliable transport protocol, the User Datagram Protocol (UDP).

The Transport layer provides end-to-end service. Transport functions, which concern such issues as quality of service and cost optimization, may include:

- Connection management
- Data verification

- Error control
- Flow control
- Sequencing
- Multiplexing

The size and complexity of a transport protocol depends on the extent of the reliability of the underlying network and network layer services. Just as there are various POTS customer classes of service, there are different classes of network transport services.

SESSION LAYER—LAYER 5

The Session layer provides a means to transfer data and control information in an organized and synchronized manner. The Session layer provides the following mechanisms:

- Dialogue discipline—Includes either half duplex or full duplex.
- Grouping—The flow of data can be marked to define groups of data.
- Recovery—The session layer can provide a checkpointing mechanism for a retransmission process.

The Session layer is responsible for establishment, management, and release of each "session" between the user and the network. Network log on and user identification is an example of a Session layer function.

PRESENTATION LAYER—LAYER 6

The Presentation layer provides an interface between the Application layer and the layers below the Presentation layer. The layers below the Presentation layer use information in a format that is useful for transmission across the network. This layer provides for translation between these two information formats.

The Presentation layer provides the services to allow the application processes to interpret the meaning of the information exchanged. This layer is responsible for the "presentation" of information to the user. The presentation layer defines the syntax used between application entities and provides for the selection and subsequent modification of the representation used. Examples of such services that may be performed at this layer include data compression, format conversion, and encryption.

APPLICATION LAYER—LAYER 7

The Application layer provides the end user with a transparent communications "window" to the network. It allows an application process to access the OSI environment and to communicate with another application process. This layer contains management functions and generally useful mechanisms to support distributed applications. In addition, general-purpose applications such as file transfer, electronic mail, and terminal access to remote computers reside at this layer.

Table 1 provides an example of the communications services as provided by the OSI protocol stack.

Table 1
OSI Protocol Stack Services

Personal Computer Protocol Stack	Services Provided
Application Layer	Application support
Presentation Layer	Translation
Session Layer	Logical connections
Transport Layer	Network addresses
Network Layer	Application ports
	Error recovery
Data Link Layer	CSMA/CD protocol
	Hardware address
	Error recovery
Physical Layer	Controller
	Transceiver
	Connectors
	Cables
	Signals

Voice over Internet Protocol

VoIP OVERVIEW

In the past, Information Technology (IT) departments did not wish to be involved in the voice service requirements for an organization. This led to splitting the telecommunications functions between a department for voice services and a department for data processing and data communications services. In today's high-technology world, it will not be long before voice and data are both delivered over the organization's data network using digital signaling technology. This service is called Voice over Internet Protocol (VoIP). As the quality of VoIP improves, as standards are developed, and as the technology matures, VoIP will be used to replace the analog telephone set. Figure 1 provides a general network overview of a VoIP environment [Douskalis, 2000].

Traditionally, voice traffic was transmitted over a circuit-switched network and data traffic was transmitted over a packet-switched network. VoIP is the transmission of voice traffic over an IP network, but it is not the same as Internet Telephony. VoIP is voice transmitted as packets over a data network, whereas Internet Telephony is voice sent as packets over the public Internet. The Internet Protocol (IP) is a standard for data transmission that is based on the packet switching technology. The three main classes of IP networks in operation today include:

- IP networks owned and managed by carriers (operators)
- open and public Internet
- IP networks for closed user groups such as WANs and Intranets

Currently, voice communication is provided primarily over circuit-switched networks, which is a technology that lends itself to services requiring high quality and minimum delays. IP networks, however, are designed for the transmission of data, where delays and occasional data loss is less critical. When deploying VoIP, vendors must therefore overcome a number of issues to ensure acceptable performance. Despite the complexities, the current trend is toward using corporate IP networks for voice and data transmission. This increased interest in VoIP on the corporate network is driven by the opportunity to combine today's data and voice networks into one single network. This combination could result in significant cost reductions, an improved ability to manage the network, and easier integration with support and maintenance systems.

VoIP HISTORY

IP telephony began in the mid-1990s when Vocaltec introduced software which made it possible to make telephone calls over the Internet. It was possible to make a telephone call from one PC to another provided both the talker and the recipient used the Internet Phone and were online simultaneously. Initially, the sound quality was poor and the conversation was often characterized by long delays. The subsequent development of IP gateways, which serve to bridge the IP and PSTN networks, made it possible to make a call from a PC to an ordinary telephone. The introduction of the IP gateway also eliminated the requirement for the call to originate from a PC. It now became possible to use the conventional telephone to call an operator who forwarded the call on to the IP network [DataPro Information Services, 1999].

TECHNOLOGY OVERVIEW

Today's networks consist of circuit-switched and packet-switched facilities. Most voice traffic is circuit switched and is transmitted over a public switched telephone network (PSTN). The speed that voice traffic is transmitted consists of an aggregate rate of 64 kbps.

Figure 1
VoIP Network Environment

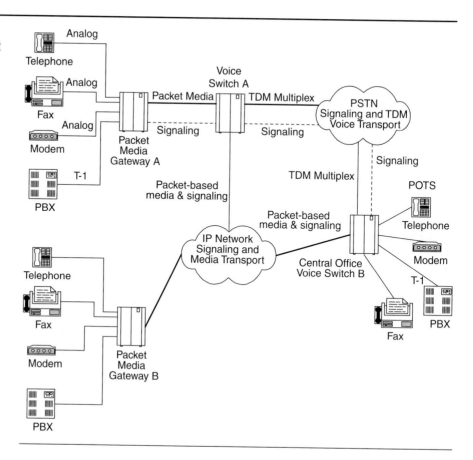

A direct connection between two connection points provides a permanent 64 kbps link for the duration of the call. This link is not available for any other purpose during this connection. PSTN provides low delay or latency and is bidirectional, and thus allows for two-way or full-duplex conversations to take place.

In a packet-switched network, data are divided into packets, each with a destination address. When the packets are transmitted through the network, the addresses are read at each network node for routing information. At the destination, the packets are reassembled and resequenced. Depending on the traffic levels and congestion in the network, packets may take different routes to the end destination. Packet switching provides a virtual circuit connection and is usually half duplex. There is no dedicated connection as is required for circuit switching, and it is therefore called a connectionless network.

The common standard for interconnection between technologically diverse networks is the TCP/IP

suite of protocols. In the TCP/IP stack, IP provides for the transportation of information; and TCP is concerned with fragmentation of the message, retransmission if required, acknowledgment of delivery, flow control, and reassembly. The layers above TCP are the application specific protocols such as Simple Mail Transfer Protocol (SMTP) for e-mail. Table 1 summarizes the functionality of the TCP/IP protocol suite.

The process for transmission of voice traffic over an IP network is as follows:

- The caller uses the PSTN and dials the access number of the IP voice gateway.
- Upon authentication of the calling party, the caller dials the number of the desired destination.
- In the gateway, the voice signal is digitized, compressed, and converted into IP packets.
- The IP packets are transmitted over an IP network that is also shared with other IP traffic.

Table 1
Functionality of TCP/IP
Suite Layers

TCP/IP Layer	Function
Application Layer	Manages the details related to specific applications. The application program elects the kind of transport needed and passes it to the Transport layer.
Transport Layer	The software segments the data to be transmitted into small packets and adds addresses. There are two transport protocols—UDP and TCP.
Network Layer (Internet)	Affects the movement of packets through the network and manages the routing of packets from node to node. Examples include SMTP and FTP.

IP TELEPHONY GATEWAY

The gateway is located between the circuit-switched and the packet-switched networks and performs the functionality for enabling voice traffic to be transmitted across different networks and technologies. Voice calls are digitized, encoded, compressed, and packetized in the originating gateway. At the destination gateway, the process is reversed. These gateways typically provide a specified number of analog or digital port interface connections on one side and a 10 Mbps or 100 Mbps Ethernet interface connection on the other side. The gateway provides the following functions:

- The interface and signaling between the networks
- Voice processing functions such as call setup and teardown
- Translation between telephone numbers and IP addresses
- Compression and decompression of voice signals
- Packetizing
- Echo control
- Silence suppression
- Forward error correction
- Jitter-buffer techniques
- QoS

Via the H323v2 protocol, the VoIP gateways are connected to a VoIP gatekeeper, which serves as a system controller, and provides for caller authentication, call accounting information, billing plans, and routing tables.

VoIP APPLICATIONS

In the near future, IP telephony will begin to offer much more than a cost-based compliment to the PSTN. The ability to integrate services over one network will become increasingly important not only from a cost perspective, but also by both enabling new advanced services and opening up new opportunities for service offerings and differentiation. It will become easier to customize and in-

tegrate voice and data solutions, thereby reducing the conformity dictated by proprietary applications. The most common applications to date are:

- Unified messaging
- Web-based call centers
- Internet call waiting
- Real-time and store-and-forward Fax over IP
- IP telephony calling cards
- IP conferencing
- Toll bypass
- International calling

VoIP ADVANTAGES

The main advantage of IP telephony comes from its complimentary function to the ordinary telephone network. The ability to bypass established players, regulations, and local loops and access charges is tempting. For the corporate community, there are additional financial advantages, and as integrated services become more readily available, the combination of lower costs and creative services will ensure that IP telephony plays a significant part of the future of telecommunications.

VoIP STANDARDS

The process of developing standards for VoIP has been evolving since the mid-1990s. Early gateways were based on proprietary protocols and were not capable of communicating with gateways on different networks. Gateways must be as robustly scalable as that which controls the PSTN. The ITU-recommended H.323 family of recommendations is the most widely used standard to bridge IP and PSTN networks, enabling interoperability between IP and PSTN gateway vendors. It provides for a common protocol that includes coding and compressing algorithms and call switching functions. Applications function seamlessly and are network, platform, and application independent. The H.323 recommendation defines how delay-sensitive data such as voice gets priority over the IP network. Version 2 includes elements for security and supplementary services such as call transfer.

The Internet Engineering Task Force (IETF) has developed the Session Initiation Protocol (SIP) as an option to H.323. SIP has several advantages over H.323 such as faster call setup, but has not managed to gain widespread support from the industry.

TECHNOLOGY PROVIDERS

This section provides a list of manufacturers of enterprise and carrier VoIP gateways and other associated communication equipment that operates in the LAN/WAN environment.

- 3Com
- Ascend
- Cisco
- Clarent
- Ericsson
- Inter-Tel
- Linkon
- Lucent Technologies
- Motorola
- MultiRech
- Netrix
- Neura
- Nokia
- Nortel Networks/Micom
- VocalTec

This Internet application is currently in the process of evolving and the requirements and environment may very well change by the time this text is published. Periodicals and up-to-date textbooks [Voice over the Net Coalition, Inc., www.von.org/, 2000] provide a wealth of information on this ever-changing subject.

Standards References

Numerous categories of documents can be utilized as a reference source in the broadband environment. This section mainly provides information on Request for Comments (RFC). There are a number of search engines available for locating these documents. The IETF Web page has a link (www.rfc-editor.org/) that provides the ability to search the Web based on a number of criteria.

REQUEST FOR COMMENTS

RFCs are documents that progress through several development stages, under the control of IETF, until they are finalized or discarded. The contents of an RFC may range from an official standardized protocol specification to research results or proposals. This list of RFCs is organized by broadband technology and related subjects and does not necessarily contain all of the relevant document references. There are numerous pages of RFCs that reference the Internet, but only a sample have been listed in this document.

ASYNCHRONOUS TRANSFER MODE

BCP0024—RSVP over ATM Implementation Guidelines
RFC2823—PPP over Simple Data Link (SDL) using SONET/SDH with ATM-like Framing
RFC2844—OSPF over ATM and Proxy-PAR
RFC2761—Terminology for ATM Benchmarking
RFC2684—Multiprotocol Encapsulation over ATM Adaptation Layer 5
RFC2682—Performance Issues in VC-Merge Capable ATM LSRs
RFC2515—Definitions of Managed Objects for ATM Management
RFC2514—Definitions of Textual Conventions and OBJECT-IDENTITIES for ATM Management
RFC2512—Accounting Information for ATM Networks
RFC2492—IPv6 over ATM Networks

RFC2417—Definitions of Managed Objects for Multicast over UNI 3.0/3.1-based ATM Networks
RFC2383—ST2+ over ATM Protocol Specification—UNI 3.1 Version
RFC2382—A Framework for Integrated Services and RSVP over ATM
RFC2381—Interoperation of Controlled–Load Service and Guaranteed Service with ATM
RFC2380—RSVP over ATM Implementation Requirements
RFC2379—RSVP over ATM Implementation Guidelines
RFC2366—Definitions of Managed Objects for Multicast over UNI 3.0/3.1, based ATM Networks
RFC2337—Intra-LIS IP multicast among Routers over ATM Using Sparse Mode PIM
RFC2331—ATM Signaling Support for IP over ATM—UNI Signaling 4.0 Update
RFC2320—Definitions of Managed Objects for Classical IP and ARP over ATM Using SMIv2 (IPOA-MIB)
RFC2269—Using the MARS Model in non-ATM NBMA Networks
RFC2226—IP Broadcast over ATM Networks
RFC2225—Classical IP and ARP over ATM
RFC2170—Application REQuested IP over ATM (AREQUIPA)
RFC2149—Multicast Server Architectures for MARS-based ATM Multicasting
RFC2098—Toshiba's Router Architecture Extensions for ATM: Overview
RFC2022—Support for Multicast over UNI 3.0/3.1-based ATM Networks
RFC1954—Transmission of Flow Labeled IPv4 on ATM Data Links Ipsilon Version 1.0
RFC1946—Native ATM Support for ST2+
RFC1932—IP over ATM: A Framework Document
RFC1926—An Experimental Encapsulation of IP Datagrams on Top of ATM

RFC1821—Integration of Real-time Services in an IP—ATM Network Architecture

RFC1755—ATM Signaling Support for IP over ATM

RFC1754—IP over ATM Working Group's Recommendations for the ATM Forum's Multiprotocol BOF Version 1

RFC1695—Definitions of Managed Objects for ATM Management Version 8.0 using SMIv2

RFC1680—IPng Support for ATM Services

RFC1626—Default IP MTU for use over ATM AAL5

RFC1577—Classical IP and ARP over ATM

RFC1483—Multiprotocol Encapsulation over ATM Adaptation Layer 5

DIGITAL SUBSCRIBER LINE

RFC2662—Definitions of Managed Objects for the ADSL Lines

FIBER DISTRIBUTED DATA INTERFACE

STD0036—Transmission of IP and ARP over FDDI Networks

RFC2467—Transmission of IPv6 Packets over FDDI Networks

RFC2019—Transmission of IPv6 Packets over FDDI

RFC1390—Transmission of IP and ARP over FDDI Networks

RFC1329—Thoughts on Address Resolution for Dual MAC FDDI Networks

RFC1285—FDDI Management Information Base

RFC1188—Proposed Standard for the Transmission of IP Datagrams over FDDI Networks

RFC1103—Proposed Standard for the Transmission of IP Datagrams over FDDI Networks

FIBRE CHANNEL

RFC2837—Definitions of Managed Objects for the Fabric Element in Fibre Channel Standard.

RFC2625—IP and ARP over Fibre Channel

FRAME RELAY

STD0055—Multiprotocol Interconnect over Frame Relay

RFC2427—Multiprotocol Interconnect over Frame Relay

RFC2115—Management Information Base for Frame Relay DTEs Using SMIv2

RFC1973—PPP in Frame Relay

RFC1604—Definitions of Managed Objects for Frame Relay Service

RFC1596—Definitions of Managed Objects for Frame Relay Service

RFC1586—Guidelines for Running OSPF over Frame Relay Networks

RFC1490—Multiprotocol Interconnect over Frame Relay

RFC1315—Management Information Base for Frame Relay DTEs

RFC1294—Multiprotocol Interconnect over Frame Relay

RFC0590—Transmission of IPv6 Packets over Frame Relay

INTEGRATED SERVICES DIGITAL NETWORK

RFC2127—ISDN Management Information Base using SMIv2

RFC1990—PPP Multilevel Protocol

RFC1356—Multiprotocol Interconnect on X.25 and ISDN in the Packet Mode

INTERNET

RFC2880—Internet Fax T.30 Features Mapping

RFC2879—Content Features Schema for Internet Fax

RFC2871—A Framework for Telephony Routing over IP

RFC2863—The Interfaces Group MIB

RFC2860—Memorandum of Understanding Concerning the Technical Work of the Internet Assigned Numbers Authority

RFC2851—Textual Conventions for Internet Network Addresses

RFC2850—Charter of the Internet Architecture Board (IAB)

MOBILE IP

RFC2794—Mobile IP Network Access Identifier Extension for IPv4

RFC2520—NHRP with Mobile NHCs

RFC2356—Sun's SKIP Firewall Traversal for Mobile IP

RFC2344—Reverse Tunneling for Mobile IP

RFC2290—Mobile—IPv4 Configuration Option for PPP IPCP

RFC2041—Mobile Network Tracing

RFC2006—The Definitions of Managed Objects for IP Mobility Support using SMIv2

RFC2002—IP Mobility Support

SWITCHED MULTIMEGABIT DATA SERVICE

STD0052—The Transmission of IP Datagrams over the SMDS Service

RFC1694—Definitions of Managed Objects for SMDS Interfaces using SMIv2

RFC1304—Definitions of Managed Objects for the SIP Interface Type

RFC1209—Transmission of IP Datagrams over the SMDS Service

SYNCHRONOUS OPTICAL NETWORK

RFC2823—PPP over Simple Data Link (SDL) using SONET/SDH with ATM-like Framing

RFC2615—PPP over SONET/SDH

RFC2558—Definitions of Managed Objects for the SONET/SDH Interface Type

RFC2175—MAPOS 16—Multiple Access Protocol over SONET/SDH with 16 Bit Addressing

RFC2171—MAPOS—Multiple Access Protocol over SONET/SDH Version 1

RFC1619—PPP over SONET/SDH

RFC1605—SONET to SONET Translation

RFC1595—Definitions of Managed Objects for the SONET/SDH Interface Type

VIRTUAL PRIVATE NETWORKS

RFC2764—A Framework for IP-based VPNs
RFC2735—NHRP Support for VPN
RFC2685—VPN Identifier
RFC2547—BGP/MPLS VPNs

WIRELESS PAGING

RFC2636—Wireless Device Configuration via ACAP
RFC2604—Wireless Device Configuration via ACAP
RFC2501—Mobile Ad hoc Networking
RFC1861—Simple Network Paging Protocol—Version 3: Two-Way Enhanced
RFC1645—Simple Network Paging Protocol—Version 2
RFC1568—Simple Network Paging Protocol—Version 1(b)

X.25

RFC1613—Cisco Systems X.25 over TCP/IP (XOT)
RFC1598—PPP in X.25
RFC1382—SNMP MIB Extension for the X.25 Packet Layer
RFC1356—Multiprotocol Interconnect on X.25 and ISDN in the Packet Mode
RFC1090—SMTP on X.25
RFC1086—ISO—TPO Bridge between TCP and X.25
RFC0874—Critique of X.25

Number Systems

This appendix includes information that will assist the reader in understanding the process for converting decimal numbers to both a binary representation and hexadecimal representation. This process is necessary for a number of network administration and management functions. A general overview will be provided in addition to a step-by-step process for accomplishing these conversions.

OVERVIEW

A number of network and data-processing applications require a conversion to binary or hexadecimal in order for the individual programs to successfully produce the desired results. As stated in Chapter 14, information received from managed network devices in the form of traps must be decoded before they can be utilized. It is also necessary for IP addresses and netmasks to be converted to binary before the "bitwise AND" process which is used to determine a network address. These network addresses are programmed in the network devices' routing tables. Network troubleshooting programs, discussed in Chapter 15, often display their information in a hexadecimal format. Sniffers and protocol analyzers display information in both binary and hexadecimal.

DECIMAL TO BINARY CONVERSION

Decimal numbers are in a base 10 format and binary numbers are in a base 2 format. There are ten different digits in decimal and two different digits in binary. To start this process, it is necessary to create a table of binary representations. Binary can be represented as 2^n, where n is a number from 0 to N. When converting IP addresses, n can vary from 0 to 7, because an IP address is

8 bits. This produces the following table for converting an IP address.

Decimal	7	6	5	4	3	2	1	0
Binary	128	64	32	16	8	4	2	1

The decimal digits of 0 through 10, therefore, can be represented in binary as follows:

Decimal	Binary
0	00000000
1	00000001
2	00000010
3	00000011
4	00000100
5	00000101
6	00000110
7	00000111
8	00001000
9	00001001
10	00001010

Additionally, when converting IP addresses, it is useful to know the binary representation of the various netmasks. The following table shows these values.

Decimal	Binary
128	10000000
192	11000000
224	11100000
240	11110000
248	11111000
252	11111100
254	11111110
255	11111111

A simple conversion of an IP address from decimal is as follows.

IP address in decimal:	10.10.10.1
IP address in binary:	00001010.00001010.00001010.00000001

The same process applied to a netmask is as follows:

Netmask in decimal:	255.240.0.0
Netmask in binary:	11111111.11110000.00000000.00000000

The first step in the conversion process is to create the 2^n heading.

The next step is to identify which of these numbers is larger than the number to be converted to binary, and then use the next smallest number as the base for the conversion. For example, use the decimal number 97 to convert to binary.

The options that are available are 1, 2, 4, 8, 16, 32, 64, 128, etc. Select 64 as it is the next smallest number from 97.

```
Binary base 2 scale 128  64  32  16  8  4  2  1
                       0   1   1   0  0  0  0  1
```

Is there a 64 in 97? **Yes.** Is there a 32 in 64? **Yes.** Is there a 16 in 1? **No.** Is there an 8 in 1? **No.** Is there a 4 in 1? **No.** Is there a 2 in 1? **No.** Is there a 1 in 1? **Yes.**

```
    97
 −  64
 =  33
 −  32
 =   1
 −   1
 =   0
```

To check the binary answer, add the elements of the binary scale that have 1s: $64 + 32 + 1 = 97$.

BINARY TO HEXADECIMAL CONVERSION

Hexadecimal characters include the digits 0 to 9 and the alpha characters A through F.

A table that provides a conversion between decimal, binary, and hexadecimal for the digits 0 through 15 follows:

Decimal	Binary	Hexadecimal
0	00000000	0
1	00000001	1
2	00000010	2
3	00000011	3
4	00000100	4
5	00000101	5
6	00000110	6
7	00000111	7
8	00001000	8
9	00001001	9
10	00001010	A
11	00001011	B
12	00001100	C
13	00001101	D
14	00001110	E
15	00001111	F

One method to convert decimal numbers to hexadecimal is to first use the previous process of converting decimal numbers to binary and then converting the binary digits to hexadecimal.

Using the previous example of a decimal 97 and a binary equivalent of 01100001, proceed as follows:

Binary 0 1 1 0 0 0 0 1 divide the binary string into bits of four digits: 0 1 1 0 | 0 0 0 1

Create a new scale of "8 4 2 1" for each four digits.

```
8  4  2  1  |  8  4  2  1

0  1  1  0  |  0  0  0  1
```

Multiply the scale by the binary digits and add.

```
0  4  2  0    0  0  0  1

      6    |          1
```

The hexadecimal representation of decimal 97 is 61.

BINARY COMPLEMENT (LOGICAL AND)

The process required to identify the network addresses that must be programmed in routers and other layer 3 devices is called "bitwise AND." Basically, the IP address and the netmask are converted to binary and the complement is taken between the two binary strings. The resulting binary string is converted back to decimal for optioning the router. When performing AND operations, each of the pairs of binary 1s and 0s are multiplied together. An example of this operation is as follows:

Binary string A	1	0	1	0	1	0	1	0
Binary string B	0	1	1	1	0	0	1	0
AND result	0	0	1	0	0	0	1	0

This process can be applied to a IPv4 32 bit address [Comer, 1999]. Step one is to convert the IP address to a binary representation. Next, convert the netmask to a binary representation. It is important that all three dots, eight 1s and 0s be properly aligned for this process to work correctly. A bit-wise AND using the IP address and the netmask produces the network address. This binary representation is then converted back into a decimal address. An example of an IP address and a netmask complement operation is as follows:

IP address	192.10.249.5	11000000.00001010.11111001.00000101
Netmask	255.255.254.0	11111111.11111111.11111110.00000000
Network Address	192.10.248.0	11000000.00001010.11111000.00000000

Broadband Case Study/Project

INSTRUCTIONS

This comprehensive enterprise network project has been designed to include most of the broadband technologies that have been presented in this textbook. There is no correct solution to this design project, only different solutions. It will be necessary to have access to the Web and a catalog of networking devices. It is also necessary to have access to pricing for carrier communication facilities. A graphics package, a word-processing application, and a spreadsheet application are also required. Access to an equipment lab will also be helpful.

Two sources of information that have been useful in the Network Design class, taught at the DeVry Technical Institute in Decatur, Georgia, are DataPro and the Black Box Corp. catalog. Technical information, connectivity, photographs, vendor information, and general network information are available from these sources. The vendor resources on the Web will also be useful in developing alternative solutions.

Information will be presented in the following categories:

- Case overview
- Geographics / locations
- User applications / dependencies
- Departmental information / security
- Employee complement / locations
- Traffic capacity / throughput requirements
- Busy hour / time zones
- Quality of service
- Cost restrictions
- Facilities

A good approach to this project is to read it thoroughly before starting the design. As in most case studies and real life, the specifications and requirements do not contain all of the information and data that are necessary to provide a viable solution. Certain details have been intentionally left out as part of the exercise; therefore, the reader should be making notes while reading the case and developing a list of questions to be answered before the design can be completed. These questions can be answered by either asking a mentor or providing personal assumptions.

Two approaches can be taken when designing this network—logical or physical. A logical design is much easier and less time consuming, but a physical design is necessary to make it work.

REQUIREMENTS

The completed case/project should include the following elements:

- A cover document that provides an overview of the solution
- A logical or physical design, or both, of the proposed solution
- A list of components with prices that make up the design
- Technical specifications of the devices and components of the design
- Review of the technologies that are utilized in the solution
- Timelines or implementation schedules
- A document that shows the flow of information and the flow of the product
- Assumptions in the design

Good designing!!

Figure 1
RST Enterprise Network

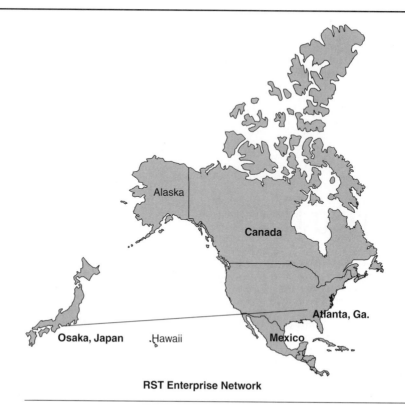

RST Enterprise Network

I. CASE OVERVIEW

RST is a multinational corporation headquartered in Osaka, Japan, which manufactures electronic components for inclusion into larger sub-assemblies. Raw materials for these components are purchased on the open market. These raw materials are not available locally. The subassemblies are shipped to final-assembly operations in the United States. RST is a billion-dollar operation and employs 800 local workers and 920 employees worldwide. The subassemblies are made to order and are not available off-the-shelf for retail sales.

The main operation for final assembly of the finished product is located on the south side of Atlanta, Georgia near the Hartsfield International Airport. The U.S. corporate office is located in downtown Atlanta. A warehouse and shipping operation is also near the airport assembly operation.

The finished product is packaged and shipped out of the airport operation to 1,000 suppliers that are located across the United States, Canada, and Mexico. These suppliers sell to the end users.

RST currently utilizes dial-up and leased-line communications services in the continental United States,

Canada, and Mexico. International telephone service is utilized between Osaka and the U.S. operations. The transmission speed currently utilized for data is 19.2 kbps. Access to the Internet is 56 kbps through local ISPs. Communication equipment consists of a mix of vendor brands. There are no broadband services deployed in this network. The ISP access to the Internet is the most sophisticated service that RST utilizes, and that is used for surfing.

A new management team has been employed to bring RST into the twenty-first century. They are soliciting for an integrated enterprise network design that will create a competitive advantage for RST. Their time frame for this proposal is thirty days. Cost is a consideration; however, RST will consider a solution that provides a positive cost/benefit relationship. This new system must be fully operational no later than six months after award of contract.

II. GEOGRAPHICS/LOCATIONS

The world headquarters for RST is located near Osaka, Japan. The manufacturing operation for the various electronic subassemblies is located in an adjacent campus environment. The U.S. headquarters is located on Peachtree Street in downtown Atlanta. The final assembly plant for the electronic devices is located south of the Atlanta Hartsfield Airport in an industrial park. The shipping and warehouse operation is also located in the same general area.

The Osaka headquarters complex includes a twenty-story building with an adjacent one-story computer complex. The manufacturing and shipping buildings are all one story.

The Atlanta headquarters are located on the top eight stories of a fifty-story tower. The computer operations are located on the top floor of the complex. The manufacturing and distribution center at the airport is housed in one-story buildings.

Completed electronic devices are shipped from the Atlanta warehouse to 1,000+ distributors across the United States, including Alaska and Hawaii. Additional distributors are located in Canada and Mexico.

III. USER APPLICATIONS/DEPENDENCIES

Computer applications at the world headquarters include the normal requirements to run a large corporation. These include management, finance, personnel, marketing and sales, research and development, manufacturing and production, and computer operations departments. Similar applications are in place in the Atlanta headquarters. There are also additional applications directed at the manufacturing, packaging, and shipping in support of the end product. These applications are resident on the large mainframe computer located at the Atlanta headquarters. The distributors have their own systems and in many cases are not compatible with the hardware and software at the Atlanta headquarters.

IV. DEPARTMENTAL INFORMATION/SECURITY

There is a Token Ring LAN at the Osaka world headquarters. Each of the departments is autonomous, but all departments share the resources that are available on the powerful mainframe computer. This computer configuration includes a very large disk farm that utilizes the RAID technology. It also includes a front end processor (FEP) and a separate communications controller for access to the various network technologies.

There is an Ethernet LAN at the Atlanta headquarters and a separate 802.3 LAN located at the airport facility. There is a client-server system located in the warehouse and in the manufacturing operation. Manufacturing and shipping are located on separate segments of the LAN, but both departments have access to the headquarters mainframe.

V. EMPLOYEE COMPLEMENT/LOCATIONS

The employee complement for the Osaka world headquarters and the Atlanta headquarters is as follows:

OSAKA:

Management	50
Finance	50
Personnel	25
Marketing and sales	100
Research and development	100
Manufacturing and production	325
Distribution	100
Computer operations department	50

ATLANTA:

Management	45
Finance	40
Personnel	25
Marketing and sales	200
Research and development	10
Manufacturing and production	400
Distribution	150
Computer operations department	40

VI. TRAFFIC CAPACITY/THROUGHPUT REQUIREMENTS

The LAN operations in Osaka and Atlanta have been experiencing throughput difficulties. These difficulties are between the headquarters operations and the manufacturing arms at both locations. Response time is also slow on the administrative LANs at both locations. The users are requesting a response time of five seconds or less.

Most of the employees have access to a desktop workstation. All employees in the manufacturing, production, and distribution departments do not require a workstation at this time. There are a number of shared printers in each department. Each department also has access to a file server. The workstation inventory is as follows:

OSAKA:

Management	50
Finance	50
Personnel	25
Marketing and sales	100
Research and development	100
Manufacturing and production	25
Distribution	10
Computer operations department	25

ATLANTA:

Management	45
Finance	40
Personnel	25
Marketing and sales	200
Research and development	10
Manufacturing and production	40
Distribution	15
Computer operations department	5

VII. BUSY HOUR/TIME ZONES

Many of the computer applications are batch in nature. These applications are normally run during the evening and night shifts. There are three shifts in computer operations for both Osaka and Atlanta. All of these applications must be completed by the morning shift of manufacturing and warehouse operations in both Osaka and Atlanta.

The Osaka and Atlanta manufacturing and shipping operations work seven days a week, twenty-four hours a day. The headquarters' personnel work schedule is 9 to 5 in both locations. Overtime often occurs at both locations.

VIII. QUALITY OF SERVICE

The quality of service operation is part of the manufacturing department. There are also departmental QoS initiatives.

Efforts are underway to improve workstation response time and reduce overtime. There are also corporate initiatives to improve product quality. Product deliveries are often late, incorrect, or misdirected. Product returns are an issue.

IX. COST RESTRICTIONS

RST is looking for a solution that is both cost effective and viable. A cost/benefit analysis will be required to satisfy the CIO and CFO.

X. FACILITIES

The Atlanta CLEC, ILEC, IXC infrastructure supports all of the broadband technologies that have been presented in this textbook. Distance limitations are an issue. There may also be RFI, EMI, and electrical power issues.

Acronyms

NUMERICS

2B1Q	Two Binary, One Quaternary

A

A/D	Analog to Digital
AAL	ATM Adaptation Layer
ABR	Available Bit Rate
ACF	Access Control Field
ACK	Acknowledge
ADM	Add/Drop Multiplexer
ADPCM	Adaptive Differential Pulse Code Modulation
ADSL	Asymmetric Digital Subscriber Line
AES	Advanced Encryption Standard
AIS	Alarm Indication Signal
AIU	Access Interface Unit
AM	Amplitude Modulation
AMI	Alternate Mark Inversion
AMPS	Advanced Mobile Phone Service
AN	Access Node
ANR	Automatic Network Routing
ANSI	American National Standards Institute
AOL	America OnLine
APS	Automatic Protection Switching
ARP	Address Resolution Protocol
ARPA	Advanced Research Projects Agency
ARPANET	Advanced Research Projects Agency Network
ARQ	Automatic Repeat Request
ASCII	American Standard Code for Information Interchange
ASK	Amplitude Shift Keying
ASN.1	Abstract Syntax Notation One
ATDM	Asynchronous Time Division Multiplexing
ATM	Asynchronous Transfer Mode
ATP	Appletalk Transaction Protocol
ATU-C	ADSL Termination Unit—Central Office
ATU-R	ADSL Termination Unit—Remote
AUX	Auxiliary
AWG	American Wire Gauge

B

B8ZS	Bipolar with 8 Zero Substitution
BA	Buffer Allocation
Bc	Committed Burst Rate
BCH	Bose Chadhuri-Hocquenghem Code
B CHANNEL	ISDN Bearer Channel
Be	Excess Burst Rate
BECN	Backward Explicit Congestion Notification
BER	Bit Error Rate
BERT	Bit Error Rate Test
BGP	Border Gateway Protocol
BICI	Broadband Intercarrier Interface
BIP8	Bit Interleave Parity 8
BISDN	Broadband Integrated Services Digital Network
BITS	Building Integrated Timing Supply
BOC	Bell Operating Companies
BOM	Beginning Of Message
Bps	Bits Per Second
BRA	Basic Rate Access
BRI	ISDN Basic Rate Interface
BRITE	Basic Rate Interface Transmission Equipment
BRZ	Bipolar Return to Zero
Btag	Beginning Tag
BUS	Broadcast Unknown Server

C

CAP	Carrierless Amplitude and Phase Modulation
CAP	Competitive Access Provider
CBR	Constant Bit Rate
CCC	Clear Channel Capability
CCS	Hundred Call Seconds Per Hour
CCS7	Common Channel Signaling 7
CCITT	Consultative Committee for International Telephony and Telegraphy
CDMA	Code Division Multiple Access
CDPD	Cellular Digital Packet Data
CEPT	Conference Europeene des Postes et Telecommunications
CERN	European Center for Nuclear Research
CIR	Committed Information Rate
CLEC	Competitive Local Exchange Carrier
CLLI	Common Language Location Identifier
CLP	Cell Loss Priority
CMI	Coded Mark Inversion
CMIP	Common Management Information Protocol
CO	Central Office
COT	Central Office Terminal
CPE	Customer Premises Equipment
CRC	Cyclic Redundancy Check
CSC	Common Signaling Channel
CSMA/CD	Carrier Sense Multiple Access/ Collision Detection
CSU	Channel Service Unit
CTIA	Cellular Telecommunications Industry Association

D

D/A	Digital to Analog
DAA	Digital Access Arrangement
D CHANNEL	ISDN director channel
DAC	Dual Attachment Concentrator
DACS	Digital Access Cross Connect Switch
DARPA	Defense Advanced Research Projects Agency
DAS	Dual Attachment Station
DAVIC	Digital Audio Visual Council
DC	Direct Current
DCE	Data Communications Equipment
DCE	Distributed Computing Environment
DCS	Digital Cross-connect System
DDD	Direct Distance Dialing
DDS	Digital Data Service
DEA	Data Encryption Algorithm
DES	Data Encryption Standard
DLC	Digital Loop Carrier
DLCI	Data Link Connection Identifier
DMT	Discrete Multitone
DNS	Domain Name System

DPSK	Differential Phase Shift Keying
DQDB	Distributed Queue Dual Bus
DSL	Digital Subscriber Line
DSLAM	Digital Subscriber Line Access Multiplexer
DSP	Digital Signal Processing
DSS	Digital Satellite System
DSU	Data Service Unit
DSX	Digital Signal Cross-connect
DTE	Data Terminal Equipment

E

EARN	European Academic Research Network
EBCDIC	Extended Binary Coded Decimal Interchange Code
ECF	Echo Frames
EDI	Electronic Data Interchange
EFS	Error-Free Seconds
EIA	Electronics Industries Association
EMI	Electromagnetic Interference
EMR	Electromagnetic Radiation
EO	End Office
EOM	End Of Message
ES	End System
ES	Errored Second
ESF	Extended Superframe
ESF	Extended Service Frames
ESS	Electronic Switching System
Etag	End Tag
ETSI	European Telecommunications Standards Institute

F

FC	Fibre Channel
FCC	Federal Communications Commission
FCS	Frame Check Sequence
FDDI	Fiber Distributed Data Interface
FDL	Facility Data Link
FDM	Frequency Division Multiplexing
FEBE	Far-End Block Error
FEC	Forward Error Correction
FECN	Forward Explicit Congestion Notification
FERF	Far-End Receive Failure
FITL	Fiber In The Loop
FM	Frequency Modulation
FOT	Fiber Optic Terminal
FR	Frame Relay
FRAD	Frame Relay Access Device
FRAD	Frame Relay Assembler/ Disassembler
FSK	Frequency Shift Keying
FTP	File Transfer Protocol
FTTC	Fiber-To-The-Curb
FTTH	Fiber-To-The-Home
FTTN	Fiber-To-The-Neighborhood
FUNI	Frame User Network Interface

G

GEO	Geosynchronous Earth Orbit
GEOS	Geosynchronous Earth Orbit Satellite
GFC	Generic Flow Control
GSM	Global System for Mobile (Communications)
GUI	Graphical User Interface

H

HDLC	High-level Data Link Control
HDSL	High-bit-rate Digital Subscriber Line
HDTV	High-Definition Television
HE	Head End
HEC	Header Error Control
HFC	Hybrid Fiber Coaxial
HSSI	High-Speed Serial Interface
HTML	HyperText Markup Language
HTTP	Hypertext Transfer Protocol

I

IANA	Internet Assigned Numbers Authority
ICI	Intercarrier Interface
ICIP	Intercarrier Interface Protocol
ICMP	Internet Control Message Protocol
IDLC	Integrated Digital Loop Carrier
IDSL	ISDN Digital Subscriber Line
IEC	Interexchange Carrier
IEEE	Institute of Electrical and Electronics Engineers
IETF	Internet Engineering Task Force
IGMP	Internet Group Management Protocol
IISP	Interim Interswitch Protocol
ILEC	Incumbent Local Exchange Carrier
ILMI	Interim Link Management Interface
IP	Internet Protocol
IPX	Internet Package eXchange
ISDN	Integrated Services Digital Network
IS	Intermediate System
ISO	International Standards Organization
ISP	Internet Service Provider
ITU-T	International Telecommunication Union—Telecom
IXC	Interexchange Carrier

K

KHz	Kilohertz

L

LAN	Local Area Network
LANE	LAN Emulation
LAPB	Link Access Procedure—Balanced
LAPD	Link Access Procedure—D Channel
LAPF	Link Access Procedure Frame
LATA	Local Access Transport Area
LEC	LAN Emulation Client
LEC	Local Exchange Carrier
LED	Light-Emitting Diode
LEO	Low Earth Orbit
LEOS	Low Earth Orbit Satellite
LER	Link Error Rate
LES	LAN Emulation Server
LLC	Logical Link Control
LMI	Local Management Interface
LOH	Line Overhead
Lpbk	Loopback Indication
LTE	Line Terminating Equipment
LTU	Line Termination Unit
LUNI	LAN Emulation User Network Interface

M

MAC	Media Access Control
MAN	Metropolitan Area Network
MAU	Media Access Unit
MAU	Multistation Access Unit
MDF	Main Distribution Frame
MEO	Medium Earth Orbit
MEOS	Medium Earth Orbit Satellite
MFJ	Modified Final Judgment
MIB	Management Information Base
MIC	Medium Interface Connector
MID	Multiplex Identification
MIME	Multipurpose Internet Mail Extension
MPEG	Motion Picture Experts Group
MSB	Most Significant Bit
MTBT	Mean Time Between Failures
MTTR	Mean Time to Repair

N

NAP	Network Access Point
NCP	Network Control Protocol
NCTE	Network Channel Terminating Equipment
NDIS	Network Driver Interface Standard
NE	Network Element
NETBEUI	NetBIOS Extended User Interface
NetBIOS	Network Basic Input/Output System
NEXT	Near-End CrossTalk
NI	Network Interface
NIC	Network Interface Card
NIF	Neighborhood Information Frame
N-ISDN	Narrowband ISDN
NIU	Network Interface Unit
NLPID	Network Layer Protocol Identifier
NN	Network Node
NNI	Network-Network Interface (Network Node Interface)
NRZ	Non-return to Zero
NRZI	Non-return to Zero Inversion
NSA	Next Station Addressing
NSAP	Network Service Access Point
NT	Network Termination
NT-1	Network Termination Type 1

NT-2	Network Termination Type 2
NTU	Network Termination Unit

O

OAM	Operations, Administration, and Management
OC	Optical Carrier
OC-1	Optical Carrier Level 1
OC-N	Optical Carrier Level N
ODI	Open Driver Interface
OH	Overhead
ONU	Optical Network Unit
OS	Operating System
OSI	Open System Interconnection
OSPF	Open Shortest Path First

P

PAD	Packet Assembler/Disassembler
PAM	Pulse Amplitude Modulation
PBX	Private Branch Exchange
PCM	Pulse Code Modulation
PCR	Peak Cell Rate
PCS	Personal Communications Service
PDH	Plesiochronous Digital Hierarchy
PDN	Public Data Network
PDU	Protocol Data Unit
PHY	Physical Layer
PLCP	Physical Layer Convergence Protocol
PLOAM	Physical Layer Operations And Management
PM	Performance Monitoring
PM	Phase Modulation
PMD	Physical Media Dependent
PMF	Parameter Management Frames
PNNI	Private Network-to-Network Interface
POH	Path Overhead
POP	Point of Presence
POTS	Plain Old Telephone Service
PPP	Point-to-Point Protocol
PRA	Primary Rate Access
PRCA	Proportional Rate Control Algorithm
PRI	Primary Rate Interface
PSDN	Public Switched Data Network
PSK	Phase Shift Keying
PSN	Packet Switched Network
PSR	Previous Segment Received
PSTN	Public Switched Telephone Network
PTE	Path Terminating Equipment
PTI	Payload Type Identifier
PTM	Packet Transfer Mode
PUB	Publication
PUC	Public Utility Commission
PVC	Permanent Virtual Circuit

Q

QAM	Quadrature Amplitude Modulation
QoS	Quality of Service
QPSK	Quaternary Phase Shift Keying

R

RADSL	Rate Adaptive Digital Subscriber Line
RAF	Resource Allocation Frames
RAM	Random Access Memory
RAM	Remote Access Multiplexer
RBOC	Regional Bell Operating Company
RDF	Resource Denied Frames
RDT	Remote Digital Terminal
RF	Radio Frequency
RFC	Request for Comment
RFI	Radio Frequency Interference
RIP	Routing Information Protocol
RMON	Remote Network Monitoring
ROM	Read-Only Memory
RSVP	Resource ReSerVation Protocol
RT	Remote Terminal
Rx	Transmit Receive
Rz	Return to Zero

S

S/N	Signal-Noise Ratio
SAC	Single Attachment Concentrator
SAP	Service Access Point
SAR	Segmentation And Reassembly
SAS	Single Attachment Station
SCR	Sustainable Cell Rate
SCSI	Small Computer System Interface
SDH	Synchronous Digital Hierarchy
SDLC	Synchronous Data Link Control
SDSL	Symmetric DSL
SDU	Service Data Unit
SEAL	Simple and Efficient Adaptation Layer
SF	Superframe Format
S/I	Signal-to-interference ratio
SIF	Station Information Frames
SIP	SMDS Interface Protocol
SLC	Subscriber Loop Carrier
SM	Service Module
SMDS	Switched Multimegabit Data Service
SMF	Single-Mode Fiber
SMT	Station Management
SMTP	Simple Mail Transfer Protocol
SN	Sequence Number
SNAP	Subnetwork Access Protocol
SNMP	Simple Network Management Protocol
SNP	Sequence Number Protection
SNR	Signal-to-noise ratio
SOF	Start of Frame
SOH	Section Overhead
SOHO	Small Office/Home Office
SONET	Synchronous Optical Network
SPANS	Simple Protocol for ATM Network Signaling
SPE	Synchronous Payload Envelope
SPID	Service Profile Identifier
SPX	Sequence Packet Exchange

SRF	Status Report Frame
SRM	Subrate Multiplexing
SRTS	Synchronous Residual Time Stamp
SS7	Signaling System 7
SSCOP	Service Specific Connection-Oriented Protocol
SSCS	Service Specific Convergence Sublayer
SSM	Single Segment Message
ST	Segment Type
STDM	Statistical Time Division Multiplexer
STE	Section Terminating Equipment
STM	Synchronous Transport Mode
STP	Shielded Twisted Pair cable
STS	Synchronous Transport Signal
STS-1	Synchronous Transport Signal Level 1
SVC	Switched Virtual Circuit
SYNTRAN	Synchronous Transmission

T

TA	Terminal Adapter
TAXI	Transparent Asynchronous Transmitter-receiver Interface
TCP	Transmission Control Protocol
TDM	Time Division Multiplexing
TDMA	Time Division Multiple Access
TE	Terminal Equipment
THT	Token Holding Timer
TOH	Transport Overhead
TP	Twisted Pair
TRT	Token Rotation Timer
TTL	Time To Live
TTRT	Target Token Rotation Timer
Tx	Transmit Send

U

UBR	Unspecified Bit Rate
UDP	User Datagram Protocol
UHF	Ultra High Frequency
ULP	Upper Layer Protocol

UME	UNI Management Entity
UNA	Upstream Neighbor Address
UNI	User Network Interface
UPC	Usage Parameter Controls
UPS	Uninterruptible Power Supply
URL	Universal Resource Locator
UTP	Unshielded Twisted Pair
UU	User to User

V

VAR	Value-added Reseller
VASSCS	Video/Audio Service Specific Convergence Sublayer
VBR	Variable Bit Rate
VC	Virtual Channel
VCC	Virtual Channel Connection
VCFC	Virtual Circuit Flow Control
VCI	Virtual Channel Identifier
VDSL	Very-high-data-rate Digital Subscriber Line
VHF	Very High Frequency
VLAN	Virtual LAN
VoIP	Voice over IP
VP	Virtual Path
VPC	Virtual Path Connection
VPI	Virtual Path Identifier
VT	Virtual Tributary

W

WAN	Wide Area Network
WDM	Wavelength Division Multiplex
WWW	World Wide Web
WYSIWYG	What You See Is What You Get

X

xDSL	x-Type Digital Subscriber Line

Z

ZCS	Zero Code Suppression

2B1Q. 2 binary, 1 quarternary. 2B1Q is a four-level pulse amplitude modulation (PAM) system used for HDSL, S-HDSL, and ISDN BRI.

8B/10B. The encoding scheme used for Fibre Channel and Gigabit Ethernet. Each 8 bits of data are converted into 10 bits for transmission. The 8B/10B scheme was developed and patented by IBM for use in its 200 megabaud ESCON interconnect system.

802.11 wireless networking standard. An IEEE standard for wireless networking; a version of the 802.11 standard appeared late in 1997.

AAL-1. Supports constant bit rate, time-dependent traffic such as voice and video.

AAL-2. Supports variable bit rate video transmission.

AAL-3/4. Supports variable bit rate, delay-tolerant data traffic requiring some sequencing and/or error detection. Includes a combination of connectionless and connection-oriented support.

AAL-5. Supports variable bit rate, delay-tolerant connection-oriented data traffic requiring minimal sequencing or error detection.

Abstract Syntax Notation One (ASN.1). LAN rules and symbols that are used to describe and define protocols and programming languages.

access line. A communications line (circuit) interconnecting a frame-relay-compatible device to a Frame Relay switch. Sometimes it is called a local loop. It is the connection between a subscriber's facility and a public network.

access point. A communications device that provides connectivity to a LAN in a wireless environment.

add drop multiplexer (ADM). A synchronous multiplexer that is used to add and drop DS1 signals into the SONET ring. The ADM is also used for ring healing in the event of a ring failure.

ADSL transceiver unit-central office (ATU-C). A splitter device located at the central office.

ADSL transceiver unit-remote (ATU-R). A splitter device located at the user's location.

Advanced Mobile Phone Service (AMPS). Another name for North American analog cellular phone system. The spectrum allocated to AMPS is shared by two cellular phone companies which are located in each geographic market.

agent. Part of the SNMP structure that is loaded onto each managed device to be monitored.

American National Standard Institute (ANSI). Sets information systems standards.

amplifier. A device used on analog circuits to strengthen and retransmit signals. It also magnifies noise and other impairments.

amplitude. A carrier wave characteristic that is manipulated to represent ones and zeros (wave height). It is the size or magnitude of a voltage or current analog waveform.

analog. A transmission mode that pertains to data in the form of continuously variable physical qualities.

ANSI T1.413. The interface standard for MDT ADSL.

antenna. A tuned electromagnetic device that can send and receive broadcast signals at particular frequencies. In wireless networking devices, an antenna is an important part of the device's sending and receiving circuitry.

arbitrated loop. A Fibre Channel topology that provides for attachments to multiple communicating ports in a loop.

ARPANET. A four-node network, named for and sponsored by the U.S. Defense Department's Advanced Research Project Agency (ARPA), that evolved into the Internet.

Asymmetrical digital subscriber line (ADSL). A method of transmitting at speeds up to 7 Mbps in one direction and up to 640 kbps in the other direction over a single copper telephone line.

asymmetrical transmission. Transmission that sends data at different rates in each direction, faster downstream than upstream.

asynchronous transfer mode (ATM). A dedicated-connection switching technology that organizes digital data into 53 byte cells or packets and transmits them over a medium using digital signal technology.

asynchronous transmission. A means of transmitting data over a network wherein each character contains a start and stop bit to keep the transmitting and receiving terminals in synchronization with each other.

ATM adaptation layer (AAL). A set of standardized protocols and formats that define support for circuit emulation, packet audio and video, and connectionless and connection-oriented data services. The process includes segmentation of the original data into the 53 octet ATM cell. See AAL-1 through AAL-5.

ATM forum. A consortium of hardware vendors, telecommunications service providers, and users who work with the ITU-T on specifications for LAN and WAN applications of ATM.

attenuation. Loss of signal strength during transmission due to resistance of the media.

authentication. In cryptography, the process of ensuring that the data come from the original source claimed.

authentication header (AH). In IPSec, the IP header used to verify that the contents of a packet have not been altered.

available bit rate (ABR). Provides leftover bandwidth whenever it is not required by the variable bit rate traffic. A flow control mechanism supports several types of feedback to control the source rate in response to changing ATM layer transfer characteristics.

backward explicit congestion notification (BECN). A frame relay flow control mechanism. It is a 1 bit field in the frame Address Field for use in congestion management.

baseline. A reference point that includes information for system comparisons when there is a malfunction in the enterprise network.

basic rate access (BRA). See basic rate interface (BRI).

basic rate interface (BRI). Consists of two 64 kbps B channels and one 16 kbps D channel. A Basic Rate user can have up to 128 kbps service.

beacon token. Used to check integrity of the primary FDDI ring following the repair of a station on that ring.

bearer (B) channel. The ISDN telecommunications channel that provides a means of transmitting and recording voice and data information in real time with changing message content. The B channel runs at 64 kbps.

bipolar 8 zero substitution (B8ZS). A technique used to accommodate the ones density requirement for digital T-carrier facilities in the public network.

bluetooth. A wireless technology that allows for the communication between mobile phones and desktop computers.

Broadband ISDN (B-ISDN). Broadband ISDN networks operating over fiber-optic links at speeds of 100 Mbps, 155 Mbps, and 600 Mbps.

Broadband network termination (BNT). Broadband node termination between the subscriber's premises network and the broadband access facility.

broadband optical telepoint network. An implementation of infrared wireless networking that supports broadband services equal to those provided by a wired network.

broadcast radio. A technique for transmitting signals, such as network data, by using a transmitter to send those signals through a communications medium. For wireless networks, this involves sending signals through the atmosphere, rather than over a wire.

browser. A program that permits its user to navigate the Web by accessing Web documents that have been coded in a language called HTML.

burstiness. In the context of a Frame Relay network, data that use bandwidth only sporadically; that is, information that does not use the total bandwidth of a circuit 100 percent of the time.

bursty. Data transmitted in short, uneven spurts.

buss. A wiring and media access strategy in which stations are wired to a common bus.

cable modem. Modem designed for use on TV coaxial circuit.

capture. The process in which an analyzer records network traffic for interpretation.

carrierless amplitude and phase modulation (CAP). A two dimensional modulation line code used in ADSL.

CDPD. See Cellular Digital Packet Data.

cell. The basic unit for ATM switching and multiplexing. Each cell consists of a 5 byte header and 48 bytes of payload.

cell loss priority (CLP). This bit in the ATM cell header indicates two levels of priority for ATM cells. CLP=0 cells are higher priority than CLP=1 cells. CLP=1 cells may be discarded during periods of congestion.

cell loss ratio (CLR). A negotiated QoS parameter; and acceptable values are network specific.

cellular digital multiple access (CDMA). Same as code division multiple access.

cellular digital packet data (CDPD). A communications technology that sends packets of digital data over unused cellular voice channels at a rate of 19.2 kbps. CDPD is one of an emerging family of mobile computing technologies.

cellular packet radio. A communications technology that sends packets of data over radio frequencies different from those used for cellular telephones. A generic term for an emerging family of mobile computing technologies.

certificate authority (CA). A trusted company or organization that will accept a user's public key, along with some proof of the user's identity, and serve as a repository of digital certificates.

CIR. See committed information rate.

circuit switching. A switching process in which physical circuits are created, maintained, and terminated for individual point-to-point or multipoint connections.

coaxial cable (coax). A cable composed of an insulated central conducting wire wrapped in another cylindrical conducting wire. This package is then wrapped in another insulating layer and an outer protecting layer. RG-59 coax is utilized in cable modem installations.

code division multiple access (CDMA). Uses digital encoding to multiplex several channels onto one line.

coder/decoder (CODEC). Used to digitize analog voice signals.

committed burst rate (B_c). Describes the maximum amount of data that a user is allowed to offer to the network during some time interval. This is established during call setup or preprovisioned with a PVC.

committed information rate (CIR). The rate at which a frame relay network agrees to transfer information under normal conditions, averaged over a minimum increment of time. Data sent above the CIR is subject to discard if the network gets congested.

common channel signaling (CCS7). A network architecture that uses Signaling System 7 (SS7) protocol for the exchange of information between telecommunications nodes and networks on an out-of-band basis.

Common Management Information Protocol (CMIP). The network management standard for OSI networks.

common part convergence sublayer (CPCS). The portion of the convergence sublayer of an AAL that remains the same regardless of traffic.

common part convergence sublayer/service data unit (CPCS-DSU). Protocol data unit to be delivered to the receiving AAL layer by the destination CPSC.

community name. Similar to a security password.

compulsory tunnel. Tunnels created without the user's consent, which may be transparent to the end user. The client-side endpoint of a compulsory tunnel typically resides on a remote access server.

compute cluster. Links multiple servers and workstations with an integrated network for processing.

computer telephony integration (CTI). A term for connecting a computer to a telephone switch and to allow the computer to issue commands to manipulate switch traffic. The classic application for CTI is in Call Centers.

concatenation. A mechanism for allocating contiguous bandwidth for transport of a payload associated with a super-rate service. The set of bits in the payload is treated as a single entity.

concentrator. A device that allows a relatively large number of circuits to share a single circuit or facility. Traffic is concentrated through a process of multiplexing.

constant bit rate (CBR). Provides a guaranteed amount of bandwidth to a given virtual path, thereby producing the equivalent of a leased T-1 or T-3.

cookie. The Internet mechanism that lets site developers place information on a client's computer for later use.

copper distributed data interface (CDDI). A version of FDDI that runs at 100 Mbps on unshielded twisted pair (UTP) instead of fiber.

customer premises equipment (CPE). Telephone and communications equipment that resides on the customer's premise.

cyclic redundancy check (CRC). A computational means to ensure the accuracy of frames transmitted between devices in a Frame Relay network. The mathematical function is computed, before the frame is transmitted, at the originating device.

data (D) channel. The telecommunications channel in ISDN that is used for signaling and control, and optionally for packet-switched user information. The D Channel speed for BRI is 16 kbps and the D Channel speed for PRI is 64 kbps.

data communications equipment (DCE). Device or connection of a communications network that comprises the network end of the user-to-network interface. Modems and interface cards are examples of DCE.

data encryption standard (DES). A block-cipher algorithm created by IBM and endorsed by the U.S. government in 1977. It uses a 56 bit key and operates on blocks of 64 bits.

data exchange interface (DXI). A variable-length frame-based ATM interface between a DTE and a special ATM CSU/DSU. The ATM CSU/DSU converts between the variable-length DXI frames and the Fixed-length ATM cells.

data link connection identifier (DLCI). A value that specifies a PVC or SVC in a frame relay network. It is a 10 bit field in the Address Field.

data service unit (DSU). A device used in digital transmission that adapts the physical interface on a DTE device to a transmission facility such as T-1 or E-1. The DSU is also responsible for functions such as timing.

data terminal equipment (DTE). The device, generally belonging to a data communications user, that provides the functional and electrical interface to the communications medium.

datagram. Refers to the Class 3 Fibre Channel service that allows data to be sent rapidly to multiple devices attached to the fabric, with no confirmation of receipt. Globally addressed message packets found in connectionless networks.

dedicated connection. A communicating circuit guaranteed and retained by the fabric for two given N_ports.

deregulation. The removal of regulatory authority to control certain activities of entrenched telephone companies. It was supposed to benefit the consumers.

diagnosis. A problem on a network that has been detected by an analyzer. The analyzer detects and alerts users to

diagnoses as it discovers them on the network to which it is attached.

digital certificate. An electronic document, issued by a certificate authority, used to establish a company's identity by verifying its public key.

digital signal. A discrete or discontinuous signal; one where various states are pulses that are discrete intervals apart.

digital subscriber line (DSL). Carries both voice and data signals at the same time, in both directions, and signaling data used for call information and customer data. DSL is another name for an ISDN BRI channel.

digital subscriber line access multiplexer (DSLAM). This technology concentrates traffic in ADSL implementations through Time Division Multiplexing (TDM) at the Central Office (CO) or remote line shelf.

direct-sequence. The form of spread-spectrum data transmission that breaks data into constant length segments called chips and transmits the data on multiple frequencies.

discard eligible (DE). The Frame Relay frames that are above the committed information rate. In case of congestion, the carrier can discard frames that have had their DE bits set on.

discrete multitone (DMT). A technology that uses digital signal processors to pump more than 6 Mbps of multimedia signals over today's existing copper wiring.

Distributed Queue Dual Bus (DQDB). The protocol used in 802.6.

divestiture. A consent decree where AT&T would divest itself of the 22 operating telephone companies.

DS-3. Consists of 672 voice circuits and 28 T-1 channels and operates at 44.736 Mbps.

DTE. See data terminal equipment.

dual attachment concentrator (DAC). A concentrator that offers two S ports for connections to the FDDI network and multiple M ports for attachment to DTE devices and other concentrators.

dual attached dual homed (DADH) station. Using Dual Attachment Stations to attach to Dual Attachment Concentrators so as to provide connectivity to both counterrotating FDDI rings, which allows for two completely separate accesses to the FDDI ring.

dual attachment station (DAS). Allows access to two separate cable systems at the same time, protecting against cable failure or damage; a device that connects to both counterrotating rings of the FDDI network.

E.164. An international numbering scheme to assign numbers to telephone lines. ATM utilizes E.164 addressing for public network addressing.

electronic data interchange (EDI). The direct computer-to-computer exchange of information normally provided on standard business documents such as invoices, bills of lading, and purchase orders.

electronic eavesdropping. The ability to listen to signals passing through some communications medium by virtue of detecting its emissions. This is especially easy to do for many wireless networking technologies because they broadcast their data into the atmosphere.

encapsulating security payload (ESP). In IPSec, an IP header that contains the encrypted contents of an IP packet.

encapsulation. A process by which an interface device places an end device's protocol-specific frames inside a Frame Relay frame.

encryption. The transformation of data from a meaningful code, called clear text, to a meaningless sequence of digits and letters that must be decrypted before it becomes meaningful again.

end device. The ultimate source or destination of data flowing through a Frame Relay network, sometimes referred to as data terminal equipment (DTE).

end station. These devices (hosts or PCs) enable the communication between ATM end stations and end stations on legacy LANs or among ATM end stations.

end system (ES). A system where an ATM connection is terminated or initiated. An originating end system initiates the ATM connection, and terminating end system terminates the ATM connection.

excess burst rate (B_e). Describes the amount of data a user may send that exceeds the committed burst rate during the time interval. The Excess Burst Rate also identifies the maximum number of bits that the network will attempt to deliver in excess of the Committed Burst Rate during this time interval.

exchange. Composed of one or more nonconcurrent sequences for a single operation.

exchange termination (ET). Typically includes the Line Termination in the CO, is responsible for multiplexing and demultiplexing the B and D Channels, and routes bit streams to the appropriate destinations within the switch.

explicit forward congestion indication (EFCI). An indication in the ATM cell header. A network element in an impending-congested state or a congested state may set EFCI so that this indication may be examined by the destination end-system.

extended LAN. Certain wireless bridges can extend the span of a LAN as far as three to twenty-five miles.

extended superframe (ESF). A DS-1 framing format in which twenty-four DS-O time slots plus a coded framing bit are organized into a frame which is repeated twenty-four times to form a superframe.

extranet. A private network that uses the Internet Protocols and the public telecommunication system to securely share part of a business's information or operations with suppliers, vendors, partners, customers, or other businesses. An extranet can be viewed as part of a company's Intranet that is extended to users outside the company.

fabric. Switched Fibre Channel network. The entity that interconnects various N_ports attached to it and handles the routing of frames.

far end receive failure (FERF). An indication returned to a transmitting Line Terminating Equipment (LTE) upon receipt of a Line AIS code or detection of an incoming line failure at the receiving LTE.

FDDI-II. An enhancement over FDDI to cope with multimedia traffic.

fiber distributed data interface (FDDI). A fiber-optic network protocol based on a dual-ring topology.

fiber-to-the curb (FTTC). Fiber is deployed to the curb, with copper from the curb to the user.

fiber-to-the-home (FTTH). Fiber is deployed all the way to the user.

fiber-to-the-neighborhood (FTTN). Fiber extends from the CO to a centrally located neighborhood. Copper runs from here to the user.

fibre channel. Provides for a practical and inexpensive means of rapidly transferring data between workstations, mainframes, storage devices, and other peripherals.

filter. An analyzer uses several varieties of filters. Determines which arriving frames will be captured for further analysis.

firewall. A computer with special software installed between the Internet and a private network for the purpose of preventing unauthorized access to the private network.

Federal Communications Commission (FCC). Regulates access to broadcast frequencies throughout the electromagnetic spectrum, including those for mobile computing and microwave transmissions.

forward explicit congestion notification (FECN). A flow control mechanism in frame relay networks. The bit in the Address Field notifies the receiving device that the network is experiencing congestion.

frame. A set of transmitted bits that define a basic Fibre Channel transport element. The message unit transmitted by a bit-oriented protocol. Each frame begins and ends with a flag.

frame check sequence (FCS). The standard 16–bit CRC used for HDLC and Frame Relay frames, error detection characters. The frame status field contains reservations and acknowledgment bits, which are used in the same manner as in Token Ring.

frame relay assembler/disassembler (FRAD). Responsible for framing data with header and trailer information prior to presentation of the frame to a Frame Relay switch. It is also referred to as a Frame Relay Access Device. One is the physical hardware device and the other is the function within the hardware device.

frame relay forum. Worldwide organization of frame relay equipment vendors, service providers, end users, and consultants working to speed the development and deployment of Frame Relay.

frames. The data link layer provides the required reliability to the physical layer transmission by organizing the bit stream into structured frames which add addressing and error checking information.

frequency. Measured in Hertz (Hz).

frequency division multiplexing (FDM). Each channel gets a portion of the available bandwidth for 100 percent of the time. This allows for carrying many conversations on one circuit.

frequency hopping. The type of spread-spectrum data transmission that switches data across a range of frequencies over time. Frequency hopping transmitters and receivers must be synchronized to hop at the same time, to the same frequencies.

full duplex. A data communication circuit over which data can be sent in both directions simultaneously.

G.lite. (also known as DSL Lite, splitterless ADSL, and Universal ADSL) Essentially a slower ADSL that does not require splitting of the line at the user end, but manages to split it for the user remotely at the telephone company. G.Lite, officially ITU-T standard G-992.2, provides a data rate from 1.544 Mbps to 6 Mbps downstream and from 128 kbps to 384 kbps upstream.

geosynchronous. A satellite in stationary orbit, synchronized with the Earth's rotation.

Global System for Mobile (GSM). A cellular system for worldwide telephone service.

graphical user interface (GUI). A graphical user interface is part of a network management system that displays options on the video screen as icons or picture symbols.

half duplex. A data communications circuit over which data can be sent in only one direction at a time.

headend. The source end of a coaxial CATV system.

hertz (Hz). A measure of broadcast frequencies, in cycles per second.

high-bit-rate digital subscriber line (HDSL). An early variation of DSL used for wideband digital transmission within a corporate site and between a telephone company and a customer.

high level data link control (HDLC). A generic link-level communications protocol developed by the ISO. HDLC manages synchronous, code-transparent, serial information transfer over a link connection.

hybrid fiber coaxial (HFC). A network that includes a FTTN with a coaxial cable to an individual home.

hybrid network. Microsoft's term for a LAN that includes both wireless and wired components.

hypertext markup language (HTML). A language used to create Web pages. These ASCII text files contain a number of tags that are interpreted by the Web browser program.

hyperText transport protocol (HTTP). Web servers run specialized Web server software, which supports HTTP in order to handle the organization of servicing the multiple Web client requests for client pages.

I.356. ITU-T specifications for traffic measurement.

I.361. B-ISDN ATM layer specification.

I.362. B-ISDN ATM layer (AAL) functional description.

I.363. B-ISDN ATM layer (AAL) specification.

I.430. Basic rate physical layer interface defined for ISDN.

I.431. Primary rate physical layer interface defined for ISDN.

I.432. ITU-T recommendation for B-ISDN User-Network interface (UNI).

interexchange carrier (IXC). A common carrier that provides long-distance service between LATAs.

intermix. A mode of service that reserves the full Fibre Channel capacity for a dedicated (Class 1) connection, but also allows connectionless (Class 2/3) traffic to share the link if the bandwidth is available.

International Standards Organization (ISO). An international organization that evaluates national recommendations for international use. The OSI Reference Model is a product of ISO.

International Telecommunications Union (ITU). The parent organization and successor to CCITT. The fundamental

standards for ATM have been defined and published by the ITU-T.

ISDN DSL (IDSL). A dedicated version of ISDN that offers 144 kbps. Since the ISDN D channel is not required for the dedicated connection, its 16 kbps is added to the 128 kbps offered by the two B channels.

ITU Q.2100. B-ISDN Signaling ATM Adaptation Layer overview.

ITU Q.2110. B-ISDN Adaptation layer, service specific connection-oriented protocol.

ITU Q.2130. B-ISDN Adaptation layer, service specific connection oriented function for support of signaling at the UNI.

infrared. That portion of the electromagnetic spectrum immediately below visible light. Infrared frequencies are popular for short- to medium-range point-to-point network connections.

Internet. A specific collection of interconnected networks encompassing most of the countries in the world.

Internet Protocol Security (IPSec). The network cryptographic protocols for protecting IP packets.

Internet Protocol (IP). See TCP/IP.

Internet service provider (ISP). Provide access to the Internet of a national, regional, or local level.

Intranet. A network of networks that is accessed by employees of an enterprise. It may consist of many interlinked LANs and also use leased lines in the WAN. Typically, an Intranet includes connections through one or more communication servers to the outside Internet. The main purpose of an Intranet is to share company information and computing resources among employees. An Intranet can also be used to facilitate working in groups and for teleconferences.

jabber. A constant bit stream sent by a malfunctioning Ethernet card. IEEE specifications limits to 15 ms.

Java. A platform-independent programming language used to create Web applications.

LAN emulation (LANE). A technique used to adapt ATM to an Ethernet network by creating a multicast network to enable pre-assigned groups of Ethernet nodes to receive transmissions.

LAPB. See Link Access Procedure-Balanced.

LAPD. See Link Access Procedure-D-Channel.

LATA. See Local Access Transport area.

latency. The delay through a device or link. It is measured from when the beginning of a frame enters the device to the time it leaves the device. The time it takes for information to get through a network, sometimes referred to as delay.

layer 2 forwarding (L2F) protocol. Supports the creation of secure virtual dial-up networks over the Internet.

leaky bucket. The term used as an analogous description of the algorithm used for conformance checking of cell flows from a user or network.

leased line. A dedicated phone circuit which bypasses CO switching equipment. It has no dial tone and is sometimes called a private line.

line. A transmission medium, together with the associated line-terminating equipment, required to provide the means of transporting information between two consecutive line-terminating Network Elements (NEs), one of which originates the line signal and the other terminates the line signal.

line alarm indication signal (AIS). A line code generated by a regenerator upon loss of input signal or loss of frame.

line code. Any method of converting digital information to analog form for transmission on a telephone line. 2B1Q, DMT, and CAP are line codes.

line-of-sight. A term that describes the requirement for narrowband, tight-beam transmitters and receivers to have an unobstructed path between the two.

line-of-sight networks. Networks that require the transmitter and receiver to have an unobstructed view, or clear line of sight, between the two devices.

line terminating equipment (LTE). Network elements that originate and/or terminate line (OC-N) signals. LTEs originate, access, modify, and/or terminate the transport overhead.

line termination (LT). Typically the location of the line card in the CO, and the network counterpart of the NT-1.

link access procedure balanced (LAPB). The most common data link control protocol used to interface X.25 DTEs with X.25.

link access procedure D-channel (LAPD). A protocol that operates at the data link layer of the OSI model. LAPD is used to convey information between layer 3 entities across the Frame Relay network.

local access transport areas (LATA). A geographic area within the franchised area of a local exchange carrier that has been established for the provision and administration of communication service. LATAs are described in the MFJ and essentially define local and long distance service areas.

local area network (LAN). A private, high-speed network used to share data, programs, or equipment within a limited geographical area.

local exchange carrier (LEC). The operating telephone company that serves a particular franchised area.

local management interface (LMI). Includes support for a keep-alive mechanism, which verifies data are flowing. It is a method for defining status information between devices such as routers.

logical link control (LLC). A protocol developed by the IEEE 802.2 committee for data-link-level transmission control. This upper sublayer of the IEEE Layer complements the MAC protocol.

low earth orbit (LEO). Designed to reduce propagation delay. Instead of being stationary above the earth, LEOs have an orbital cycle that causes them to appear over a given area of the earth for a specific time period.

management information base (MIB). Database of network performance information that is stored on an agent.

maximum burst size (MBS). The maximum number of cells that can be sent at the peak cell rate (PCR).

media. Most often the conduit or link that carries transmissions. Transport media include coaxial cable, copper wire, radio waves, and fiber.

media access control (MAC). IEEE specifications for the lower half of the data link layer that defines topology-dependent access control protocols for IEEE LAN specifications.

media interface connector (MIC). FDDI de facto standard connector.

metropolitan area network (MAN). As described in 802.6, essentially a LAN covering a large geographical area operating at speeds consistent with data rates available on the public network to allow easier access.

micro-segmentation. Reducing the amount of users on a segment by continually dividing the number of users. Usually for every division by one half of users, available bandwidth goes up by a factor of two.

micron. One thousandth of a millimeter.

minimum cell rate (MCR). The minimum guaranteed cell rate.

mobile computing. A form of wireless networking that uses common carrier frequencies to permit networked devices to move around freely within the broadcast coverage area yet remain connected to the network.

mobile telephone switching office (MTSO). This central office houses the field monitoring and relay stations for switching calls between the cellular and wire-based central office. The MTSO controls the entire operations of a cellular system.

modem (modulator/demodulator). A device that modulates and demodulates signals for data transmission over analog transmission facilities.

Modified Final Judgement (MFJ). The agreement reached on January 8, 1982, between the U.S. DOJ and AT&T and approved by the courts on August 24, 1982, which settled the 1974 antitrust case of the United States v. AT&T.

monitor port. A port on a switch or device that can be connected to a network analyzer.

nanometer (nm). One billionth of a meter.

narrowband radio. A type of broadcast-based networking technology that uses a single, specific radio frequency to send and receive data. Low powered narrowband implementations do not usually require FCC approval.

narrowband sockets. An emerging programming interface designed to facilitate communication between cellular networks and the Internet.

network access server (NAS). See Remote Access Server.

network channel terminating equipment (NCTE). Equipment that is designed for terminating a telephone circuit or facility at the customer's premise.

network element (NE). An ATM NE may be realized as either a standalone device or a geographically distributed system.

network interface (NI). The point of interconnection between telephone company communications facilities and the terminal equipment at a subscriber's premise.

network management. A general term describing the protocols and applications used to manage networks.

network service access point (NSAP). OSI generic standard for a network address consisting of twenty octets. ATM has specified E.164 for public network addressing and the NSAP address structure for private network addresses.

network-to-network interface (NNI). Defines standards for interoperability between the network vendor's equipment. It describes the connection between two public Frame Relay services.

network termination (NT). Equipment that provides the functions necessary for a terminal to gain access to the network. Network termination provides essential functions for signal transmission.

node. A collection of one or more fibre channel N_ports.

non-return-to-zero alternate mark inversion (NRZI). Digital encoding format. This scheme incorporates error detect capabilities by a process known as polarity violation.

OC-1. The optical standard; STS-1 is the copper standard. Both equal 51.840 Mbps. Calculated by the following formula: 90 * 9 * 8 * 8,000 = 51,840,000 bps.

OC-3. A multiple of OC-1 (51.840 Mbps); it operates at 155 Mbps.

octet. A term for 8 bits that is sometimes used interchangeably with byte to mean the same thing.

Open Systems Interconnection (OSI). A telecommunications standards framework recommended by the International Organization for Standardization (ISO) to ensure interoperability between different products and services.

operation. A set of one or more, possibly concurrent, exchanges that is associated with a logical construct above the Fibre Channel FC-2 layer.

operations administration and maintenance (OAM). A group of network management functions that provide network fault indication, performance information, and data and diagnosis functions.

optical carrier (OC). A SONET optical signal.

optical carrier level 1 (OC-1). The optical signal that results from an optical conversion of an electrical STS-1 signal. This signal forms the basis of the interface.

optical carrier level N (OC-N). The optical signal that results from an optical conversion of an electrical STS-N signal.

optical fiber. A communications medium made of very thin glass or plastic fiber that conducts light waves.

originator. The logical function associated with a Fibre Channel N_port that initiates an exchange.

overhead. In data communications, all information found on the network at a given time. This includes control, routing, and error-checking characters, in addition to user-transmitted data.

packet. A group of fixed-length binary digits, including the data and call control signals, which are transmitted through a X.25 packet switching network as a composite whole.

packet assembler/disassembler (PAD). A device used on a packet switched network that assembles information into packets and converts received packets into a continuous data stream.

packet switching. As opposed to circuit switching, user's data shares physical circuits with data from numerous other users.

packet switching network (PSN). A network that divides a message into fixed-size packets and routes them to the destination.

packets. Network layer protocols are responsible for providing network layer (end-to-end) addressing schemes and for enabling internetwork routing of network layer data packets. See frames.

passband. The bandwidth of a link is determined by the range of frequencies it can carry. Local loops will limit the range to the voice passband, around 300 Hz to 3,300 Hz. Any frequency outside this range is not carried on the link.

patch panel. A device in which temporary connections can be made between incoming and outgoing lines. This can be used for modifying or reconfiguring a communications system or for connecting test instruments.

path. Logical connection between the point at which a standard frame format for the signal at the given rate is assembled, and the point at which the standard frame format for the signal is disassembled.

path overhead (POH). Overhead assigned to and transported with the payload until the payload is demultiplexed. It is used for functions that are necessary to transport the payload.

payload pointer. Indicates the location of the beginning of the Synchronous Payload Envelope.

payload type (PT). A 3 bit field in the ATM cell header that discriminates between a cell carrying management information and one carrying user information.

peak cell rate (PCR). The maximum rate at which cells may be created and sent.

permanent virtual circuit (PVC). A virtual circuit that is permanently established. PVCs save bandwidth associated with circuit establishment and tear down in situations where certain virtual circuits must exist all the time. A PVC uses a fixed logical channel to maintain a permanent association between the DTEs.

personal communications service (PCS). A new, lower powered, higher frequency technology that is competitive to cellular. Whereas cellular typically operates in the 800 MHz–900 MHz range, PCS operates in the 1.5 GHZ to 1.8 GHz range.

phase. One characteristic of a wave which can be manipulated in phase modulation schemes in order to represent ones and zeros (offset). An attribute of an analog signal that describes its relative position measured in degrees.

physical Layer (PHY). The OSI Layer that includes all electrical and mechanical aspects relating to the connection of a device to a transmission medium.

physical medium dependent (PMD). This sublayer defines the parameters at the lowest level, such as bit speeds on the media.

ping. A diagnostic utility that sends ICMP Echo Request messages to a specific IP address on the network.

plesiochronous. Two signals of corresponding significant instants that occur at nominally the same rate.

point of presence (POP). The Inter-Exchange Carrier (IXC) equivalent of a local phone company's central office. The POP is the location at which the IXC terminates a long-distance call and passes the call to the local phone company's lines.

point to point protocol (PPP). A protocol that allows a computer to connect to the Internet with a standard dial-up telephone line and a high-speed modem. PPP features error detection, data compression, and other elements of modem communications protocols that SLIP lacks.

point to point tunneling protocol (PPTP). A new protocol that enables virtual private networking. PPTP will help companies deploy remote access to employees more quickly, using fewer resources, by allowing them to take advantage of existing enterprise network infrastructures such as the Internet for remote access.

port. The hardware entity within a node that performs data communications over a Fibre Channel link. An N_port is a port at the end-system end of a link; an F_port is an access point of the fabric.

portal. A site on the Internet that the owner positions as an entrance to other sites on the Internet.

primary rate access (PRA). See PRI.

primary rate interface (PRI). Consists of twenty-three B channels and one 64 kbps D channel in the United States or thirty B channels and one D channel in Europe.

private network-network interface (PNNI). A routing information protocol that enables extremely scalable, full function, dynamic multivendor ATM switches to be integrated in the same network.

protocol. A set of rules and procedures that permit the orderly exchange of information within and across a network.

protocol data unit (PDU). A message of a given protocol comprising payload and protocol-specific control information, typically contained in a header. Each of the OSI layers generates a uniquely formatted message intended for the corresponding layer in the receiving station.

proxy server. A type of firewall that employs a store-and-forward approach to protecting crucial data and applications. A proxy server terminates the incoming connection from the source and initiates a second connection to the destination, ensuring that the incoming user has appropriate access rights to use the data requested from the destination before passing that data on to the user.

public key certificate. Specially formatted data blocks that report the value of a public key, the name of the key's owner, and a digital signature of the issuing organization. These certificates are used to identify the owner of a particular public key.

Public Switched Telephone Network (PSTN). Refers to the local and long-distance telephone system that is used in everyday personal and business activities.

pulse code modulation (PCM). A voice digitization technique that digitizes voice in to 64 kbps by assigning voice levels to one of 256 8-bit codes.

Q.921. See LAPD.

Q.922 Annex A (Q.922A). The international draft standard that defines the structure of frame relay frames.

Q.931. A network layer protocol for ISDN call setup and teardown. It is used by the D channel to provide out-of-band call control via messages instead of the "tone-and-click" method used for POTS.

quadrature amplitude modulation (QAM). A two-dimensional modulation used for ADSL, cable modems, and proposed for VDSL. In QAM, a single-carrier frequency is modulated in both sine and cosine components.

quality of service (QoS). ATM quality of service objectives set by the carriers as class of service 1, 2, 3, and 4. QoS is defined on an end-to-end basis in terms of Cell Loss Ratio, Cell Transfer Delay, and Cell Delay Variation.

rate-adaptive DSL (RADSL). An ADSL technology that relies on software to determine the rate at which signals can be transmitted on a given customer phone line. The data rate is adjusted according to varying line conditions. RADSL can deliver from 640 kbps to 2.2 Mbps downstream and from 272 kbps to 1.088 Mbps upstream over an existing line.

receiver. A data communications device designed to capture and interpret signals broadcast at one or more frequencies in the electromagnetic spectrum.

reflective wireless network. An infrared wireless networking technology that uses a central optical transceiver to relay signals between end stations.

remote access server (RAS). Used by ISPs to allow their customers access into their networks. They typically measure how many simultaneous dial-in users they can handle.

repeater. A device used by carriers on digital transmission lines to regenerate digital signals over long distances.

Request for Comment (RFC). Documents that progress through several development stages, under the control of IETF, until they are finalized or discarded.

responder. The logical function in a Fibre Channel N_port responsible for supporting an exchange initiated by an originator.

ring. A wiring and MAC strategy in which each station is logically connected to two adjacent stations.

rotation time. Access algorithm for the FDDI network. All stations on the ring negotiate for rotation time based on their required network bandwidth.

router. A sophisticated device that divides networks into logical software-oriented subnetworks, enabling data traffic to be more efficiently routed around a network.

satellite microwave. A microwave transmission system that uses geosynchronous satellites to send and relay signals between the sender and receiver.

scatter infrared network. An infrared LAN technology that uses flat reflective surfaces such as walls and ceilings to bounce wireless transmissions between the sender and receiver.

section. The portion of a transmission facility, including terminating points, between a terminal Network Element (NE) and a regenerator or two regenerators. A terminating point is the point after signal regeneration at which performance is done.

security gateway. A device that sits between public and private networks, preventing unauthorized intrusions into the private network.

segment. A physical group of LAN nodes that communicates with the same domain.

segmentation and Reassembly (SAR). Method of breaking up arbitrarily sized packets.

sequence. A set of one or more data frames with a common sequence ID, transmitted unidirectionally over Fibre Channel from one N_port to another N_port, with a corresponding response, if applicable, transmitted in response to each data frame.

Serial Line Internet Protocol (SLIP). One of the most common protocols used by individual subscribers to connect to ISPs.

service access point (SAP). There are several groups of SAPs that are specified as valid for native ATM services.

service access point identifier (SAPI). Identifies a logical point at which layer 2 services are provided by a data link layer entity to a layer 3 entity.

service level agreement (SLA). An agreement between a user and a service provider, defining the nature of the service provided and establishing a set of metrics to be used to measure the quality provided against the agreed level of service.

S-HDSL. Single pair transmission using HDSL technology, normally 2B1Q.

Signaling System 7 (SS7). Common Channel Signaling System 7: the current international standard for out-of-band call control signaling on digital networks.

Simple Mail Transport Protocol (SMTP). The TCP/IP protocol governing electronic mail transmissions and receptions. It is an application layer protocol that supports text-oriented e-mail.

Simple Network Management Protocol (SNMP). A protocol used to enable network agents to gather statistics that can be requested by management applications.

simplex. Operating a communications channel in one direction, with no ability to operate in the other direction.

single attachment concentrator (SAC). Offers one S port for connection to the FDDI ring and multiple M ports for DTE device attachments.

single attachment station (SAS). A device that connects directly to only one counter-rotating ring of the FDDI network.

single-frequency radio. A form of wireless networking technology that passes data using only a single broadcast frequency.

splitter. Splits the incoming bit stream into voice and data. Typically the telephone company would install this at a specified site.

splitterless. A xDSL device that splits the bit stream into voice and data, but the telephone company does not have to come to the site to install it.

star. A wiring and media access strategy in which each station is wired directly to a central server or switch.

station management (SMT). The part of FDDI that manages stations on a ring.

statistical time division multiplexing (STDM). An advanced form of TDM that seeks to overcome TDM inefficiencies by dynamically adapting polling of channels. Time on a communications channel is assigned to terminals only when they have data to transport.

store and forward. A characteristic of a network that uses packet switches to forward packets. Each packet switch along a transmission path receives a packet and temporarily stores it in memory. The switch continuously selects a packet from the queue in memory, and routes the packet to the next appropriate hop.

subscriber's premises network (SPN). Multiplexes broadband ISDN (B-ISDN) service traffic.

sustainable cell rate (SCR). An upper bound on the conforming average rate of an ATM connection over time scales, which are long relative to those for which the peak cell rate (PCR) is defined.

switched line. Connected to a central office switch, provides dial tone, and reaches different destinations by the dialing of different phone numbers.

switched virtual channel connection (SVCC). One that is established and taken down dynamically through control signaling. A virtual channel connection (VCC) is an ATM connection where switching is performed on the VPI/VCI fields of each cell.

switched virtual circuit (SVC). A virtual circuit that is dynamically established on demand and is torn down when transmission is complete. SVCs are used in situations where transmission is sporadic. It is the datacom equivalent of a dial-up call; the specific path provided is determined on a call-by-call basis and in consideration of the endpoints and the congestion level in the network.

symmetrical transmission. Transmission in which a channel sends and receives data with the same signaling rate.

Symmetrical DSL (SDSL). Similar to HDSL in that both support data rates approximating T-1. SDSL uses a single-pair connection, whereas HDSL uses a two-pair local loop connection. A newer version of HDSL called HDSL-2, requires only a single-pair local loop connection.

synchronous network. The synchronization of synchronous transmission systems with synchronous payloads to a master (network) clock that can be traced to a reference clock.

synchronous optical network (SONET). Consists of physical layer communication facilities, using fiber optics. OC-1 provides for a 51.840 Mbps data rate and OC-3 provides for a 155.520 Mbps data rate. Networks based on SONET can deliver voice, data, and video communications.

synchronous transport signal level 1 (STS-N). The basic logical building block electrical signal with a rate of 51.840 Mbps.

synchronous transmission. A method of transmitting data over a network wherein the sending and receiving stations are kept in synchronization with each other by a clock signal embedded in the data.

STS envelope capacity. Bandwidth within, and assigned to, the STS Frame which carries the STS Synchronous Payload Envelope (SPE).

STS path overhead (STS POH). Nine evenly distributed POH bytes per 125 ms starting at the first byte of the STS SPE. STS POH provides for communication between the point of creation of a STS SPE and its point of disassembly.

STS path terminating equipment (STS PTE). Network Elements that multiplex/demultiplex the STS payload. STS PEs originate, access, modify, and/or terminate the STS POH necessary to transport the STS payload.

STS payload capacity. The maximum bandwidth within the STS Synchronous Payload Envelope that is available for payload.

STS synchronous payload envelope (STS PE). Network Elements that multiplex/demultiplex the STS payload. STS PEs originate, access, modify, and/or terminate the STS POH necessary to transport the STS payload.

target token rotation time (TTRT). A FDDI token travels along the network ring from node to node. If a node does not need to transmit, it passes the token to the next node. If it has a need to transmit, it can send as many frames as desired for a fixed amount of time.

terminal adapter (TA). A device to interconnect present non-ISDN terminal equipment with the digital environment.

terminal endpoint identifier (TEI). Used to identify a specific connection endpoint with a service access point.

terminal equipment (TE). TE-1 (ISDN compatible), TE-2 (non-ISDN compatible).

terrestrial microwave. A wireless microwave networking technology that uses line-of-sight communications between pairs of earth-based transmitter and receivers to relay information.

time division multiplexing (TDM). A method of combining several communication channels by dividing a channel into time increments and assigning each channel to a time slot.

token. A ring type of LAN in which a supervisory frame, or token, must be received by an attached terminal before that device can start transmitting.

Transmission Control Protocol/Internet Protocol—(TCP/IP). A networking protocol that provides communication across interconnected networks between computers with diverse hardware architectures and various operating systems.

transponder. Satellite repeater that receives at one frequency and transmits at another.

transport overhead (TOH). The overhead added to the STS SPE for transport purposes. Transport Overhead consists of Line and Sections Overhead.

trigger. An analyzer feature that allows a user to define an event after which the analyzer will stop capturing.

trunks. A circuit that carries user channels between telephone switching centers. (2) The communications circuit between two network nodes or switches.

tunneling. Used to implement VPNs. Basically, information from a LAN is encapsulated within IP packets and encrypted for transmission to a specific corporate location through the public Internet or a carrier's shared IP network. Encapsulation is the wrapping of data in a particular protocol header. As an example, Ethernet data are wrapped in a specific Ethernet header before network transit.

U interface. An ISDN interface standard for connecting NT-1 network channel termination equipment to the network.

user channel. This is allocated to the user for input of information such as data communication for use in maintenance activities and remoting of alarms external to the span equipment, in a proprietary fashion.

user-to-network interface (UNI). Defines standards for interoperability between end-user equipment and the network equipment. It is the physical, electrical, and functional demarcation point between the user and the public network service provider.

unspecified bit rate (UBR). An ATM service category that does not specify traffic-related service guarantees.

variable bit rate (VBR). Provides a guaranteed minimum threshold amount of constant bandwidth below that which the available bandwidth will not drop.

very high data rate DSL (VDSL). A developing technology that promises much higher data rates over relatively short distances (between 51 Mbps and 55 Mbps over lines up to 1,000 ft).

virtual channel (VC). A generic term used to describe unidirectional transport of ATM cells associated by a common unique identifier value.

virtual channel connection (VCC). A concatenation of VC links that extends between two points where ATM service users access the ATM layer.

virtual channel identifier (VCI). A field within the ATM cell header that is used to switch virtual channels.

virtual circuit. A circuit that is established between two terminal devices by assigning a logical path over which data can flow. A virtual circuit can either be permanent, in which terminals are assigned a permanent path, or switched, in which case the circuit is established each time a terminal has data to send.

virtual docking. Numerous point-to-point wireless infrared technologies exist that permit laptops to exchange data with desktop machines or permit data exchange between a computer and a handheld device or a printer.

virtual path (VP). A generic term used to describe the unidirectional transport of ATM cells belonging to virtual channels that are associated with a common unique identifier value.

virtual path connection (VPC). A concatenation of VP links that extends between the point where the VCI values are assigned and the point where those values are translated or removed.

virtual path identifier (VPI). A field within the ATM cell header that is used to switch virtual paths, defined as groups of virtual channels.

Virtual Private Network (VPN). A private data network that makes use of the public telecommunication infrastructure, maintaining privacy through the use of a tunneling protocol and security procedures. A VPN can be contrasted with a system of owned or leased lines that can only be used by one company.

virtual tributary (VT). A structure designed for transport and switching of sub-STS-1 payloads. There are currently four sizes of VT.

voluntary tunnel. A vehicle set up at the request of the end user. The client-side endpoint of a voluntary tunnel resides on the user's computer.

VT group. A 9-row X 12-column structure (108 bytes) that carries one or more VTs of the same size. Seven VT groups are byte interleaved within the VT-organized SPE.

VT path terminating equipment (VT PTE). Network Elements that multiplex/demultiplex the VT payload. PTEs originate access, modify and/or terminate the VT Path Overhead necessary to transport the VT payload.

wide area network (WAN). A network that covers a large geographical area. Contrast with LAN.

wireless. Indicates that a network connection depends on transmission at some kind of electromagnetic frequency through the atmosphere to carry data transmissions from one networked device to another.

wireless bridge. Consists of a pair of devices, typically narrowband and tight-beam, that are used to relay network traffic from one location to another. Wireless bridges that use spread-spectrum radio, infrared, and laser technologies are available.

World Wide Web (WWW or the Web). The universe of accessible information that is available on many computers spread across the world and attached to a very large computer network called the Internet. The Web has a body of software, a set of protocols, and a set of defined conventions for accessing the information on the Web.

X.509. A specification for public key certificates, originally developed as part of the CCITT's X.500 directory specification.

References and Other Resources

Access Technologies Forum (ACTEF). Vendor's ISDN Association (VIA). www.via-isdn.org/index.htm, 2000.

ATM Forum. Asynchronous Transfer Mode (ATM) support organization. www.atmforum.com/, 2000.

ATM Forum. "The ATM Forum Glossary." World Wide Web: http://www.atmforum.com/atmforum/library/glossary/glosspage.html, March 16, 2000.

Badgett, T., M. J. Palmer, and N. Jonker, 1999. *A Guide to Operating Systems: Troubleshooting and Problem Solving.* Cambridge, Mass.: Course Technology.

Bates, R. and D. W. Gregory, 1997. *Voice and Data Communications Handbook.* New York: McGraw-Hill.

Beyda, W. 2000. *Data Communications: From Basics to Broadband.* Upper Saddle River, N.J.: Prentice Hall.

Black, U. 1997. *Emerging Communications Technologies.* Upper Saddle River, N. J.: Prentice Hall.

Bluetooth Special Interest Group (SIG). Bluetooth Overview. World Wide Web: http://www.bluetooth.com/developer/specification/overview.asp, Nov. 27, 2000.

CDMA Development Group. Code-Division Multiple Access (CDMA) technology. World Wide Web: www.cdg.org/frame_tech.html, 2000.

Cellular Digital Packet Data (CDPD). "Your Wireless Data Guide." World Wide Web: http://www.cdpd.org/cdpd/, Aug. 15, 2000.

Cellular Telecommunications Industry Association (CTIA). Cellular telecommunications support organization. World Wide Web: www.wow-com.com/, 2000.

Cisco Systems, Inc. *Troubleshooting Overview: Symptoms, Problems, and Solutions.* World Wide Web: http://www.cisco.com/univercd/home/home.htm, Jan. 1, 2000.

Cole, M. 2000. *Introduction to Telecommunications: Voice, Data, and the Internet.* Upper Saddle River, N.J.: Prentice Hall.

Comer, D. 1999. *Computer Networks and Internets.* Upper Saddle River, N.J.: Prentice Hall.

DataPro Information Services. *An Introduction to Internet Telephone (or Voice over IP).* World Wide Web: http://www.datapro.com/, March 16, 2000.

Douskalis, B. 2000. *IP Telephony: The Integration of Robust VoIP Services.* Upper Saddle River, N.J.: Prentice Hall.

DSL Forum. Digital Subscriber Line (DSL) support organization. World Wide Web: www.adsl.com/dsl_forum.html, 2000.

Elahi, A. 2001. *Network Communications Technology.* Albany: Delmar Thomson Learning.

Enterprise Computer Telephony Forum. Computer telephony support organization. World Wide Web: www.ectf.org/, 2000.

Fibre Channel Industry Association (FCIA). Fibre Channel support organization. "Fibre Channel—Overview of the Technology." World Wide Web: http://www.fibrechannel.com, Aug. 15, 2000.

Fiber Optics Association, Inc. Fiber optics support organization. World Wide Web: www.std.com/fotec/foa.htm., 2000.

FitzGerald, J. and D. Alan. 1999. *Business Data Communications and Networking*. New York: John Wiley & Sons, Inc.

Frame Relay Forum (The). Frame Relay support organization. "Basic Guide to Frame Relay Networking." World Wide Web: http://www.frforum.com/basicsguide.html, Aug. 15, 2000.

Garg, V. K. and J. E. Wilkes. 1996. *Wireless and Personal Communications Systems*. Upper Saddle River, N.J.: Prentice Hall, Inc.

Gelber, S. 1997. *Data Communications Today: Network, The Internet, and The Enterprise*. Upper Saddle River, N.J.: Prentice Hall, Inc.

Gigabit Ethernet Alliance. Gigabit Ethernet support organization. "Gigabit Ethernet Technical Overview." World Wide Web: http://www.gigabit-ethernet.org/technology, Aug. 15, 2000.

Goldman, J. E. 1998. *Applied Data Communications: A Business-Oriented Approach*. New York: John Wiley & Sons, Inc.

Goralski, W. 1998. *ADSL and DSL Technologies*. New York: McGraw Hill.

Green, J. H. 1996. *The Irwin Handbook of Telecommunications Management*. New York: McGraw Hill.

Guide to TCP/IP on Microsoft Windows NT 4.0 (A). 1998. Cambridge, Mass.: Course Technology.

Haynal, R. 2000. Russ Haynal's ISP Page. "Major Internet Backbones". World Wide Web: http://www.navigators.com/isp.html, Aug. 15, 2000.

Held, G. 1996. *Dictionary of Communications Technology*. New York: John Wiley & Sons. Ltd.

Hioki, W. 1998. *Telecommunications*. Upper Saddle River, N.J.: Prentice Hall.

HP OpenView SNMP Agent Administrator's Guide. 1995. Hewlett-Packard Co.

Hudson, K. 2000. *CCNA Guide to Cisco Networking Fundamentals*. Albany: Thomson Learning.

Huitema, C. 1999. *Routing in the Internet*. Upper Saddle River, N.J.: Prentice Hall.

Huntington-Lee, Terplan, Kornel, and J.Gibson. 1997. *HP Openview: A Manager's Guide*. McGraw Hill.

Infrared Data Association. Wireless technology support organization. World Wide Web: www.irda.org/products/index.asp, 2000.

Institute of Electrical and Electronics Engineers, Inc. (IEEE). See www.ieee.org/ for the latest standards.

Internet Engineering Task Force (IETF). Request for Comments (RFCs). See www.rfc-editor.org/ for the latest standards and draft standards.

Internet Engineering Task Force (IETF). Support organization. World Wide Web: www.ietf.cnri.reston.va.us/, 2000.

Internetworking Technologies Handbook, 2d ed. 1998.Cisco Systems, Inc.

ISO: X.25 Technical Help Forum. X.25 support organization. World Wide Web: www.tek-tips.com/. 2000.

Kessler, G.C. and P. Southwick. 1997. *ISDN—Concepts, Facilities and Services*. New York: McGraw Hill.

Martin, J. 1996. *Enterprise Networking—Data Link Subnetworks*. Upper Saddle River, N.J.: Prentice Hall.

McDysan, D. E. and D. L. Spohn. 1998. *Hands-On ATM*. New York: McGraw Hill.

Miller, M.A. 2000. *Data and Network Communications*. Albany: Delmar Thomson Learning.

National ISDN Council. National ISDN support organization. World Wide Web: www.nationalisdncouncil.com/, 2000.

Newton, H. 1999. *Newton's Telecom Dictionary*. New York: Miller Freeman, Inc.

Network General Corporation. 1996. *Troubleshooting with the Sniffer Network Analyzer—Troubleshooting Techniques.*

North American ISDN User's Forum. ISDN support organization. World Wide Web: www.niuf.nist.gov, 2000.

Ohio State University. ATM Products. World Wide Web: http://www.cis.ohio-state.edu/Research/research.html, Nov. 25, 2000.

Oppenheimer, P. 1999. *Top-Down Network Design.* Indianapolis, Ind: Macmillan Technical Publishing.

Palmer, M. and R. B. Sinclair. 1999. *A Guide to Designing and Implementing Local and Wide Area Networks.* Cambridge, Mass.: *Course Technology.*

Panko, R. 1999. *Business Data Communications and Networking.* Upper Saddle River, N.J.: Prentice Hall.

Rose, M. T. 1996. *The Simple Book: An Introduction to Networking Management.* Upper Saddle River, N.J.: Prentice Hall.

Schneiderman, R. (ed.) 2000. IEEE Spectrum. Bluetooth Technology. World Wide Web: www.spectrum.ieee.org/, Nov. 29, 2000.

Shay, W. A. 1999. *Understanding Data Communications & Networks.* Brooks/Cole Publishing Co.

SONET Interoperability Forum (SIF). Synchronous Optical NETwork (SONET) support organization. World Wide Web: www.atis.org/atis/sif/sifhom.htm/, 2000.

Stallings, W. 1997. *Data & Computer Communications,* 5th ed. Upper Saddle River, N.J.: Prentice Hall.

Stallings, W. 1997. *Local and Metropolitan Area Networks.* Upper Saddle River, N.J.: Prentice Hall.

Stallings, W. 2000. *Data & Computer Communications,* 6th ed. Upper Saddle River, N.J.: Prentice Hall.

Stamper, D. A. 1999. *Business Data Communications.* Addison-Wesley Longman, Inc.

Sveum, M. E. 2000. *Data Communications: An Overview.* Upper Saddle River, N.J.: Prentice Hall.

Telecommunications Industry Association (TIA). Telecommunications support organization. "The Voice of Manufacturers & Suppliers of Communications & Information Technology Products & Services." World Wide Web: www.tiaonline.org/, 2000.

Thompson, A. 2000. *Understanding Local Area Networks: A Practical Approach.* Upper Saddle River, N.J.: Prentice Hall.

Thurwachter, Jr., C. N. 2000. *Data and Telecommunications: Systems and Applications.* Upper Saddle River, N.J.: Prentice Hall.

Tittel, E. and D. Johnson. 1998. *A Guide to Networking Essentials.* Cambridge, Mass.: Course Technology.

Voice over the Net Coalition, Inc. (VON). IP Telephony support organization. World Wide Web: www.von.org/, 2000.

VPN Consortium. Virtual Private Network (VPN) support organization. World Wide Web: www.vpnc.org/, 2000.

Wesel, E. K. 1998. *Wireless Multimedia Communications: Networking Video, Voice, and Data.* Addison-Wesley Longman, Inc.

World Wide Web Consortium. WWW support organization. www.w3.org/, 2000.

INDEX